AF361892

PI3K/AKT/mTOR Signaling Network in Human Health and Diseases

PI3K/AKT/mTOR Signaling Network in Human Health and Diseases

Guest Editors

Jean Christopher Chamcheu
Claudia Bürger
Shile Huang

Basel • Beijing • Wuhan • Barcelona • Belgrade • Novi Sad • Cluj • Manchester

Guest Editors

Jean Christopher Chamcheu
Department of Biological
Sciences and Chemistry
Southern University
Baton Rouge
United States

Claudia Bürger
Department of Dermatology
Goethe University Frankfurt
Frankfurt am Main
Germany

Shile Huang
Department of Biochemistry
and Molecular Biology
Louisiana State University
Health Sciences Center
Shreveport
United States

Editorial Office
MDPI AG
Grosspeteranlage 5
4052 Basel, Switzerland

This is a reprint of the Special Issue, published open access by the journal *Cells* (ISSN 2073-4409), freely accessible at: www.mdpi.com/journal/cells/special_issues/PI3K_AKT_mTOR.

For citation purposes, cite each article independently as indicated on the article page online and using the guide below:

Lastname, A.A.; Lastname, B.B. Article Title. *Journal Name* **Year**, *Volume Number*, Page Range.

ISBN 978-3-7258-2774-9 (Hbk)
ISBN 978-3-7258-2773-2 (PDF)
https://doi.org/10.3390/books978-3-7258-2773-2

Contents

About the Editors

Jean Christopher Chamcheu

Jean Christopher Chamcheu, Ph.D., is an associate professor in the Department of Biological Sciences and Chemistry at Southern University and A&M College (SUBR) and an associate professor of Viral Oncology at the Louisiana State University School of Veterinary Medicine (LSU SVM). He earned his Ph.D. in Dermatology and Venereology from Uppsala University, Sweden, and completed postdoctoral training at the University of Wisconsin–Madison, where he advanced to become an assistant staff scientist and instructor in Dermatology. In 2017, he became an assistant professor of Pharmacology at the University of Louisiana at the Monroe College of Pharmacy. Dr. Chamcheu joined SUBR and the LSU SVM as an associate professor in 2024. His research focuses on central mTOR signaling in skin inflammation and carcinogenesis, exploring therapeutic interventions using bioactive small molecules, natural dietary ingredients, synthetic scaffolds, and oncolytic viruses.

Claudia Bürger

Claudia Bürger, Ph.D., is heading a research lab clinic of dermatology at a university hospital in Frankfurt, Germany. Her research investigates molecular mechanisms and, in particular, the signal transduction cascades that contribute to the maintenance of a healthy skin barrier and are involved in inflammatory dermatoses such as psoriasis.

Shile Huang

Shile Huang, Ph.D., is a professor in the Department of Biochemistry and Molecular Biology and adjunct professor in the Department of Hematology and Oncology, Louisiana State University Health Sciences Center in Shreveport, LA, USA. His research focuses on the role of mTOR signaling in tumorigenesis and metastasis, the mechanisms of the anticancer action of small molecules, and the mechanisms of the neurotoxicity of cadmium.

Preface

Global health challenges, driven by genetic alterations and environmental factors, have led to a rise in human diseases. Understanding the molecular basis of these diseases is crucial for developing targeted treatments. Signal transduction, a fundamental life process, allows for communication within and between cells, maintaining cellular integrity and initiating physiological responses. The dysregulation of these pathways is linked to various diseases, including cancers, infections, chronic inflammation, and neurological, developmental, metabolic, and cardiovascular disorders.

The PI3K/AKT/mTOR pathway regulates key cellular processes such as metabolism, growth, proliferation, and survival. It has gained attention for its role in the development and progression of diseases like cancers, cardiovascular diseases, and autoimmune disorders. Understanding this pathway is essential for developing targeted therapies that maintain normal tissue homeostasis.

This Special Issue features original research, communications, and reviews exploring the PI3K/AKT/mTOR pathway in both physiological and pathological settings, addressing knowledge gaps. Topics include its involvement in diseases such as breast and liver cancers, psoriasis, and myocardial infarction, as well as its potential as a therapeutic target for treatments like stem cell therapies and oxidative stress interventions. The issue highlights the importance of advancing our knowledge to improve disease management and treatment.

The reprint compiles sixteen peer-reviewed publications, comprising twelve original research articles, three reviews, and an editorial. It examines the PI3K/AKT/mTOR pathway's role in various diseases and therapeutic approaches, emphasizing its relevance across multiple contexts and investigating the impact of PIK3CA gene mutations on cancer. Novel technologies for studying AKT1 biology and treatment mechanisms are also highlighted.

We sincerely appreciate the contributions of the authors and reviewers and acknowledge Dr. Tolulope O. Omolekan for his support in drafting and editing. We extend our gratitude to the MDPI book staff and the editorial team of *Cells*, especially Ms. Eleanor Wang.

Jean Christopher Chamcheu, Claudia Bürger, and Shile Huang
Guest Editors

Editorial

PI3K/AKT/mTOR Signaling Network in Human Health and Diseases

Tolulope O. Omolekan [1,2,3], Jean Christopher Chamcheu [1,2,4,*], Claudia Buerger [5,*] and Shile Huang [6,7,8,*]

[1] Department of Biological Sciences and Chemistry, College of Sciences and Engineering, Southern University and A&M College, Baton Rouge, LA 70813, USA; tolulope.omolekan@bowen.edu.ng
[2] Department of Pathological Sciences, School of Veterinary Medicine, Louisiana State University, Baton Rouge, LA 70803, USA
[3] Department of Biochemistry, Bowen University, Iwo 232101, Nigeria
[4] Department of Pathology and Translational Pathobiology, Louisiana State University Health Sciences Center, Shreveport, LA 71103, USA
[5] Department of Dermatology, University Hospital Frankfurt, Goethe University Frankfurt, Theodor-Stern-Kai 7, 60590 Frankfurt am Main, Germany
[6] Department of Biochemistry and Molecular Biology, Louisiana State University Health Sciences Center, 1501 Kings Highway, Shreveport, LA 71103, USA
[7] Department of Hematology and Oncology, Louisiana State University Health Sciences Center, 1501 Kings Highway, Shreveport, LA 71103, USA
[8] Feist-Weiller Cancer Center, Louisiana State University Health Sciences Center, 1501 Kings Highway, Shreveport, LA 71103, USA
* Correspondence: jchamcheu@lsu.edu or jeanchristopher.c@sus.edu (J.C.C.); cbuerger@med.uni-frankfurt.de (C.B.); shile.huang@lsuhs.edu (S.H.)

check for updates

Citation: Omolekan, T.O.; Chamcheu, J.C.; Buerger, C.; Huang, S. PI3K/AKT/mTOR Signaling Network in Human Health and Diseases. *Cells* **2024**, *13*, 1500. https://doi.org/10.3390/cells13171500

Received: 28 August 2024
Accepted: 29 August 2024
Published: 6 September 2024

1. Introduction

Transduction of molecular signaling is a fundamental mechanism that allows a living cell to communicate internally with other cells and its environment through chemical or physical signals, thereby maintaining its structural integrity and triggering physiological responses. The human body, a complex multicellular entity, has an extensive and coordinated network of signaling pathways necessary for its health, survival, and functionality. These pathways enable the body to maintain homeostasis in response to a range of internal and external stimuli, both under normal and disease conditions, throughout different stages of life. Dysregulation of cell signaling is associated with the onset of many diseases, including cancer, infections, chronic inflammation, as well as neurological, developmental, metabolic, and cardiovascular disorders. Such dysregulations arise from a variety of factors, such as alterations of genes, transcription factors, splicing and chromatin regulators, and abnormal levels of signaling molecules, leading to disruption of the regulatory networks essential for cell function and communication [1]. Since Claude Bernard first introduced the concept of signaling in 1855, research into the molecular complexities of cell signaling in health and disease has spurred the discovery of disease biomarkers, new drug targets, and the development of innovative therapeutic strategies. The PI3K/AKT/mTOR pathway, a highly conserved intracellular pathway in eukaryotic cells, plays a vital role in cell metabolism and regulates various cellular events such as cell growth, proliferation, survival, motility, adhesion, and differentiation [2]. Frequent dysregulation of this pathway in numerous diseases has made it a focus of research to identify biomarkers and define therapeutic targets associated with this signaling cascade. The phosphoinositide 3-kinase (PI3K) can be activated by receptor tyrosine kinases (RTKs), such as the platelet-derived growth factor receptor (PDGFR) or epidermal growth factor receptor (EGFR), which promote cell proliferation and migration, the insulin-like growth factor receptor (IGFR), which stimulates cell growth and survival, and the insulin receptor (IR), which maintains metabolic homeostasis. To synchronize the various cellular responses to multiple external stimuli, PI3K effectors modify multiple physiological aspects of the cell. For example, signals that propel cell cycle progression are synchronized with those that increase the need for metabolic pathways to generate the necessary energy and macromolecules for cell growth and division. PI3K mediates the synthesis of phosphatidylinositol (3,4,5)-trisphosphate (PIP3) in the cells, which acts as a lipid second messenger and recruits AKT (also known as protein kinase B (PKB)) and phosphoinositol-dependent kinase (PDK1) to the membrane. PDK1 can then activate AKT by phosphorylation at Thr308. Complete activation of AKT also requires phosphorylation at Ser473 by mTOR complex 2 (mTORC2) [3]. Fully activated AKT then phosphorylates a variety of signaling molecules with diverse functions and coordinates a complex set of metabolic responses to meet these needs while maintaining homeostasis. Through a feedback loop, the network is regulated, preventing abnormal cell proliferation under nutrient scarcity and other stress conditions. The mTOR complex 1 (mTORC1) phosphorylates and stabilizes the growth factor receptor-bound protein 10 (GRB10), an adaptor protein that binds and inhibits IGFR and IR [4]. Mutations in *GRB10*, *PIK3CA* genes, and other genes in the PI3K-AKT-mTOR pathway result in complex pathologies in humans like type 2 diabetes, congenital lipomatous overgrowth, vascular malformations, epidermal nevis, spinal/skeletal anomalies/scoliosis (CLOVES) syndrome, immune-mediated inflammatory conditions, and hyperproliferative disorders [5]. Understanding the molecular and physiological intricacies of the PI3K/AKT/mTOR signaling network could facilitate the development of new therapies that target this network in pathological conditions without disrupting normal tissues. Such drugs specifically target mutant forms of oncogenic proteins (e.g., p110α with mutant H1047R), thus sparing endogenous signaling molecules, which are critical for the maintenance of normal homeostasis.

Therapeutic strategies targeting this signaling network have been extensively explored, especially in conditions like cancer, neurodegenerative disorders, inflammation, autoimmune diseases, obesity, and diabetes. These strategies include the use of nutraceuticals, synthetic small-molecule inhibitors, and combinations with drugs to enhance efficacy. Preclinical and clinical studies are ongoing to develop these targeted therapies [6]. The U.S. Food and Drug Administration has approved certain inhibitors of PI3K, AKT, and mTOR for some types of cancer, including breast cancer with specific genetic mutations and chronic lymphocytic leukemia [7]. However, it is important to note that while some drugs have been approved, others are still under evaluation or have not received approval for the treatment of any human disease.

This Special Issue comprises published original research and high-quality reviews of scientific literature that deepen our understanding of the PI3K/AKT/mTOR pathway and its biology. In particular, it encompasses investigations into the roles of this pathway in various human diseases (e.g., cancer, psoriasis, viral infection, and myocardial infarction) and treatments, as well as the delivery of programmed phospho-variants of AKT1 into cells. The findings not only underscore its involvement in various disease conditions but also examine the diverse effects of hotspot mutations in the *PIK3CA* gene on cancer progression. Furthermore, it highlights novel technologies for selectively studying AKT1 biology and the mechanisms of action of different treatment modalities.

2. An Overview of Published Articles

Siddika et al., Ghodsinia et al., Ferreri et al., and George et al. independently reported the biology, activity, genome-wide transcriptomic changes, and the consequences of mutations of important effectors of the signaling network in disease conditions. Siddika et al. (contribution 1) described the development of a novel and effective delivery system for programmed AKT1 phospho-variants into human cells. This system selectively induces the phosphorylation of its substrate (glycogen synthase kinase 3α) and downstream effector (ribosomal protein S6) at Ser9 and Ser240/244, respectively, in the PI3K/AKT/mTOR signaling pathway. Until now, traditional methods of phosphorylating and activating AKT via growth factors or insulin have resulted in the activation of multiple kinases, complicating the selective study of AKT1. These variants were generated by fusing AKT1 with an N-terminal cell penetrating peptide tag derived from the human immunodeficiency virus trans-activator of transcription (TAT) protein, and then expressed and purified from *E. coli*. The tag did not alter AKT1 kinase activity but facilitated the efficient and rapid delivery of AKT1 phospho-protein variants into human cells. These findings demonstrate an efficient delivery system for programmed AKT1 phospho-variants into human cells, thus providing a novel cell-based model system for specifically investigating AKT1 signaling activity.

Ghodsinia et al. (contribution 2) characterized two novel non-hotspot mutants at Q661K (exon 13) and C901R (exon 19) of the *PIK3CA* gene in NIH3T3 and HCT116 cells. This gene plays a crucial role in colorectal cancer (CRC) as it encodes the p110α catalytic subunit of PI3K, which is involved in cell growth and migration. Well-characterized mutations in hotspots of the *PIK3CA* gene, such as E545K (exon 9) and H1047R (exon 20), contribute differently to CRC progression. E545K promotes a wide range of oncogenic behaviors, while H1047R has a more limited scope, highlighting the importance of understanding the specific effects of each mutation within the gene. Their findings offer insights into the diversity of mutation effects within the same gene, affecting cellular functions such as proliferation, apoptosis resistance, and cytoskeletal reorganization, all crucial for cancer development and metastasis. These mutations may disrupt the interaction between the p110α catalytic subunit and the p85α regulatory subunit, leading to increased PI3K signaling and consequent cancer progression. Additionally, they influence RNA expression levels, tumor microenvironment, and the distribution of immune cells within the tumor. While many *PIK3CA* mutations in CRC occur in hotspot regions, they are not hereditary. The findings contribute to the growing body of knowledge that will help refine the prognostic and predictive value of *PIK3CA* mutations in CRC, with significant implications for improving patient outcomes.

The activity of mTOR complex 1 (mTORC1) in the proliferative layer of healthy skin prevents differentiation of keratinocytes and is suppressed by the tuberous sclerosis complex (TSC) through its inhibition of the small GTPase, Ras homolog enriched in brain (Rheb), in other tissues [8]. However, in inflammatory conditions, cytokines disrupt this regulatory mechanism and prevent proper keratinocyte differentiation. Therefore, Ferreri et al. (contribution 3) investigated the role of proinflammatory cytokines in the regulation of TSC and the pathogenesis of psoriasis. Proinflammatory cytokines, such as tumor necrosis factor-α (TNF-α) and interleukin-1β (IL-1β), contribute to the pathogenesis of psoriasis. They found that TNF-α and IL-1β could induce phosphorylation of TSC2 at S939 via the PI3K/AKT and mitogen-activated protein kinase (MAPK) pathways in HaCaT keratinocytes. Surprisingly, phosphorylation of TSC2 S939 was not detected in the lesional psoriatic skin of the patients. Further in vitro studies showed that proinflammatory cytokines induce the dissociation of TSC2 from lysosomes, leading to its destabilization and degradation. This in turn results in chronic mTORC1 hyperactivation and impaired keratinocyte differentiation, thus contributing to the phenotypical changes seen in psoriatic epidermis.

Overexpression or hyperactivation of the AKT pathway is a common event that contributes to the progression of breast cancer by promoting the survival and proliferation of breast cancer cells. However, little is known about the precise genome-wide transcriptomic changes associated with the AKT pathway in breast cancer cells. To clarify this, George et al. (contribution 4) conducted a genome-wide RNA-sequencing analysis after selectively knocking down AKT1 using specific siRNAs or inhibiting its activity with a pan-AKT inhibitor VIII in breast cancer cells. Alterations in AKT1 cellular levels impact the expression of a set of differentially expressed genes, which in turn affects its cellular functions. Surprisingly, AKT1 also plays an intrinsic role in suppressing the expression of a subset of genes in both unstimulated and growth factor-stimulated breast cancer cells. Consistently, the expression of this subset of genes also increases in breast tumors when AKT1 is depleted or not undetectable and decreases in breast tumors when AKT1 levels are elevated. Real-time polymerase chain reaction (qRT-PCR) analysis validated the RNA-sequencing data from two breast cancer cell lines and breast cancer patient-derived cells. The results provide insights into the AKT1-dependent modulation of gene expression in breast cancer cells and emphasize its importance in cellular function and potential therapeutic targets.

Wilczek et al. (contribution 5) studied the consequences of JCPyV infection on progressive multifocal leukoencephalopathy (PML). Using RNA sequencing and chemical inhibitors of PI3K, AKT, and mTOR, they showed that the PI3K/AKT/mTOR pathway is essential for JCPyV infection in primary astrocytes but not in transformed cell lines. The findings indicate that the pathway could be a potential target for developing therapeutics against PML and open possibilities for repurposing anticancer drugs that inhibit the PI3K/AKT/mTOR pathway for treatment of PML.

Osteonecrosis of the femoral head (ONFH) induced by glucocorticoid use is a challenging condition to manage. Ma et al. (contribution 6) demonstrated a therapeutic potential of extracellular vesicles derived from bone marrow stem cells (BMSC-EVs) in the treatment of glucocorticoid-induced ONFH. They found that glucocorticoid inhibited autophagy by activating the PI3K/AKT/mTOR pathway, resulting in a decrease in cell viability and angiogenesis capacity and an increase in apoptosis of bone microvascular endothelial cells (BMECs). BMSC-EVs prevented glucocorticoid-induced injury of BMECs by suppressing the glucocorticoid-activated PI3K/AKT/mTOR pathway, promoting autophagy.

The PDK1 signaling plays a crucial role in the radiotherapy resistance (IR resistance) observed in hepatocellular carcinoma (HCC), contributing to treatment failure. Bamodu et al. (contribution 7) suggest that PDK1 is pivotal in promoting IR resistance by enhancing DNA damage repair and facilitating post-IR relapse in aggressive HCC cells. The aberrant expression of PDK1 in poorly differentiated HCC cells, in contrast to well-differentiated or normal liver cells, underscores its role in activating the PI3K/AKT/mTOR signaling pathway, which is significant. This activation enables cells to evade IR toxicity, leading to enhanced survival, proliferation, and a dedifferentiated phenotype that is highly resistant to radiotherapy. Molecular ablation of PDK1 function increased the susceptibility of HCC cells to IR and suppressed PI3K/AKT/mTOR signaling. Therefore, targeting PDK1 could potentially enhance radiosensitivity in HCC treatment. The positive correlation between PDK1-driven IR resistance and factors such as cell motility, invasiveness, and stemness marker expression further supports its critical role in this resistance development. Upregulation of stemness markers like ALDH1A1, PROM1, SOX2, KLF4, and POU5F1, increased tumor sphere-formation efficiency, and suppressed biomarkers of DNA damage suggest a complex network of signaling, contributing to the radioresistant phenotype. This preclinical evidence implicates PDK1 as an active driver of IR resistance through the activation of PI3K/AKT/mTOR signaling pathway, modulation of cancer stemness signaling, and suppression of DNA damage. This underscores the potential of PDK1-targeted therapies to enhance radiosensitivity in HCC treatment. These findings provide valuable insights into the molecular mechanisms underlying IR resistance in HCC and open new avenues for therapeutic intervention. It will be intriguing to see translational research in clinical settings to determine whether PDK1-targeted therapies can indeed improve outcomes for patients with IR-resistant HCC.

Inhibition of Notch signaling through γ-secretase inhibitor (GSI) treatment activates the AKT/mTOR signaling pathways, critical for survival, differentiation, myogenesis, and muscle protein synthesis (MPS) in myotubes. Huot et al. (contribution 8) explored the potential dependency of GSI's impact on myogenesis and MPS via the AKT/mTOR signaling in C2C12 cells by assessing myotube formation, anabolic signaling, and MPS after exposing C2C12 cells to GSI along with rapamycin and API-1, inhibitors of mTOR and AKT, respectively. Rapamycin and API-1 counteracted GSI-mediated effects on myotube formation and fusion in C2C12 cells, whereas GSI treatment rescued MPS and GSK3β Ser9 phosphorylation in C2C12 cells treated with rapamycin and API-1. These findings suggest that GSI treatment rescues MPS in C2C12 myotubes independently of AKT/mTOR signaling, possibly by modulating the phosphorylation of GSK3β.

Luo et al. (contribution 9) investigated the mechanism of anticancer effects of dihydroartemisinin (DHA), an anti-malarial drug, using rhabdomyosarcoma (RMS) cells as an experimental model. They found that DHA inhibits the mTORC1 signaling pathway in RMS cells but not in normal cells. Mechanistically, DHA does not directly bind to mTOR or

FKBP12, nor does it inhibit IGFR, PI3K, and ERK1/2 pathways, or activate phosphatase and tensin homolog (PTEN). Instead, DHA activates AMP-activated protein kinase (AMPK). Inhibition of AMPK, either pharmacologically or genetically, attenuates DHA's inhibitory effect on mTORC1. Also, DHA causes the dissociation of raptor from mTOR, crucial for inhibiting mTORC1 activity. Oral treatment with artesunate, a prodrug of DHA, effectively suppresses the tumor growth of RMS xenografts by activating AMPK and inhibiting mTORC1. These findings support the idea that DHA has a great potential for treatment of RMS. The study underscores the importance of understanding the molecular mechanisms of drugs, enabling their repositioning for treatments beyond their original indications.

mTORC2 is essential for cellular growth and metabolism through the phosphorylation of AGC family kinases such as AKT and protein kinase C (PKC) at their hydrophobic motif (HM), which is crucial for their activation. The mitogen activated protein kinase/extracellular signal-regulated kinase (MAPK/ERK) pathway also significantly contributes to cell proliferation and differentiation by phosphorylating p90 ribosomal S6 kinase (RSK) at Ser380, a conserved HM site within the AGC kinase family. The study by Chou et al. (contribution 10) demonstrates the complexity of these interactions by showing that RSK phosphorylation at Ser380 can occur independently of mTOR's catalytic activity, although optimal phosphorylation of RSK at this site requires an intact mTORC2. This finding suggests that mTORC2 may primarily serve as a scaffold to facilitate this process. This phosphorylation event at Ser380 is critical as it potentially influences the substrate specificity of RSK and indicates how RSK can respond to alterations in nutrient availability. This example highlights the fascinating ability of cells to integrate multiple signals to maintain equilibrium and adapt to environmental changes.

Hsueh et al. (contribution 11) elucidated the protective effects of ascorbic acid (AA) against oxidative stress in corneal endothelial cells. Using in vitro and in vivo models, they demonstrated the mechanisms by which AA mitigates damage from paraquat-induced reactive oxygen species (ROS) and improves corneal health. The downregulation of AKT phosphorylation in response to oxidative stress and its attenuation through AA pretreatment is a significant finding, highlighting the potential role of the PI3K/AKT pathway in protecting corneal endothelium. The in vivo rabbit corneal damage model further supports the topical application of AA as a viable perioperative strategy to enhance corneal clarity and prevent oxidative damage. These findings provide a valuable contribution to ophthalmology, especially for patients undergoing procedures like phacoemulsification who are at risk of oxidative stress-induced corneal endothelial decompensation. This represents a promising step towards improving surgical outcomes and patient vision post-surgery.

Galectin-1 (GAL1) is a β-galactoside-binding protein involved in multiple aspects of tumorigenesis. Su et al. (contribution 12) highlight the significant role of GAL1 in upper tract urothelial carcinoma (UTUC). GAL1 expression is associated with worse outcomes in UTUC, indicated by poorer recurrence-free survival (RFS) and cancer-specific survival (CSS) in patients with higher GAL1 levels. In vitro studies conducted on four urothelial carcinoma (UC) cell lines (BFTC-909, T24, RT4, and J82) suggest that GAL1 promotes tumor invasiveness and migration, potentially through its effects on proteins like MMP-2, MMP-9, and TIMP-1, which are involved in the breakdown and remodeling of the extracellular matrix—a key process in cancer metastasis. Overexpression of GAL1 activates the FAK/PI3K/AKT/mTOR pathway, which correlates with disease progression and patients' survival in upper urinary urothelial carcinoma. These suggest GAL1 as a potential therapeutic target, given that its modulation could affect crucial pathways involved in cancer progression. Moreover, leveraging public genomic data from TCGA and GSE32894 for comparison adds a valuable dimension, broadening the understanding of GAL1's role in UTUC. It is evident that GAL1's function in cancer biology is complex and multifaceted, influencing various aspects of tumor growth and metastasis. Hence, this study contributes significantly to the growing body of evidence highlighting GAL1's importance in cancer and may lead to novel therapeutic strategies in UTUC.

The treatment and recovery from myocardial infarction (MI) are significantly influenced by the PI3K/AKT signaling pathway. Activation of this pathway is crucial as it provides cardioprotection, enhances cell survival, and mitigates the negative consequences of post-infarction myocardial remodeling. Walkowski et al. (contribution 13) reviewed the role of the PI3K/AKT pathway in each step of ischemia and subsequent left ventricular remodeling. They described a notable impairment in the function of the PI3K/AKT pathway in diabetes, which exacerbates adverse changes in the myocardium following an infarct. This impairment is a notable concern, given the frequent coexistence of diabetes and cardiovascular diseases in patients, potentially compromising the efficacy of treatments targeting the PI3K/AKT pathway. They also discuss some cardiac and antidiabetic drugs, which can activate or inhibit different components of the PI3K/AKT pathway, influencing myocardial ischemia and left ventricular remodeling. It is important to note that while there are promising substances and drugs that interact with the PI3K/AKT pathway, more investigation is needed before they can be introduced into clinical practice. This includes understanding the molecular mechanisms of their effects and conducting extensive clinical trials to ensure safety and efficacy for patients with MI, especially those with diabetes. The interplay between diabetes, cardiac conditions, and the PI3K/AKT pathway is complex, and personalized medicine approaches may be required to optimize treatment strategies for individual patients.

In a review article, Loissell-Baltazar and Dokudovskaya (contribution 14) gave a summary of the key points regarding the Seh 1 associated (SEA) complex. The SEA complex in yeast and its homologue in humans (the GATOR complex) are crucial upstream regulators of mTORC1 that integrate various signals from amino acids, growth factor, oxygen, and DNA-damaging agents to regulate cellular responses to stress. The complex plays a pivotal role in nutrient sensing and response, influencing the cell's decision to pursue metabolic pathways, which are critical for cell growth, metabolism, and autophagy. Mutations in this complex disrupt the normal functioning of the mTORC1 pathway, leading to pathological conditions. The ongoing research in the SEA/GATOR complex not only enhances our comprehension of cellular metabolism and growth but also underscores the complex's potential as a therapeutic target for related diseases. It is a prime example of how fundamental research can lead to breakthroughs in understanding and potentially treating complex diseases.

Transcription factor EB (TFEB) is a master regulator of autophagy and lysosomal biogenesis; thus, its ability to clear intracellular pathogens makes it a suitable therapeutic target for pathologies related to autophagy dysfunction. The identification of TFEB agonists and their progression into preclinical and clinical studies mark significant advancements in this area. In a review article, Chen et al. (contribution 15) discussed TFEB regarding its structure, regulatory mechanisms, and implications in diseases, as well as a variety of TFEB agonists. This review underscores the importance of TFEB in managing diseases associated with autophagy dysregulation and suggests that modulating TFEB activity could be a promising therapeutic approach. As research continues, we can expect to see further developments in the use of TFEB agonists for the management of human diseases.

3. Conclusions

Overall, the 15 papers published in this Special Issue highlight the importance of the PI3K/AKT/mTOR pathway in human health and diseases. In summary, a comprehensive understanding of the physiological and pathophysiological functions of this pathway is of great importance for the development of effective strategies to better manage human health and diseases.

Funding: We acknowledge the financial support for the research conducted in the laboratories of Buerger, Chamcheu, and Huang, which is partly enabled by grants from various funding agencies. The Buerger research laboratory was specifically supported by the Deutsche Forschunggemeinschaft (DFG-German Research Foundation) BU1840/5-3, UfIB/NatLifE2020 German Federal Ministry of Education and Research 031B0716, as well as the Rolf. M. Schwiete Foundation (Germany). The

Chamcheu research laboratory was partly supported by a full project award from an LSU Center for Pre-Clinical Cancer Research COBRE funded NIH/NIGMS grant 1P20 GM135000-01A1, a full project award from an NIH/NIGMS IDeA LBRN grant P20 GM103424-21, a Louisiana Board of Regents support fund grant LEQSF (2021-24)-RD-A-22, and an NIH/NCI R15 award grant 1R15 CA290568-01. The Huang research laboratory is supported by the Louisiana Cancer Research Center (LCRC) Seed Award, the LSU Collaborative Cancer Research Initiative (CCRI) Award, and LSU Health Shreveport-Ochsner Collaborative Intramural Research Program (CIRP) Award.

Acknowledgments: We extend our heartfelt gratitude to all the authors for their exceptional contributions to this Special Issue (SI) "PI3K/AKT/mTOR Signaling Network in Human Health and Diseases". We are immensely thankful to the reviewers for their insightful and constructive critiques. Our sincere appreciation to the Editors-in-Chief and the Assistant Editor for their unwavering support throughout the inception and development of this SI project. This SI underscores the dire need for continued research into the PI3K/AKT/mTOR Signaling Network and its pivotal role in diseases, injuries, infections, and treatments.

Conflicts of Interest: The authors declare no conflicts of interest.

List of Contributions

1. Siddika, T.; Balasuriya, N.; Frederick, M.I.; Rozik, P.; Heinemann, I.U.; O'Donoghue, P. Delivery of Active AKT1 to Human Cells. *Cells* **2022**, *11*, 3834.
2. Ghodsinia, A.A.; Lego, J.-A.M.T.; Garcia, R.L. Mutation-Associated Phenotypic Heterogeneity in Novel and Canonical PIK3CA Helical and Kinase Domain Mutants. *Cells* **2020**, *9*, 1116.
3. Ferreri, A.; Lang, V.; Kaufmann, R.; Buerger, C. mTORC1 Activity in Psoriatic Lesions Is Mediated by Aberrant Regulation through the Tuberous Sclerosis Complex. *Cells* **2022**, *11*, 2847.
4. George, B.; Gui, B.; Raguraman, R.; Paul, A.M.; Nakshatri, H.; Pillai, M.R.; Kumar, R. AKT1 Transcriptomic Landscape in Breast Cancer Cells. *Cells* **2022**, *11*, 2290.
5. Wilczek, M.P.; Armstrong, F.J.; Mayberry, C.L.; King, B.L.; Maginnis, M.S. PI3K/AKT/mTOR Signaling Pathway Is Required for JCPyV Infection in Primary Astrocytes. *Cells* **2021**, *10*, 3218.
6. Ma, J.; Shen, M.; Yue, D.; Wang, W.; Gao, F.; Wang, B. Extracellular Vesicles from BMSCs Prevent Glucocorticoid-Induced BMECs Injury by Regulating Autophagy via the PI3K/Akt/mTOR Pathway. *Cells* **2022**, *11*, 2104.
7. Bamodu, O.A.; Chang, H.-L.; Ong, J.-R.; Lee, W.-H.; Yeh, C.-T.; Tsai, J.-T. Elevated PDK1 Expression Drives PI3K/AKT/MTOR Signaling Promotes Radiation-Resistant and Dedifferentiated Phenotype of Hepatocellular Carcinoma. *Cells* **2020**, *9*, 746.
8. Huot, J.R.; Thompson, B.; McMullen, C.; Marino, J.S.; Arthur, S.T. GSI Treatment Preserves Protein Synthesis in C2C12 Myotubes. *Cells* **2021**, *10*, 1786.
9. Luo, J.; Odaka, Y.; Huang, Z.; Cheng, B.; Liu, W.; Li, L.; Shang, C.; Zhang, C.; Wu, Y.; Luo, Y.; et al. Dihydroartemisinin Inhibits mTORC1 Signaling by Activating the AMPK Pathway in Rhabdomyosarcoma Tumor Cells. *Cells* **2021**, *10*, 1363.
10. Chou, P.-C.; Rajput, S.; Zhao, X.; Patel, C.; Albaciete, D.; Oh, W.J.; Daguplo, H.Q.; Patel, N.; Su, B.; Werlen, G.; et al. mTORC2 Is Involved in the Induction of RSK Phosphorylation by Serum or Nutrient Starvation. *Cells* **2020**, *9*, 1567.
11. Hsueh, Y.-J.; Meir, Y.-J.J.; Yeh, L.-K.; Wang, T.-K.; Huang, C.-C.; Lu, T.-T.; Cheng, C.-M.; Wu, W.-C.; Chen, H.-C. Topical Ascorbic Acid Ameliorates Oxidative Stress-Induced Corneal Endothelial Damage via Suppression of Apoptosis and Autophagic Flux Blockage. *Cells* **2020**, *9*, 943.
12. Su, Y.-L.; Luo, H.-L.; Huang, C.-C.; Liu, T.-T.; Huang, E.-Y.; Sung, M.-T.; Lin, J.-J.; Chiang, P.-H.; Chen, Y.-T.; Kang, C.-H.; et al. Galectin-1 Overexpression Activates the FAK/PI3K/AKT/mTOR Pathway and Is Correlated with Upper Urinary Urothelial Carcinoma Progression and Survival. *Cells* **2020**, *9*, 806.
13. Walkowski, B.; Kleibert, M.; Majka, M.; Wojciechowska, M. Insight into the Role of the PI3K/Akt Pathway in Ischemic Injury and Post-Infarct Left Ventricular Remodeling in Normal and Diabetic Heart. *Cells* **2022**, *11*, 1553.
14. Loissell-Baltazar, Y.A.; Dokudovskaya, S. SEA and GATOR 10 Years Later. *Cells* **2021**, *10*, 2689.
15. Chen, M.; Dai, Y.; Liu, S.; Fan, Y.; Ding, Z.; Li, D. TFEB Biology and Agonists at a Glance. *Cells* **2021**, *10*, 333.

References

1. Nisar, S.; Hashem, S.; Macha, M.A.; Yadav, S.K.; Muralitharan, S.; Therachiyil, L.; Segeena, G.; Al-Naemi, H.; Haris, M.; Bhat, A.A. Exploring dysregulated signaling pathways in cancer. *Curr. Pharm. Des.* **2020**, *26*, 429–445. [CrossRef] [PubMed]
2. Glaviano, A.; Foo, A.S.C.; Lam, H.Y.; Yap, K.C.H.; Jacot, W.; Jones, R.H.; Eng, H.; Nair, M.G.; Makvandi, P.; Geoerger, B.; et al. PI3K/AKT/mTOR signaling transduction pathway and targeted therapies in cancer. *Mol. Cancer* **2023**, *22*, 138. [CrossRef] [PubMed]
3. Hoxhaj, G.; Manning, B.D. The PI3K-AKT network at the interface of oncogenic signalling and cancer metabolism. *Nat. Rev. Cancer* **2020**, *20*, 74. [CrossRef] [PubMed]
4. Yu, Y.; Yoon, S.O.; Poulogiannis, G.; Yang, Q.; Ma, X.M.; Villén, J.; Kubica, N.; Hoffman, G.R.; Cantley, L.C.; Gygi, S.P.; et al. Phosphoproteomic analysis identifies Grb10 as an mTORC1 substrate that negatively regulates insulin signaling. *Science* **2011**, *332*, 1322–1326. [CrossRef] [PubMed]
5. Roy, T.; Boateng, S.T.; Uddin, M.B.; Banang-Mbeumi, S.; Yadav, R.K.; Bock, C.R.; Folahan, J.T.; Siwe-Noundou, X.; Walker, A.L.; King, J.A.; et al. The PI3K-Akt-mTOR and Associated Signaling Pathways as Molecular Drivers of Immune-Mediated Inflammatory Skin Diseases: Update on Therapeutic Strategy Using Natural and Synthetic Compounds. *Cells* **2023**, *12*, 1671. [CrossRef] [PubMed]
6. Tostivint, E.P.; Thibault, B.; Guibert, J.G. Targeting PI3K Signaling in Combination Cancer Therapy. *Trends Cancer* **2017**, *3*, 454–469. [CrossRef] [PubMed]
7. Chamcheu, J.C.; Roy, T.; Uddin, M.B.; Banang-Mbeumi, S.; Chamcheu, R.-C.N.; Walker, A.L.; Liu, Y.-Y.; Huang, S. Role, and Therapeutic Targeting of the PI3K/Akt/mTOR Signaling Pathway in Skin Cancer: A Review of Current Status and Future Trends on Natural and Synthetic Agents Therapy. *Cells* **2019**, *8*, 803. [CrossRef] [PubMed]
8. Wang, J.; Cui, B.; Chen, Z.; Ding, X. The regulation of skin homeostasis, repair, and the pathogenesis of skin diseases by spatiotemporal activation of epidermal mTOR signaling. *Front. Cell Dev. Biol.* **2022**, *10*, 950973. [CrossRef] [PubMed]

Article

Delivery of Active AKT1 to Human Cells

Tarana Siddika [1], Nileeka Balasuriya [1], Mallory I. Frederick [1], Peter Rozik [1], Ilka U. Heinemann [1,*] and Patrick O'Donoghue [1,2,*]

[1] Department of Biochemistry, The University of Western Ontario, London, ON N6A 5C1, Canada
[2] Department of Chemistry, The University of Western Ontario, London, ON N6A 5C1, Canada
* Correspondence: ilka.heinemann@uwo.ca (I.U.H.); patrick.odonoghue@uwo.ca (P.O.)

Abstract: Protein kinase B (AKT1) is a serine/threonine kinase and central transducer of cell survival pathways. Typical approaches to study AKT1 biology in cells rely on growth factor or insulin stimulation that activates AKT1 via phosphorylation at two key regulatory sites (Thr308, Ser473), yet cell stimulation also activates many other kinases. To produce cells with specific AKT1 activity, we developed a novel system to deliver active AKT1 to human cells. We recently established a method to produce AKT1 phospho-variants from *Escherichia coli* with programmed phosphorylation. Here, we fused AKT1 with an N-terminal cell penetrating peptide tag derived from the human immunodeficiency virus trans-activator of transcription (TAT) protein. The TAT-tag did not alter AKT1 kinase activity and was necessary and sufficient to rapidly deliver AKT1 protein variants that persisted in human cells for 24 h without the need to use transfection reagents. TAT-pAKT1^{T308} induced selective phosphorylation of the known AKT1 substrate GSK-3α, but not GSK-3β, and downstream stimulation of the AKT1 pathway as evidenced by phosphorylation of ribosomal protein S6 at Ser240/244. The data demonstrate efficient delivery of AKT1 with programmed phosphorylation to human cells, thus establishing a cell-based model system to investigate signaling that is dependent on AKT1 activity.

Keywords: cell penetrating peptide; cellular signaling; kinase; protein kinase B (AKT1); phosphoinositide-dependent kinase (PDK1); recombinant protein; trans-activator of transcription (TAT)

Citation: Siddika, T.; Balasuriya, N.; Frederick, M.I.; Rozik, P.; Heinemann, I.U.; O'Donoghue, P. Delivery of Active AKT1 to Human Cells. *Cells* **2022**, *11*, 3834. https://doi.org/10.3390/cells11233834

Academic Editors: Jean Christopher Chamcheu, Claudia Bürger and Shile Huang

Received: 19 October 2022
Accepted: 28 November 2022
Published: 29 November 2022

1. Introduction

Protein kinase B (PKB or AKT1) is a serine/threonine kinase and a key mediator of cell growth and survival processes, including glucose metabolism, apoptosis, transcription, cell proliferation, and migration [1,2]. The AKT protein family encompasses three isoforms: AKT1, AKT2, and AKT3 [3]. The AKT isoforms have both overlapping substrates and unique functions that are not compensated for by the other isoforms due to their individual subcellular localization [4]. Overexpression of AKT isoforms is associated with multiple human cancers [5–7], and AKT1 is a prime drug target [8] that is hyper-phosphorylated and overactive in most human cancers [9]. AKT1 is activated by phosphorylation at two key regulatory sites (Thr308 and Ser473) that are both used as clinical diagnostic markers [10]. Activated AKT1 phosphorylates downstream targets that stimulate cell survival and inhibit apoptosis [11] (Figure 1). Thr308 and Ser473 phosphorylation is also regulated by phosphatases. Protein phosphatase 2A (PP2A) dephosphorylates AKT1 at Thr308, and PH domain leucine-rich repeat phosphatase (PHLPP) dephosphorylates Ser473 [12,13].

AKT1 contains three conserved domains, namely the N-terminal pleckstrin homology domain (PH), followed by the central kinase catalytic (CAT) domain and the C-terminal extension (EXT) domain containing the hydrophobic motif (HM) and proline-rich sequences [14,15]. The PH domain plays an important role in co-localization with upstream kinases and translocation of AKT1 from the cytoplasm to the cell membrane where it is activated. In response to growth factor, the phosphoinositide 3-kinase (PI3K) phosphorylates phosphatidylinositol (4,5)-bisphosphate (PIP2) to generate phosphatidylinositol (3,4,5)-trisphosphate (PIP3). As a result, PH domain-containing proteins, including AKT1, bind to PIP3 and transiently anchor to the plasma membrane [16,17].

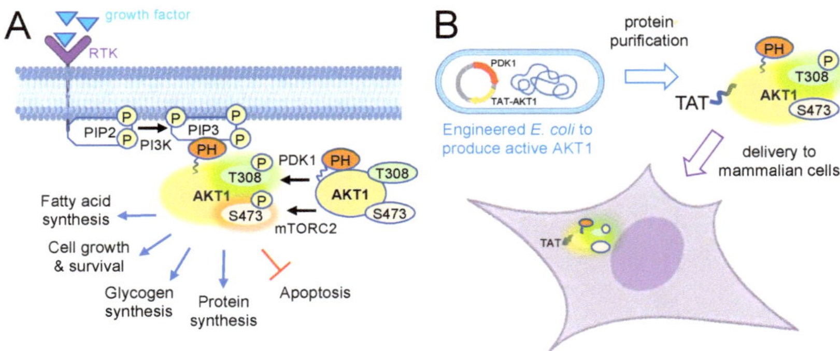

Figure 1. AKT1 activation pathway and delivery of site-specifically phosphorylated AKT1 into human cells. (**A**) Protein kinase B (AKT1) is normally activated in response to growth factors that bind to a receptor tyrosine kinase (RTK) in the membrane, activating phosphoinositide 3-kinase (PI3K). PI3K subsequently phosphorylates phosphatidylinositol-4,5-bisphosphate (PIP2) to phosphatidylinositol-3,4,5-trisphosphate (PIP3). When the pleckstrin homology (PH) domain of AKT1 binds to the membrane phospholipid PIP3, the auto-inhibitory effect of the PH domain is released from AKT1, exposing the activation sites of AKT1 for phosphorylation by the up-stream kinases mTOR complex 2 (mTORC2) at Ser473 and phosphoinositide-dependent kinase 1 (PDK1) phosphorylates Thr308. The activation of AKT1 initiates a phosphorylation cascade, leading to the activation of many downstream pathways controlling cell growth and protein synthesis as well as inhibition of apoptosis. (**B**) Trans-Activator of Transcription (TAT) is a cell penetrating peptide facilitating protein delivery to mammalian cells. TAT-tagged AKT1 was expressed and purified from *E. coli* cells. PDK1 was co-expressed in *E. coli* to facilitate phosphorylation of AKT1 at Thr308 during protein production, yielding active, phosphorylated TAT-tagged AKT1. TAT-tagged AKT1 variants were then delivered to human cells.

At the cell membrane, another PH-domain containing protein, phosphoinositide-dependent kinase 1 (PDK1) binds to PIP3 and phosphorylates AKT1 in the catalytic domain at Thr308 [18]. Activated by receptor tyrosine kinases (RTK), mammalian target of rapamycin complex 2 (mTORC2) phosphorylates AKT1 in the HM domain at Ser473 [15] (Figure 1). Fully activated AKT1 has many downstream substrates [19]. Phosphorylated AKT1 plays an important role in glucose metabolism by inhibiting glycogen synthase kinase-3 (GSK-3) isoforms through phosphorylation of GSK-3α at Ser21 and GSK-3β at Ser9 [20,21]. AKT1-dependent phosphorylation of the tuberous sclerosis 2 (TSC2) protein counteracts its inhibitory effect on mammalian target of rapamycin complex 1 (mTORC1), a regulator of protein synthesis that affects many cellular processes including cell survival and growth. Stimulated mTORC1 activates the p70 ribosomal protein S6 kinase which leads to phosphorylation of ribosomal protein S6 at Ser240/244 [22–24] and stimulation of protein synthesis [8].

Traditional approaches to investigate AKT1 biology are based on the over-expression of AKT1 followed by stimulation of cells with insulin or epidermal growth factor (EGF) leading to phosphorylation and activation of AKT1 [25,26]. EGF treatment activates AKT1 as well as Raf-1 kinase [27], mitogen activated protein kinases [28], and many other protein kinases and signaling cascades [29,30]. We previously developed an approach to generate recombinant and active human AKT1 from bacterial cells with site-specific phosphorylation [19,31–34]. Here, we fused the cell penetrating TAT peptide to recombinant AKT1 constructs and produced and purified both inactive (TAT-AKT1) and active (TAT-pAKT1^{T308}) AKT1 variants. We demonstrated the delivery of TAT-AKT1 and TAT-pAKT1^{T308} to human cells in culture. We found that active TAT-pAKT1^{T308} but not inactive TAT-AKT1 delivery led to increased phosphorylation of AKT1 targets and activated AKT1 signaling pathways in human cells.

2. Materials and Methods

2.1. Molecular Cloning and Gene Synthesis

The human AKT1 gene, which was acquired from the Harvard PlasmidID repository service (Boston, MA, USA), was previously cloned [33] using *NcoI*/*NotI* restriction sites into the pCDF-Duet-1 vector in multiple-cloning site 1 (MCS1). The unphosphorylated AKT1 was produced from this pCDF-Duet-1 vector, with no gene cloned into multiple-cloning site 2 (MCS2). For pAKT1^{T308} production, the MCS2 contains human PDK1 as described before [33]. Briefly, the human PDK1 gene, which was also purchased from the Harvard PlasmidID repository, was cloned using *KpnI*/*NdeI* into MCS2 of pCDF-Duet-1 as detailed before. For TAT tagged gene constructs, the TAT-tag (YGRKKRRQRRR) and TAT-mCherry-tagged AKT1 constructs were codon optimized for *Escherichia coli* expression and synthesized (Azenta Life Sciences, Chelmsford, MA, USA). The His$_6$-TAT tag DNA sequence was derived from the pTAT-HA vector, which was a kind gift from Steven Dowdy (Addgene, Watertown, MA, USA, plasmid #35612) [35]. The synthetic genes were flanked by *NcoI*/*NotI* restriction sites and cloned at Azenta Life Sciences into pCDF-Duet-1 and separately into the pCDF-Duet-1-PDK1 vector noted above. Each of the resulting AKT1 constructs contain a His$_6$ tag for affinity purification. Sequences of each construct were verified by DNA sequencing at Azenta Life Sciences and are available in the Supplementary Material Section S.2.

2.2. Protein Production and Purification

All AKT1 protein variants were expressed in *E. coli* BL21(DE3) as His-tag fusion proteins and the six AKT1 variants (AKT1, pAKT1^{T308}, TAT-AKT1, TAT-pAKT1^{T308}, TAT-mCherry-AKT1, and TAT-mCherry-pAKT1^{T308}) were purified by affinity chromatography followed by size exclusion and anion exchange chromatography (see below). For protein production, *E. coli* BL21(DE3) chemically competent cells were transformed with the indicated pCDF-Duet-1 vector and selected on Luria-Bertani (LB) agar with 50 μg/mL spectinomycin. Independent protein preparations were initiated with a single transformed *E. coli* colony that was used to inoculate 100 mL of liquid LB media with spectinomycin at 50 μg/mL and incubated overnight at 37 °C with shaking. From the starter culture, a 10 mL inoculum was added to 4 × 1 L of LB with spectinomycin at 50 μg/mL. The preparative cultures were grown with shaking at 37 °C until A$_{600}$ = 0.8, when protein production was induced by adding 300 μM of isopropyl β-D-1-thiogalactopyranoside (IPTG). Bacterial cultures were then incubated at 16 °C for 18 h with shaking. Cells were pelleted by centrifugation at 5500× *g* and stored at −80 °C until further analysis. Cell pellets were suspended in lysis buffer (20 mM 4-(2-Hydroxyethyl)-1-piperazine ethanesulfonic acid (HEPES) pH 7.4, 150 mM NaCl, 5 mM β-mercaptoethanol, 1 mM phenylmethylsulfonyl fluoride (PMSF), 1 tablet of ethylenediaminetetraacetic acid (EDTA)-free mini protease inhibitor cocktail (Roche Canada, Mississauga, ON, Canada) and 10 mM imidazole). For the pAKT1^{T308} protein variants, phosphatase inhibitors (1 mM Na$_3$VO$_4$, 5 mM NaF) were added to the lysis buffer. Following sonication, cell lysates were centrifuged at 170,000× *g* for 1 h. The cell free extract was filtered using a sterile 0.45 μm filter.

2.3. Affinity Column Chromatography

Ni^{2+}-nitrilotriacetic acid (NTA) affinity column chromatography (HisTrap FF 5 mL column volume, #17525501, Sigma-Aldrich Canada, Oakville, ON, Canada) was used to purify all His-tagged AKT1 protein variants using an ÄKTA Pure L1 FPLC system (Cytiva Life Sciences, Shrewsbury, MA, USA). Columns were equilibrated with buffer A (20 mM

HEPES pH 7.4, 150 mM NaCl, 5 mM β-mercaptoethanol, 15 mM imidazole), and cell lysates were loaded onto the column. Unbound proteins were washed with 10 column volumes of buffer A, and 8 column volumes of buffer B (20 mM HEPES pH 7.4, 150 mM NaCl, 5 mM β-mercaptoethanol, 30 mM imidazole). The His_6-tagged proteins were eluted using 6 column volumes of elution buffer (20 mM HEPES pH 7.4, 150 mM NaCl, 5 mM β-mercaptoethanol, and 70 mM imidazole). The $pAKT1^{T308}$ proteins were eluted in elution buffer containing 200 mM imidazole. Elution fractions of 1 mL were taken over the total 5 mL elution volume. Samples from each fraction were separated by 10% sodium dodecyl sulfate (SDS)-polyacrylamide gel electrophoresis (PAGE) and visualized with Coomassie staining to identify fractions containing the His_6-tagged AKT1 proteins.

2.4. Size Exclusion Chromatography

Affinity purified AKT1 proteins variants were further purified by size exclusion chromatography using a 13 mL Superdex 200 (Cytiva Life Sciences) gel filtration column on the ÄKTA Pure L1 FPLC system. The column was equilibrated with 3 column volumes of buffer (20 mM HEPES pH 7.4, 300 mM NaCl, 5 mM β-mercaptoethanol). The pooled and concentrated fractions from the affinity chromatography purification were loaded onto the column with a flow rate of 0.5 mL/min and 1 mL elution fractions were collected. Elution fractions were separated by 10% SDS-PAGE to identify fractions containing purified proteins and pure fractions were pooled and concentrated using Vivaspin 6 spin-column (5 mL, 10 kDa molecular weight cutoff (MWCO), #14558502, Sartorius Canada, Oakville, ON, Canada). Concentrated AKT1 proteins were dialyzed into storage buffer (20 mM HEPES pH 7.4, 150 mM NaCl, 5 mM β-mercaptoethanol, 10% glycerol) and stored at $-80\,^{\circ}$C until further use.

2.5. Anion Exchange Chromatography

$pAKT1^{T308}$ protein variants were further purified by anion exchange chromatography using the ÄKTA Pure L1 FPLC system (Cytiva Life Sciences) with a HiTrap Q HP anion exchange 1 mL column. The column was equilibrated with 5 column volumes of buffer A1 (20 mM HEPES pH 7.4, 150 mM NaCl, 5 mM β-mercaptoethanol). Unbound proteins were washed with 10 column volumes of buffer A1, and pAKT1 proteins were eluted with a gradient of 0 to 100% buffer B1 (20 mM HEPES pH 8.0, 300 mM NaCl, 5 mM β-mercaptoethanol). The flow rate was maintained at 0.5 mL/min and 1 mL fractions were collected. Fractions were separated by 10% SDS-PAGE and fractions containing pure pAKT protein were pooled and concentrated using a Vivaspin 20 concentrator (25 mL, 10 kDa MWCO, Sartorius, #14558502). Purified pAKT1 proteins were dialyzed into storage buffer (20 mM HEPES pH 7.4, 150 mM NaCl, 5 mM β-mercaptoethanol, 10% glycerol) and stored at $-80\,^{\circ}$C until further use.

2.6. AKT1 Kinase Activity Assay

To assure equal loading in all assays, purified AKT protein concentrations were measured using a Bradford assay (Biorad Canada, Mississauga, ON, Canada). The activity of each AKT1 and TAT-AKT1 variant was characterized by performing kinase assays in the presence of 200 µM substrate peptide CKRPRAASFAE (SignalChem, Vancouver, BC, Canada) derived from the natural AKT1 substrate GSK-3β. Kinase assays were performed in a buffer containing 25 mM 3-(N-morpholino)propanesulfonic acid (MOPS) pH 7.0, 12.5 mM β-glycerolphosphate, 25 mM $MgCl_2$, 5 mM ethylene glycolbis(β-aminoethyl ether)-N,N,N',N'-tetraacetic acid (EGTA) pH 8.0, 2 mM EDTA, 0.2 mM ATP, and 0.4 µCi (33 nM) [γ-^{32}P]-ATP in a 30 µL reaction volume. Reactions were incubated at 37 $^{\circ}$C with agitation and time points were taken over a 15 min time course. Reactions were initiated by the addition of 18 pmol of the specific AKT1 variant for a final enzyme concentration of 600 nM and reactions were quenched by spotting 3 µL aliquots of reaction solution on P81 paper at 5, 10, and 15 min. Following washes with 1% phosphoric acid (3 $\times$ 10 min) and 95% ethanol (1 $\times$ 5 min), the P81 paper was air-dried. Addition of radioactive ^{32}P to the GSK-3β peptide was detected by phosphorimaging and imaged using a Storm 860 Molecular Imager (Molecular Dynamics, Sunnyvale, CA, USA). Spot intensity was quantitated using ImageQuant TL software (Cytiva Life Sciences).

2.7. Protein Incubation with Cells

HEK 293T cells were maintained in Dulbecco's modified eagle's medium (DMEM, Cellgro, ThermoFisher Scientific, Ottawa, ON, Canada) containing 10% fetal bovine serum and 1% penicillin/streptomycin at 37 °C with 5% CO_2. Equal numbers of cells were seeded onto 6 well plates and cultured for 2–4 days until 90% confluence. AKT1 or TAT-AKT1 protein variants were added to a final concentration of 1 μM. After 1 h incubation, cells were washed twice with phosphate-buffered saline (PBS), harvested, and stored at −80 °C until further use.

2.8. Transfection of HEK 293T Cells

A pcDNA3.1 plasmid encoding mCherry-AKT1 [33] was used to genetically over-express AKT1 in HEK 293T cells. The same number of cells were plated onto 6 well plates, and cells were transfected with 1 μg of pcDNA3.1 plasmid using Lipofectamine 3000 (ThermoFisher Scientific, #L3000015) according to the manufacturer's instructions at 70% confluence. At 24 h after transfection, cells were stimulated with epidermal growth factor (EGF) for 10 min, harvested, and stored at −80 °C until further use.

2.9. Western Blotting

For cells incubated with AKT1 or TAT-AKT1 protein variants or for cells transfected with pCDNA3.1-AKT1, protein extraction was initiated by suspending harvested cells in cold lysis buffer (50 mM Na_2HPO_4, 1 mM $Na_4P_2O_7$, 20 mM NaF, 2 mM EDTA, 2 mM EGTA, 1 mM dithiothreitol, 300 μM PMSF, and protease inhibitor cocktail (Roche Canada, #04693159001)) and incubated on ice for 10 min with vortexing every 2 min. Lysates were cleared by centrifugation at 15,000× g for 15 min at 4 °C. Protein concentration was measured by Bradford assay and equal concentration of proteins (50 μg) were separated in 10% SDS-PAGE gel for each sample.

For western blotting of the purified recombinant AKT1 proteins, proteins were denatured using 3 × SDS sample loading buffer and extracted proteins from HEK 293T cells were diluted to separate 50 μg of protein by 10% SDS-PAGE. Proteins were transferred to a polyvinylidene fluoride (PVDF) membrane using a Turbo-Blot Turbo transfer system (Biorad Canada). The membranes were blocked with either 5% bovine serum albumin (BSA) in 1 × tris buffered saline with 0.1% Tween-20 (TBST) or 5% milk in 1 × PBS with 0.1% Tween-20 (PBST) for 1 h at room temperature. The membrane was immunoblotted with the indicated primary antibody overnight at 4 °C followed by 3 × 10 min washes in TBST or PBST. Next, the membrane was incubated with a secondary antibody for 1 h at room temperature and subsequently washed 3 × 10 min in TBST. Membranes were stored in TBS or PBS for further analysis. The blots were visualized by fluorescence on a LiCOR imaging system (LiCOR Biosciences, Lincoln, NE, USA). The following primary antibodies were used: AKT1, Cell Signaling Technology, Danvers, MA, USA #2938; pAKT (Thr308), Cell Signaling Technology #4060; GSK-3α/β, Santa Cruz Biotechnology sc-7291 (0011-A); pGSK-3α/β, Cell Signaling Technology #9331S; ribosomal protein S6, Cell Signaling Technology #2317S; and pS6 (S240/S244), Cell Signaling Technology #2215S. A glyceraldehyde 3-phosphate dehydrogenase (GAPDH) antibody (Abcam, Waltham, MA, USA, #ab8245) was used as loading control. The secondary fluorescent antibodies were Goat-anti-Mouse (Sigma-Aldrich Canada, #AQ127) and Donkey-anti-Rabbit (Sigma-Aldrich Canada, #AP182P).

2.10. Microscopy and Cell Imaging

HEK 293T cells were cultured in 6 well dishes containing DMEM (Cellgro, ThermoFisher Scientific) supplemented with 10% fetal bovine serum and 1% penicillin/streptomycin at 37 °C with 5% CO_2. For cell imaging experiments, cells were cultured in 24 well plates and 0.5 µM of each AKT1 protein variant was added to wells when cells reached 90% confluence and then incubated for 24 h. Following incubation with the AKT1 protein variants, cells were washed twice with DMEM prior to imaging. For HEK 293T cells incubated with mCherry-AKT1 protein variants or for cells transfected with plasmid encoded mCherry-AKT1, the cells were imaged at 24 h after incubation or transfection with an EVOS FL Auto 2 imaging system (ThermoFisher) using brightfield and fluorescence imaging. To detect mCherry fluorescence, cells incubated with TAT-mCherry-AKT1 protein variants or cells transfected with the mCherry-AKT1 plasmid were imaged with the RFP filter cube (excitation 542 ± 20 nm, emission 593 ± 40 nm). All images were captured with the EVOS 4× objective (fluorite, PH, long-working distance, 0.13 numerical aperture/10.58 mm working distance). Fluorescent images from wells representing three biological replicates for each condition were quantified using Image J software (National Institutes of Health, National Institute of Mental Health, Bethesda, MD, USA). Image J was used to count the total number of cells in brightfield images and then used to determine the total number of red fluorescing cells to calculate transfection efficiency. Statistical analysis was performed using GraphPad Prism software (GraphPad, San Diego, CA, USA). Details for confocal imaging are provided in the Supplementary Materials (S.3 Supplementary Methods).

2.11. Cytotoxicity Assay

Equal numbers of HEK 293T cells were seeded in 96 well dishes containing DMEM (Cellgro, ThermoFisher Scientific) supplemented with 10% fetal bovine serum and 1% penicillin/streptomycin at 37 °C with 5% CO_2. At 90% confluence, 0.5 µM of TAT-AKT1 protein variants were added to each well and incubated for 24 h. The CytoTox-Glo cytotoxicity assay kit (Promega, #G9290, Madison, WI, USA) was used to measure the total cell and dead cell numbers by following the supplied protocol. Each experiment was performed in three biological replicates.

2.12. Trypan Blue and Sytox Blue Assays

Equal numbers of HEK 293T cells were cultured in 96 well plates containing DMEM (Cellgro) supplemented with 10% fetal bovine serum and 1% penicillin/streptomycin at 37 °C in 5% CO_2. When cell confluence reached 90%, TAT-AKT1 and TAT-pAKT1^{T308} proteins were added at a concentration of 1 µM and incubated for 1 h or 24 h. An aliquot of 100 µL of cells from each well was transferred to a microcentrifuge tube and 100 µL of trypan blue dye was added. Viable and dead cells were counted under a light microscope after 1 h or 24 h of protein delivery. Sytox blue (ThermoFisher, #S34857) was used to measure the fluorescence of dead cells [36] at 24 h after TAT-AKT1 protein incubation. Cell images were taken with a EVOS FL Auto 2 cell imaging system using a CFP filter cube (emission 445 ± 45 nm, excitation 510 ± 42 nm). Dead cell fluorescence was quantified using image J software. Each experiment was performed in three biological replicates.

2.13. Quantification and Statistical Analysis

For all experiments, at least three biological replicates were used and indicated on all graphs. Graphs were generated in Microsoft Excel. Western blots were quantified using Image Lab (Biorad) and the data were compiled and normalized in Microsoft Excel. *p*-values were calculated by one-way analysis of variance (ANOVA), two-way ANOVA, or Student's two-tailed *t*-test as indicated.

3. Results

3.1. Production and Purification of Recombinant AKT1 and Site-Specifically Phosphorylated AKT1 Variants

TAT peptide tags were shown by us [37] and others [38–40] to enable rapid and efficiency delivery of recombinant proteins to mammalian cells. To determine if a TAT-peptide fusion will enable delivery of recombinant AKT1 to human cells (Figure 1B), we first constructed, produced, and purified AKT1 variants with a TAT tag as well as AKT1 variants lacking the TAT-tag as before [19,33,34]. Full-length human AKT1 variants were expressed and produced in *E. coli* to generate unphosphorylated AKT1 or TAT-tagged AKT1. We also produced pAKT1^{T308} variants with or without the TAT-tag by co-expressing each variant with PDK1 to produce site-specific phosphorylated pAKT1^{T308} and TAT-pAKT1^{T308}. Using a similar approach to that we established previously [33], all variants were initially purified using Ni^{2+}-NTA affinity chromatography relying on the His$_6$-tag included in each construct (see Supplementary Material Section S.2). The unphosphorylated variants were purified further by size exclusion chromatography leading to pure fractions (Figure S1A,C). Adhering to our previous approach [33], the phosphorylated variants were purified by size exclusion and then further purified by anion exchange chromatography to yield pure fractions of pAKT1^{T308} (Figure S1B) and TAT-pAKT1^{T308} (Figure S1D).

To confirm incorporation of pThr308 in both the wild-type and TAT-tagged AKT1 variants we performed western blotting on the purified proteins (Figure S2). We blotted the purified proteins with both an AKT1 antibody (Figure S2A) and an antibody specific for pAKT^{T308} (Figure S2B). The AKT1 antibody confirmed equivalent loading of each of the purified AKT1 and TAT-AKT1 variants, and we found that the pAKT^{T308} specific antibody displayed a robust response to only the pAKT1^{T308} and TAT-pAKT1^{T308} variants. The blots indicate an equivalent level of phosphorylation on the pAKT1^{T308} and TAT-pAKT1^{T308} proteins (Figure S2C). We previously used multiple mass spectrometry approaches to characterize the pAKT1^{T308} variant and found stoichiometric incorporation of phosphate at the Thr308 site [19,33,34]. Here, we performed liquid chromatography tandem mass spectrometry (LC-MS/MS) analysis of the TAT-pAKT1^{T308} protein and independently confirmed incorporation of pThr308 (Figure S2D). In agreement with our previous studies of pAKT1^{T308} [33], the TAT-pAKT1^{T308} protein showed no evidence of phosphorylation at the other regulatory site, Ser473. Indeed, we observed only the Ser473 peptide (Figure S2E).

3.2. Enzymatic Activity of AKT1 and TAT-AKT1 Variants

To determine if addition of the TAT-tag affected AKT1 activity, we tested the enzymatic activity of AKT1 variants with and without a TAT-tag. The activity of each AKT1 variant was investigated using a standard kinase assay, as before [19,33,34]. Briefly, we monitored the transfer of a radiolabeled γ-phosphate from [γ-^{32}P]-ATP to a substrate peptide based on the natural AKT1 substrate, GSK-3β. We measured phospho-peptide production over a 15 min time course. While both the unphosphorylated AKT1 and TAT-AKT1 proteins showed no activity that was distinguishable from background measurements, we found that both pAKT1^{T308} and TAT-pAKT1^{T308} showed robust kinase activity that increased linearly during the time course (Figure 2A and Figure S3). There was no significant difference in the phosphorylation rate catalyzed by pAKT1^{T308} or TAT-pAKT1^{T308} (Figures 2B and S3), indicating that the TAT-tag does not perturb the activity of the enzyme. The activity data agree with our finding of a similar level of Thr308 phosphorylation in both the pAKT1^{T308} and TAT-pAKT1^{T308} variants (Figure S2B,C).

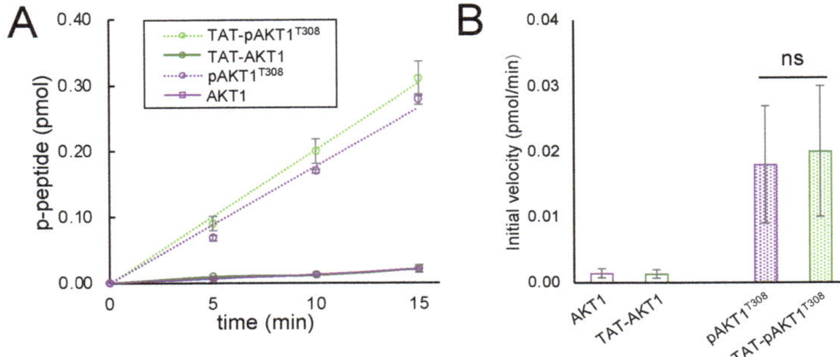

Figure 2. Enzymatic activity of AKT1 is dependent on phosphorylation at Thr308 and was not affected by fusion of AKT1 to the cell penetrating TAT-tag. (**A**) Time courses and (**B**) initial velocity of three independent in vitro kinase activity assays. AKT1 variants were incubated with a GSK-3β substrate peptide and [γ-^{32}P]-ATP and spotted on filter paper. Unreacted [γ-^{32}P]-ATP was washed away and reaction products visualized by phosphorimaging (Figure S3) and quantified. AKT1 (purple) and TAT-AKT1 (green) did not show significant activity above background, whereas pAKT1^{T308} (purple dotted) and TAT-pAKT1^{T308} (green dotted) were enzymatically active. The TAT-tag did not significantly alter pAKT1^{T308} activity. Significant differences were calculated by two-tailed *t*-test (ns—not significant).

3.3. Delivery of TAT-AKT1 Variants to Human Cells

We used the HEK 293T cell line as a model system to characterize TAT-tag dependent delivery of AKT1 to human cells. In future and on-going studies beyond the scope of this work, we will also deliver TAT-AKT1 variants to other mammalian cell types. HEK 293T cells are well established cell biological model systems that display robust growth and high transfection efficiencies [41]. This aspect was important for our comparison of TAT-dependent AKT1 delivery to more traditional transient transfection approaches described below (see Section 3.6). AKT1 is also highly expressed in many human tissues, including the kidney [42], making HEK 293T cells an appropriate model system to study AKT1 signaling. Indeed, some of the initial characterizations of AKT1 activity [43] as well as more recent studies to screen AKT1 inhibitors [44] and to define new roles for AKT1 in diverse cellular processes including RNA metabolism [45,46], miRNA regulation [31], and non-sense mediated decay [47] each relied on HEK 293 cells as an appropriate cell line to investigate AKT1 biology.

Cells were grown to 90% confluence and then incubated with one of the AKT1 variants characterized above. After a 1-h incubation of the cells with AKT1, pAKT1^{T308}, TAT-AKT1, or TAT-pAKT1^{T308}, we used western blotting to demonstrate the TAT-tag mediated cellular uptake and to monitor the phosphorylation status of endogenous AKT, TAT-AKT1 and TAT-pAKT1^{T308} using specific antibodies for AKT1, pAKT1^{T308}, and GAPDH as a loading control. The AKT1 and pAKT1^{T308} constructs contain a His$_6$-tag derived from pCDF-Duet-1 as described before [33] that have a relative molecular weight of 3.1 kDa greater than the endogenous AKT1. The TAT-tagged AKT1 variants have a larger N-terminal tag, including both His$_6$ and the TAT peptide, derived from the pTAT-HA plasmid [35] (see Supplementary Material Section S.2). Thus, the TAT-tagged AKT1 variants were easily differentiated from endogenous AKT1 in the western blots due to their size difference of 6 kDa (Figure 3).

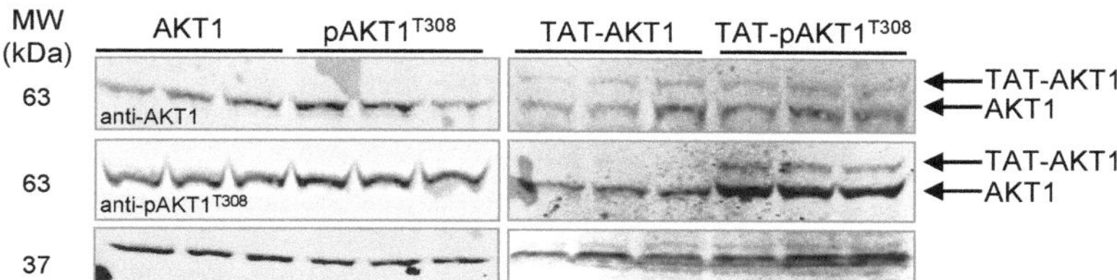

Figure 3. Delivery of TAT-AKT1 protein variants to HEK 293T cells. HEK 293T cells were incubated with AKT1, pAKT1^{T308}, TAT-AKT1, or TAT-pAKT1^{T308} for 1 h in three biological replicates. Cell extracts were separated by SDS-PAGE and immunoblotted with AKT1, pAKT^{T308}, and GAPDH specific antibodies. Cells incubated with AKT1 or pAKT1^{T308} showed a single band in all blots corresponding to endogenous AKT1. Two bands, corresponding to endogenous AKT1 and TAT-AKT1 separated by size, were apparent in cells transfected with TAT-tagged AKT1 variants, demonstrating effective delivery of both TAT-tagged AKT1 variants into cells. TAT-AKT1 delivery did not lead to phosphorylation of TAT-AKT1 in cells, as no phosphorylated TAT-AKT1 was detected by the pThr308 specific antibody after 1 h. The delivered TAT-pAKT1^{T308} was readily detected by phospho-specific antibody for pAKT^{T308}.

Following the protein incubation period, the cells were extensively washed with fresh media to remove any protein remaining outside of the cells. In western blots, we clearly observed a band in the anti-AKT1 blots corresponding to the higher molecular weight of the TAT-AKT1 proteins relative to endogenous AKT (Figure 3). In contrast, in cells incubated with the AKT1 variants lacking the TAT-tag, we observed only the endogenous AKT1 and no additional bands corresponding to the His$_6$-AKT1 or His$_6$-pAKT1^{T308} proteins. In agreement with our characterization of the purified TAT-AKT1 and TAT-pAKT1^{T308} proteins (Figure S3), only the TAT-pAKT1^{T308} protein was detected by the pAKT^{T308} specific antibody (Figure 3). Together the data show that both the TAT-AKT1 and TAT-pAKT1^{T308} proteins were delivered to the interior of the cells, and further, that the TAT-pAKT1^{T308} retained phosphorylation at Thr308 while the TAT-AKT1 protein did not acquire Thr308 phosphorylation in the cells (Figure 3).

To further verify that the AKT1 proteins lacking the TAT-tag were not taken up by the cells, we performed an independent experiment where HEK 293T cells were incubated with buffer only, with AKT1, or with pAKT1 proteins lacking the TAT-tag, and we found no change in the level of AKT1 compared to a GAPDH loading control in any of these conditions (Figure S4). Thus, the TAT-tag was necessary and sufficient to delivery TAT-AKT1 and TAT-pAKT1^{T308} variants to human cells.

3.4. Selective Phosphorylation of GSK-3α by TAT-pAKT1^{T308}

GSK-3 was the first identified AKT1-dependent substrate [48]. GSK-3 has two paralogs that are often referred to as isoforms, including a higher molecular weight GSK-3α and a lower molecular weight GSK-3β protein [49]. A typical approach to promote AKT activity in cells involves genetic expression of a myristoylated-AKT1 variant (myr-AKT1) and subsequent stimulation of cells with insulin or growth factors such as EGF [50] or platelet derived growth factor (PGDF) [51]. The myr-tag anchors AKT1 to the plasma membrane. Studies with PGDF show myr-AKT1 is activated and causes increased phosphorylation of GSK-3α at Ser21 (SGRARTSsFAEPGGG) and GSK-3β at Ser9 (SGRPRTTsFAESCKP) [51]. The AKT1 target peptides in GSK-3α and GSK-3β are homologous but not identical, and both can be visualized using anti-GSK-3 and anti-pGSK-3 specific antibodies.

To determine if TAT-pAKT1^{T308} is active once delivered to cells, we used western blotting to probe cells for GSK-3 and pGSK-3 levels following incubation of cells with AKT1, pAKT1^{T308}, TAT-AKT1, or TAT-pAKT1^{T308} proteins (Figure 4). Following incubation of HEK 293T cells with each of the indicated AKT1 variants, we identified no significant changes in the levels of GSK-3α or GSK-3β in comparison to a GAPDH loading control (Figure 4A,B). We next compared the levels of pGSK-3α and pGSK-3β to the GAPDH loading control (Figure 4A,C), and we also determined the ratio of phosphorylated GSK-3α and GSK-3β to the levels of the respective un-phosphorylated proteins (Figure 4D). In agreement with our findings above, neither of the AKT1 variants lacking the TAT-tag led to any stimulation of GSK-3 phosphorylation. In addition, the unphosphorylated TAT-AKT1 protein did not alter GSK-3 phosphorylation levels as anticipated. Strikingly, TAT-pAKT1^{T308} protein delivered to cells showed a robust 3-fold and significant increase in GSK-3α phosphorylation but no stimulation of GSK-3β phosphorylation was detected (Figure 4C,D). The data demonstrate that increased pGSK-3α levels were dependent on both the TAT-tag and the phosphorylation status of the delivered AKT1 protein.

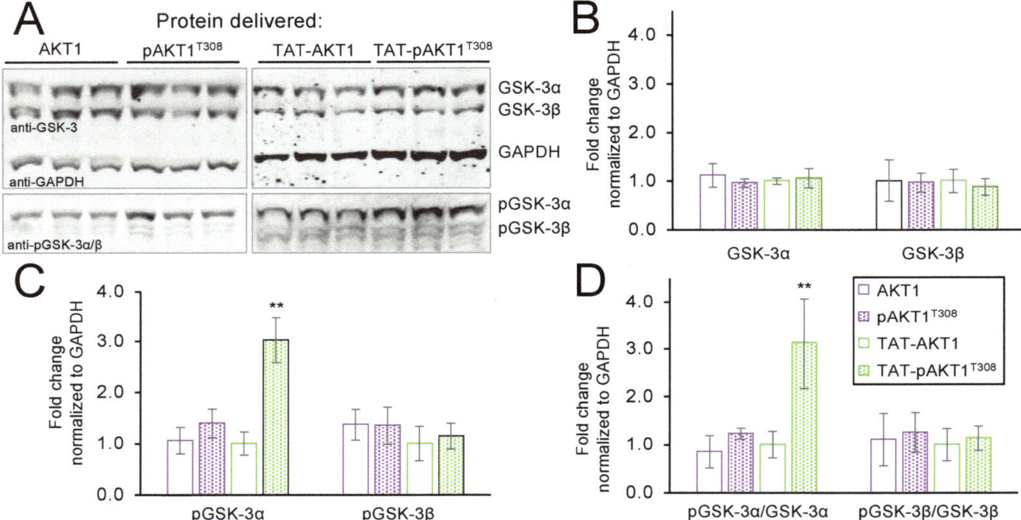

Figure 4. TAT-pAKT1^{T308} protein delivery stimulates AKT1 signaling. HEK 293T cells were incubated with AKT1, pAKT1^{T308}, TAT-AKT1, or TAT-pAKT1^{T308} for 1 h in three biological replicates. (**A**) Cell extracts were separated by SDS-PAGE and immunoblotted with GSK-3, pGSK-3, and GADPH specific antibodies. GSK-3 homologs GSK-3α and GSK-3β are direct substrates of AKT1. (**B**) Quantification of western blots showed no change in GSK-3α or GSK-3β protein abundance. Changes in phosphorylation of GSK-3α or GSK-3β were quantified and normalized to (**C**) GAPDH or (**D**) GSK-3α and GSK-3β, respectively, showing that TAT-pAKT1^{T308} specifically stimulates phosphorylation of GSK-3α, but not GSK-3β. Significant differences were calculated by two-tailed *t*-test and are indicated by asterisks (** $p < 0.01$).

3.5. TAT-pAKT1^{T308} Stimulates Downstream AKT1 Signaling to Ribosomal Protein S6

Although AKT1 is thought to regulate many cellular substrates directly and signaling pathways indirectly, one of the best characterized downstream pathways involves AKT1-dependent stimulation of protein synthesis via indirect phosphorylation of the ribosomal protein S6 at Ser240/Ser244 (reviewed in [52]). AKT1 directly phosphorylates the mammalian target of rapamycin (mTOR), which in turn phosphorylates and activates p70 S6 kinase (p70S6K) that then targets and phosphorylates ribosomal protein S6 at Ser240/Ser244 [53,54]. To determine if TAT-pAKT1^{T308} stimulates a downstream signaling pathway of AKT1, we used western blotting to measure S6 and pS6 levels in cells incubated with AKT1, pAKT1^{T308}, TAT-AKT1, or TAT-pAKT1^{T308} proteins (Figure 5A). We monitored S6 and pS6 levels relative to the GAPDH loading control (Figure 5A,B), and we determined the level of pS6 compared to the level of S6 protein in each condition (Figure 5A,C). We observed no significant changes in S6 protein levels in each condition (Figure 5B), and we found no significant changes in pS6 levels in cells treated with AKT1, pAKT1^{T308}, or TAT-AKT1. We did observe a significant 1.5-fold stimulation of S6 phosphorylation at Ser240/Ser244 only in cells treated TAT-pAKT1^{T308} (Figure 5B,C). The data indicate that TAT-pAKT1^{T308} stimulates signaling pathways downstream of AKT1.

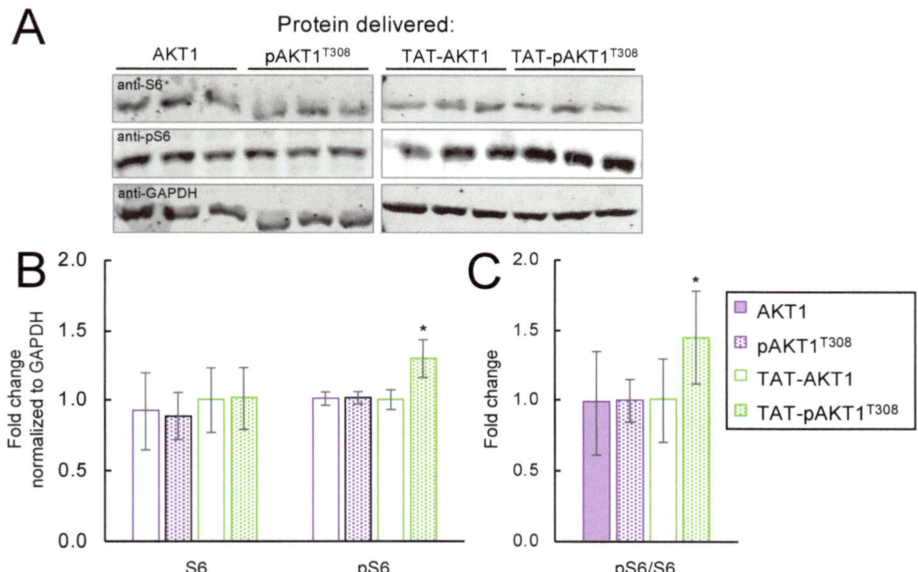

Figure 5. TAT-pAKT1^{T308} protein delivery stimulates downstream AKT1 signaling to ribosomal protein S6. HEK 293T cells were incubated with AKT1, pAKT1^{T308}, TAT-AKT1, and TAT-pAKT1^{T308} for 1 h in three biological replicates. (**A**) Western blots and (**B,C**) quantification of blots of cell extracts separated by SDS-PAGE and immunoblotted with ribosomal protein S6, pS6, and GADPH specific antibodies. S6 phosphorylation was significantly increased after incubation with TAT-pAKT1^{T308}, but not following incubation of cells with AKT1 or pAKT1 lacking the TAT tag or with unphosphorylated TAT-AKT protein variants. Significant differences were calculated by two-tailed *t*-test and are indicated by asterisks (* $p < 0.05$).

3.6. Genetic Over-Expression Model of AKT1 Activity in Stimulated Cells

As noted above, traditional approaches to generate cells with active AKT1 often involve addition of EGF to induce AKT1 phosphorylation and stimulate AKT signaling pathways [55]. To provide a direct comparison of this approach to our novel method to deliver active AKT1 to cells, we conducted a set of independent experiments to over-express AKT1 from a plasmid transiently transfected to HEK 293T cells (Figure 6). We used a well-established over-expression model of AKT1, in which an mCherry-tagged AKT1 protein is produced from a pCDNA3.1 backbone vector in the cells [33,55]. We used western blotting to show the level of endogenous AKT1 as well as to demonstrate expression of the plasmid encoded mCherry-AKT1 (Figure 6A). The data confirm over-expression of mCherry-AKT1 in cells transfected with the relevant plasmid and show a similar level of mCherry-AKT1 to the level of endogenous AKT1 observed in the same cells. We note that the plasmid-based mCherry-AKT1 expression leads to a similar level of additional AKT1 in the cells compared to our experiments with delivered TAT-tagged AKT1 variants (Figure 3).

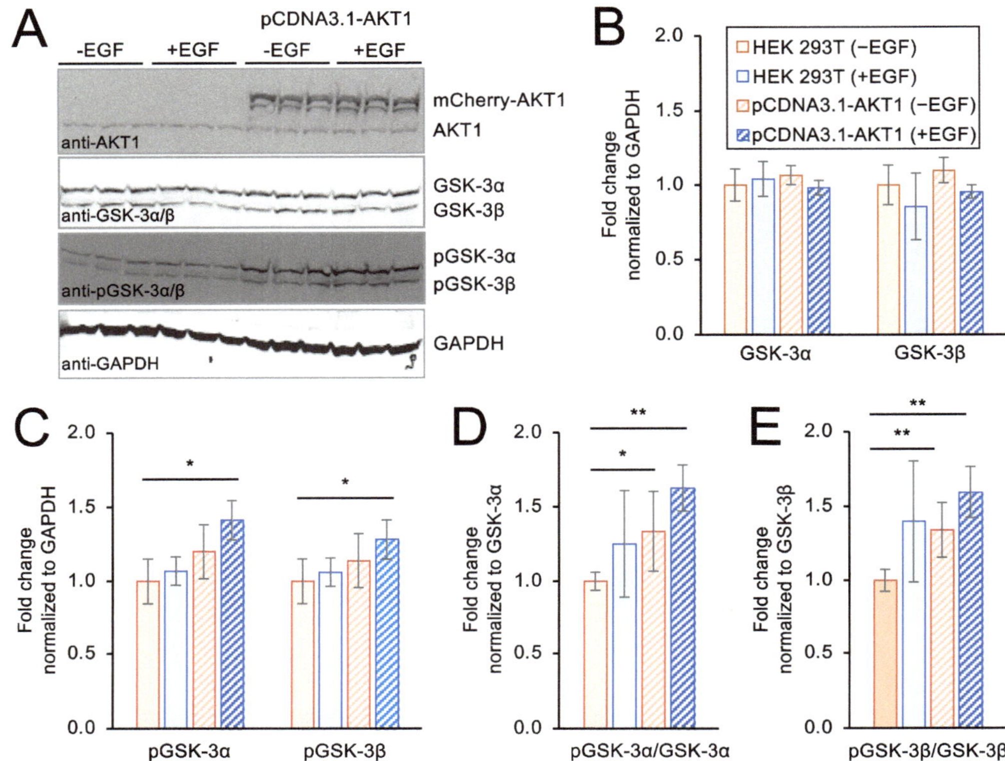

Figure 6. AKT1 over-expression and growth factor stimulation activates AKT1 signaling in HEK 293T cells. HEK 293T cells were transfected with an mCherry-AKT1 over-expression plasmid. At 24 h after transfection, cells were stimulated with EGF for 10 min. (**A**) Cell extracts were separated by SDS-PAGE and immunoblotted with AKT1, GSK-3, pGSK-3, and GAPDH specific antibodies in three biological replicates. (**B**) Quantification of western blots showed no change in GSK-3α or GSK-3β protein abundance. (**C**) GSK-3α and GSK-3β phosphorylation normalized to GAPDH was significantly increased in cells over-expressing AKT1 after EGF stimulation, but not significantly changed in EGF stimulated cells alone or unstimulated cells over-expressing AKT1. (**D**) GSK-3α and (**E**) GSK-3β phosphorylation was also normalized to GSK-3α and GSK-3β levels, showing increased phosphorylation of both GSK-3 isoforms in response to AKT1 over-expression, which was further increased upon EGF stimulation. Error bars represent the standard deviation of the mean. *p*-values were calculated by two-tailed *t*-test and are indicated by asterisks (* $p < 0.05$, ** $p < 0.01$).

We used western blotting to monitor the levels of GSK-3α and GSK-3β as well as pGSK-3α and pGSK-3β in untreated cells and in transfected cells with or without addition of EGF. None of the conditions caused any significant change in GSK-3α or GSK-3β levels relative to a GAPDH control (Figure 6A,B). As above, we monitored pGSK-3α and pGSK-3β levels relative to the GAPDH control (Figure 6C) as well as the levels of pGSK-3α and pGSK-3β compared to the levels of un-phosphorylated GSK-3α and GSK-3β, respectively (Figure 6D,E). Both approaches led to a consistent observation of a significant and ~1.5-fold increase in pGSK-3α and pGSK-3β levels in cells over-expressing AKT1 and stimulated with EGF relative to unstimulated control cells lines lacking AKT1 over-expression. These findings agree with previously published studies [50,51] where growth factor stimulation of cells with over-expressed AKT1 causes an increase in phosphorylation of both GSK-3α and GSK-3β.

3.7. Delivery Efficiency and Localization of TAT-Tagged AKT1

To measure the efficiency of TAT-tagged AKT1 delivery to live human cells, we constructed, produced, and purified AKT1 and pAKT1^{T308} variants in *E. coli* that contain the TAT-tag followed by an N-terminal mCherry tag (see Supplementary Material Section S.2). Since traditional approaches to study AKT1 signaling in cells often rely on over-expression of mCherry-tagged AKT1 in transiently transfected cells [33,55–57], we determined the transfection efficiency of plasmid-borne mCherry-AKT1 compared to delivery of the TAT-mCherry-tagged AKT1 variants. At 24 h after transfection with pCDNA3.1 encoded mCherry-AKT1 or 24 h after incubation of cells with TAT-mCherry-AKT1 variants, the cells were imaged by fluorescence and brightfield microscopy. Based on the images, delivery efficiency was calculated for the TAT-mCherry-tagged proteins and plasmid-based transfection efficiency was also determined (Figure 7). In HEK 293T cells, we found that both the plasmid transfection and the TAT-tagged protein delivery were similarly efficient. The TAT-mCherry-AKT1 entered the cells with ~90% delivery efficiency, while in our plasmid-based transfection we recorded a mean of ~80% transfection efficiency. Statistical analysis showed that TAT-mCherry tagged AKT1 protein delivery was not significantly different from the plasmid-based transfection efficiency. Thus, the TAT-tagged protein transfection is similarly efficient compared to traditional approaches. In the case of the plasmid-based experiment, lipofectamine transfection reagent was employed, while the TAT-tagged proteins required no transfection or other additional reagents to induce efficient cellular uptake.

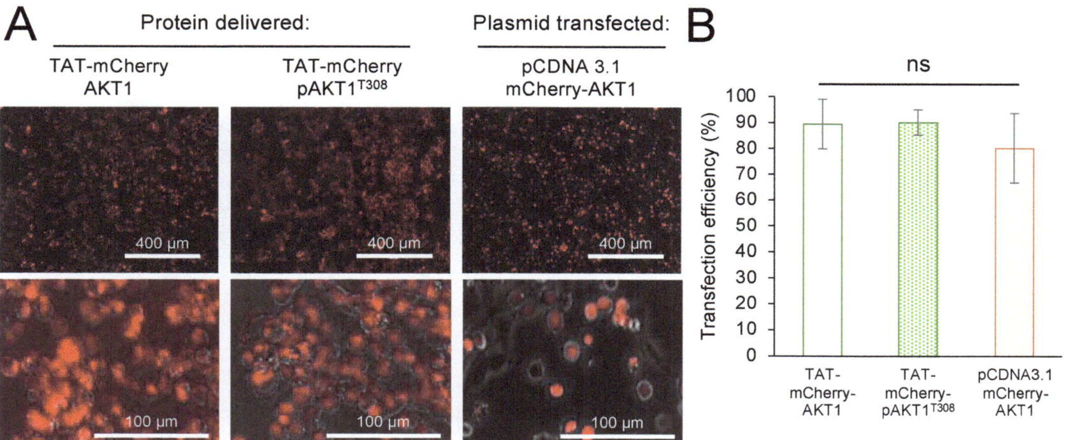

Figure 7. Delivery of TAT-mCherry-AKT1 variants compared to plasmid-based transfection efficiency of mCherry-AKT1. (**A**) Overlay of brightfield and fluorescent (excitation 531 nm, emission 593 nm) images (above) and zoomed-in images (below) of cells 24 h after incubation with TAT-mCherry-AKT1, TAT-mCherry-pAKT1^{T308}, or lipofectamine mediated transfection with an mCherry-AKT1 expressing plasmid. (**B**) Quantification of delivery or transfection efficiency from three biological replicates was calculated as the ratio of transfected cells/total cells and by determining fluorescence of the mCherry-AKT1 fusion proteins in cells. Protein delivery and plasmid transfection efficiency was not significantly different. The data are based on three biological replicates, error bars show ±1 standard deviation, and significance was calculated by one-way ANOVA (ns—not significant).

To observe the localization of TAT-AKT1, we included zoomed-in images of cells incubated with the TAT-mCherry-AKT1 variants as well as the plasmid-transfected cells. These images show that the TAT-tagged AKT1 variants and the plasmid expressed AKT1 are well-distributed in the interior of the cells (Figure 7). To get a more definitive image of TAT-AKT1 delivered to cells, we also conducted confocal fluorescence microscopy (see Supplementary Material Section S.3) of HEK 293T cells incubated with TAT-AKT1 (Figure S5A,B) or TAT-pAKT1^{T308} (Figure S5C,D) variants. The TAT-AKT1 proteins were visualized with an anti-TAT antibody and fluorescent secondary antibody, while the nuclei were stained with 4′,6-diamidino-2-phenylindole (DAPI) and the membranes were stained with an AlexaFluor-linked phalloidin (Figure S5). The confocal images agree with our TAT-mCherry-AKT1 studies and demonstrate that the TAT-AKT1 variants are well-distributed in the cytoplasm.

3.8. Cellular Viability and Toxicity

To quantify the impact on cell viability following protein delivery of TAT-AKT1 protein variants, we conducted trypan blue assays on cells incubated with TAT-AKT1 or TAT-pAKT1^{T308} protein for 1 h (Figure S6A) or 24 h (Figure S6B). In the trypan blue assay, dead and live cells were counted using brightfield imaging. Cells treated with TAT-AKT1 or TAT-pAKT1^{T308} showed >94% cell survival compared to the control cells incubated with buffer only (Figure S6A,B).

Next, dead cell nuclei were stained with the fluorescent Sytox blue probe and images were taken by fluorescence microscopy at 24 h after incubation with no protein (buffer only), with TAT-AKT1, or with TAT-pAKT1^{T308} (Figure 8). There was no significant difference in the fluorescence from dead cells in comparing cells transfected with either TAT-AKT1 variant or without protein.

Finally, a cytotoxicity assay was conducted to measure potential toxicity associated with either of the TAT-tagged AKT1 variants. The same number of HEK 293T cells were seeded into the 96 well plates in culture media. Following a 24-h incubation of the cells with buffer only (no protein), with TAT-AKT1, or with TAT-pAKT1^{T308} proteins, the live-cell impermeant peptide substrate aminoluciferin (AAF-Glo) was added into each well to assess the percentage of dead cells in a microplate reader. Following cell lysis, the luminescent signal was used to count the total number of cells. Cells transfected with no protein (buffer only), TAT-AKT1, or TAT-pAKT1^{T308} each showed no significant difference in cytotoxicity at 24 h after protein incubation (Figure S6C). Together the data show that TAT-tagged AKT1 variants are not toxic to cells, nor do they reduce cell viability.

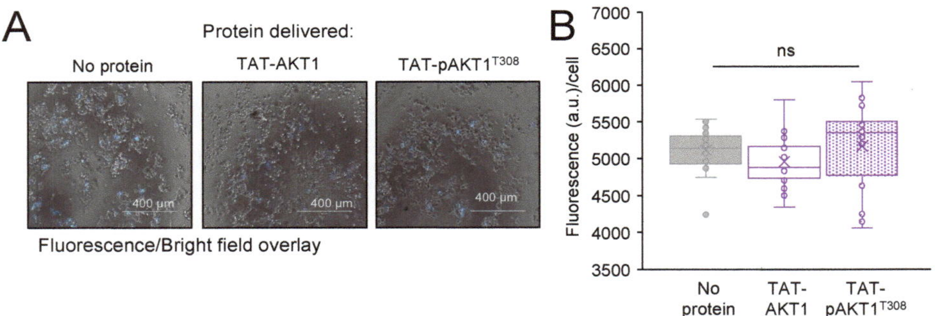

Figure 8. TAT-AKT1 protein delivery is non-toxic to HEK 293T cells. Cytotoxicity was measured after a 24-h incubation with no protein (buffer only), TAT-AKT1, or TAT-pAKT1^{T308} by staining dead cells with Sytox blue in three biological replicates. (**A**) Images of cells with Sytox blue staining of dead cells. (**B**) Quantification of dead cells shows no significant impact of TAT-AKT1 and TAT-pAKT1^{T308} protein delivery compared to cells treated with buffer only (no protein). Significance was calculated by one-way ANOVA (ns—not significant).

4. Discussion

4.1. Fusing AKT1 with the Cell Penetrating Peptide TAT

We engineered AKT1 and pAKT1 variants to include an N-terminal TAT tag to facilitate their cellular uptake into mammalian cells. The TAT peptide is a member of a diverse and growing collection of cell penetrating peptides (CPPs) that have been used to deliver small molecules [58], proteins [39], and even mRNAs [59] to mammalian cells. The TAT peptide (YGRKKRRQRRR) is a fragment from a larger TAT protein that is essential for HIV replication in cells [60]. Early studies identified a region of the TAT protein (residues 37–72) that is responsible for cellular uptake [38]. That same segment of TAT was also fused to a variety of other proteins, such as β-galactosidase and horseradish peroxidase, to facilitate their uptake into mammalian cells [40]. Recently, we fused the TAT-tag to the human selenocysteine-containing thioredoxin reductase 1, and demonstrated that the TAT-tag was necessary and sufficient to enable cytosolic delivery of an active selenoprotein to human cells for the first time [37].

Previous studies have appended His-tags [33], mCherry [55], and glutathione tags [61] to the N-terminus of AKT1, and each study found that the N-terminally tagged AKT1 retained kinase activity. Because of the important role of the C-terminal HM domain in AKT1 activation [15], C-terminal tagging is not appropriate for AKT1. Since TAT-fusions of AKT1 had not been created before, we characterized the biochemical activity of AKT1 and pAKT1^{T308} with and without the TAT-tag. We confirmed that the addition of a TAT-tag at the N-terminus does not alter catalytic activity of the AKT1 protein variants using in vitro kinase assays with a substrate peptide derived from GSK-3β.

4.2. Impact of TAT-Tagged Protein Delivery on Cell Fitness

One report suggested that TAT-tag molecules, such as TAT-NEMO binding domain (NBD), can have a cytotoxic effect at high concentration in treated cells. Incubation with a TAT-tagged NBD peptide led to decreased cell numbers at concentration of 100 μM [62]. We found that following incubation with human cells in culture, TAT-AKT1 and TAT-pAKT1^{T308} proteins at concentrations of 0.5 to 1 μM provided a level of delivered TAT-AKT1 that was similar to the level of endogenous AKT1 and that did not cause significant changes in cell viability or cytotoxicity according to multiple independent assays. More than 94% cells were viable following incubation with no protein or with either TAT-tagged AKT1 variant. As the cell survival rate was similar to the untreated cells, we measured cytotoxicity in cells harboring TAT-tagged AKT1 variants. We used a sensitive luminescent cytotoxicity assay to determine the relative number of dead cells in which the intensity of luminescence directly associates with the number of cells undergoing cytotoxic stress [63,64]. Compared to untreated HEK 293T cells, cells incubated with TAT-AKT1 variants did not show significant increases in dead cell numbers or in cells undergoing cytotoxic stress. The data underline the effectiveness of the cell penetrating TAT-tag for rapid and efficient delivery of active or inactive recombinant human AKT1 through the cell membrane without significantly compromising cellular viability or toxicity.

4.3. Delivery of Active TAT-pAKT1^{T308} Stimulates AKT1 Signaling in Human Cells

We utilized the TAT-mCherry-AKT1 to visualize protein delivery and calculate delivery efficiency. Independently, we utilized the TAT-pAKT1 variants lacking the mCherry fusion protein to investigate AKT1-dependent signaling. TAT-tagged protein delivery efficiency was ~90% after 24 h demonstrating efficient protein uptake by the cells. In comparison, transfection of cells with a plasmid bearing mCherry-AKT1 showed a similar (~80%) and not significantly different level of transfection efficiency. Together the data suggest that TAT-tagged AKT1 delivery is of similar efficiency to traditional plasmid-based approaches in HEK 293T cells, however, the TAT-tagged protein was delivered to cells without additional transfection reagents and can be applied to any recombinantly produced protein, including proteins with non-canonical amino acids [37].

AKT-mediated downstream regulation of effector proteins is directly linked to glycogenosis, glucose import, and protein synthesis [65–67]. GSK-3 is a highly conserved regulatory enzyme, and inhibition of GSK-3 activity by AKT-dependent phosphorylation was implicated in cancer progression, neuronal disease, aging, and metabolic disorders such as diabetes [66]. The GSK-3α and 3β isoforms have distinct regulatory roles and they do not compensate for each other's functions. GSK-3 is a critical downstream target of the PI3K-AKT cell signaling pathway, and GSK-3 activity is controlled in insulin or growth factor stimulated cells by AKT-dependent phosphorylation of GSK-3α at Ser21 and of GSK-3β at Ser9 [48]. The substrate specificity and recognition capabilities of GSK-3α and 3β also vary between and within tissues. In brain cells, deletion of each GSK-3 isoform produces distinct substrate phosphorylation of Collapsin response mediator protein (CRMP). CRMP Thr509, Thr514, and Ser518 phosphorylation is observed in GSK-3α depleted cells, but not GSK-3β depleted cells [68]. A knockdown of GSK-3α in mice showed increased sensitivity towards insulin, suggesting the significant role of GSK-3α in glucose synthesis and as a therapeutic target for diabetes [69]. Therefore, GSK-3 isoforms show a clear variation in their functions, and the distinct roles of GSK-3α and GSK-3β need to be delineated to understand the mechanism underlying various disorders.

In cells that we transfected with an AKT1 over-expression construct and then stimulated with growth factor, we reproduced the well-established result that AKT1 caused increased phosphorylation of its target substrates on GSK-3α and GSK-3β. In contrast, cells incubated with TAT-pAKT1^{T308} showed a strong and selective induction of phosphorylation on GSK-3α only. We observed evidence that TAT-pAKT1^{T308} stimulates direct downstream AKT1 substrates, such as GSK-3α, and we also observed evidence of downstream stimulation of ribosomal protein S6 phosphorylation by TAT-pAKT1^{T308} likely due to the well-established role of AKT1 in activation of the mTOR pathway [22–24].

We showed before that peptides representing the GSK-3α and GSK-3β (Figure 2) AKT1 phosphorylation sites are both competent substrates for pAKT1^{T308} [19]. We previously observed that pAKT1^{T308} is significantly more active with a GSK-3α peptide compared to a doubly phosphorylated ppAKT1^{T308,S473} [19], while pAKT1^{T308} and ppAKT1^{T308,S473} showed similar and robust activity with a GSK-3β peptide [19]. Taken together with the data presented here, our observations suggest that either doubly phosphorylated ppAKT1^{T308,S473} or growth factor stimulation of other kinases are responsible for phosphorylation of GSK-3β in cells. In the context of cells, scaffolding effects [70] and the role of phosphatases [71] are important factors for the accumulation of phosphorylation on specific sites that may also explain the apparent selective activity of TAT-pAKT1^{T308} on GSK-3α in cells. Fascinatingly, the phosphatase laforin is selective for de-phosphorylation of GSK-3β at Ser9 and has no effect on GSK-3α phosphorylation at Ser21 [72]. Insulin stimulation represses laforin protein levels [73] and enhances GSK-3β pSer9, while the same site is hyper-phosphorylated in laforin knock-out mice [73]. Over-expression of laforin is sufficient to abrogate GSK-3β phosphorylation even in insulin stimulated HEK 293 cells [74]. It is, therefore, feasible that TAT-pAKT1^{T308} phosphorylates both GSK-3α and GSK-3β, and GSK-3β is subsequently dephosphorylated by laforin. We will test this hypothesis in future work.

4.4. TAT-Fusion of Peptides and Proteins to Modulate AKT1 Signaling

The TAT-peptide was used in previous studies of AKT1, but no other studies fused TAT to the full-length or to the activated AKT1 as we have demonstrated here. Previously, the TAT-peptide was fused to the PH domain of AKT1 (residues 1 to 147) to generate a dominant negative and truncated version of AKT1 lacking the kinase domain [75]. The TAT-PH protein was successfully delivered to cells and was found to inhibit the anti-apoptotic effects of estrogen in vascular endothelial cells [75]. TAT was also fused to other peptides, such as a segment of the angiotensin II type I receptor to partially inhibit AKT signaling in HEK 293 cells [76], or to a fragment of the NF-κB essential modulator-binding domain to generate a 33 amino acid peptide that was shown to reduce expression of AKT [77].

TAT-peptides were also conjugated to nanoparticles in approaches to improve delivery of chemotherapeutics that disrupt AKT1 signaling [78,79]. There are many examples in the literature of fusing TAT to peptides or other compounds that inhibit human kinases, including inhibitors of the EGF receptor [80], protein kinase C [80], and casein kinase 2 [81].

Interestingly, a larger version of the TAT-peptide (residues 48–60) itself was shown to inhibit AGC family kinases [82]. The TAT-peptide (residues 48–60, GRKKRRQRRRAHQ) was found to be a μM level inhibitor of protein kinase A. The report also showed that a concentration of 10 μM of the TAT_{48-60} peptide was required to have a substantial impact on AKT kinase activity, and a concentration of 30 μM TAT_{48-60} peptide was needed to reduce ERK2 phosphorylation in HeLa cells. These levels are 30 to 60-fold above the concentrations in which we were working, and we used a TAT-peptide fused to AKT rather than a TAT-peptide in isolation. Furthermore, our TAT-fusion peptide is composed of residues 47–57 (YGRKKRRQRRR), which is a different peptide and may not retain the inhibitory effect observed previously [82].

Some previous studies have found that TAT-tagged proteins can remain trapped in endosomes. For example, studies of TAT-tagged green fluorescent protein (GFP) indicated inefficient endosomal escape at a concentration of 10 μM [83], yet we operated at a concentration 10 to 20-folder lower with TAT-tagged AKT1 (Figure S5). The same study [83] found that fusion of a haemagglutinin (HA) tag produced a TAT-HA-GFP that was well-distributed in the cytosol according to confocal imaging. According to fluorescence microscopy and confocal imaging experiments, our TAT-AKT1 variants showed similar cytosolic distribution to the TAT-HA-GFP previously reported [83] without the need of an additional HA tag. Thus, the nature of the protein cargo is a critical factor to achieve cytosolic delivery of TAT-tagged proteins [84]. It is also well-established that in the absence of inhibitors of endosomal release, TAT-tagged proteins are released from endosomes and delivered to the cytosol and nucleus [85].

To our knowledge, ours is the first report of fusing the TAT peptide to a full-length and active human kinase and successful demonstration of delivering a recombinant human kinase to human cells. Here, we generated novel TAT-tagged AKT1 and pAKT1 proteins using an approach that we previously developed to generate differentially active variants of full-length human AKT1 from *E. coli* [33]. The Thr308 phosphorylation is achieved by co-expression of AKT1 with the natural upstream Thr308 kinase PDK1 in *E. coli*. While we [19,32,34] and others [61] have also used this approach to produce a truncated AKT1 lacking the N-terminal PH domain, here and previously [33] we found this same approach works well to generate full-length AKT1 containing stoichiometric phosphorylation at Thr308. In our previous studies [19,31–34], we combined this approach with genetic code expansion to incorporate pSer473 in response to a UAG stop codon at the second regulatory site in AKT1, and thus produced AKT1 with phosphorylation at either or both regulatory sites. We also showed that pThr308 is necessary and sufficient to stimulate maximal AKT1 signaling in mammalian cells [33], and we employed a large scale chemoproteomic approach to determine that pSer473 functions to tune the substrate selectivity of AKT1 [19,32]. In future and on-going studies, beyond the scope of the current work, we are generating TAT-tagged version of AKT1 with phosphorylation at only Ser473 or at both activating sites.

5. Conclusions

We demonstrated a novel approach using a TAT fusion peptide to deliver both inactive and activated versions of the human kinase AKT1 to human cells. In our future work with this approach, we will be able to deliver different phospho-AKT1 variants to mammalian cells and study their potential to activate AKT1 signaling and to tune AKT1 substrate selectivity in cells. We anticipate that our approach will serve as a blueprint to deliver other isoforms and phospho-forms of AKT to human cells as well as a wide variety of other human kinases to provide novel methods to study the human kinome in the homologous context of human cells.

Supplementary Materials: The following supporting information can be downloaded at: https://www.mdpi.com/article/10.3390/cells11233834/s1, S.1 Supplementary Figures: Figure S1. Purification of AKT1 and TAT-AKT1 variants; Figure S2. Immunoblotting of purified AKT1 and TAT-AKT1 variants and mass spectrometry of TAT-pAKT1^{T308}; Figure S3. Autoradiography images of enzymatic activity assays of AKT1 and TAT-AKT1 variants; Figure S4. AKT1 and pAKT1^{T308} lacking the TAT-tag were not delivered to the interior of HEK 293T cells; Figure S5. Confocal images showing localization of TAT-AKT1 and TAT-pAKT1^{T308} in HEK 293T cells; Figure S6. TAT-AKT1 delivery is not toxic to HEK 293T cells. S.2 TAT-tagged DNA and protein sequences for AKT1 constructs; S.3 Supplementary Methods.

Author Contributions: Conceptualization, I.U.H. and P.O.; methodology, T.S., N.B., M.I.F., P.R. and P.O.; data analysis, T.S. and N.B.; investigation, T.S., N.B. and M.I.F.; resources, I.U.H. and P.O.; writing—original draft preparation, T.S. and P.O.; writing—review and editing, T.S., N.B., M.I.F., I.U.H. and P.O.; visualization, T.S., N.B. and M.I.F.; supervision, I.U.H. and P.O.; funding acquisition, I.U.H. and P.O. All authors have read and agreed to the published version of the manuscript.

Funding: This work was supported from the Natural Sciences and Engineering Research Council of Canada [04776 to I.U.H; 04282 to P.O.]; Canada Research Chairs [232341 to P.O.]; and the Canadian Institutes of Health Research [165985 to P.O.]; Ontario Ministry of Research and Innovation [ER-18-14-183 to I.U.H].

Informed Consent Statement: Not applicable.

Data Availability Statement: All data are available in the Figures and Supplementary Material.

Acknowledgments: We are grateful to David Wright, Gary Shaw, Yasmeen Shamiya for technical assistance and critical discussions. We thank TuKiet Lam at the Mass Spectrometry & Proteomics Resource of the W.M. Keck Foundation Biotechnology Resource Laboratory at Yale University for his assistance with mass spectrometry. We also thank Karen Nygard and Marc Courchesne at the Biotron Integrated Microscopy Facility at Western University for their technical assistance with confocal imaging.

Conflicts of Interest: The authors declare no conflict of interest.

References

1. Vogiatzi, P.; Giordano, A. Following the tracks of AKT1 gene. *Cancer Biol. Ther.* **2007**, *6*, 1521–1524. [CrossRef]
2. Chen, W.S.; Xu, P.-Z.; Gottlob, K.; Chen, M.-L.; Sokol, K.; Shiyanova, T.; Roninson, I.; Weng, W.; Suzuki, R.; Tobe, K. Growth retardation and increased apoptosis in mice with homozygous disruption of the Akt1 gene. *Genes Dev.* **2001**, *15*, 2203–2208. [CrossRef]
3. Karege, F.; Perroud, N.; Schürhoff, F.; Meary, A.; Marillier, G.; Burkhardt, S.; Ballmann, E.; Fernandez, R.; Jamain, S.; Leboyer, M. Association of AKT1 gene variants and protein expression in both schizophrenia and bipolar disorder. *Genes Brain Behav.* **2010**, *9*, 503–511. [PubMed]
4. Santi, S.A.; Lee, H. The Akt isoforms are present at distinct subcellular locations. *Am. J. Physiol. -Cell Physiol.* **2010**, *298*, C580–C591. [CrossRef]
5. Staal, S.P. Molecular cloning of the akt oncogene and its human homologues AKT1 and AKT2: Amplification of AKT1 in a primary human gastric adenocarcinoma. *Proc. Natl. Acad. Sci. USA* **1987**, *84*, 5034–5037. [CrossRef]
6. Bellacosa, A.; De Feo, D.; Godwin, A.K.; Bell, D.W.; Cheng, J.Q.; Altomare, D.A.; Wan, M.; Dubeau, L.; Scambia, G.; Masciullo, V. Molecular alterations of the AKT2 oncogene in ovarian and breast carcinomas. *Int. J. Cancer* **1995**, *64*, 280–285. [CrossRef]
7. Cheng, J.Q.; Ruggeri, B.; Klein, W.M.; Sonoda, G.; Altomare, D.A.; Watson, D.K.; Testa, J.R. Amplification of AKT2 in human pancreatic cells and inhibition of AKT2 expression and tumorigenicity by antisense RNA. *Proc. Natl. Acad. Sci. USA* **1996**, *93*, 3636–3641. [CrossRef]
8. Martini, M.; De Santis, M.C.; Braccini, L.; Gulluni, F.; Hirsch, E. PI3K/AKT signaling pathway and cancer: An updated review. *Ann. Med.* **2014**, *46*, 372–383. [CrossRef] [PubMed]
9. Hoxhaj, G.; Manning, B.D. The PI3K-AKT network at the interface of oncogenic signalling and cancer metabolism. *Nat. Rev. Cancer* **2020**, *20*, 74–88. [CrossRef]
10. Xing, Y.; Lin, N.U.; Maurer, M.A.; Chen, H.; Mahvash, A.; Sahin, A.; Akcakanat, A.; Li, Y.; Abramson, V.; Litton, J.; et al. Phase II trial of AKT inhibitor MK-2206 in patients with advanced breast cancer who have tumors with PIK3CA or AKT mutations, and/or PTEN loss/PTEN mutation. *Breast Cancer Res.* **2019**, *21*, 78. [CrossRef] [PubMed]
11. Zhang, X.; Tang, N.; Hadden, T.J.; Rishi, A.K. Akt, FoxO and regulation of apoptosis. *Biochim. Biophys. Acta (BBA) Mol. Cell Res.* **2011**, *1813*, 1978–1986. [CrossRef]

12. Gao, T.; Furnari, F.; Newton, A.C. PHLPP: A phosphatase that directly dephosphorylates Akt, promotes apoptosis, and suppresses tumor growth. *Mol. Cell* **2005**, *18*, 13–24. [CrossRef]
13. Ugi, S.; Imamura, T.; Maegawa, H.; Egawa, K.; Yoshizaki, T.; Shi, K.; Obata, T.; Ebina, Y.; Kashiwagi, A.; Olefsky, J.M. Protein phosphatase 2A negatively regulates insulin's metabolic signaling pathway by inhibiting Akt (protein kinase B) activity in 3T3-L1 adipocytes. *Mol. Cell. Biol.* **2004**, *24*, 8778–8789. [CrossRef] [PubMed]
14. Datta, S.R.; Brunet, A.; Greenberg, M.E. Cellular survival: A play in three Akts. *Genes Dev.* **1999**, *13*, 2905–2927. [CrossRef] [PubMed]
15. Kumar, C.C.; Madison, V. AKT crystal structure and AKT-specific inhibitors. *Oncogene* **2005**, *24*, 7493–7501. [CrossRef] [PubMed]
16. Andjelkovic, M.; Alessi, D.R.; Meier, R.; Fernandez, A.; Lamb, N.J.; Frech, M.; Cron, P.; Cohen, P.; Lucocq, J.M.; Hemmings, B.A. Role of translocation in the activation and function of protein kinase B. *J. Biol. Chem.* **1997**, *272*, 31515–31524. [CrossRef] [PubMed]
17. Vivanco, I.; Sawyers, C.L. The phosphatidylinositol 3-kinase–AKT pathway in human cancer. *Nat. Rev. Cancer* **2002**, *2*, 489–501. [CrossRef]
18. Alessi, D.R.; James, S.R.; Downes, C.P.; Holmes, A.B.; Gaffney, P.R.; Reese, C.B.; Cohen, P. Characterization of a 3-phosphoinositide-dependent protein kinase which phosphorylates and activates protein kinase Bα. *Curr. Biol.* **1997**, *7*, 261–269. [CrossRef] [PubMed]
19. Balasuriya, N.; Davey, N.E.; Johnson, J.L.; Liu, H.; Biggar, K.K.; Cantley, L.C.; Li, S.S.-C.; O'Donoghue, P. Phosphorylation-dependent substrate selectivity of protein kinase B (AKT1). *J. Biol. Chem.* **2020**, *295*, 8120–8134. [CrossRef]
20. Manning, B.D.; Cantley, L.C. AKT/PKB signaling: Navigating downstream. *Cell* **2007**, *129*, 1261–1274. [CrossRef]
21. Stambolic, V.; Woodgett, J.R. Functional distinctions of protein kinase B/Akt isoforms defined by their influence on cell migration. *Trends Cell Biol.* **2006**, *16*, 461–466. [CrossRef] [PubMed]
22. Rosner, M.; Hengstschlager, M. Nucleocytoplasmic localization of p70 S6K1, but not of its isoforms p85 and p31, is regulated by TSC2/mTOR. *Oncogene* **2011**, *30*, 4509–4522. [CrossRef]
23. Rosner, M.; Hengstschlager, M. Evidence for cell cycle-dependent, rapamycin-resistant phosphorylation of ribosomal protein S6 at S240/244. *Amino Acids* **2010**, *39*, 1487–1492. [CrossRef] [PubMed]
24. Roux, P.P.; Shahbazian, D.; Vu, H.; Holz, M.K.; Cohen, M.S.; Taunton, J.; Sonenberg, N.; Blenis, J. RAS/ERK signaling promotes site-specific ribosomal protein S6 phosphorylation via RSK and stimulates cap-dependent translation. *J. Biol. Chem.* **2007**, *282*, 14056–14064. [CrossRef] [PubMed]
25. Tsuchiya, A.; Kanno, T.; Nishizaki, T. PI3 kinase directly phosphorylates Akt1/2 at Ser473/474 in the insulin signal transduction pathway. *J. Endocrinol.* **2014**, *220*, 49. [CrossRef] [PubMed]
26. Okano, J.-i.; Gaslightwala, I.; Birnbaum, M.J.; Rustgi, A.K.; Nakagawa, H. Akt/protein kinase B isoforms are differentially regulated by epidermal growth factor stimulation. *J. Biol. Chem.* **2000**, *275*, 30934–30942. [CrossRef] [PubMed]
27. App, H.; Hazan, R.; Zilberstein, A.; Ullrich, A.; Schlessinger, J.; Rapp, U. Epidermal growth factor (EGF) stimulates association and kinase activity of Raf-1 with the EGF receptor. *Mol. Cell. Biol.* **1991**, *11*, 913–919. [PubMed]
28. Wu, J.; Dent, P.; Jelinek, T.; Wolfman, A.; Weber, M.J.; Sturgill, T.W. Inhibition of the EGF-activated MAP kinase signaling pathway by adenosine 3′,5′-monophosphate. *Science* **1993**, *262*, 1065–1069. [CrossRef]
29. Schlessinger, J. Common and distinct elements in cellular signaling via EGF and FGF receptors. *Science* **2004**, *306*, 1506–1507. [CrossRef]
30. Komurov, K.; Padron, D.; Cheng, T.; Roth, M.; Rosenblatt, K.P.; White, M.A. Comprehensive mapping of the human kinome to epidermal growth factor receptor signaling. *J. Biol. Chem.* **2010**, *285*, 21134–21142. [CrossRef]
31. Frederick, M.I.; Siddika, T.; Zhang, P.; Balasuriya, N.; Turk, M.A.; O'Donoghue, P.; Heinemann, I.U. miRNA-Dependent Regulation of AKT1 Phosphorylation. *Cells* **2022**, *11*, 821. [CrossRef] [PubMed]
32. McKenna, M.; Balasuriya, N.; Zhong, S.; Li, S.S.; O'Donoghue, P. Phospho-Form Specific Substrates of Protein Kinase B (AKT1). *Front. Bioeng. Biotechnol.* **2020**, *8*, 619252. [CrossRef] [PubMed]
33. Balasuriya, N.; Kunkel, M.T.; Liu, X.; Biggar, K.K.; Li, S.S.-C.; Newton, A.C.; O'Donoghue, P. Genetic code expansion and live cell imaging reveal that Thr-308 phosphorylation is irreplaceable and sufficient for Akt1 activity. *J. Biol. Chem.* **2018**, *293*, 10744–10756. [CrossRef] [PubMed]
34. Balasuriya, N.; McKenna, M.; Liu, X.; Li, S.S.; O'Donoghue, P. Phosphorylation-dependent inhibition of Akt1. *Genes* **2018**, *9*, 450. [CrossRef] [PubMed]
35. Nagahara, H.; Vocero-Akbani, A.M.; Snyder, E.L.; Ho, A.; Latham, D.G.; Lissy, N.A.; Becker-Hapak, M.; Ezhevsky, S.A.; Dowdy, S.F. Transduction of full-length TAT fusion proteins into mammalian cells: TAT-p27Kip1 induces cell migration. *Nat. Med.* **1998**, *4*, 1449–1452. [CrossRef]
36. Truernit, E.; Haseloff, J. A simple way to identify non-viable cells within living plant tissue using confocal microscopy. *Plant Methods* **2008**, *4*, 1–6. [CrossRef]
37. Wright, D.E.; Siddika, T.; Heinemann, I.U.; O'Donoghue, P. Delivery of the selenoprotein thioredoxin reductase 1 to mammalian cells. *Front. Mol. Biosci.* **2022**, *9*, 1031756. [CrossRef]
38. Green, M.; Loewenstein, P.M. Autonomous functional domains of chemically synthesized human immunodeficiency virus tat trans-activator protein. *Cell* **1988**, *55*, 1179–1188. [CrossRef]
39. Kurrikoff, K.; Vunk, B.; Langel, U. Status update in the use of cell-penetrating peptides for the delivery of macromolecular therapeutics. *Expert Opin. Biol. Ther.* **2021**, *21*, 361–370. [CrossRef]

40. Fawell, S.; Seery, J.; Daikh, Y.; Moore, C.; Chen, L.L.; Pepinsky, B.; Barsoum, J. Tat-mediated delivery of heterologous proteins into cells. *Proc. Natl. Acad. Sci. USA* **1994**, *91*, 664–668. [CrossRef]
41. Thomas, P.; Smart, T.G. HEK293 cell line: A vehicle for the expression of recombinant proteins. *J. Pharmacol. Toxicol. Methods* **2005**, *51*, 187–200. [CrossRef] [PubMed]
42. Meier, R.; Alessi, D.R.; Cron, P.; Andjelkovic, M.; Hemmings, B.A. Mitogenic activation, phosphorylation, and nuclear translocation of protein kinase Bbeta. *J. Biol. Chem.* **1997**, *272*, 30491–30497. [CrossRef] [PubMed]
43. Dufner, A.; Andjelkovic, M.; Burgering, B.M.; Hemmings, B.A.; Thomas, G. Protein kinase B localization and activation differentially affect S6 kinase 1 activity and eukaryotic translation initiation factor 4E-binding protein 1 phosphorylation. *Mol. Cell. Biol.* **1999**, *19*, 4525–4534. [CrossRef] [PubMed]
44. Myohanen, T.T.; Mertens, F.; Norrbacka, S.; Cui, H. Deletion or inhibition of prolyl oligopeptidase blocks lithium-induced phosphorylation of GSK3b and Akt by activation of protein phosphatase 2A. *Basic Clin. Pharmacol. Toxicol.* **2021**, *129*, 287–296. [CrossRef] [PubMed]
45. Chung, C.Z.; Balasuriya, N.; Manni, E.; Liu, X.; Li, S.S.; O'Donoghue, P.; Heinemann, I.U. Gld2 activity is regulated by phosphorylation in the N-terminal domain. *RNA Biol.* **2019**, *16*, 1022–1033. [CrossRef] [PubMed]
46. Chung, C.Z.; Balasuriya, N.; Siddika, T.; Frederick, M.I.; Heinemann, I.U. Gld2 activity and RNA specificity is dynamically regulated by phosphorylation and interaction with QKI-7. *RNA Biol.* **2021**, *18*, 397–408. [CrossRef]
47. Palma, M.; Leroy, C.; Salome-Desnoulez, S.; Werkmeister, E.; Kong, R.; Mongy, M.; Le Hir, H.; Lejeune, F. A role for AKT1 in nonsense-mediated mRNA decay. *Nucleic Acids Res.* **2021**, *49*, 11022–11037. [CrossRef]
48. Cross, D.A.; Alessi, D.R.; Cohen, P.; Andjelkovich, M.; Hemmings, B.A. Inhibition of glycogen synthase kinase-3 by insulin mediated by protein kinase B. *Nature* **1995**, *378*, 785–789. [CrossRef]
49. Beurel, E.; Grieco, S.F.; Jope, R.S. Glycogen synthase kinase-3 (GSK3): Regulation, actions, and diseases. *Pharmacol. Ther.* **2015**, *148*, 114–131.
50. Burgering, B.M.; Coffer, P.J. Protein kinase B (c-Akt) in phosphatidylinositol-3-OH kinase signal transduction. *Nature* **1995**, *376*, 599–602. [CrossRef]
51. Zhang, H.; Zha, X.; Tan, Y.; Hornbeck, P.V.; Mastrangelo, A.J.; Alessi, D.R.; Polakiewicz, R.D.; Comb, M.J. Phosphoprotein analysis using antibodies broadly reactive against phosphorylated motifs. *J. Biol. Chem.* **2002**, *277*, 39379–39387. [CrossRef] [PubMed]
52. Meyuhas, O. Ribosomal Protein S6 Phosphorylation: Four Decades of Research. *Int. Rev. Cell Mol. Biol.* **2015**, *320*, 41–73. [PubMed]
53. Gerasimovskaya, E.V.; Tucker, D.A.; Weiser-Evans, M.; Wenzlau, J.M.; Klemm, D.J.; Banks, M.; Stenmark, K.R. Extracellular ATP-induced proliferation of adventitial fibroblasts requires phosphoinositide 3-kinase, Akt, mammalian target of rapamycin, and p70 S6 kinase signaling pathways. *J. Biol. Chem.* **2005**, *280*, 1838–1848. [CrossRef] [PubMed]
54. Weichhart, T.; Hengstschlager, M.; Linke, M. Regulation of innate immune cell function by mTOR. *Nat. Rev. Immunol.* **2015**, *15*, 599–614.
55. Kunkel, M.T.; Ni, Q.; Tsien, R.Y.; Zhang, J.; Newton, A.C. Spatio-temporal dynamics of protein kinase B/Akt signaling revealed by a genetically encoded fluorescent reporter. *J. Biol. Chem.* **2005**, *280*, 5581–5587. [CrossRef]
56. Toulany, M.; Maier, J.; Iida, M.; Rebholz, S.; Holler, M.; Grottke, A.; Jüker, M.; Wheeler, D.L.; Rothbauer, U.; Rodemann, H.P. Akt1 and Akt3 but not Akt2 through interaction with DNA-PKcs stimulate proliferation and post-irradiation cell survival of K-RAS-mutated cancer cells. *Cell Death Discov.* **2017**, *3*, 1–10. [CrossRef]
57. Bessière, L.; Todeschini, A.-L.; Auguste, A.; Sarnacki, S.; Flatters, D.; Legois, B.; Sultan, C.; Kalfa, N.; Galmiche, L.; Veitia, R.A. A hot-spot of in-frame duplications activates the oncoprotein AKT1 in juvenile granulosa cell tumors. *eBioMedicine* **2015**, *2*, 421–431. [CrossRef]
58. Tian, Y.; Zhou, S. Advances in cell penetrating peptides and their functionalization of polymeric nanoplatforms for drug delivery. *Wiley Interdiscip. Rev. Nanomed. Nanobiotechnol.* **2021**, *13*, e1668. [CrossRef]
59. Yokoo, H.; Oba, M.; Uchida, S. Cell-Penetrating Peptides: Emerging Tools for mRNA Delivery. *Pharmaceutics* **2021**, *14*, 78. [CrossRef]
60. Arya, S.K.; Guo, C.; Josephs, S.F.; Wong-Staal, F. Trans-activator gene of human T-lymphotropic virus type III (HTLV-III). *Science* **1985**, *229*, 69–73. [CrossRef]
61. Klein, S.; Geiger, T.; Linchevski, I.; Lebendiker, M.; Itkin, A.; Assayag, K.; Levitzki, A. Expression and purification of active PKB kinase from *Escherichia coli*. *Protein Expr. Purif.* **2005**, *41*, 162–169. [CrossRef] [PubMed]
62. Jones, S.W.; Christison, R.; Bundell, K.; Voyce, C.J.; Brockbank, S.M.; Newham, P.; Lindsay, M.A. Characterisation of cell-penetrating peptide-mediated peptide delivery. *Br. J. Pharmacol.* **2005**, *145*, 1093. [CrossRef] [PubMed]
63. Riss, T.; Niles, A.; Moravec, R.; Karassina, N.; Vidugiriene, J. Cytotoxicity assays: In vitro methods to measure dead cells. In *Assay Guidance Manual*; Eli Lilly & Company and the National Center for Advancing Translational Sciences: Bethesda, MD, USA, 2019.
64. Harms, J.S.; Khan, M.; Hall, C.; Splitter, G.A.; Homan, E.J.; Bremel, R.D.; Smith, J.A. Brucella peptide cross-reactive major histocompatibility complex class I presentation activates SIINFEKL-specific T cell receptor-expressing T cells. *Infect. Immun.* **2018**, *86*, e00281-18. [CrossRef] [PubMed]
65. Sheng, S.; Qiao, M.; Pardee, A.B. Metastasis and AKT activation. *J. Cell. Physiol.* **2009**, *218*, 451–454. [CrossRef] [PubMed]
66. McCubrey, J.A.; Steelman, L.S.; Bertrand, F.E.; Davis, N.M.; Sokolosky, M.; Abrams, S.L.; Montalto, G.; D'Assoro, A.B.; Libra, M.; Nicoletti, F. GSK-3 as potential target for therapeutic intervention in cancer. *Oncotarget* **2014**, *5*, 2881. [PubMed]

67. Mailleux, A.A.; Overholtzer, M.; Schmelzle, T.; Bouillet, P.; Strasser, A.; Brugge, J.S. BIM regulates apoptosis during mammary ductal morphogenesis, and its absence reveals alternative cell death mechanisms. *Dev. Cell* **2007**, *12*, 221–234. [CrossRef]
68. Soutar, M.P.; Kim, W.Y.; Williamson, R.; Peggie, M.; Hastie, C.J.; McLauchlan, H.; Snider, W.D.; Gordon-Weeks, P.R.; Sutherland, C. Evidence that glycogen synthase kinase-3 isoforms have distinct substrate preference in the brain. *J. Neurochem.* **2010**, *115*, 974–983. [CrossRef]
69. MacAulay, K.; Doble, B.W.; Patel, S.; Hansotia, T.; Sinclair, E.M.; Drucker, D.J.; Nagy, A.; Woodgett, J.R. Glycogen synthase kinase 3α-specific regulation of murine hepatic glycogen metabolism. *Cell Metab.* **2007**, *6*, 329–337. [CrossRef]
70. Hu, J.; Ahuja, L.G.; Meharena, H.S.; Kannan, N.; Kornev, A.P.; Taylor, S.S.; Shaw, A.S. Kinase regulation by hydrophobic spine assembly in cancer. *Mol. Cell. Biol.* **2015**, *35*, 264–276. [CrossRef]
71. Antal, C.E.; Newton, A.C. Spatiotemporal dynamics of phosphorylation in lipid second messenger signaling. *Mol. Cell. Proteom.* **2013**, *12*, 3498–3508. [CrossRef]
72. Liu, R.; Wang, L.; Chen, C.; Liu, Y.; Zhou, P.; Wang, Y.; Wang, X.; Turnbull, J.; Minassian, B.A.; Liu, Y.; et al. Laforin negatively regulates cell cycle progression through glycogen synthase kinase 3beta-dependent mechanisms. *Mol. Cell. Biol.* **2008**, *28*, 7236–7244. [CrossRef] [PubMed]
73. Vernia, S.; Heredia, M.; Criado, O.; Rodriguez de Cordoba, S.; Garcia-Roves, P.M.; Cansell, C.; Denis, R.; Luquet, S.; Foufelle, F.; Ferre, P.; et al. Laforin, a dual specificity phosphatase involved in Lafora disease, regulates insulin response and whole-body energy balance in mice. *Hum. Mol. Genet.* **2011**, *20*, 2571–2584. [CrossRef] [PubMed]
74. Liu, Y.; Wang, Y.; Wu, C.; Liu, Y.; Zheng, P. Dimerization of Laforin is required for its optimal phosphatase activity, regulation of GSK3beta phosphorylation, and Wnt signaling. *J. Biol. Chem.* **2006**, *281*, 34768–34774. [CrossRef] [PubMed]
75. Koga, M.; Hirano, K.; Hirano, M.; Nishimura, J.; Nakano, H.; Kanaide, H. Akt plays a central role in the anti-apoptotic effect of estrogen in endothelial cells. *Biochem. Biophys. Res. Commun.* **2004**, *324*, 321–325. [CrossRef]
76. Yu, J.; Taylor, L.; Mierke, D.; Berg, E.; Shia, M.; Fishman, J.; Sallum, C.; Polgar, P. Limiting angiotensin II signaling with a cell-penetrating peptide mimicking the second intracellular loop of the angiotensin II type-I receptor. *Chem. Biol. Drug Des.* **2010**, *76*, 70–76. [CrossRef]
77. Liu, R.; Xi, L.; Luo, D.; Ma, X.; Yang, W.; Xi, Y.; Wang, H.; Qian, M.; Fan, L.; Xia, X.; et al. Enhanced targeted anticancer effects and inhibition of tumor metastasis by the TMTP1 compound peptide TMTP1-TAT-NBD. *J. Control. Release* **2012**, *161*, 893–902. [CrossRef]
78. Fan, Y.X.; Liang, Z.X.; Liu, Q.Z.; Xiao, H.; Li, K.B.; Wu, J.Z. Cell penetrating peptide of sodium-iodide symporter effect on the I-131 radiotherapy on thyroid cancer. *Exp. Ther. Med.* **2017**, *13*, 989–994. [CrossRef]
79. He, L.; Lai, H.; Chen, T. Dual-function nanosystem for synergetic cancer chemo-/radiotherapy through ROS-mediated signaling pathways. *Biomaterials* **2015**, *51*, 30–42. [CrossRef]
80. Konoeda, H.; Yang, H.; Yang, C.; Gower, A.; Xu, C.; Zhang, W.; Liu, M. Protein Kinase C-delta Inhibitor Peptide Formulation using Gold Nanoparticles. *J. Vis. Exp.* **2019**, e58741. [CrossRef]
81. Zanin, S.; Sandre, M.; Cozza, G.; Ottaviani, D.; Marin, O.; Pinna, L.A.; Ruzzene, M. Chimeric peptides as modulators of CK2-dependent signaling: Mechanism of action and off-target effects. *Biochim. Biophys. Acta (BBA) -Proteins Proteom.* **2015**, *1854*, 1694–1707. [CrossRef]
82. Ekokoski, E.; Aitio, O.; Tornquist, K.; Yli-Kauhaluoma, J.; Tuominen, R.K. HIV-1 Tat-peptide inhibits protein kinase C and protein kinase A through substrate competition. *Eur. J. Pharm. Sci.* **2010**, *40*, 404–411. [CrossRef] [PubMed]
83. Patel, S.G.; Sayers, E.J.; He, L.; Narayan, R.; Williams, T.L.; Mills, E.M.; Allemann, R.K.; Luk, L.Y.P.; Jones, A.T.; Tsai, Y.H. Cell-penetrating peptide sequence and modification dependent uptake and subcellular distribution of green florescent protein in different cell lines. *Sci. Rep.* **2019**, *9*, 6298. [CrossRef] [PubMed]
84. El-Andaloussi, S.; Jarver, P.; Johansson, H.J.; Langel, U. Cargo-dependent cytotoxicity and delivery efficacy of cell-penetrating peptides: A comparative study. *Biochem. J.* **2007**, *407*, 285–292. [CrossRef] [PubMed]
85. Potocky, T.B.; Menon, A.K.; Gellman, S.H. Cytoplasmic and nuclear delivery of a TAT-derived peptide and a beta-peptide after endocytic uptake into HeLa cells. *J. Biol. Chem.* **2003**, *278*, 50188–50194. [CrossRef]

Article

mTORC1 Activity in Psoriatic Lesions Is Mediated by Aberrant Regulation through the Tuberous Sclerosis Complex

Antonio Ferreri, Victoria Lang, Roland Kaufmann and Claudia Buerger *

Department of Dermatology, University Hospital, Goethe University Frankfurt, 60590 Frankfurt am Main, Germany
* Correspondence: claudia.buerger@kgu.de

Abstract: In the basal, proliferative layer of healthy skin, the mTOR complex 1 (mTORC1) is activated, thus regulating proliferation while preventing differentiation. When cells leave the proliferative, basal compartment, mTORC1 signaling is turned off, which allows differentiation. Under inflammatory conditions, this switch is hijacked by cytokines and prevents proper differentiation. It is currently unknown how mTORC1 is regulated to mediate these effects on keratinocyte differentiation. In other tissues, mTORC1 activity is controlled through various pathways via the tuberous sclerosis complex (TSC). Thus, we investigated whether the TS complex is regulated by proinflammatory cytokines and contributes to the pathogenesis of psoriasis. TNF-α as well as IL-1β induced the phosphorylation of TSC2, especially on S939 via the PI3-K/AKT and MAPK pathway. Surprisingly, increased TSC2 phosphorylation could not be detected in psoriasis patients. Instead, TSC2 was strongly downregulated in lesional psoriatic skin compared to non-lesional skin of the same patients or healthy skin. In vitro inflammatory cytokines induced dissociation of TSC2 from the lysosome, followed by destabilization of the TS complex and degradation. Thus, we assume that in psoriasis, inflammatory cytokines induce strong TSC2 phosphorylation, which in turn leads to its degradation. Consequently, chronic mTORC1 activity impairs ordered keratinocyte differentiation and contributes to the phenotypical changes seen in the psoriatic epidermis.

Keywords: psoriasis; inflammation; mTORC1; tuberous sclerosis complex; cytokines

1. Introduction

Psoriasis is a common, chronic inflammatory skin disease that affects 2–3% of the population and is associated with a reduced quality of life and a shortened life expectancy due to the association with the metabolic syndrome and cardiovascular pathologies [1]. Clinically, psoriasis presents with red, scaly plaques, which mostly affect predilection sites such as the extensor surfaces of forearms and shins, umbilical, perianal, retro-auricular regions, and the scalp [2]. These plaques are characterized by epidermal hyper-proliferation with impaired keratinocyte differentiation, extravasation of lymphocytes, and angio(neo)genesis.

In healthy skin, keratinocytes are subject to a strict control between proliferation by asymmetric cell division in the basal layer and ordered terminal differentiation and maturation into corneocytes, forming a tight epidermal barrier. In psoriasis, certain trigger factors (trauma, drugs, infections), together with dysregulated expression of the antimicrobial peptide LL37, are thought to induce sustained activation of plasmacytoid dendritic cells. This process, which normally controls the defense against infection in injured skin, promotes the maturation of myeloid dendritic cells, which induce the differentiation of T cells into Th17 cells through the secretion of interleukin-12 (IL-12) and IL-23. These activated Th17 cells produce effector cytokines such as IL-17A, IL-17F and IL-22, which stimulate keratinocytes to increase proliferation, while differentiation is impaired [3,4]. Activated keratinocytes, in turn, produce important pro-inflammatory cytokines, including TNF-α and IL-1β, which induces a "vicious cycle" of exuberant immune response, epidermal hyper-proliferation and neovascularization, leading to the complex clinical presentation of psoriasis [5].

Citation: Ferreri, A.; Lang, V.; Kaufmann, R.; Buerger, C. mTORC1 Activity in Psoriatic Lesions Is Mediated by Aberrant Regulation through the Tuberous Sclerosis Complex. *Cells* **2022**, *11*, 2847. https://doi.org/10.3390/cells11182847

Academic Editor: Stephen Yarwood

Received: 15 June 2022
Accepted: 7 September 2022
Published: 13 September 2022

Publisher's Note: MDPI stays neutral with regard to jurisdictional claims in published maps and institutional affiliations.

We previously found that AKT [6], as well as the mTOR kinase and its downstream signaling molecules [7,8] are hyper-activated in psoriatic skin. In healthy skin, mTORC1 signaling is only active in the basal layer and regulates proliferation while preventing differentiation. When cells leave the proliferative compartment, mTORC1 signaling is switched off, which allows for ordered epidermal differentiation. However, pro-inflammatory cytokines activate the PI3K/Akt pathway, which promotes aberrant proliferation [8]. At the same time persistent mTORC1 activation inhibits proper differentiation [8,9].

Beyond this model, it is currently unknown how mTORC1 activity is regulated to promote these effects on keratinocyte differentiation. In order to coordinate cell growth and proliferation, the tuberous sclerosis complex (TSC) integrates extracellular cues (e.g., nutrients, energy, oxygen and growth factors) through different kinases such as AKT [10], RSK1 [11] or ERK [12], which regulate mTORC1 activity [13]. This multiprotein complex consists of TSC1/hamartin, TSC2/tuberin and the TBC1 domain family member TBC1D7 [14]. Mutations in TSC2 cause non-malignant tumors in various organs, including the skin, leading to the clinical manifestation of tuberos sclerosis [15]. Under favorable conditions, the mentioned kinases directly phosphorylate TSC at specific serin (S) and threonine (T) residues [16], leading to its dissociation from the lysosome and destabilization of the protein complex [17]. Thus, TSC2, the catalytic subunit of the complex, can no longer act as a GTPase activating protein (GAP) towards the small GTPase RHEB, which remains GTP-loaded and is able to activate mTORC1 on the lysosome [18–20]. Activated mTORC1, in turn, regulates key anabolic processes, such as protein biosynthesis through the phosphorylation of S6 kinase-1 (S6K-1) and eukaryotic initiation factor 4E (eIF-4E) binding protein-1 (4E-BP1), and lipid and nucleotide synthesis, and inhibits catabolic processes such as autophagy [9]. In contrast, under unfavorable extracellular conditions i.e., in the absence of growth factors or amino acids, TSC is not phosphorylated, thus remaining at the lysosomal membrane [17,21] and promoting GTP cleavage by RHEB. RHEB is thus unable to activate mTORC1, and anabolic, mTOR-dependent processes fail to occur.

We hypothesized that the TS complex might also serve as a critical switch to control the mTORC1 function during epidermal maturation. We showed that TSC2 is strongly phosphorylated on S939 by pro-inflammatory cytokines such as TNF-α as well as IL-1β. Blocking the PI3-K/AKT or the MAPK pathway with kinase inhibitors not only reduced TSC Ser939 phosphorylation, but also blocked mTORC1 activity. Surprisingly, we could not find evidence for increased TSC2 phosphorylation in psoriasis patients. Instead, we discovered that TSC2 is significantly downregulated in lesional psoriatic skin when compared to healthy skin or non-lesional skin of the same patients. Additionally, we uncovered, that persistent exposure to inflammatory cytokines such as in psoriasis dislocates the TS complex, inducing its destabilization and degradation, which results in the hyperactivation of mTORC signaling.

2. Materials and Methods

2.1. Antibodies and Chemicals

P-TSC2 S939 (LS-C358381) was from Biozol (Sontheim, Germany). Antibodies specific for TSC2/tuberin (#4308), P-AKT S473 (#4060), AKT (#4691), P-S6K T389 (#9234), P-S S235/6 (#2211), S6 (#2217), P-ERK T202/Y204 (#4370), ERK (#4696), P-4E-BP T37/46 (#2855), as well as MG132 and cyclohexamide were from Cell Signaling (Danvers, MA, USA). Involucrin (SY8; ab20202) was from Abcam (Cambridge, UK), Actin antibody (A1978) was from Sigma-Aldrich (St. Louis, MO, USA) and LAMP-2 antibody (sc-18822) from SantaCruz Biotechnology (Dallas, TX, USA). LY294002, U0126 were from were from Calbiochem (San Diego, CA, USA).

2.2. Immunohistochemistry

The study was approved by the ethics committee of the Clinic of the Goethe-University (116/11); the Declaration of Helsinki protocols were followed. A total of 10 psoriasis patients between 23–60 years with a confirmed diagnosis of severe plaque-type psoriasis vulgaris for at least 6 months and no current systemic anti-inflammatory therapy gave written informed consent, as did 5 healthy individuals. Punch biopsies (6 mm) from the lesional and non-lesional skin of patients or normal skin of healthy individuals were taken. Specimens were fixed in 4% PFA, and paraffin embedded and 4 μm sections were processed routinely. Primary antibodies were applied overnight, and Histofine Simple Stain AP Multi (Nichirei Bioscience, Tokyo, Japan) was used for detection. Nuclei were stained with hematoxylin. Images were acquired using a Nikon Eclipse Ci microscope (Nikon Europe, Amstelveen, The Netherlands). For semi-quantitative analysis, IHC staining intensity was estimated by two independent investigators on a scale between 0 to 3. Mean values were calculated, and statistical significance was evaluated by one-way ANOVA and uncorrected Fisher's LSD test.

2.3. Cell Culture and Western Blotting

HaCaT keratinocytes [22] were cultured in DMEM with 10% FCS (Life Technologies, Carlsbad, CA, USA). Cells were treated as indicated and lysed in RIPA lysis buffer (Cell Signaling Technology, Dancers, MA, USA). Lysates were adjusted for equal protein amounts, subjected to SDS–PAGE and blotted onto PVDF membranes. After blocking in 5% milk/TBS-T, membranes were probed with the indicated antibodies overnight. Bound antibodies were visualized with HRP-conjugated secondary antibodies using ECL Substrate (Bio-Rad, Hercules, CA, USA).

2.4. Immunofluorescence Staining

HaCaT cells were grown on glass coverslips, treated as indicated and formalin fixed. Cells were permeabilized with 0.2% Triton X-100/PBS and blocked with 5% normal goat serum/0.2% Triton/PBS. Primary antibody was applied overnight at 4 °C, and staining with Alexa 488 or Alexa594 labelled secondary antibodies (Life Technologies, Carlsbad, CA, USA) was performed for 1 h at room temperature. Images were acquired through a 100× oil immersion objective on a Nikon Eclipse Ci microscope (Nikon Europe, Amstelveem, The Netherlands). Colocalization of TSC2 and LAMP-2 was quantified using the Coloc2 plugin of Fiji software 2.3.0 (doi:10.1038/nmeth.2019). For each condition, the Manders' co-localization coefficient using automatic Costes thresholding was calculated from 8 to 14 separate images, each containing between 3 to 10 cells. Mean values were calculated, and the statistical significance was evaluated by one-way ANOVA and uncorrected Fisher's LSD test.

2.5. Statistical Analysis

The mean ± standard error of the mean (SEM) was depicted in the diagrams. Graph Pad Prism 9.4.1 (GraphPad Software, LLC, San Diego, CA, USA) was used for statistical analysis. Multiple groups were compared by ordinary one-way ANOVA followed by an uncorrected Fisher's LSD test. A p-value ≤ 0.05 was considered statistically significant with (* $p < 0.05$, ** $p < 0.01$).

3. Results

3.1. Th1 Cytokines Induce TSC2 Phosphorylation via the PI3-K/AKT and MAPK Pathway

To investigate whether cytokine-dependent hyperactivation of mTORC1 is regulated by the tuberous sclerosis complex, we first examined whether TSC2 can be phosphorylated by cytokines involved in the pathogenesis of psoriasis. As we could previously show that AKT is hyper-activated in psoriatic skin, we focused on residues in TSC2, which are mainly regulated by AKT. For four out of the five residues, phosphorylated by AKT (S981, S1130, S1132, T1462), no reliable antibodies could be acquired. Thus, we investigated S939, which is one of the more important residues to control TSC functions towards mTORC1 signaling [10,23,24]. HaCaT keratinocytes were treated with Th1 cytokines with a known effect on keratinocytes during the psoriatic inflammation. As shown previously, IL-17A and IL-22 induced mild phosphorylation of AKT S473, while IL-1β, TNF-α as well as the Th1 mix consisting of IL-1β, IL-17A and TNF-α conferred robust activation of AKT, resulting in strong mTORC1 activation measured by the phosphorylation of the mTORC1 target proteins S6K, S6 and 4E-BP1. At the same time especially TNF-α, as well as the Th1 mix, mediated strong phosphorylation of TSC2 on S939 (Figure 1a).

To verify that cytokine-dependent phosphorylation of TSC2 is mediated via PI3-K/AKT, HaCaT cells were pre-treated with the PI3-K inhibitor LY294002, which not only inhibited TNF-α-induced AKT activity, but also completely abolished TSC2 phosphorylation. In addition, the MAPK pathway was blocked using the MEK inhibitor U0126, which inhibited TNF-α-induced ERK1 activation, but had only a mild effect on TSC2 phosphorylation (Figure 1b). These results show that pro-inflammatory cytokines of the Th1 spectrum can confer TSC2 phosphorylation via the PI3-K/AKT pathway, and to some degree via the MAPK pathway.

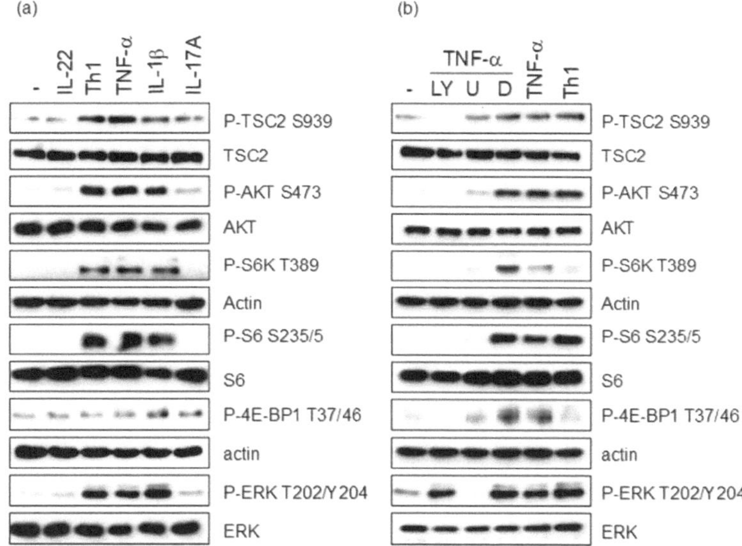

Figure 1. (**a**) HaCaT cells were starved overnight and stimulated with the indicated cytokines (20 ng/mL) or a mix of Th1 cytokines (IL-1β, IL-17A and TNF-α; 20 ng/mL each) for 30 min. (**b**) HaCaT cells were starved overnight and treated for 30 min with the indicated inhibitors (L = 50 μM LY294002; U = 10 μM U0126; D = DMSO), followed by stimulation with 20 ng/mL TNF-α or the Th1 mix (20 ng/mL each). Protein lysates were prepared and analyzed by Western blotting with the indicated antibodies. Equal protein loading was confirmed using total protein antibodies or actin antibody.

3.2. TSC2 Is Downregulated in Lesional Psoriatic Skin

To test whether these pro-inflammatory cytokines also mediate the phosphorylation of TSC2 in vivo, non-lesional and lesional psoriatic skin was investigated for TSC2 activation and expression. While healthy skin (NN) displayed the phosphorylation of TSC2 S939 throughout all epidermal layers (Figure 2c,f), non-lesional psoriatic skin (PN) showed comparable levels of TSC2 phosphorylation (Figure 2b,e). Surprisingly, we found that in lesional psoriatic skin (PP), TSC2 did not display enhanced phosphorylation at S939 as expected from the in vitro data, but rather showed similar (Figure 2d) or slightly reduced activation (Figure 2a) compared to non-lesional skin from the same patient. Quantitative analysis of ten patients, underlined this finding, showing slightly but not significantly reduced TSC2 phosphorylation in lesional psoriatic skin (Figure 2g).

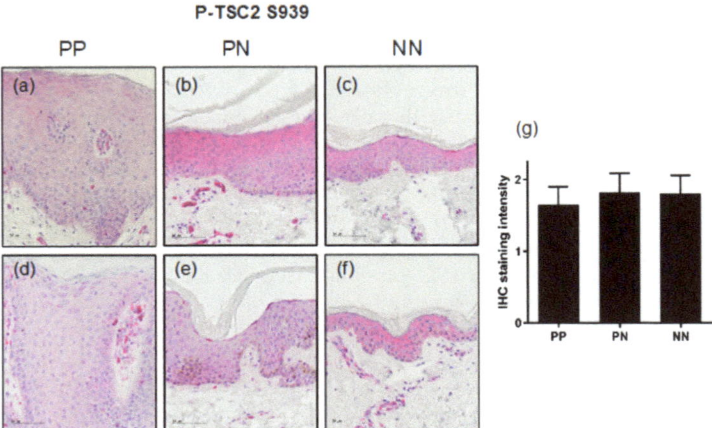

Figure 2. (**a–f**) IHC staining with P-TSC2-S939-specific antibodies of two representative biopsies from lesional (PP; **a,d**) or non-lesional (PN; **b,e**) skin of psoriasis vulgaris patients or healthy donors (NN; **c,f**). Bars represent 50 μm. (**g**) The graph represents quantification of IHC staining intensities from 10 psoriasis patients and 3 healthy individuals (mean ± SEM).

As accumulating phosphorylation of the TS complex on different residues results in the dissociation of TSC from the lysosome and ubiquitin-dependent degradation. We next investigated whether TSC is destabilized in psoriatic skin. We found that healthy skin (NN; Figure 3c,f) as well as non-lesional skin (PN; Figure 3b,e), displayed homogeneous expression of TSC2 throughout the epidermis. However, lesional psoriatic skin (PP; Figure 3a,d) showed a reduced expression of TSC2 in all epidermal layers compared to non-lesional skin from the same patient (Figure 3b,e). This finding was further substantiated by quantifying the staining intensities of all ten psoriasis samples. Psoriatic skin showed a significantly reduced expression of TSC2 in comparison to non-lesional or healthy skin (Figure 3g). Unfortunately, the level of TSC1 being in a complex with TSC2 in the epidermis could not be determined due to the lack of a TSC1-specific, IHC-compatible antibody.

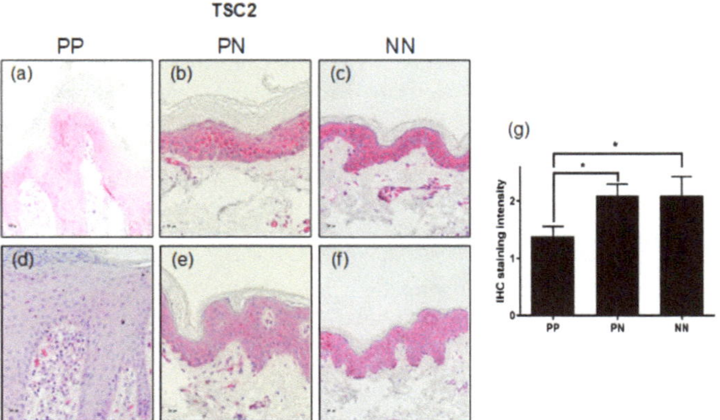

Figure 3. (**a–f**) IHC staining with TSC2-specific antibodies of two representative biopsies from lesional (PP; **a,d**) or non-lesional (PN; **b,e**) skin of psoriasis vulgaris patients or healthy donors (NN; **c,f**). Bars represent 50 μm. (**g**) The graph represents the quantification of IHC staining intensities from 10 psoriasis patients and 3 healthy individuals (mean ± SEM). Statistical significance was calculated with one-way ANOVA and Fisher's LSD test (* $p < 0.05$).

3.3. Pro-Inflammatory Cytokines Regulate TSC Function by Spatial Re-Distribution and Degradation

To test whether pro-inflammatory cytokines impair TSC function in psoriatic skin by delocalization and degradation as known from other tissues, HaCaT keratinocytes were stimulated with the Th1 mix or TNF-α, and the localization of TSC2 was visualized by immunofluorescence staining. In starved cells, TSC2 was localized to the lysosomal membrane as indicated by co-staining with the lysosomal marker LAMP2 (Figure 4a). Thus, TSC2 remained in the vicinity of RHEB, supporting its GTPase activity, so that mTORC1 was not activated. However, Th1 cytokines (Figure 4b) or TNF-α (Figure 4d) stimulated dissociation from the lysosomal surface, indicated by diffuse red TSC2 staining, and less yellow co-localization could be detected. This effect was mediated via the cytokine-dependent stimulation of the PI3-K/AKT axis, as the inhibition of PI3-K with LY294002 prevented AKT-mediated TSC2 phosphorylation, and TSC2 remained at the lysosome (Figure 4c,e). Co-localization analysis of several similar images confirmed re-localization of TSC2 into the cytoplasm after an inflammatory stimulus, as the Manders coefficient decreased under these conditions (Figure 4f). Thus, we propose that the psoriatic inflammation drives TSC2 phosphorylation, resulting in its spatial re-distribution, which contributes to TSC2 degradation.

To investigate this, we first examined the stability of the TSC2 protein by treating HaCaT cells with cycloheximide (CHX), which blocks general protein translation. We could show that TSC2 levels are starting to decline after two to four hours of treatment and are completely absent after eight hours. In contrast, other proteins of the pathway, such as AKT, are unaffected by CHX treatment, thus not being regulated though protein stability (Figure 5a). As we hypothesized that proinflammatory cytokines impair TSC2 stability, cells were treated for these time points with Th1 cytokines, but no impact on TSC2 levels could be seen. However, chronic cytokine treatment for 72 h or longer led to massively reduced TSC2 levels (Figure 5b). Taking the reduced levels of TSC2 in the cytokine-treated samples into account, TSC2 was hyperphosphorylated even after this prolonged treatment. As chronic inflammation interferes with epidermal maturation, the differentiation marker involucrin was assessed to monitor the effect of the cytokines. During the course of the experiment, involucrin accumulated in the untreated samples as keratinocytes grew post-confluently, leading to differentiation, while chronic cytokine exposure blocked the expression of the differentiation marker (Figure 5b). To verify that this cytokine-mediated effect on TSC2 is due to reduced protein stability, we inhibited the 26S proteasome using MG132, which efficiently reduces the degradation of ubiquitin-conjugated proteins. As keratinocytes did not tolerate this inhibitor for the time periods needed for cytokine-induced TSC2 degradation, we established a protocol where keratinocytes were serum-starved and then treated with CHX as well as the Th1 mix. Under these conditions, pro-inflammatory cytokines were able to induce massive degradation of TSC2 already after two and four hours of treatment (Figure 5c). If, under these conditions, the proteasome was inhibited by MG132, a rescue of cytokine-induced TSC2 degradation could be detected (Figure 5d). This provides evidence that inflammation-induced TSC2 phosphorylation likely leads to its ubiquitinylation and degradation via the proteasome.

In summary, we could show that the TS complex is significantly involved in mTORC1 hyperactivation in inflammatory skin diseases, as pro-inflammatory cytokines induce the phosphorylation of TSC2 at S939, which in turn favors its dislocation from the lysosome, followed by degradation. This is underlined by the finding that in psoriatic skin, reduced TSC2 levels could be detected.

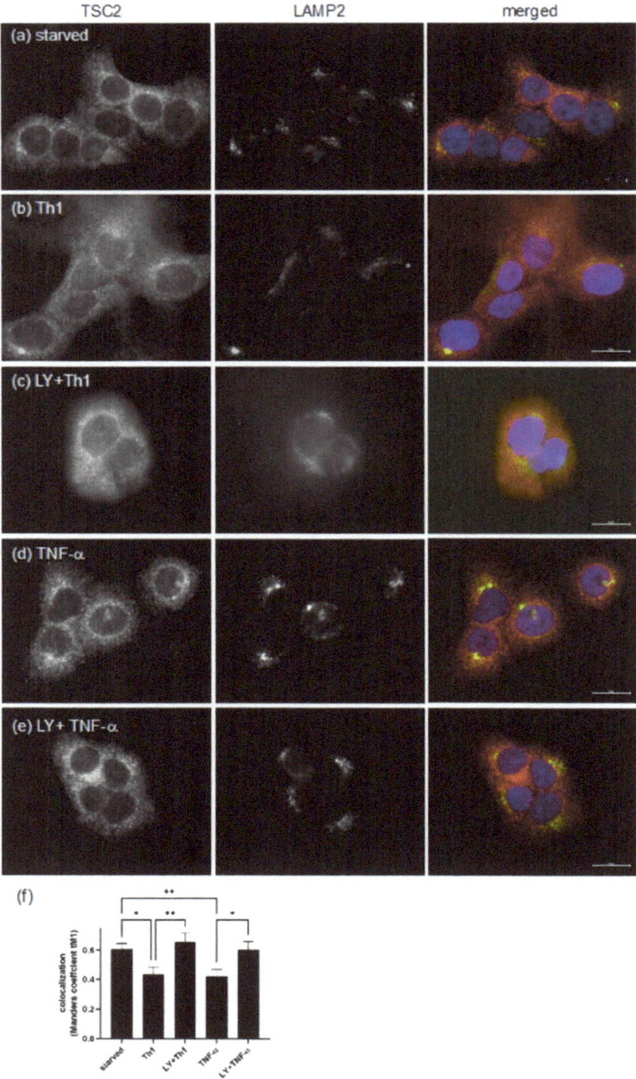

Figure 4. HaCaT cells were serum-starved (**a**) and treated with a cytokine cocktail consisting of IL-1β, IL-17A and TNF-α (20 ng/mL each) (**b,c**) or TNF-α alone (20 ng/mL) (**d,e**) for 30 min. If indicated, cells were pre-treated with 50 μM LY294002 (LY) for 30 min (**c,e**). Cells were fixed and stained for TSC2 (red) and LAMP2 (green), and nuclei were visualized with DAPI (blue). Monochrome images of the red and green channel as well as merged images are shown. Bars indicate 20 μm. (**f**) Colocalization analysis of TSC2 and LAMP2 (thresholded Manders coefficient) is shown as mean ± SEM of 8–14 images from three individual experiments. Statistical significance was assessed with one-way ANOVA and Fisher's LSD test (* $p < 0.05$, ** $p < 0.01$).

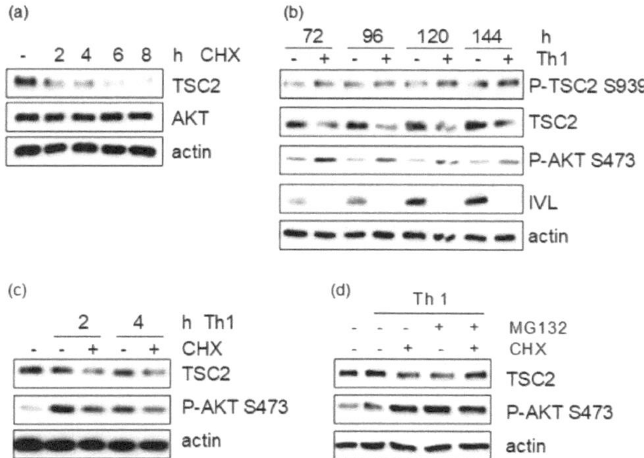

Figure 5. (**a**) HaCaT cells were treated with 10 μg/mL cycloheximide (CHX) for the indicated time points. (**b**) Cells were treated with a Th1 cytokine cocktail (IL-1β, IL-17A, TNF-α 20 ng/mL each) for the indicated time points. (**c**) HaCaT cells were serum-starved overnight and treated with Th1 cytokine and 10 μg/mL cycloheximide (CHX) as indicated for 2 or 4 h. (**d**) Serum-starved cells were treated with Th1 cytokine mix, 10 μg/mL cycloheximide (CHX) or 10 μM MG132 as indicated for 4 h. Protein lysates were prepared and subjected to Western blotting, followed by detection of the indicated proteins.

4. Discussion

We provide evidence that the TS complex might serve as a crucial switch during epidermal maturation. Under inflammatory conditions such as in psoriasis, it is significantly downregulated, leading to persistent mTORC1 activation, which in turn hampers epidermal maturation.

The TSC1/2 complex has emerged as a central sensor and signal integrator of mTORC1 signaling, being a target for several protein kinases that respond to extracellular cues such as energy, nutrients, growth signals and stress [16]. Especially well described is the regulation through AKT and ERK signaling. We could show that both pathways are activated by cytokines, with a proven role in psoriasis, leading to TSC2 phosphorylation on S939. We and others could show that this residue is a direct target of AKT [24]. In addition, ERK activates RSK1, which has overlapping target sites with AKT such as TSC2 S939 [16]. Thus, chronic exposure to pro-inflammatory cytokines in psoriasis leads to enhanced phosphorylation, which represents a negative signal towards mTORC1. It has been recently found that mTORC1 signaling events are localized at the lysosomal surface [25]. In the absence of any anabolic signals, such as growth factors or amino acids, the TS complex translocates to the lysosome, where TSC2 serves as a GAP for the small GTPase RHEB, thus blocking mTORC1 activation [17,21]. We could show that psoriatic cytokines are able to direct TSC2 away from the lysosomal surface, thus allowing constant mTORC1 activation. Lysosomes are especially important for the formation of a protective epidermal barrier and the formation of corneocytes. Corneocytes are dead cells that mainly serve as a shell for highly insoluble cornified envelop proteins. Thus, keratinocytes eliminate intracellular material and organelles via the lysosomal pathway [26,27]. Interestingly, in HaCaT cells treated with a similar mix of psoriatic cytokines, as well as in psoriatic skin, an altered number of lysosomes could be detected [28], and lysosomal function was impaired [29]. This could imply that under inflammatory conditions detrimental changes in the lysosomal signal platform also affect mTORC1 signaling.

As an additional regulatory mechanism, the phosphorylated TS complex is not only recruited away from the lysosomal mTORC1 signaling complex but is also regulated by protein stability. TSC2 can be bound by the ubiquitin-ligases HERC1 [30], PAM [31], E6AP [32] or TRIM6 [33], becoming highly ubiquitinated and targeted for proteasomal degradation [34]. Furthermore, TSC stability is regulated by the phosphorylation-dependent binding of 14-3-3 [35]. Shumway et al., found that the binding of 14-3-3β inhibits TSC function [36], while it was also described that 14-3-3β binding to the TS complex member TBC1D7 prevents binding of the E3-ubiquitin-ligase β-TrCP1 and thus stabilizes the protein. However, this interaction is only relevant when TBC1D7 is not bound by TSC1, as this interaction stabilizes the whole complex [14,37]. Interestingly, the expression of 14-3-3β and 14-3-3ζ was found to be lower in psoriatic lesions than in healthy skin [38], which would result in increased TSC activity and decreased mTORC1 signaling, which contradicts our previous findings [7,8]. A different report describes 14-3-3β upregulation in psoriatic skin only on the RNA level, but not on the protein level [39]. Thus, a thorough expression analysis of 14-3-3 isoforms in psoriasis and their impact on TSC functions is needed to fully elucidate these mechanisms.

In addition to the presented modes of TSC regulation through phosphorylation, recent work suggests that TSC is also regulated by other post-translational modifications. Gen et al. show that TSC2 is methylated at R1457 and R1459 by the protein arginine methyltransferase 1 (PRMT1), which affects TSC2 stability and interferes with phosphorylation at Thr1462 [40]. In addition, Garcia-Aguilar et al., demonstrate that TSC2 is also acetylated on lysine residues, which promotes its ubiquitination, thus stimulating mTORC1 activity [41]. These secondary modifications are also worth investigating in psoriatic samples. Besides post-translational regulation of the TS complex, the psoriatic inflammation also seems to influence TSC transcription, as gene expression analysis showed the downregulation of TSC1 and 2 in lesional psoriatic skin compared to non-lesional skin [42,43].

Thus, an inflammatory milieu as existing in the psoriatic epidermis affects TSC function through different mechanisms, which in turn leads to aberrant mTORC1 activation. mTORC1 hyperactivation not only contributes to enhanced keratinocyte proliferation, leading to epidermal thickening [8], but also interferes with ordered differentiation through the altered expression of keratin 6 [44] and filaggrin [45] as well as reduced nucleophagy [27]. Independent from its role in regulating mTORC1 signaling, the TS complex also contributes to epidermal barrier function by regulating tight junction (TJ) formation. Ablation of TSC1 in mice disrupts TJs and causes a psoriasis-like phenotype. At the same time, junctional TSC1 was markedly reduced in psoriatic skin, suggesting that TSC1 deficiencies underlies epidermal barrier deficiencies due to TJ impairment [46].

Hence, the TS complex seems to have divergent levels of regulation that contribute to its localization, stability and activity towards mTORC1 signaling and beyond.

In summary, we provide the first evidence that the TS complex serves as an important regulator in epidermal differentiation and maturation. Under healthy conditions, in the absence of a signal that stimulates AKT or ERK, TSC2 resides on the lysosomal surface and serves as a GAP for RHEB. This prevents activation of mTORC1 and allows proper keratinocyte differentiation (Figure 6 top panel). Under inflammatory conditions, such as in psoriasis, AKT and ERK become highly activated, leading to the phosphorylation of TSC2, which relocates from the lysosome and is targeted for degradation. This allows for the chronic activation of mTORC1 signaling, which prevents ordered epidermal differentiation and maturation, leading to the phenotypic changes seen in psoriatic skin (Figure 6 lower panel). Consequently, TSC represents as a major switch during epidermal homeostasis, so that its deregulation contributes significantly to the development of skin diseases. Thus, our study supports the notion to further explore the AKT/TSC/mTORC1 pathway as a therapeutic target for the treatment of inflammatory skin diseases. In the past, the mTORC1 inhibitor rapamycin/sirolimus has been used sporadically for systemic therapy due to its immunosuppressive properties, but has been associated with significant side effects [47–49]. A single, small clinical trial [50] as well as a mouse study [51], used rapamycin as a topical treatment, resulting in a significant improvement in the clinical score, which support the assumption that the anti-proliferative and differentiation-promoting effects of rapamycin are more important for the therapeutic outcome than its immune-modulatory properties.

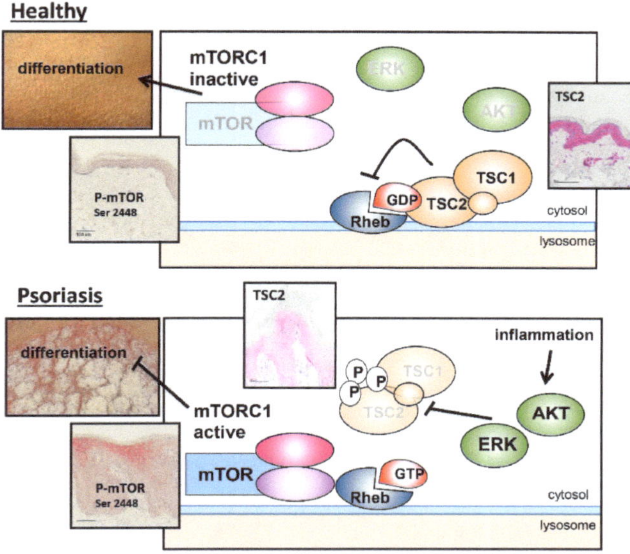

Figure 6. Mechanistic scheme on the regulation of the TS complex in healthy and psoriatic skin.

Author Contributions: Conceptualization, C.B. and R.K.; methodology and investigation, A.F. and V.L.; data analysis and figure preparation, A.F. and C.B.; writing—original draft preparation, C.B.; writing—review and editing, C.B. and R.K. All authors have read and agreed to the published version of the manuscript.

Funding: This research was funded by German Research Foundation (DFG), BU 1840/5-3 to C.B.

Institutional Review Board Statement: The study was conducted in accordance with the Declaration of Helsinki and approved by the Ethics Committee of the University Hospital Frankfurt, Goethe University (116/11; 21 October 2015).

Informed Consent Statement: Informed consent was obtained from all subjects involved in the study.

Data Availability Statement: Not applicable.

Acknowledgments: The author would like to thank Sandra Diehl for excellent technical support and Magdalena Jahn for critical reading of the manuscript.

Conflicts of Interest: The authors declare no conflict of interest.

References

1. Boehncke, W.H. Systemic Inflammation and Cardiovascular Comorbidity in Psoriasis Patients: Causes and Consequences. *Front. Immunol.* **2018**, *9*, 579. [CrossRef] [PubMed]
2. Boehncke, W.H.; Schon, M.P. Psoriasis. *Lancet* **2015**, *386*, 983–994. [CrossRef]
3. Lowes, M.A.; Bowcock, A.M.; Krueger, J.G. Pathogenesis and therapy of psoriasis. *Nature* **2007**, *445*, 866–873. [CrossRef] [PubMed]
4. Schakel, K.; Schon, M.P.; Ghoreschi, K. [Pathogenesis of psoriasis]. *Hautarzt* **2016**, *67*, 422–431. [CrossRef] [PubMed]
5. Lowes, M.A.; Russell, C.B.; Martin, D.A.; Towne, J.E.; Krueger, J.G. The IL-23/T17 pathogenic axis in psoriasis is amplified by keratinocyte responses. *Trends Immunol.* **2013**, *34*, 174–181. [CrossRef] [PubMed]
6. Buerger, C.; Richter, B.; Woth, K.; Salgo, R.; Malisiewicz, B.; Diehl, S.; Hardt, K.; Boehncke, S.; Boehncke, W.H. Interleukin-1beta interferes with epidermal homeostasis through induction of insulin resistance: Implications for psoriasis pathogenesis. *J. Investig. Dermatol.* **2012**, *132*, 2206–2214. [CrossRef] [PubMed]

7. Buerger, C.; Malisiewicz, B.; Eiser, A.; Hardt, K.; Boehncke, W.H. Mammalian target of rapamycin and its downstream signalling components are activated in psoriatic skin. *Br. J. Dermatol.* **2013**, *169*, 156–159. [CrossRef]

8. Buerger, C.; Shirsath, N.; Lang, V.; Berard, A.; Diehl, S.; Kaufmann, R.; Boehncke, W.H.; Wolf, P. Inflammation dependent mTORC1 signaling interferes with the switch from keratinocyte proliferation to differentiation. *PLoS ONE* **2017**, *12*, e0180853. [CrossRef]

9. Buerger, C. Epidermal mTORC1 Signaling Contributes to the Pathogenesis of Psoriasis and Could Serve as a Therapeutic Target. *Front. Immunol.* **2018**, *9*, 2786. [CrossRef]

10. Potter, C.J.; Pedraza, L.G.; Xu, T. Akt regulates growth by directly phosphorylating Tsc2. *Nat. Cell Biol.* **2002**, *4*, 658–665. [CrossRef]

11. Roux, P.P.; Ballif, B.A.; Anjum, R.; Gygi, S.P.; Blenis, J. Tumor-promoting phorbol esters and activated Ras inactivate the tuberous sclerosis tumor suppressor complex via p90 ribosomal S6 kinase. *Proc. Natl. Acad. Sci. USA* **2004**, *101*, 13489–13494. [CrossRef] [PubMed]

12. Ma, L.; Chen, Z.; Erdjument-Bromage, H.; Tempst, P.; Pandolfi, P.P. Phosphorylation and functional inactivation of TSC2 by Erk implications for tuberous sclerosis and cancer pathogenesis. *Cell* **2005**, *121*, 179–193. [CrossRef] [PubMed]

13. Huang, J.; Manning, B.D. A complex interplay between Akt, TSC2 and the two mTOR complexes. *Biochem. Soc. Trans.* **2009**, *37*, 217–222. [CrossRef] [PubMed]

14. Dibble, C.C.; Elis, W.; Menon, S.; Qin, W.; Klekota, J.; Asara, J.M.; Finan, P.M.; Kwiatkowski, D.J.; Murphy, L.O.; Manning, B.D. TBC1D7 is a third subunit of the TSC1-TSC2 complex upstream of mTORC1. *Mol. Cell* **2012**, *47*, 535–546. [CrossRef] [PubMed]

15. Uysal, S.P.; Sahin, M. Tuberous sclerosis: A review of the past, present, and future. *Turk. J. Med. Sci.* **2020**, *50*, 1665–1676. [CrossRef]

16. Huang, J.; Manning, B.D. The TSC1-TSC2 complex: A molecular switchboard controlling cell growth. *Biochem. J.* **2008**, *412*, 179–190. [CrossRef]

17. Menon, S.; Dibble, C.C.; Talbott, G.; Hoxhaj, G.; Valvezan, A.J.; Takahashi, H.; Cantley, L.C.; Manning, B.D. Spatial control of the TSC complex integrates insulin and nutrient regulation of mTORC1 at the lysosome. *Cell* **2014**, *156*, 771–785. [CrossRef]

18. Castro, A.F.; Rebhun, J.F.; Clark, G.J.; Quilliam, L.A. Rheb binds tuberous sclerosis complex 2 (TSC2) and promotes S6 kinase activation in a rapamycin- and farnesylation-dependent manner. *J. Biol. Chem.* **2003**, *278*, 32493–32496. [CrossRef]

19. Garami, A.; Zwartkruis, F.J.; Nobukuni, T.; Joaquin, M.; Roccio, M.; Stocker, H.; Kozma, S.C.; Hafen, E.; Bos, J.L.; Thomas, G. Insulin activation of Rheb, a mediator of mTOR/S6K/4E-BP signaling, is inhibited by TSC1 and 2. *Mol. Cell* **2003**, *11*, 1457–1466. [CrossRef]

20. Inoki, K.; Li, Y.; Xu, T.; Guan, K.L. Rheb GTPase is a direct target of TSC2 GAP activity and regulates mTOR signaling. *Genes Dev.* **2003**, *17*, 1829–1834. [CrossRef]

21. Demetriades, C.; Doumpas, N.; Teleman, A.A. Regulation of TORC1 in response to amino acid starvation via lysosomal recruitment of TSC2. *Cell* **2014**, *156*, 786–799. [CrossRef] [PubMed]

22. Boukamp, P.; Petrussevska, R.T.; Breitkreutz, D.; Hornung, J.; Markham, A.; Fusenig, N.E. Normal keratinization in a spontaneously immortalized aneuploid human keratinocyte cell line. *J. Cell Biol.* **1988**, *106*, 761–771. [CrossRef] [PubMed]

23. Inoki, K.; Li, Y.; Zhu, T.; Wu, J.; Guan, K.L. TSC2 is phosphorylated and inhibited by Akt and suppresses mTOR signalling. *Nat. Cell Biol.* **2002**, *4*, 648–657. [CrossRef]

24. Manning, B.D.; Tee, A.R.; Logsdon, M.N.; Blenis, J.; Cantley, L.C. Identification of the tuberous sclerosis complex-2 tumor suppressor gene product tuberin as a target of the phosphoinositide 3-kinase/akt pathway. *Mol. Cell* **2002**, *10*, 151–162. [CrossRef]

25. Sancak, Y.; Bar-Peled, L.; Zoncu, R.; Markhard, A.L.; Nada, S.; Sabatini, D.M. Ragulator-Rag complex targets mTORC1 to the lysosomal surface and is necessary for its activation by amino acids. *Cell* **2010**, *141*, 290–303. [CrossRef]

26. Monteleon, C.L.; Agnihotri, T.; Dahal, A.; Liu, M.; Rebecca, V.W.; Beatty, G.L.; Amaravadi, R.K.; Ridky, T.W. Lysosomes Support the Degradation, Signaling, and Mitochondrial Metabolism Necessary for Human Epidermal Differentiation. *J. Investig. Dermatol.* **2018**, *138*, 1945–1954. [CrossRef] [PubMed]

27. Akinduro, O.; Sully, K.; Patel, A.; Robinson, D.J.; Chikh, A.; McPhail, G.; Braun, K.M.; Philpott, M.P.; Harwood, C.A.; Byrne, C.; et al. Constitutive Autophagy and Nucleophagy during Epidermal Differentiation. *J. Investig. Dermatol.* **2016**, *136*, 1460–1470. [CrossRef] [PubMed]

28. Bochenska, K.; Moskot, M.; Malinowska, M.; Jakobkiewicz-Banecka, J.; Szczerkowska-Dobosz, A.; Purzycka-Bohdan, D.; Plenkowska, J.; Slominski, B.; Gabig-Ciminska, M. Lysosome Alterations in the Human Epithelial Cell Line HaCaT and Skin Specimens: Relevance to Psoriasis. *Int. J. Mol. Sci.* **2019**, *20*, 2255. [CrossRef]

29. Klapan, K.; Frangez, Z.; Markov, N.; Yousefi, S.; Simon, D.; Simon, H.U. Evidence for Lysosomal Dysfunction within the Epidermis in Psoriasis and Atopic Dermatitis. *J. Investig. Dermatol.* **2021**, *141*, 2838–2848.e4. [CrossRef]

30. Chong-Kopera, H.; Inoki, K.; Li, Y.; Zhu, T.; Garcia-Gonzalo, F.R.; Rosa, J.L.; Guan, K.L. TSC1 stabilizes TSC2 by inhibiting the interaction between TSC2 and the HERC1 ubiquitin ligase. *J. Biol. Chem.* **2006**, *281*, 8313–8316. [CrossRef]

31. Han, S.; Witt, R.M.; Santos, T.M.; Polizzano, C.; Sabatini, B.L.; Ramesh, V. Pam (Protein associated with Myc) functions as an E3 ubiquitin ligase and regulates TSC/mTOR signaling. *Cell Signal.* **2008**, *20*, 1084–1091. [CrossRef] [PubMed]

32. Zheng, L.; Ding, H.; Lu, Z.; Li, Y.; Pan, Y.; Ning, T.; Ke, Y.; Liu, W.; Yi, Y.; Zhang, C.; et al. E3 ubiquitin ligase E6AP-mediated TSC2 turnover in the presence and absence of HPV16 E6. *Genes Cells* **2008**, *13*, 285–294. [CrossRef] [PubMed]

33. Liu, W.; Yi, Y.; Zhang, C.; Zhou, B.; Liao, L.; Liu, W.; Hu, J.; Xu, Q.; Chen, J.; Lu, J. The Expression of TRIM6 Activates the mTORC1 Pathway by Regulating the Ubiquitination of TSC1-TSC2 to Promote Renal Fibrosis. *Front. Cell Dev. Biol.* **2020**, *8*, 616747. [CrossRef] [PubMed]

34. Plas, D.R.; Thompson, C.B. Akt activation promotes degradation of tuberin and FOXO3a via the proteasome. *J. Biol. Chem.* **2003**, *278*, 12361–12366. [CrossRef]
35. Li, Y.; Inoki, K.; Yeung, R.; Guan, K.L. Regulation of TSC2 by 14-3-3 binding. *J. Biol. Chem.* **2002**, *277*, 44593–44596. [CrossRef]
36. Shumway, S.D.; Li, Y.; Xiong, Y. 14-3-3beta binds to and negatively regulates the tuberous sclerosis complex 2 (TSC2) tumor suppressor gene product, tuberin. *J. Biol. Chem.* **2003**, *278*, 2089–2092. [CrossRef]
37. Madigan, J.P.; Hou, F.; Ye, L.; Hu, J.; Dong, A.; Tempel, W.; Yohe, M.E.; Randazzo, P.A.; Jenkins, L.M.M.; Gottesman, M.M.; et al. The tuberous sclerosis complex subunit TBC1D7 is stabilized by Akt phosphorylation-mediated 14-3-3 binding. *J. Biol. Chem.* **2018**, *293*, 16142–16159. [CrossRef]
38. Man, X.; Zhang, X.; Tang, J.; Chen, Y.; Li, H.; Xu, B.; Pan, L. Downregulation of 14-3-3beta and 14-3-3zeta in lesions of psoriasis vulgaris. *Clin. Exp. Dermatol.* **2013**, *38*, 390–395. [CrossRef]
39. Raaby, L.; Otkjaer, K.; Salvskov-Iversen, M.L.; Johansen, C.; Iversen, L. A Characterization of the expression of 14-3-3 isoforms in psoriasis, basal cell carcinoma, atopic dermatitis and contact dermatitis. *Dermatol. Rep.* **2010**, *2*, e14. [CrossRef]
40. Gen, S.; Matsumoto, Y.; Kobayashi, K.I.; Suzuki, T.; Inoue, J.; Yamamoto, Y. Stability of tuberous sclerosis complex 2 is controlled by methylation at R1457 and R1459. *Sci. Rep.* **2020**, *10*, 21160. [CrossRef]
41. Garcia-Aguilar, A.; Guillen, C.; Nellist, M.; Bartolome, A.; Benito, M. TSC2 N-terminal lysine acetylation status affects to its stability modulating mTORC1 signaling and autophagy. *Biochim. Biophys. Acta* **2016**, *1863*, 2658–2667. [CrossRef] [PubMed]
42. Ding, J.; Gudjonsson, J.E.; Liang, L.; Stuart, P.E.; Li, Y.; Chen, W.; Weichenthal, M.; Ellinghaus, E.; Franke, A.; Cookson, W.; et al. Gene expression in skin and lymphoblastoid cells: Refined statistical method reveals extensive overlap in cis-eQTL signals. *Am. J. Hum. Genet.* **2010**, *87*, 779–789. [CrossRef] [PubMed]
43. Swindell, W.R.; Xing, X.; Stuart, P.E.; Chen, C.S.; Aphale, A.; Nair, R.P.; Voorhees, J.J.; Elder, J.T.; Johnston, A.; Gudjonsson, J.E. Heterogeneity of inflammatory and cytokine networks in chronic plaque psoriasis. *PLoS ONE* **2012**, *7*, e34594. [CrossRef]
44. Hickerson, R.P.; Leake, D.; Pho, L.N.; Leachman, S.A.; Kaspar, R.L. Rapamycin selectively inhibits expression of an inducible keratin (K6a) in human keratinocytes and improves symptoms in pachyonychia congenita patients. *J. Dermatol. Sci.* **2009**, *56*, 82–88. [CrossRef]
45. Naeem, A.S.; Tommasi, C.; Cole, C.; Brown, S.J.; Zhu, Y.; Way, B.; Willis Owen, S.A.; Moffatt, M.; Cookson, W.O.; Harper, J.I.; et al. A mechanistic target of rapamycin complex 1/2 (mTORC1)/V-Akt murine thymoma viral oncogene homolog 1 (AKT1)/cathepsin H axis controls filaggrin expression and processing in skin, a novel mechanism for skin barrier disruption in patients with atopic dermatitis. *J. Allergy Clin. Immunol.* **2017**, *139*, 1228–1241. [CrossRef]
46. Lai, M.; Zou, W.; Han, Z.; Zhou, L.; Qiu, Z.; Chen, J.; Zhang, S.; Lai, P.; Li, K.; Zhang, Y.; et al. Tsc1 regulates tight junction independent of mTORC1. *Proc. Natl. Acad. Sci. USA* **2021**, *118*, e2020891118. [CrossRef]
47. Wei, K.C.; Lai, P.C. Combination of everolimus and tacrolimus: A potentially effective regimen for recalcitrant psoriasis. *Dermatol. Ther.* **2015**, *28*, 25–27. [CrossRef]
48. Reitamo, S.; Spuls, P.; Sassolas, B.; Lahfa, M.; Claudy, A.; Griffiths, C.E. Efficacy of sirolimus (rapamycin) administered concomitantly with a subtherapeutic dose of cyclosporin in the treatment of severe psoriasis: A randomized controlled trial. *Br. J. Dermatol.* **2001**, *145*, 438–445. [CrossRef]
49. Frigerio, E.; Colombo, M.D.; Franchi, C.; Altomare, A.; Garutti, C.; Altomare, G.F. Severe psoriasis treated with a new macrolide: Everolimus. *Br. J. Dermatol.* **2007**, *156*, 372–374. [CrossRef]
50. Ormerod, A.D.; Shah, S.A.; Copeland, P.; Omar, G.; Winfield, A. Treatment of psoriasis with topical sirolimus: Preclinical development and a randomized, double-blind trial. *Br. J. Dermatol.* **2005**, *152*, 758–764. [CrossRef]
51. Burger, C.; Shirsath, N.; Lang, V.; Diehl, S.; Kaufmann, R.; Weigert, A.; Han, Y.Y.; Ringel, C.; Wolf, P. Blocking mTOR Signalling with Rapamycin Ameliorates Imiquimod-induced Psoriasis in Mice. *Acta Derm. Venereol.* **2017**, *97*, 1087–1094. [CrossRef] [PubMed]

Article

AKT1 Transcriptomic Landscape in Breast Cancer Cells

Bijesh George [1,2], **Bin Gui** [3,†], **Rajeswari Raguraman** [1,†], **Aswathy Mary Paul** [1,2], **Harikrishna Nakshatri** [4], **Madhavan Radhakrishna Pillai** [1,*] and **Rakesh Kumar** [1,5,6,7,*]

[1] Cancer Research Program, Rajiv Gandhi Centre for Biotechnology, Trivandrum 695014, India; bijeshgeorge@rgcb.res.in (B.G.); rajraghuraman@rgcb.res.in (R.R.); aswathym@rgcb.res.in (A.M.P.)

[2] Graduate Degree Program, Manipal Academy of Higher Education, Manipal, Udupi 576104, India

[3] Biological Sciences, Ribon Therapeutics, Inc., Cambridge, MA 02140, USA; easter0731@gmail.com

[4] Department of Surgery, Indiana University of School of Medicine, Indianapolis, IN 46202, USA; hnakshat@iupui.edu

[5] Cancer Research Institute, Himalayan Institute of Medical Sciences, Swami Rama Himalayan University, Dehradun 248016, India

[6] Division of Hematology and Oncology, Department of Medicine, Rutgers New Jersey Medical School, Newark, NJ 07103, USA

[7] Department of Human and Molecular Genetics, School of Medicine, Virginia Commonwealth University, Richmond, VA 23298, USA

* Correspondence: mrpillai@gmail.com (M.R.P.); rakeshkumar@srhu.edu.in (R.K.)

† These authors contributed equally to this work.

Abstract: Overexpression and hyperactivation of the serine/threonine protein kinase B (AKT) pathway is one of the most common cellular events in breast cancer progression. However, the nature of AKT1-specific genome-wide transcriptomic alterations in breast cancer cells and breast cancer remains unknown to this point. Here, we delineate the impact of selective AKT1 knock down using gene-specific siRNAs or inhibiting the AKT activity with a pan-AKT inhibitor VIII on the nature of transcriptomic changes in breast cancer cells using the genome-wide RNA-sequencing analysis. We found that changes in the cellular levels of AKT1 lead to changes in the levels of a set of differentially expressed genes and, in turn, imply resulting AKT1 cellular functions. In addition to an expected positive relationship between the status of AKT1 and co-expressed cellular genes, our study unexpectedly discovered an inherent role of AKT1 in inhibiting the expression of a subset of genes in both unstimulated and growth factor stimulated breast cancer cells. We found that depletion of AKT1 leads to upregulation of a subset of genes—many of which are also found to be downregulated in breast tumors with elevated high AKT1 as well as upregulated in breast tumors with no detectable AKT expression. Representative experimental validation studies in two breast cancer cell lines showed a reasonable concurrence between the expression data from the RNA-sequencing and qRT-PCR or data from ex vivo inhibition of AKT1 activity in cancer patient-derived cells. In brief, findings presented here provide a resource for further understanding of AKT1-dependent modulation of gene expression in breast cancer cells and broaden the scope and significance of AKT1 targets and their functions.

Keywords: breast cancer; AKT1; RNA-sequencing; transcriptome; emerging functions and targets; cancer therapeutics

Citation: George, B.; Gui, B.; Raguraman, R.; Paul, A.M.; Nakshatri, H.; Pillai, M.R.; Kumar, R. AKT1 Transcriptomic Landscape in Breast Cancer Cells. *Cells* **2022**, *11*, 2290. https://doi.org/10.3390/cells11152290

Academic Editors: Jean Christopher Chamcheu, Claudia Bürger and Shile Huang

Received: 28 May 2022
Accepted: 11 July 2022
Published: 25 July 2022

Publisher's Note: MDPI stays neutral with regard to jurisdictional claims in published maps and institutional affiliations.

1. Introduction

Breast cancer (BC) is a polygenic and heterogeneous disease, which accounts for more than 24% of all cancers in women and about 684,996 deaths in 2020 [1]. The primary BC subtypes are stratified on the basis of the levels of estrogen receptor-alpha, progesterone receptor, and human epidermal growth factor 2 (HER2) receptor as well as on the basis of genomic, transcriptomic, epigenetic, morphological, and metabolic alterations. These alterations largely contribute to the noticed heterogeneity among more than 20 subtypes of breast cancers [2].

Cancer progression to more invasive phenotypes involves coordinated action of growth factors and oncogenes to counteract the activities of growth inhibitory pathways and tumor suppressors, in addition to other regulatory pathways. The process of oncogenesis is generally associated with dysregulated regulatory signaling, including mitogenic growth factors [3–5]. Mitogenic growth factors in conjunction with chromatin remodeling machinery (and other pathways) stimulate the proliferation, survival, motility, and invasive signaling pathways and resulting phenotypes [3–7]. One of such dysregulated signaling pathways in human cancer is the serine/threonine protein kinase B (AKT) pathway, which could be stimulated by multiple upstream molecules, i.e., insulin, platelet-derived growth factor (PDGF), insulin-like growth factor 1 (IGF1), epidermal growth factor (EGF), cytokines, nutrients, etc. [8]. The PI-3 kinase-AKT signaling pathway regulates cell cycle progression, survival, DNA repair, RNA export, differentiation, and tumorigenesis in several cancer cell types [4,8]. Accordingly, constitutive activation of this pathway has been also explored as a promising anticancer therapeutic strategy [9].

Activation of AKT signaling by growth factors, such as EGF, engages numerous downstream signaling cascades, leading to improved cell survival and proliferation in diverse cell types, including mammary epithelial cells [10,11]. The human AKT family of kinases consists of three distinct genes encoded on different loci, i.e., AKT1, AKT2, and AKT3, on chromosomes 14, 19, and 1, respectively. A large volume of initial studies in the field were conducted using AKT1 as a prototype of the AKT family, and conclusions drawn were initially presumed to also be implied for other AKT isoforms. However, a large body of work over the years involving either the gain- or loss-functions of AKT and AKT isoforms in the mouse and human model systems, respectively, has revealed differentiating biology of the AKT isoforms and their roles in the development and involution of the mammary gland as well as in the development and progression of breast cancer [12]. Previous studies have also shown that AKT1 mutations are found in ~1% of all cancers, and the most prevalent mutant AKT1(E17K) leads to its localization to the plasma membrane, invoking a consistent activation of AKT signaling in cancer cells [13,14].

All three AKT isoforms have been reported to be upregulated in human cancer and act as oncogenes and promote tumor proliferation at different levels [12,15,16]. In general, AKT1 knockdown leads to inhibition of tumor growth via blocking the cell-cycle progression and/or promoting apoptosis in breast cancer model systems [12,17,18]. Similarly, overexpression and/or constitutive activation of Akt1 in the mammary epithelial cells inhibits the pro-apoptotic signals as well as activates the survival signals to support the process of tumorigenesis [19,20]. Studies from transgenic mice suggested that Akt1 plays an important role in the initiation, development, and progression of breast tumors [20,21], whereas Akt2 has no major involvement in the process of tumor initiation but contributes to the process of tumor growth [20]. Consistent with these findings, hyperactivated AKT1 pathways are highly correlated with the initiation and development of breast cancer [21,22].

Results from knockout murine studies revealed that individual knockout of any one of the three isoforms was not lethal but contributed to growth retardation [23,24]. However, the double knockout murine studies involving Akt1 and Akt2 or Akt2 and Akt3 but not Akt1 and Akt3 were shown to be lethal in nature [25–27]. In the context of breast cancer, studies involving Akt1 knockout mice revealed that the loss of Akt1 suppresses ErbB2-induced mammary carcinogenesis [28] and mammary adenocarcinomas in mouse mammary tumor virus (MMTV)-ErbB2/neu, MMTV-polyoma middle T transgenic mice [29] and the growth of A2780 ovarian tumors in xenograft models [30]. AKT2 knockdown inhibited the chemotaxis of breast cancer cells [31], whereas knockdown of AKT3 resulted in reduced expression of HER2 and HER3 and upregulation of ER-alpha, resulting in an increased responsiveness of murine model cells to antiestrogen [32]. Ablation of AKT1 or AKT2 in murine breast cancer models and of AKT1 or AKT2 in human breast cell lines was associated with suppression of tumor progression and cell-cycle progression, increased apoptosis, and an overall reduced metastatic potential of target cells. In contrast, ablation of AKT3 has been shown to be associated with no major effect on the tumor growth but significantly decreases the tumorigenic potential of triple-negative breast cancer cells [33]. AKT2 has been shown to be involved in the maintenance of the tumorigenic characteristics of cells, as its knockdown was associated with tumor inhibition [33]. More recently, a circular AKT3 transcript has been shown to exert tumor suppressive function in glioblastoma cells, presumably by inhibiting the PI-3 kinase signaling [34]. In addition, AKT chemical inhibitors have been shown to inhibit the growth of model tumors through phosphorylation of downstream substrates in breast and other cancer cell types [35]. In addition, allosteric and competitive AKT inhibitors have been shown to prevent cancerous growth in a limited clinical study [15]. Because of structural similarities between the isoforms, many of the past experimental studies have utilized pan-AKT inhibitors such as MK2206, AZD5363, Ipatasertib, and perifosine [36,37].

The above studies suggest that in spite of a large body of work, the effect of the prototypic family member, AKT1, on the whole genome transcriptome in breast cancer cells remains unknown until this point, and hence, is examined in the present study.

2. Materials and Methods

Cell culture: Breast cancer cells MCF-7 and SKBR-3 were grown in Dulbecco's modified eagle medium (DMEM) (Invitrogen, Waltham, MA, USA) containing 10% fetal bovine serum (FBS) (Invitrogen) and 1% penicillin-streptomycin antibiotic (Hi-media). The cells were maintained in a humidified atmosphere with 5% CO_2 at 37 °C and used for experiments after they attained 70% confluence.

Transfection: Cells (4×10^5 cells/mL) were seeded onto six-well plates and incubated for 24 h at 37 °C and later transfected with 50 nm AKT1 specific sure siRNA (Cat no. sc-29195) and non-specific siRNA (Cat no. sc-37007) from Santa Cruz Biotechnology, Inc., Dallas, TX, USA, in serum-free medium for 36 h. For EGF+ condition, cells were treated with 120 ng/mL of hEGF (Cat no. E9644, Sigma-Aldrich® Solutions, St. Louis, MI, USA) for 8 h. Cells were harvested, and total RNA was isolated with the RNeasy extraction kit (Qiagen India Pvt Ltd., New Delhi, India).

RNA isolation: The total RNA from direct and indirect coculture assays was isolated with RNeasy kit (Qiagen India Pvt Ltd., New Delhi, India) following manufacturer's instructions. The quality of RNA was checked with a Biospec nano spectrophotometer (Eppendorf, Hamburg, Germany), and 1 μg of total RNA was used for the cDNA conversion.

Single cell sequencing-based data analysis: AKT1, AKT2, and AKT3 expression data were extracted from single cell RNA-seq data of the healthy breast tissues using Loupe Browser (10X Genomics) as reported recently [38]. Markers used for cell type annotation are also described in the publication. The details of the samples and methods used to generate the referred single-cell sequencing atlas are described in Bhat-Nakshatri et al. [38].

RNA sequencing and resulting data processing: Total RNA samples were subjected to the whole transcriptome RNA-sequencing by Beckman Coulter Genomics, Newton, MA, USA. The vendor used the standard time-tested methodology for removing large and small ribosomal RNA, quality control, cDNA synthesis, DNA library preparation, paired-end sequencing with 2×100 bp using Illumina HiSeq 2000, read alignment to the reference hg19 (Ensembl GRCh37.75 build) genome using Tophat [39] version 2.0.9 in conjunction with Bowtie [40] version 1.0.0, quality control, and transcript assembly. Prior to mapping, reads are inspected and trimmed for adapter sequence with Flexbar [41] version 2.4. Thus, only reads not mapping to the transcriptome are attempted directly on the genome, allowing for prediction of novel exons, isoforms, and genes. Reads mapped to the transcriptome are documented with their genome-equivalent coordinates. 'Proper' read pairs either fall entirely within exons or hit adjacent exons. Singleton reads do not have their mate-read mapped on the genome due to sequence quality of the mate or to the incompleteness of the genome reference.

Read counting and differential expression analysis: Reads were counted using HTSeq-count [42] version 0.6.0, and multiple alignments were excluded. Gene counting was performed for genes and transcripts. Differential expression analysis was performed using

DESeq package [43]. The samples were compared at gene level for all six experimental conditions to find the differentially expressed genes that are regulated by AKT isoforms. Together we have studied the transcription changes regulated by AKT gene by using pan-AKT inhibitor (Inhibitor VIII, a widely used AKT pan-inhibitor VIII) as well as its isoforms by knocking down AKT1 specific siRNA.

Complementary DNA (cDNA) conversion: The total RNA isolated from MCF 7 cells was converted to cDNA using the script cDNA synthesis kit (Bio-Rad). Briefly, 1 μg of total RNA was converted to DNA following manufacturer's instructions. The mixture was incubated under the cycling conditions (25 °C for 5 min, 42 °C for 30 min, and 85 °C 95 for 5 min and 4 °C hold) in a PCR machine (Eppendorf) and were stored at −20 °C until use.

Quantitative reverse transcriptase polymerase chain reaction (qRT-PCR): The effect of knocking down AKT1 using siRNA or inhibiting AKT kinases by AKT inhibitor VIII on the levels of test genes in breast cancer cells, stimulated or unstimulated by EGF, was determined using qRT-PCR. The qRT-PCR reaction mixture containing 30 ng of cDNA was prepared with 5 μL SYBR-Green 2xmaster mix (TAKARA BIO INC., Kusatsu, Japan) and 0.4 μM each of forward and reverse primers (Sigma-Aldrich® Solutions). The PCR reaction was carried out in an Applied Biosystem Quant Studio 7 plus real time PCR machine. Relative quantification of the gene expression (siNON v/s siAKT1) was determined using the 2-$\Delta\Delta$Ct method [44], and relative expression values (log 2-fold change (FC)) were normalized to GAPDH endogenous control values. The primer sequences for genes were commercially procured from Sigma Aldrich. The experiments were performed in triplicate for each sample.

Splice variation analysis: BAM files resulting from the processed RNA-Seq data alignment were analyzed for splice variations. BAM files were subjected to percent spliced-in (PSI) calculation using psi.sh as per the protocol described, and the exon inclusion counts were obtained [45,46]. Count files were fed to the DEXSeq [47] package in R with metadata to identify potential exon usage by each condition. DEXSeq provides differential expression analysis for a set of experimental conditions with a common denominator. Splicing events with p-value < 0.05 and p-adj < 0.1 were selected based on the higher exon usage coefficient to identify the highly abundant transcripts, and exons showing higher dispersion between AKT1 silenced samples and fold change are reported.

Comparative analysis with gene expression omnibus and TCGA datasets: Comparative analysis was performed for the differential expressed genes using AKT silenced datasets from GEO for the accession numbers GSE71900 [8] and GSE98078 [48]. Statistically significant (at least a fold change of 1.5 with a p-value < 0.05) genes were compared with the list of genes identified in our experiment to check the overlap, and the results are included. Gene expression analysis of breast cancer data are available from The Cancer Genome Atlas (TCGA) and Molecular Taxonomy of Breast Cancer International Consortium [49–51] samples. Samples are categorized into AKT1 high and low expressing samples using Onco-Query Language provided as per cBioPortal datasets [52,53].

Gene ontology and gene enrichment analysis: Gene ontology analysis was performed using the Funrich tool [54].

3. Results

3.1. Expression AKTs in Breast Cancer Cells

RNA interference (RNAi) technology coupled with gene expression analysis is widely used to map the regulatory network by inhibiting specific targeted mRNA. This approach is a powerful strategy to identify the transcriptomic variations, and, in turn, gain clues about the nature of the dysregulated pathways [55]. High throughput RNA sequencing allows us to capture the transcriptomic profile and compare test profiles to identify the alterations between distinct experimental settings.

To understand the significance of AKT1 in the mammary epithelial cells, first we examined the status of AKT isoforms by single cell sequencing in major mammary epithelium cell types, i.e., basal cells, luminal progenitor, and mature luminal cells, isolated from healthy women [38]. We found that AKT1 is highly expressed in the basal cells, luminal progenitor, and mature luminal cells. In contrast, the expression of AKT2 somewhat overlapped with that of AKT1 (Figure 1A), suggesting that resulting phenotypes in mammary epithelial cells and, perhaps, in breast cancer, could be differentially affected by AKTs. We next examined the expression of AKT1, 2, and 3 in Breast Cancer METABRIC TCGA datasets [49–51,56] (Figure 1B). We found that each of AKT isoforms had a distinct overexpression pattern and that AKT1 and AKT2 are largely overexpressed as well as amplified, whereas AKT3 is predominantly amplified (and not overexpressed). Based on these observations and the results from the previous gain and loss of functional studies in the field, showing the role of AKT1 in the cell survival and growth regulation [12,15]—the focus of the present study—we decided to examine the impact of selective depletion of AKT1 and the inhibiting of the AKT's activity by inhibitor VIII on the genome-wide transcriptome of breast cancer cells using breast cancer MCF-7 cells as a model system, as these cells express abundant levels of AKT1 and are widely used for a large number of genome-wide discovery studies. Results obtained were validated in MCF-7 and SKBR-3 breast cancer cell lines as well as in publicly available databases wherein human specimens were treated with the AKT inhibitor VIII.

3.2. Analysis of AKT1 Transcriptome in Breast Cancer Cells

The expression of endogenous AKT1 in MCF-7 cells was silenced using selective siRNAs directed against AKT1. In addition, in certain experiments, we used a pan-AKT inhibitor VIII, 5 nm for 30 min, which has been widely used to inhibit the activities of AKTs in multiple previous studies [57,58] (Supplementary Figure S1A). As the AKT pathway has been shown to be stimulated in cancer cells by growth factors, we chose to use EGF as a mitogen to stimulate the AKT1 pathway in MCF-7 cells. Cells were stimulated with or without 120 ng/mL EGF for 8 h after treating the cells with selective or control 50 nm siRNAs for 36 h (Figure 1C). Samples were prepared and subjected to paired-end RNA-sequencing [44], and we observed an average Pearson correlation of 0.93 between the replicates (Supplementary Table S1). Data analysis was performed using several commonly used, open-source algorithms and tools, schematically depicted in Figure 1C. An average of 78 million reads were generated for each sample, out of which an average of 87% of reads were aligned to human reference genome (Ensemble GRCh37.75 build, hg19) for each condition; 87.09% of reads were aligned as proper pairs, 7.22% were aligned as long pairs, and 5.69% were aligned as singletons (Figure 1D and Supplementary Table S2).

An average of 15,500 genes were identified per sample using Ensemble annotations. Genes with at least 10 aligned reads were considered for subsequent analyses (Supplementary Table S3), and a highly abundant 10 transcripts were considered as stable abundance and analyzed across the samples (Supplementary Table S4). Initial analysis on read mapping confirmed that the read mapping was proper and thus excluded the possibility in the variation in read distribution among the treatment conditions. In brief, these studies accurately mapped the high quality paired end reads to the human genome and, thus, appropriateness of the read depth and coverage for further analysis.

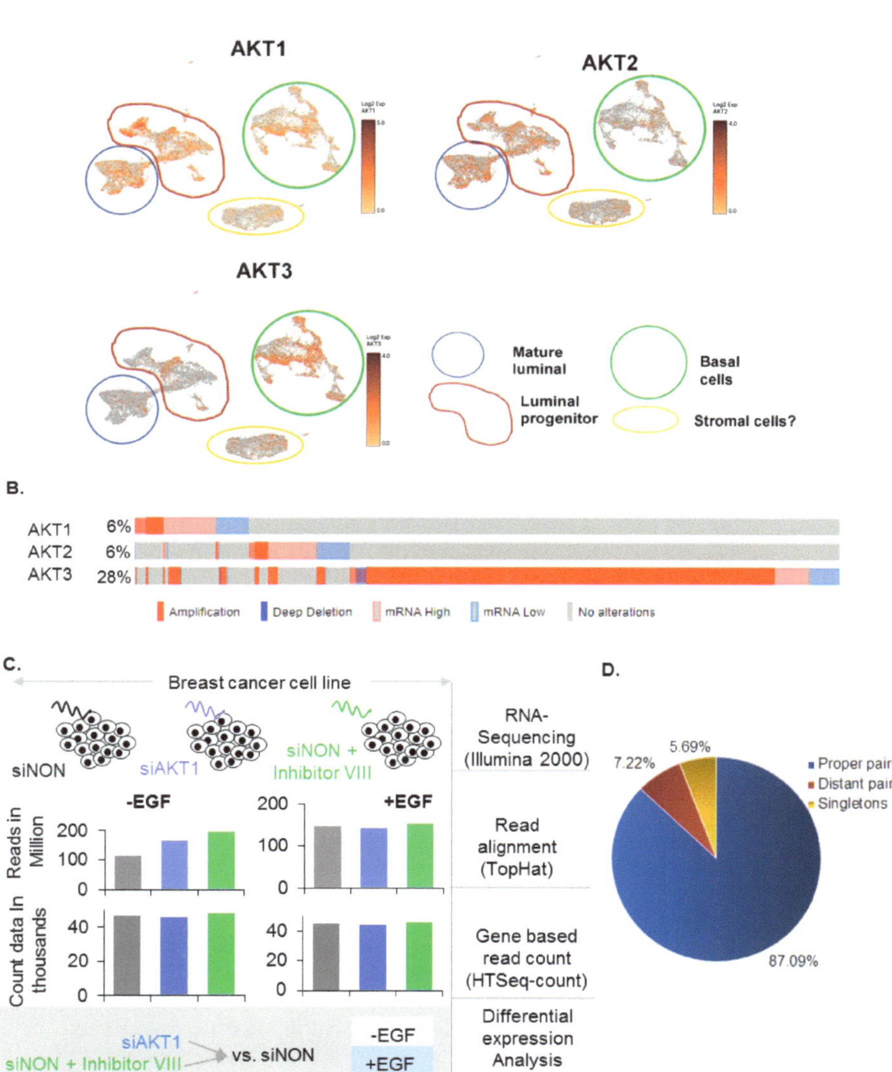

Figure 1. Expression of AKT isoforms and experimental strategy. (**A**) AKT1, AKT2, and AKT3 mRNA at the single-cell level of breast tissue. The image was generated using a recently published healthy breast atlas [38]. Expression of AKT1 and AKT2 was found to be widespread compared to AKT3. For example, AKT3 was expressed at a higher level in many subclusters of basal cells and in subcluster 4 of the luminal progenitor, but least in the mature luminal cells. The identity of the cells in yellow was suspected to be stromal in nature. (**B**) Expression of AKT isoforms in METABRIC datasets. (**C**) An overview of the analysis workflow, including the steps and the method followed to analyze the data and the numeric figures related to each step in the workflow. (**D**) Pie chart shows the average percentage of reads mapped to the human genome (hg19, Ensemble Grch37).

3.3. Influence of AKT1 on the Status of Growth Factor Induced Genes

Differentially expressed genes were obtained for each experimental condition with respect to control and with and without EGF stimulation. To assess the effectiveness of the selective AKT1-siRNA used, we first determined the levels of AKT1 mRNA (and AKT2 and AKT3 mRNA as controls) in the processed datasets, in addition to initial examination of the AKT1 protein (Supplementary Figure S1A). As expected, use of AKT1-siRNA was accompanied by a reduced expression of AKT1 (not AKT3 as a negative control) (Supplementary Figure S1B). For obtaining an overall larger view of gene distribution among experimental conditions and for performing an initial assessment of AKT1 isoform specific changes in the transcriptome, the comparison of the differentially expressed genes was performed before applying the statistical threshold in Figure 2. This was followed by implementation of the quality control measures to select differentially expressed genes with at least a 1.5-fold change and a p-value less than 0.05 for further analysis. Statistically significant genes with p-value < 0.05 and >1.5-fold change were identified using a negative binomial test. A total of 3898 and 2908 genes were found to be differentially expressed, respectively, for AKT1 knockdown and cells treated with inhibitor VIII with at least a 2-fold change over the cells treated with the control siRNA (siNON). Comparative analysis of differentially expressed genes found sets of 2653 and 1663 genes to be uniquely regulated by siAKT1 and Inhibitor VIII (AKT-VIII), respectively, in unstimulated breast cancer cells (Figure 2A). Upon EGF stimulation, these affected gene numbers were changed to 5325 (siAKT1) and 2740 (AKT-VIII) differentially expressed genes. As the goal of the study was to determine the nature of AKT1-dependent modulation of transcriptome, we found a total of 4202 and 1619 genes were uniquely regulated by siAKT1 and inhibitor VIII, respectively (Figure 2C). The number of differentially expressed genes in each chromosome was analyzed to observe the choice of the preferred target gene genomic loci. We noticed that chromosomes 1 and 19 represent relatively higher fractions of altered genes in both unstimulated and EGF-stimulated breast cancer cells, whereas differentially expressed genes on chromosome 2 were observed only upon EGF stimulation (Figure 2C,D). In brief, we observed that AKT1 could regulate specific set of genes and thus could influence the nature of breast cancer transcriptome.

3.4. EGF Modulation of AKT-Dependent Transcriptome

As we were interested in understanding the effect of EGF stimulation of AKT1-dependent transcriptome, we found that a total of 2519, 4299, and 2643 genes were altered, with at least 2-fold change in expression, in cells treated with siNON, siAKT1, and VIII, respectively. Among these differentially expressed genes, 1343, 3078, and 1453 genes were found to be unique to the referred experimental conditions (Figure 2E). Chromosome-wise distribution showed that EGF stimulation in the absence of AKT1 alters a higher number of genes across the genome (Figure 2F).

We next analyzed the fold-change distributions against statistical significance, i.e., differentially expressed genes with at least 1.5-fold change with p-value < 0.05 for further analysis (Figure 3A). Differentially expressed genes were categorized based on the coding potential; the protein coding genes were found highly altered followed by lncRNAs and antisense RNAs (Figure 3B, Supplementary Table S5). A set of 1624 genes were found to be significantly affected by the knockdown of AKT1 with respect to siNON in the absence or presence of EGF stimulation (Figure 3C). In brief, we describe a set of differentially expressed genes which are preferentially modulated by the levels of AKT1.

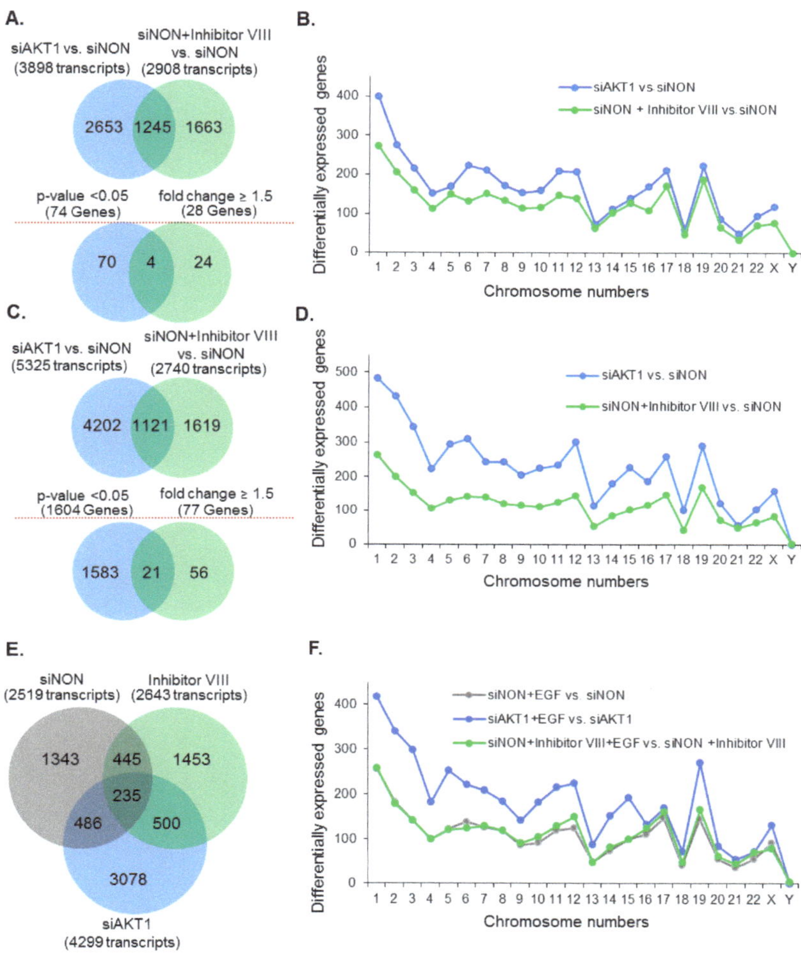

Figure 2. Effect of AKT1 on the status of differentially expressed genes in breast cancer cells. (**A**) Venn diagram showing the shared and uniquely expressed differentially expressed genes upon depletion of AKT1 or treatment with AKT inhibitor VIII as compared to cells treated with siNON. The total number of differentially expressed transcripts with 2-fold change are represented in parentheses. Results of comparative analysis of statistically significant (*p*-value < 0.05 and fold change > 1.5) differentially expressed genes are shown in the lower panels; (**B**) ≥chromosome-wise distribution of differentially expressed transcripts corresponding to the preceding upper panel (**A**); (**C**) Venn diagram showing the comparison of the number of differentially expressed transcripts with 2-fold change in EGF-stimulated breast cancer cells. Results of comparative analysis of statistically significant (*p*-value < 0.05 and fold change > 1.5) differentially expressed genes are shown in the lower panels; (**D**) chromosome-wise distribution of transcripts corresponding to genes in upper panel (**C**); (**E**) genes altered (>2-fold change) in the indicated experimental conditions in EGF-stimulated breast cancer cells; and (**F**) chromosome-wise distribution of differentially expressed genes in EGF-stimulated breast cancer cells in Panel (**E**).

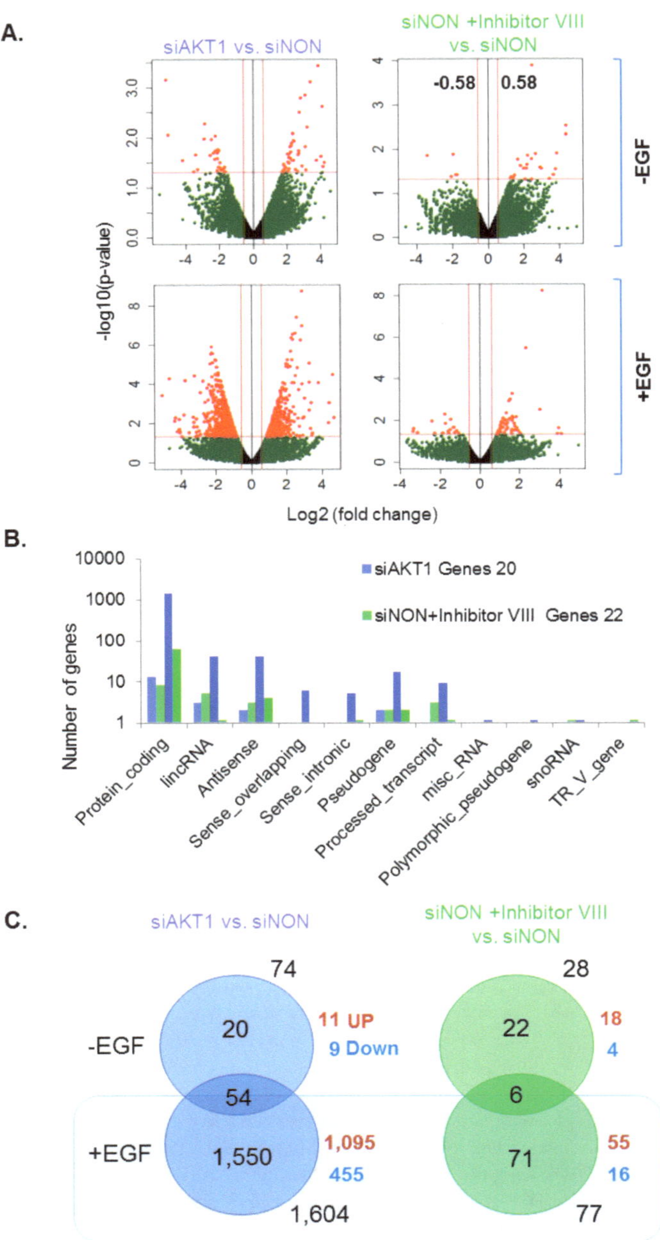

Figure 3. Status of statistically significant differentially expressed genes. (**A**) Statistical distribution of differentially expressed genes; *p*-value 0.05 threshold is marked on the Y-axis (red) and 2-fold change (green) are marked on the X-axis; (**B**) overall distribution of the coding potential of the statistically significant genes across experimental conditions; and (**C**) pictorial representation of differentially expressed genes across various experimental conditions.

3.5. Identification of AKT1 Specific Regulatory Pathways

We next compared the genes affected by the status of AKT1 and identified a total of 1624 genes to be regulated by AKT1 (Figure 4A). Gene ontology analysis of 1624 genes showed an enrichment of these genes in protein metabolism, metabolism, energy pathways, cell growth and/or maintenance, transport etc. (Figure 4B, Supplementary File S2). A set of 99 AKT regulated genes were functionally involved in regulation of the immune system process, female pregnancy, negative regulation of endopeptidase activity, triglyceride catabolic process, etc. (Figure 4D, Supplementary File S2). Comparative analysis of functions of AKT1 regulated genes versus genes regulated when all the AKTs were inhibited by the Pan-AKT Inhibitor VIII revealed sharing of only six predicted functions, namely, signal transduction, immune response, cell cycle, cell adhesion, cell proliferation, and cell differentiation (Supplementary Figure S2, Supplementary File S2). A set of 1104 genes found downregulated were functionally involved in protein metabolism, metabolism, energy pathways, cell growth and/or maintenance, transport, cell communication, signal transduction, etc. (Figure 4D, Supplementary File S2). Interestingly, 466 AKT-dependent upregulated genes were also functionally annotated and found to be largely involved in the biological processes like protein metabolism, metabolism, energy pathways, cell growth and/or maintenance, transport, cell communication, etc. (Figure 4E, Supplementary File S2). A total of 466 AKT-knockdown associated upregulated genes might be important, as these genes were presumably inhibited by the presence of an active AKT pathway. An overall gene ontology analysis [59–61] of these 466 genes showed that a large number of these genes are involved in metabolic related functions (Figure 4D). Among the 466 genes upregulated in the absence of AKT1 (and hence, these genes are expected to be downregulated by AKT1), 25 genes are downregulated in breast cancer with high AKT1 expression [62]. In brief, our study strengthens the notion of AKT1-specific functions and discovered an unexpected role of AKT1 signaling in inhibitory transcriptomic changes.

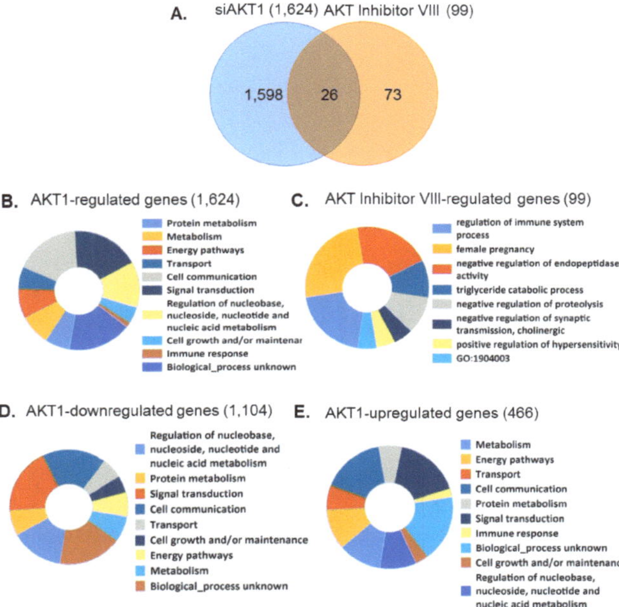

Figure 4. Pathway and functional analysis of AKT1-dependent differentially altered genes. (**A**) Venn diagram showing the shared and unique genes regulated by AKT1 and/or AKT inhibitor VIII;

(**B**) doughnut chart showing the functional annotation of differentially expressed genes upon AKT1 silencing using siAKT1; (**C**) doughnut chart showing the functional annotation of differentially expressed genes upon using the Pan-AKT Inhibitor VIII; (**D**) doughnut chart showing the functional annotation of downregulated genes upon AKT1 silencing; and (**E**) doughnut chart showing the functional annotation of upregulated genes upon AKT1 silencing.

3.6. Validation of AKT1 Regulated Significant Genes

Next, we validated a set of selected genes of interest from the RNA-seq analysis using qRT-PCR in two different breast cancer MCF-7 and SKBR-3 cell lines. First, MCF-7 cells were treated with siAKT1 or control siNON, followed by stimulation with or without EGF for 8 h. In general, we noticed a similar trend of increased or reduced expression of 11 tested transcripts, out of 21 selected genes, belonging to the pathways of interest to the laboratory, between the RNA-sequencing and qRT-PCR results in cells with siAKT1 (Figure 5A, Supplementary Figure S3) and among them 7 (without EGF) and 6 (EGF stimulated samples) genes were also found to exhibit the same pattern of expression in another breast cancer SKBR3 cell line (Supplementary Figure S3).

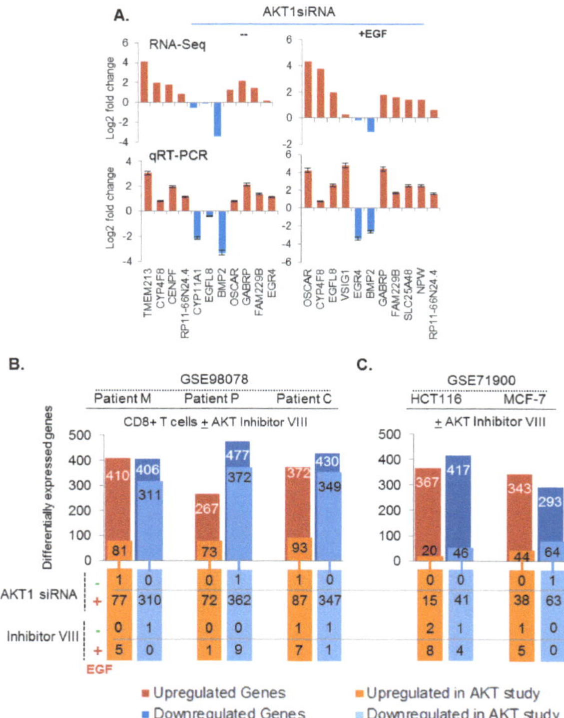

Figure 5. Validation of AKT-dependent modulation of transcriptomes. (**A**) Examples of genes with a similar expression pattern in the RNA-seq data and RT-PCR validation studies; (**B**) comparative analysis of RNA sequencing data presented here with AKT inhibitor VIII-modulated transcriptome from the accession number GSE98078. The numbers on the top of the bar show the overlap with upregulated and downregulated genes as compared to the individual sample; the lower part of the diagram shows the number of gene with unique overlap with AKT inhibitor VIII treated samples with and without EGF stimulation; and (**C**) the comparative analysis of the RNA-seq data presented here with the effect of AKT inhibitor VIII on the gene expression in colorectal cancer HCT116 and breast cancer MCF-7 cells under the accession number GSE71900. Only the genes showing at least 1.5 fold change with a *p*-value < 0.05 are considered from all the studies.

To further validate the noticed modulation of AKT1 status-dependent transcriptome in patient-derived biomaterial, we attempted to search for datasets from human cancer or other studies. We found two AKT-silencing based transcriptome datasets from the Gene Expression Omnibus (GEO). The first dataset reported mRNA expression profiles of CD8+ T cells, derived from an unrelated study involving three acute lymphocytic leukemia (ALL) patients—designated as patient M, P, and C, and ex vivo treated with a pan-AKT inhibitor VIII [63]. We observed an overlap of 816 (410 upregulated, 406 downregulated), 744 (267 upregulated, 477 downregulated), 802 (372 upregulated, 430 downregulated) genes between the results from MCF-7 cells and patients M, P, and C, respectively (Figure 5B). When we analyzed the status of upregulated and downregulated genes individually upon silencing AKT1, we observed a higher number of overlaps among the downregulated genes (Figure 5C, blue bars). Upon comparing the levels of overlapped genes between AKT1-silenced MCF-7 cells with the AKT inhibitor VIII-treated CD8+ T cells (Figure 5B, lower panel), we observed a substantial overlap between two differentially regulated gene sets, especially under conditions of EGF stimulation. For example, we noticed an overlap of 77 upregulated genes between EGF-stimulated MCF-7 cells with siAKT1 and 81 upregulated genes in cells from patient M. Similarly, there was an overlap of 310 downregulated genes between EGF-stimulated MCF-7 cells with siAKT1 and 311 downregulated genes in patient M-derived cells treated with pan-AKT inhibitor. Using the second dataset, we performed a comparative analysis of our results with previously reported differentially expressed genes in AKT inhibitor VIII-treated HCT116 colorectal cancer cells and MCF-7 breast cancer cells [8]. As illustrated in Figure 5C, cancer cell lines also exhibited a similar pattern in the levels of differentially expressed genes with that of patients (Figure 5C, blue bars).

As elucidated in Figure 5A, one of the interesting observations of the present study is it revealed an unexpected role of active AKT1 signaling in inhibiting the expression of certain genes through an undefined mechanism at the moment—as silencing of AKT1 resulted in elevated expression of such genes. Examples of such validated genes include transmembrane protein 213 (TMEM213), upregulated in cells treated with siAKT1; Cytochrome P450 Family 4 Subfamily F Member 8 (CYP4F8), Gamma-aminobutyric acid receptor subunit pi (GABRP) and osteoclast-associated immunoglobulin-like receptor (OSCAR)—upregulated in cells treated with siAKT1, etc. Consistent with these observations, we noticed that overexpression of AKT1 mRNA in breast tumors in the TCGA dataset is generally accompanied by a substantial reduction in the levels of TMEN213, VS1G1, CYP4F8, HAS3, and OSCAR (Supplementary Figure S4). In brief, these results validated the notion that the nature of AKT1-dependent transcriptomic shows changes in cellular and patient-derived ex vivo cellular models and revealed a set of gene expression that might be negatively affected by AKT1 signaling.

3.7. Influence of the Endogenous Status of AKT1 on Splice Variation

Transcript variants are the number of different mRNAs reported for a single gene from a single transcription start site (TSS) that might contribute to genomic heterogenicity [62,64] and are also reported to be important to splicing in breast cancer [65]. AKT signaling is known to be involved in alternative splicing through phosphorylation of its downstream substrates and, in turn, modulation of the post-transcriptional regulation of target genes [66,67]. Having observed an effect of the selective depletion of AKT1 on the genome-wide transcriptome of MCF-7 cells, we reasoned that AKT, being a kinase, might also influence the functionality of the splicing machinery through yet-to-be defined mechanisms and, hence, lead to splice variants. We next attempted to examine the status of splice variance among highly abundant transcripts across different experimental conditions in the presence or absence of EGF stimulation. Splice variation was estimated using the number of unique reads mapped to the spliced region when compared to read abundance for each exon for a single gene.

Splice variation analysis was performed on transcriptome of AKT1 silenced MCF-7 cells and compared to control siNON with and without EGF stimulation using publicly available tools and algorithms (Figure 6A). A total of 3236 statistically significant (*p*-value < 0.05 and fold change > 2) splicing events were identified from all the comparisons: the splicing was counted only if more than 10 reads mapped to the target region. We found that AKT1 knockdown exhibited an overall higher number of splicing events with a similar trend of expression than that of differential expression of the full-length transcripts. A total of 1592 splice variations were identified in MCF-7 cells with AKT1 knockdown and 257 splice variances in cells treated with Inhibitor VIII in cells without EGF stimulation. Interestingly, EGF stimulation of breast cancer cells leads to 945 splice variances in cells with AKT1 knockdown and 442 splice variances in Inhibitor VIII-treated cells. These findings suggested the status of AKT1 might have an effect on the magnitude of influence on the genome-wide splice variance (Figure 6B,C). We next analyzed the top 10 ranked exon splicing events that were either highly abundant or dispersed between the cells with AKT1 knockdown. One of such examples is the ribosomal protein family, of which six out of ten highly abundant splice variations are identified to be affected by the status of AKT1 (this study, Supplementary Tables S6 and S7).

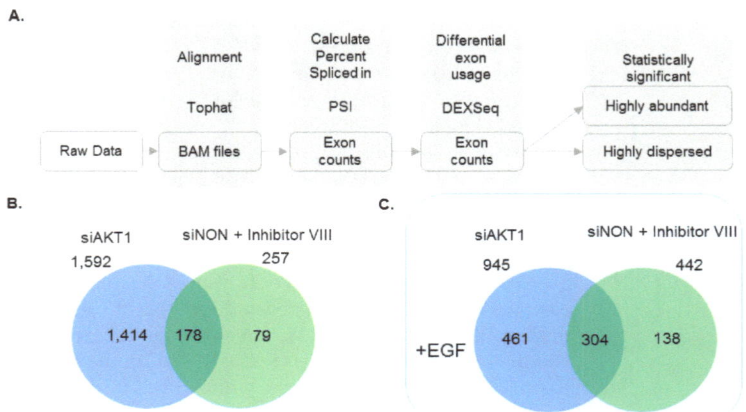

Figure 6. Identification of splice variants in AKT1-dependent transcriptome. (**A**) Tools and pipeline to identify the splice variants; and (**B,C**) comparative analysis of splice variants identified in each experimental conditions without (**B**) or with (**C**) EGF stimulation. Only the spliced genes showing at least 2-fold change with a *p*-value < 0.05 are considered from all the studies.

3.8. Top 10 Highly Dispersed Splicing Events between the AKT1 Knockdown

We found 13 examples of spliced variants downregulated upon AKT1 silencing. Splice variants related to four genes (DBNDD2, DAP, ITPK1, and ROGDI) previously reported upregulated in breast/other human cancers were found downregulated in AKT1 silenced samples compared to control (siNoN) (Supplementary Table S6). Data in Figure 7A show the expression levels of top 10 spliced variants. For example, the Fascin-3 variant FSCN3:2/exon 2 was reported differentially upregulated in metastasis [68]. Dysbindin domain-containing protein 2 variant DBNDD2:34 that was upregulated in BRCA1 mutated breast cancer [63] was found downregulated in the AKT1 silenced sample compared to control (siNoN). Inositol-tetrakisphosphate 1-kinase variant ITPK1:27 was found over expressed in breast cancer [69] but was also found downregulated in the AKT1 silenced sample compared to control (siNoN). Protein rogdi homolog variant ROGDI:43–47, upregulated and reported as a target molecule in cervical cancer [70], was found downregulated in the AKT1 silenced sample compared to control (siNoN). However, upon EGF stimulation, we observed 10 examples of spliced variants downregulated upon AKT1 silencing, among them many were reported with an alteration related to breast/human cancers (Figure 7B, Supplementary Table S7). For example, Nucleolysin TIA-1 isoform p40 variant TIA1:61 was found downregulated in the AKT1 silenced sample compared to control (siNoN), which was reported downregulated in human cancers [71]. PH and SEC7 domain-containing protein 3 variant PSD3:18 found downregulated in the AKT1 silenced sample compared to control was reported downregulated in breast cancer [72]. Interestingly, we also observed examples of AKT1-dependent splicing of two validated genes, TMEM213 and HAS3.

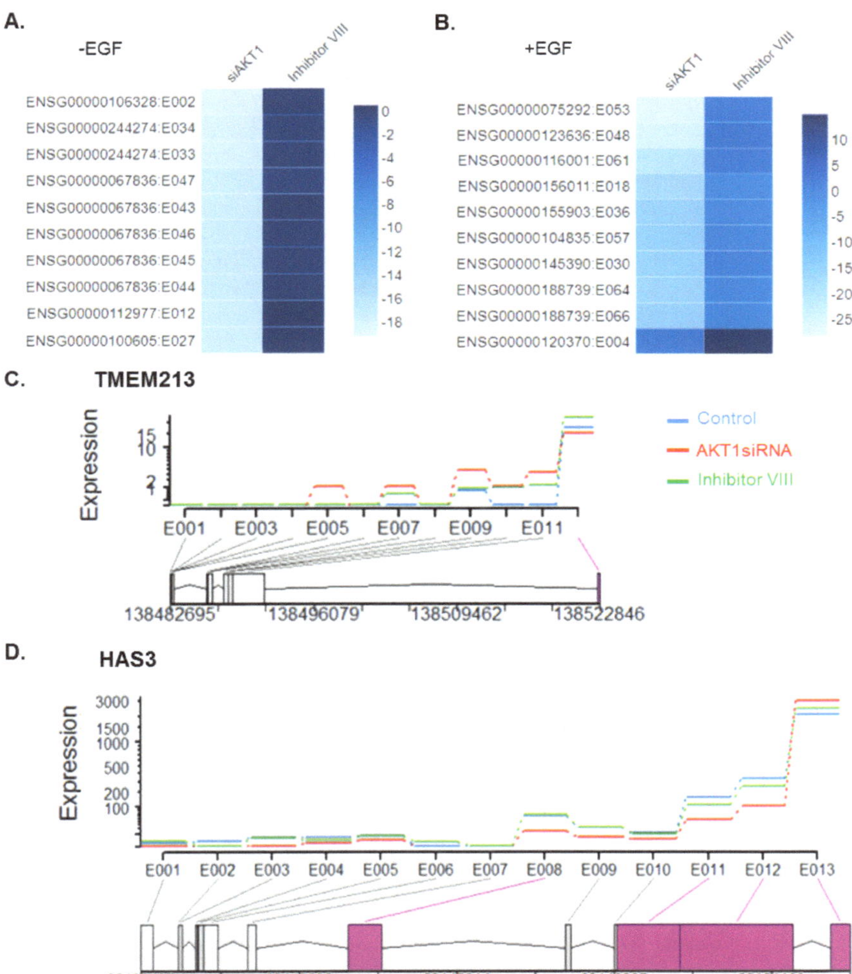

Figure 7. Highly dispersed splicing events observed between AKT1 silenced cells upon EGF stimulation. (**A,B**) Top 10 differential spliced variants in AKT1 silenced samples with or without EGF stimulation; (**C**) exon-based splicing of TMEM213 showing differential splicing of exon 12 under indicated experimental conditions; and (**D**) exon-based splicing pattern of HAS3 showing differential splicing of exon 8, 11, 12, and 13. Spliced exons are denoted in pink.

4. Discussion

Many extrinsic signals influence the pro-survival and invasive phenotypes of cancer cells by stimulating the AKT signaling pathway and its downstream cellular processes feeding into cancer phenotypes. In addition, the functionality of signal-dependent cellular events is profoundly determined by the functionality of transcriptomic heterogenicity, which, in part, is influenced by differencing splicing in addition to other regulatory steps of transcriptomic and post-transcriptional mechanisms of gene expression. In this context, here we uncovered the effect of the status of AKT1 as well as inhibition of AKT's kinase activity on the genome-wide transcriptomic and differential splicing events in breast cancer cells. Our experimental strategy involved selectively knocking down the endogenous AKT1 as well as treating the model breast cancer cells with a pan-AKT activity Inhibitor VIII. This was followed by stimulation of cells with EGF before subjecting them to genome-wide RNA-sequencing.

We observed that silencing of the endogenous AKT1 and/or inhibiting AKTs could alter the expression of up- and downregulated differentially expressed transcripts in breast cancer cells. As AKT1 has been shown to be overexpressed and/or hyperactivated in breast tumors, our results imply that many downstream phenotypic effects of AKT are not merely mediated by AKT1-signaling dependent phosphorylation of its direct substrates but also by genomic effects of AKT1. It remains unclear how exactly AKT1 contributes to the noticed genomic changes. The highest population of AKT status-dependent gene alterations were found on chromosome 1 (the longest chromosome with the highest number of genes) and chromosome 19. As chromosome 19 has been shown to exhibit extremely high incidence of loss of heterozygosity linked with breast cancer metastasis [73], it is interesting to observe that AKT1 signaling also preferentially modulates the expression of genes on chromosome 19—the underlying basis of which remains unknown at this time. As cancer cells are exposed to a variety of mitogenic growth factors, one of our experimental strategies was to also reveal the nature of transcriptomic changes in cells stimulated with epidermal growth factor. We noticed that only a portion of AKT-responsive transcripts undergoes further alterations in its expression in growth-factor stimulated cells, presumably due to stimulated hyperactivated AKT1 signaling by EGF. These findings imply that the nature of AKT1-responsive pathways is not only affected by the levels of AKT1 transcripts but also by the presence of mitogenic signals feeding into AKT1 signaling.

As expected, we observed a positive relationship between the status of AKT1 expression and many differentially expressed genes related to cellular processes. This could be attributed to the ability of AKT1 kinase to phosphorylate its substrate and/or cascade effects on the transcriptome. The top ten AKT1-dependent highly abundant transcripts identified in the present study—TFF1 (ENST00000291527), EEF2 (ENST00000309311), SCD (ENST00000370355), SPTSSB (ENST00000359175), KRT81 (ENST00000327741), LAPTM4A (ENST00000175091), NUCKS1 (ENST00000367142), LAPTM4B (ENST00000445593), PERP (ENST00000421351), and KRT19 (ENST00000361566)—are also found to be upregulated in human cancers, and nine of them are highly significant cancer-associated genes in breast cancer (Supplementary Table S4). This suggests that biological effects of generally noticed AKT1-hyperactivation in human cancers, including breast cancer, might be impacted by co-overexpression of AKT1 status-dependent expression of a subset of genes. To examine the validity of this hypothesis, we performed a multivariate analysis of AKT1 overexpression in conjunction with 10 highly abundant AKT-dependent transcripts on the overall disease-free survival of patients with breast tumors. We found that, indeed, co-overexpression of test genes further shortens the duration of overall survival of patients as compared to AKT1 alone (Supplementary Figure S5).

As many components of the cellular splicing machinery are phosphoproteins and the process of differential splicing has been shown to be regulated by upstream signaling [73], the present study also sheds new light on the effect of the AKT1 status on the splicing events, which, in turn, also contributes to tumor heterogeneity. Our results revealed that the cellular status of AKT1 might determine the magnitude of splice variance (Figure 6). Interestingly, many of the examples of the top 10 highly dispersed splicing events affected by the status of AKT1 have been previously implicated in cancerous phenotypes [63,68–72,74–81]. Though not in the top 10, we also found examples of AKT1-dependent transcripts that also underwent differential splicing in a manner dependent on the presence or silencing of AKT1. Examples of such genes include TMEM213, and HAS3 (Figure 7C,D).

Our observation that the absence of AKT1 leads to upregulation of about 466 genes represents another notable unexpected finding, as it uncovered a potential role of AKT1 signaling on the expression of a subset of genes—the mechanism of which is yet to be defined. A broader significance of this finding might be that 25 of such loss-of-AKT1-associated upregulated genes, implying that these might be normally inhibited by AKT1, have been shown to be downregulated in breast cancer [62]. Examples of the noticed unexpected upregulation of candidate genes upon AKT1 silencing include: cytochrome P450 family 4 subfamily F member 8 (CYP4F8)—involved in drug metabolism and biosynthesis of lipids and cholesterol; osteoclast-associated receptor (OSCAR)—involved in adaptive and innate immunity; and two poorly studied genes—transmembrane protein 213 (TMEM213) and V-Set and Immunoglobulin Domain Containing 1 (VSIG1) (see below).

Transmembrane protein 213 (TMEM213) is a poorly studied protein-coding gene with a predicted protein localization in the endoplasmic reticulum. TMEM213 has been shown to be downregulated in clear cell renal cell carcinoma with a predicted association with invasion and metastasis [82], whereas it is upregulated in lung adenocarcinoma and contributes to a longer survival of patients [83]. Bioinformatic studies have largely linked TMEM213 to pathways with roles in drug metabolism and transporters [83]. The levels of TMEM213 were found to be upregulated upon silencing AKT1 (this study), and an inverse correlation exists between the levels of AKT1 and TMEM213 in breast tumors (Supplementary Figure S4a). Additionally, overexpression of TMEM213 in breast tumors was found to be associated with better survival (Supplementary Figure S4b); we suggest that the generally observed increased expression and/or hyperactivation of PI-3 kinase/AKT signaling might impair the expression of TMEM213 and the resulting role in cancer progression.

V-Set and Immunoglobulin Domain Containing 1 (VSIG1) protein is a recently discovered member of the junctional adhesion family and has been widely dysregulated in human cancer. Earlier studies suggest that the levels of VSIG1 have been implicated in prometastatic and EMT: its reduced expression correlates with a poor prognosis [84] and differentiation [85] of certain cancer types [85,86]. As expression of VSIG1 might be repressed by AKT expression/signaling (this study) and the fact that VSIG1 is predicted to be localized in the plasma membrane (https://www.genecards.org/cgi-bin/carddisp.pl?gene=VSIG1, accessed on 12 November 2021); these findings raise the possibility that some of the recognized cancerous phenotypes of AKT1 might be mediated by its influence on the levels of VSIG1 through an undefined mechanism at this point.

In brief, results presented here shed new insights on the significance of AKT1 signaling on the genome-wide transcriptome and differential splicing in breast cancer cells. In addition to broadening the scope of AKT1-dependent positive regulation of gene expression, our study unexpectedly discovered that an active AKT1 signaling could also inhibit gene expression—many of which are widely known to be downregulated during cancer progression. These initial findings have raised several follow-up issues, including further experimental validation in multiple cellular models and delineating the fine mechanistic details through which AKT1 contributes to gene expression [49–53,63,68–72,77,86–93].

Supplementary Materials: The following supporting information can be downloaded at: https://www.mdpi.com/article/10.3390/cells11152290/s1, Figure S1: Expression levels of AKTs; Figure S2: Comparative analysis of functions regulated by AKT1 regulated genes versus the functions inhibited by pan-AKT inhibitor VIII; Figure S3: Status of the fold-change values from the RNA-sequencing and RT-PCR assays for selected genes using MCF-7 and SKBR-3 cell lines; Figure S4: The expression levels of selected genes in breast cancer datasets along with the changes in patient survival; Figure S5: Gene expression and survival analysis of select highly abundant genes; Table S1: The Pearson correlation coefficient between the replicates; Table S2: Details of the reads mapped to the genes for each sample; Table S3: The number of genes identified from different experimental conditions; Table S4: The number of 10 highly abundant transcripts across experimental conditions; Table S5: The number of potential differentially expressed genes according to their coding potential; Table S6: The 10 selected and highly spliced genes compared to control when AKTs are silenced from samples without EGF stimulation; Table S7: The 10 selected and highly spliced genes compared to control when AKTs are silenced from EGF stimulated samples; Supplementary File S2: Complete results of the functional analysis of differentially expressed genes.

Author Contributions: Conceptualization, design, direction, and overall supervision of the study, R.K.; methodology, B.G. (Bijesh George), B.G. (Bin Gui) and R.K.; sample preparation, B.G. (Bin Gui), B.G. (Bijesh George) and R.K.; single cell sequencing, H.N.; validation studies, B.G. (Bijesh George) and R.R.; software, B.G. (Bijesh George) and A.M.P.; data analysis, B.G. (Bijesh George), A.M.P., M.R.P. and R.K.; data curation, B.G. (Bijesh George); writing, reviewing, and editing, B.G. (Bijesh George), R.K. and M.R.P.; visualization, B.G. (Bijesh George) and R.K.; resources, R.K. and M.R.P.; project administration, R.K. All authors have read and agreed to the published version of the manuscript.

Funding: The authors acknowledge partial funding from the Department of Science & Technology, Government of India to M.R.P. (Sanction Number: VI-D&P/535/2015-16/TDT(G)).

Institutional Review Board Statement: Not applicable.

Informed Consent Statement: Not applicable.

Data Availability Statement: The data presented in this study are available in Supplementary Material and sequencing raw files deposited to the NCBI SRA and available under the accession PRJNA859811.

Acknowledgments: The authors thank the Rajiv Gandhi Center for Biotechnology for anticipated support of the publication cost.

Conflicts of Interest: The authors declare no competing financial interests.

References

1. International Agency for Research on Cancer. Cancer Today. Available online: https://gco.iarc.fr/today/online-analysis-pie?v=2020&mode=cancer&mode_population=continents&population=900&populations=900&key=total&sex=2&cancer=39&type=0&statistic=5&prevalence=0&population_group=0&ages_group%5B%5D=0&ages_group%5B%5D=17&nb_items=7&group (accessed on 15 June 2021).

2. Eroles, P.; Bosch, A.; Pérez-Fidalgo, J.A.; Lluch, A. Molecular Biology in Breast Cancer: Intrinsic Subtypes and Signaling Pathways. *Cancer Treat. Rev.* **2012**, *38*, 698–707. [CrossRef] [PubMed]

3. Hanahan, D.; Weinberg, R.A. The Hallmarks of Cancer. *Cell* **2000**, *100*, 57–70. [CrossRef]

4. Vadlamudi, R.K.; Kumar, R. P21-Activated Kinases in Human Cancer. *Cancer Metastasis Rev.* **2003**, *22*, 385–393. [CrossRef] [PubMed]

5. Manavathi, B.; Acconcia, F.; Rayala, S.K.; Kumar, R. An inherent role of microtubule network in the action of nuclear receptor. *Proc. Natl. Acad. Sci. USA* **2006**, *103*, 15981–15986. [CrossRef] [PubMed]

6. Nair, S.S.; Mishra, S.K.; Yang, Z.; Balasenthil, S.; Kumar, R.; Vadlamudi, R.K. Potential Role of a Novel Transcriptional Coactivator PELP1 in Histone H1 Displacement in Cancer Cells. *Cancer Res.* **2004**, *64*, 6416–6423. [CrossRef] [PubMed]

7. Nair, S.S.; Kumar, R. Chronatin remodeling in cancer: A gateway to regulate gene transcription. *Mol. Oncol.* **2012**, *6*, 611–619. [CrossRef]

8. Salony; Solé, X.; Alves, C.P.; Dey-Guha, I.; Ritsma, L.; Boukhali, M.; Lee, J.H.; Chowdhury, J.; Ross, K.N.; Haas, W.; et al. AKT Inhibition Promotes Nonautonomous Cancer Cell Survival. *Mol. Cancer Ther.* **2016**, *15*, 142–153. [CrossRef]

9. Santi, S.A.; Douglas, A.C.; Lee, H. The Akt Isoforms, Their Unique Functions and Potential as Anticancer Therapeutic Targets. *Biomol. Concepts* **2010**, *1*, 389–401. [CrossRef]

10. Burgering, B.M.; Coffer, P.J. Protein Kinase B (c-Akt) in Phosphatidylinositol-3-OH Kinase Signal Transduction. *Nature* **1995**, *376*, 599–602. [CrossRef]

11. Okano, J.; Gaslightwala, I.; Birnbaum, M.J.; Rustgi, A.K.; Nakagawa, H. Akt/Protein Kinase B Isoforms Are Differentially Regulated by Epidermal Growth Factor Stimulation. *J. Biol. Chem.* **2000**, *275*, 30934–30942. [CrossRef]

12. Hinz, N.; Jücker, M. Distinct Functions of AKT Isoforms in Breast Cancer: A Comprehensive Review. *Cell Commun. Signal.* **2019**, *17*, 154. [CrossRef] [PubMed]

13. Carpten, J.D.; Faber, A.L.; Horn, C.; Donoho, G.P.; Briggs, S.L.; Robbins, C.M.; Hostetter, G.; Boguslawski, S.; Moses, T.Y.; Savage, S.; et al. A Transforming Mutation in the Pleckstrin Homology Domain of AKT1 in Cancer. *Nature* **2007**, *448*, 439–444. [CrossRef]

14. Landgraf, K.E.; Pilling, C.; Falke, J.J. Molecular Mechanism of an Oncogenic Mutation That Alters Membrane Targeting: Glu17Lys Modifies the PIP Lipid Specificity of the AKT1 PH Domain. *Biochemistry* **2008**, *47*, 12260–12269. [CrossRef]

15. Brown, J.S.; Banerji, U. Maximising the Potential of AKT Inhibitors as Anti-Cancer Treatments. *Pharmacol. Ther.* **2017**, *172*, 101–115. [CrossRef] [PubMed]

16. Polytarchou, C.; Iliopoulos, D.; Hatziapostolou, M.; Kottakis, F.; Maroulakou, I.; Struhl, K.; Tsichlis, P.N. Akt2 Regulates All Akt Isoforms and Promotes Resistance to Hypoxia through Induction of MiR-21 upon Oxygen Deprivation. *Cancer Res.* **2011**, *71*, 4720–4731. [CrossRef] [PubMed]

17. Watson, K.L.; Moorehead, R.A. Loss of Akt1 or Akt2 Delays Mammary Tumor Onset and Suppresses Tumor Growth Rate in MTB-IGFIR Transgenic Mice. *BMC Cancer* **2013**, *13*, 375. [CrossRef] [PubMed]
18. Gargini, R.; Cerliani, J.P.; Escoll, M.; Antón, I.M.; Wandosell, F. Cancer Stem Cell-like Phenotype and Survival Are Coordinately Regulated by Akt/FoxO/Bim Pathway. *Stem Cells* **2015**, *33*, 646–660. [CrossRef]
19. Hutchinson, J.; Jin, J.; Cardiff, R.D.; Woodgett, J.R.; Muller, W.J. Activation of Akt (Protein Kinase B) in Mammary Epithelium Provides a Critical Cell Survival Signal Required for Tumor Progression. *Mol. Cell. Biol.* **2001**, *21*, 2203–2212. [CrossRef]
20. Dillon, R.L.; Marcotte, R.; Hennessy, B.T.; Woodgett, J.R.; Mills, G.B.; Muller, W.J. Akt1 and Akt2 Play Distinct Roles in the Initiation and Metastatic Phases of Mammary Tumor Progression. *Cancer Res.* **2009**, *69*, 5057–5064. [CrossRef]
21. Riggio, M.; Polo, M.L.; Blaustein, M.; Colman-Lerner, A.; Lüthy, I.; Lanari, C.; Novaro, V. PI3K/AKT Pathway Regulates Phosphorylation of Steroid Receptors, Hormone Independence and Tumor Differentiation in Breast Cancer. *Carcinogenesis* **2012**, *33*, 509–518. [CrossRef]
22. Liu, H.; Radisky, D.C.; Nelson, C.M.; Zhang, H.; Fata, J.E.; Roth, R.A.; Bissell, M.J. Mechanism of Akt1 Inhibition of Breast Cancer Cell Invasion Reveals a Protumorigenic Role for TSC2. *Proc. Natl. Acad. Sci. USA* **2006**, *103*, 4134–4139. [CrossRef] [PubMed]
23. Chen, W.S.; Xu, P.Z.; Gottlob, K.; Chen, M.L.; Sokol, K.; Shiyanova, T.; Roninson, I.; Weng, W.; Suzuki, R.; Tobe, K.; et al. Growth Retardation and Increased Apoptosis in Mice with Homozygous Disruption of the Akt1 Gene. *Genes Dev.* **2001**, *15*, 2203–2208. [CrossRef] [PubMed]
24. Garofalo, R.S.; Orena, S.J.; Rafidi, K.; Torchia, A.J.; Stock, J.L.; Hildebrandt, A.L.; Coskran, T.; Black, S.C.; Brees, D.J.; Wicks, J.R.; et al. Severe Diabetes, Age-Dependent Loss of Adipose Tissue, and Mild Growth Deficiency in Mice Lacking Akt2/PKB Beta. *J. Clin. Investig.* **2003**, *112*, 197–208. [CrossRef] [PubMed]
25. Peng, X.-D.; Xu, P.-Z.; Chen, M.-L.; Hahn-Windgassen, A.; Skeen, J.; Jacobs, J.; Sundararajan, D.; Chen, W.S.; Crawford, S.E.; Coleman, K.G.; et al. Dwarfism, Impaired Skin Development, Skeletal Muscle Atrophy, Delayed Bone Development, and Impeded Adipogenesis in Mice Lacking Akt1 and Akt2. *Genes Dev.* **2003**, *17*, 1352–1365. [CrossRef]
26. Yang, Z.-Z.; Tschopp, O.; Di-Poï, N.; Bruder, E.; Baudry, A.; Dümmler, B.; Wahli, W.; Hemmings, B.A. Dosage-Dependent Effects of Akt1/Protein Kinase Balpha (PKBalpha) and Akt3/PKBgamma on Thymus, Skin, and Cardiovascular and Nervous System Development in Mice. *Mol. Cell. Biol.* **2005**, *25*, 10407–10418. [CrossRef]
27. Dummler, B.; Tschopp, O.; Hynx, D.; Yang, Z.-Z.; Dirnhofer, S.; Hemmings, B.A. Life with a Single Isoform of Akt: Mice Lacking Akt2 and Akt3 Are Viable but Display Impaired Glucose Homeostasis and Growth Deficiencies. *Mol. Cell. Biol.* **2006**, *26*, 8042–8051. [CrossRef]
28. Ju, X.; Katiyar, S.; Wang, C.; Liu, M.; Jiao, X.; Li, S.; Zhou, J.; Turner, J.; Lisanti, M.P.; Russell, R.G.; et al. Akt1 Governs Breast Cancer Progression in Vivo. *Proc. Natl. Acad. Sci. USA* **2007**, *104*, 7438–7443. [CrossRef]
29. Maroulakou, I.G.; Oemler, W.; Naber, S.P.; Tsichlis, P.N. Akt1 Ablation Inhibits, Whereas Akt2 Ablation Accelerates, the Development of Mammary Adenocarcinomas in Mouse Mammary Tumor Virus (MMTV)-ErbB2/Neu and MMTV-Polyoma Middle T Transgenic Mice. *Cancer Res.* **2007**, *67*, 167–177. [CrossRef]
30. Bilodeau, M.T.; Balitza, A.E.; Hoffman, J.M.; Manley, P.J.; Barnett, S.F.; Defeo-Jones, D.; Haskell, K.; Jones, R.E.; Leander, K.; Robinson, R.G.; et al. Allosteric Inhibitors of Akt1 and Akt2: A Naphthyridinone with Efficacy in an A2780 Tumor Xenograft Model. *Bioorg. Med. Chem. Lett.* **2008**, *18*, 3178–3182. [CrossRef]
31. Wang, J.; Wan, W.; Sun, R.; Liu, Y.; Sun, X.; Ma, D.; Zhang, N. Reduction of Akt2 Expression Inhibits Chemotaxis Signal Transduction in Human Breast Cancer Cells. *Cell. Signal.* **2008**, *20*, 1025–1034. [CrossRef]
32. Grabinski, N.; Möllmann, K.; Milde-Langosch, K.; Müller, V.; Schumacher, U.; Brandt, B.; Pantel, K.; Jücker, M. AKT3 Regulates ErbB2, ErbB3 and Estrogen Receptor α Expression and Contributes to Endocrine Therapy Resistance of ErbB2$^+$ Breast Tumor Cells from Balb-NeuT Mice. *Cell. Signal.* **2014**, *26*, 1021–1029. [CrossRef]
33. Santi, S.A.; Lee, H. Ablation of Akt2 Induces Autophagy through Cell Cycle Arrest, the Downregulation of P70S6K, and the Deregulation of Mitochondria in MDA-MB231 Cells. *PLoS ONE* **2011**, *6*, e14614.
34. Xia, X.; Li, X.; Li, F.; Wu, X.; Zhang, M.; Zhou, H.; Huang, N.; Yang, X.; Xiao, F.; Liu, D.; et al. A novel tumor suppressor protein encoded by circular AKT3 RNA inhibits glioblastoma tumorigenicity by competing with active phosphoinositide-dependent Kinase-1. *Mol. Cancer* **2019**, *18*, 131. [CrossRef] [PubMed]
35. Barnett, S.F.; Defeo-Jones, D.; Fu, S.; Hancock, P.J.; Haskell, K.M.; Jones, R.E.; Kahana, J.A.; Kral, A.M.; Leander, K.; Lee, L.L.; et al. Identification and Characterization of Pleckstrin-Homology-Domain-Dependent and Isoenzyme-Specific Akt Inhibitors. *Biochem. J.* **2005**, *385*, 399–408. [CrossRef] [PubMed]
36. Gills, J.J.; Dennis, P.A. Perifosine: Update on a Novel Akt Inhibitor. *Curr. Oncol. Rep.* **2009**, *11*, 102–110. [CrossRef] [PubMed]
37. Nitulescu, G.M.; Margina, D.; Juzenas, P.; Peng, Q.; Olaru, O.T.; Saloustros, E.; Fenga, C.; Spandidos, D.A.; Libra, M.; Tsatsakis, A.M. Akt Inhibitors in Cancer Treatment: The Long Journey from Drug Discovery to Clinical Use (Review). *Int. J. Oncol.* **2016**, *48*, 869–885. [CrossRef]
38. Bhat-Nakshatri, P.; Gao, H.; Sheng, L.; McGuire, P.C.; Xuei, X.; Wan, J.; Liu, Y.; Althouse, S.K.; Colter, A.; Sandusky, G.; et al. A Single-Cell Atlas of the Healthy Breast Tissues Reveals Clinically Relevant Clusters of Breast Epithelial Cells. *Cell Rep. Med.* **2021**, *2*, 100219. [CrossRef]
39. Kim, D.; Pertea, G.; Trapnell, C.; Pimentel, H.; Kelley, R.; Salzberg, S.L. TopHat2: Accurate Alignment of Transcriptomes in the Presence of Insertions, Deletions and Gene Fusions. *Genome Biol.* **2013**, *14*, R36. [CrossRef]

40. Langmead, B.; Trapnell, C.; Pop, M.; Salzberg, S.L. Ultrafast and Memory-Efficient Alignment of Short DNA Sequences to the Human Genome. *Genome Biol.* **2009**, *10*, R25. [CrossRef]

41. Dodt, M.; Roehr, J.T.; Ahmed, R.; Dieterich, C. FLEXBAR-Flexible Barcode and Adapter Processing for Next-Generation Sequencing Platforms. *Biology* **2012**, *1*, 895–905. [CrossRef]

42. Anders, S.; Pyl, P.T.; Huber, W. HTSeq—A Python Framework to Work with High-Throughput Sequencing Data. *Bioinformatics* **2015**, *31*, 166–169. [CrossRef] [PubMed]

43. Anders, S.; Huber, W. Differential Expression Analysis for Sequence Count Data. *Genome Biol.* **2010**, *11*, R106. [CrossRef] [PubMed]

44. Eswaran, J.; Cyanam, D.; Mudvari, P.; Reddy, S.D.N.; Pakala, S.B.; Nair, S.S.; Florea, L.; Fuqua, S.A.W.; Godbole, S.; Kumar, R. Transcriptomic Landscape of Breast Cancers through MRNA Sequencing. *Sci. Rep.* **2012**, *2*, 264. [CrossRef] [PubMed]

45. Schafer, S.; Miao, K.; Benson, C.C.; Heinig, M.; Cook, S.A.; Hubner, N. Alternative Splicing Signatures in RNA-Seq Data: Percent Spliced in (PSI). *Curr. Protoc. Hum. Genet.* **2015**, *87*, 11.16.1–11.16.14. [CrossRef] [PubMed]

46. Eswaran, J.; Horvath, A.; Godbole, S.; Reddy, S.D.; Mudvari, P.; Ohshiro, K.; Cyanam, D.; Nair, S.; Fuqua, S.A.W.; Polyak, K.; et al. RNA Sequencing of Cancer Reveals Novel Splicing Alterations. *Sci. Rep.* **2013**, *3*, 1689. [CrossRef] [PubMed]

47. Anders, S.; Reyes, A.; Huber, W. Detecting Differential Usage of Exons from RNA-Seq Data. *Genome Res.* **2012**, *22*, 2008–2017. [CrossRef]

48. Klebanoff, C.A.; Crompton, J.G.; Leonardi, A.J.; Yamamoto, T.N.; Chandran, S.S.; Eil, R.L.; Sukumar, M.; Vodnala, S.K.; Hu, J.; Ji, Y.; et al. Inhibition of AKT Signaling Uncouples T Cell Differentiation from Expansion for Receptor-Engineered Adoptive Immunotherapy. *JCI Insight* **2017**, *2*, e95103. [CrossRef]

49. Pereira, B.; Chin, S.F.; Rueda, O.M.; Vollan, H.K.; Provenzano, E.; Bardwell, H.A.; Pugh, M.; Jones, L.; Russell, R.; Sammut, S.J.; et al. The Somatic Mutation Profiles of 2,433 Breast Cancers Refines Their Genomic and Transcriptomic Landscapes. *Nat. Commun.* **2016**, *7*, 11479. [CrossRef]

50. Rueda, O.M.; Sammut, S.-J.; Seoane, J.A.; Chin, S.-F.; Caswell-Jin, J.L.; Callari, M.; Batra, R.; Pereira, B.; Bruna, A.; Ali, H.R.; et al. Dynamics of Breast-Cancer Relapse Reveal Late-Recurring ER-Positive Genomic Subgroups. *Nature* **2019**, *567*, 399–404. [CrossRef]

51. Curtis, C.; Shah, S.P.; Chin, S.-F.; Turashvili, G.; Rueda, O.M.; Dunning, M.J.; Speed, D.; Lynch, A.G.; Samarajiwa, S.; Yuan, Y.; et al. The Genomic and Transcriptomic Architecture of 2,000 Breast Tumours Reveals Novel Subgroups. *Nature* **2012**, *486*, 346–352. [CrossRef]

52. Cerami, E.; Gao, J.; Dogrusoz, U.; Gross, B.E.; Sumer, S.O.; Aksoy, B.A.; Jacobsen, A.; Byrne, C.J.; Heuer, M.L.; Larsson, E.; et al. The CBio Cancer Genomics Portal: An Open Platform for Exploring Multidimensional Cancer Genomics Data. *Cancer Discov.* **2012**, *2*, 401–404. [CrossRef]

53. Gao, J.; Aksoy, B.A.; Dogrusoz, U.; Dresdner, G.; Gross, B.; Sumer, S.O.; Sun, Y.; Jacobsen, A.; Sinha, R.; Larsson, E.; et al. Integrative Analysis of Complex Cancer Genomics and Clinical Profiles Using the CBioPortal. *Sci. Signal.* **2013**, *6*, pl1. [CrossRef] [PubMed]

54. Fonseka, P.; Pathan, M.; Chitti, S.V.; Kang, T.; Mathivanan, S. FunRich Enables Enrichment Analysis of OMICs Datasets. *J. Mol. Biol.* **2021**, *433*, 166747. [CrossRef] [PubMed]

55. Franceschini, A.; Meier, R.; Casanova, A.; Kreibich, S.; Daga, N.; Andritschke, D.; Dilling, S.; Rämö, P.; Emmenlauer, M.; Kaufmann, A.; et al. Specific Inhibition of Diverse Pathogens in Human Cells by Synthetic MicroRNA-like Oligonucleotides Inferred from RNAi Screens. *Proc. Natl. Acad. Sci. USA* **2014**, *111*, 4548–4553. [CrossRef] [PubMed]

56. Turner, K.M.; Sun, Y.; Ji, P.; Granberg, K.J.; Bernard, B.; Hu, L.; Cogdell, D.E.; Zhou, X.; Yli-Harja, O.; Nykter, M.; et al. Genomically Amplified Akt3 Activates DNA Repair Pathway and Promotes Glioma Progression. *Proc. Natl. Acad. Sci. USA* **2015**, *112*, 3421–3426. [CrossRef]

57. Halacli, S.O.; Dogan, A.L. FOXP1 Regulation via the PI3K/Akt/P70S6K Signaling Pathway in Breast Cancer Cells. *Oncol. Lett.* **2015**, *9*, 1482–1488. [CrossRef]

58. Zhu, L.; Derijard, B.; Chakrabandhu, K.; Wang, B.-S.; Chen, H.-Z.; Hueber, A.-O. Synergism of PI3K/Akt Inhibition and Fas Activation on Colon Cancer Cell Death. *Cancer Lett.* **2014**, *354*, 355–364. [CrossRef]

59. Mi, H.; Muruganujan, A.; Ebert, D.; Huang, X.; Thomas, P.D. PANTHER Version 14: More Genomes, a New PANTHER GO-Slim and Improvements in Enrichment Analysis Tools. *Nucleic Acids Res.* **2018**, *47*, D419–D426. [CrossRef]

60. The Gene Ontology Consortium. The Gene Ontology Resource: Enriching a GOld Mine. *Nucleic Acids Res.* **2021**, *49*, D325–D334. [CrossRef]

61. Ashburner, M.; Ball, C.A.; Blake, J.A.; Botstein, D.; Butler, H.; Cherry, J.M.; Davis, A.P.; Dolinski, K.; Dwight, S.S.; Eppig, J.T.; et al. Gene Ontology: Tool for the Unification of Biology. The Gene Ontology Consortium. *Nat. Genet.* **2000**, *25*, 25–29. [CrossRef]

62. Bhat-Nakshatri, P.; Wang, G.; Appaiah, H.; Luktuke, N.; Carroll, J.S.; Geistlinger, T.R.; Brown, M.; Badve, S.; Liu, Y.; Nakshatri, H. AKT Alters Genome-Wide Estrogen Receptor Alpha Binding and Impacts Estrogen Signaling in Breast Cancer. *Mol. Cell. Biol.* **2008**, *28*, 7487–7503. [CrossRef] [PubMed]

63. Privat, M.; Rudewicz, J.; Sonnier, N.; Tamisier, C.; Ponelle-Chachuat, F.; Bignon, Y.-J. Antioxydation and Cell Migration Genes Are Identified as Potential Therapeutic Targets in Basal-Like and BRCA1 Mutated Breast Cancer Cell Lines. *Int. J. Med. Sci.* **2018**, *15*, 46–58. [CrossRef] [PubMed]

64. Bhat-Nakshatri, P.; Song, E.-K.; Collins, N.R.; Uversky, V.N.; Dunker, A.K.; O'Malley, B.W.; Geistlinger, T.R.; Carroll, J.S.; Brown, M.; Nakshatri, H. Interplay between Estrogen Receptor and AKT in Estradiol-Induced Alternative Splicing. *BMC Med. Genom.* **2013**, *6*, 21. [CrossRef] [PubMed]

65. Horvath, A.; Pakala, S.B.; Mudvari, P.; Reddy, S.D.N.; Ohshiro, K.; Casimiro, S.; Pires, R.; Fuqua, S.A.W.; Toi, M.; Costa, L.; et al. Novel Insights into Breast Cancer Genetic Variance through RNA Sequencing. *Sci. Rep.* **2013**, *3*, 2256. [CrossRef] [PubMed]

66. Jiang, K.; Patel, N.A.; Watson, J.E.; Apostolatos, H.; Kleiman, E.; Hanson, O.; Hagiwara, M.; Cooper, D.R. Akt2 Regulation of Cdc2-like Kinases (Clk/Sty), Serine/Arginine-Rich (SR) Protein Phosphorylation, and Insulin-Induced Alternative Splicing of PKCbetaII Messenger Ribonucleic Acid. *Endocrinology* **2009**, *150*, 2087–2097. [CrossRef]

67. Yea, S.; Narla, G.; Zhao, X.; Garg, R.; Tal-Kremer, S.; Hod, E.; Villanueva, A.; Loke, J.; Tarocchi, M.; Akita, K.; et al. Ras Promotes Growth by Alternative Splicing-Mediated Inactivation of the KLF6 Tumor Suppressor in Hepatocellular Carcinoma. *Gastroenterology* **2008**, *134*, 1521–1531. [CrossRef]

68. Whitsett, T.G.; Inge, L.J.; Dhruv, H.D.; Cheung, P.Y.; Weiss, G.J.; Bremner, R.M.; Winkles, J.A.; Tran, N.L. Molecular Determinants of Lung Cancer Metastasis to the Central Nervous System. *Transl. Lung Cancer Res.* **2013**, *2*, 273–283.

69. Speers, C.; Tsimelzon, A.; Sexton, K.; Herrick, A.M.; Gutierrez, C.; Culhane, A.; Quackenbush, J.; Hilsenbeck, S.; Chang, J.; Brown, P. Identification of Novel Kinase Targets for the Treatment of Estrogen Receptor-Negative Breast Cancer. *Clin. Cancer Res.* **2009**, *15*, 6327–6340. [CrossRef]

70. Chen, Y.-F.; Cho, J.J.; Huang, T.-H.; Tseng, C.-N.; Huang, E.-Y.; Cho, C.-L. Downregulation of a Novel Human Gene, ROGDI, Increases Radiosensitivity in Cervical Cancer Cells. *Cancer Biol. Ther.* **2016**, *17*, 1070–1078. [CrossRef]

71. Hong, S. RNA Binding Protein as an Emerging Therapeutic Target for Cancer Prevention and Treatment. *J. Cancer Prev.* **2017**, *22*, 203–210. [CrossRef]

72. Thomassen, M.; Tan, Q.; Kruse, T.A. Gene Expression Meta-Analysis Identifies Chromosomal Regions and Candidate Genes Involved in Breast Cancer Metastasis. *Breast Cancer Res. Treat.* **2009**, *113*, 239–249. [CrossRef] [PubMed]

73. Meng, Q.; Rayala, S.K.; Gururaj, A.E.; Talukder, A.H.; O'Malley, B.W.; Kumar, R. Signaling-Dependent and Coordinated Regulation of Transcription, Splicing, and Translation Resides in a Single Coregulator, PCBP1. *Proc. Natl. Acad. Sci. USA* **2007**, *104*, 5866–5871. [CrossRef] [PubMed]

74. Ghosh, D.K.; Roy, A.; Ranjan, A. Aggregation-Prone Regions in HYPK Help It to Form Sequestration Complex for Toxic Protein Aggregates. *J. Mol. Biol.* **2018**, *430*, 963–986. [CrossRef] [PubMed]

75. Li, Y.; Huang, J.; Sun, J.; Xiang, S.; Yang, D.; Ying, X.; Lu, M.; Li, H.; Ren, G. The Transcription Levels and Prognostic Values of Seven Proteasome Alpha Subunits in Human Cancers. *Oncotarget* **2017**, *8*, 4501–4519. [CrossRef] [PubMed]

76. Takei, N.; Yoneda, A.; Sakai-Sawada, K.; Kosaka, M.; Minomi, K.; Tamura, Y. Hypoxia-Inducible ERO1α Promotes Cancer Progression through Modulation of Integrin-B1 Modification and Signalling in HCT116 Colorectal Cancer Cells. *Sci. Rep.* **2017**, *7*, 9389. [CrossRef]

77. Sarajlić, A.; Filipović, A.; Janjić, V.; Coombes, R.C.; Pržulj, N. The Role of Genes Co-Amplified with Nicastrin in Breast Invasive Carcinoma. *Breast Cancer Res. Treat.* **2014**, *143*, 393–401. [CrossRef]

78. Dolezal, J.M.; Dash, A.P.; Prochownik, E.V. Diagnostic and Prognostic Implications of Ribosomal Protein Transcript Expression Patterns in Human Cancers. *BMC Cancer* **2018**, *18*, 275. [CrossRef]

79. Kikuchi, M.; Katoh, H.; Waraya, M.; Tanaka, Y.; Ishii, S.; Tanaka, T.; Nishizawa, N.; Yokoi, K.; Minatani, N.; Ema, A.; et al. Epigenetic Silencing of HOPX Contributes to Cancer Aggressiveness in Breast Cancer. *Cancer Lett.* **2017**, *384*, 70–78. [CrossRef]

80. Drew, B.G.; Hamidi, H.; Zhou, Z.; Villanueva, C.J.; Krum, S.A.; Calkin, A.C.; Parks, B.W.; Ribas, V.; Kalajian, N.Y.; Phun, J.; et al. Estrogen Receptor (ER)α-Regulated Lipocalin 2 Expression in Adipose Tissue Links Obesity with Breast Cancer Progression. *J. Biol. Chem.* **2015**, *290*, 5566–5581. [CrossRef]

81. Satih, S.; Chalabi, N.; Rabiau, N.; Bosviel, R.; Fontana, L.; Bignon, Y.-J.; Bernard-Gallon, D.J. Gene Expression Profiling of Breast Cancer Cell Lines in Response to Soy Isoflavones Using a Pangenomic Microarray Approach. *OMICS* **2010**, *14*, 231–238. [CrossRef]

82. Wrzesiński, T.; Szelag, M.; Cieślikowski, W.A.; Ida, A.; Giles, R.; Zodro, E.; Szumska, J.; Poźniak, J.; Kwias, Z.; Bluyssen, H.A.R.; et al. Expression of Pre-Selected TMEMs with Predicted ER Localization as Potential Classifiers of CcRCC Tumors. *BMC Cancer* **2015**, *15*, 518. [CrossRef] [PubMed]

83. Zou, J.; Li, Z.; Deng, H.; Hao, J.; Ding, R.; Zhao, M. TMEM213 as a Novel Prognostic and Predictive Biomarker for Patients with Lung Adenocarcinoma after Curative Resection: A Study Based on Bioinformatics Analysis. *J. Thorac. Dis.* **2019**, *11*, 3399–3410. [CrossRef]

84. Chen, Y.; Pan, K.; Li, S.; Xia, J.; Wang, W.; Chen, J.; Zhao, J.; Lü, L.; Wang, D.; Pan, Q.; et al. Decreased Expression of V-Set and Immunoglobulin Domain Containing 1 (VSIG1) Is Associated with Poor Prognosis in Primary Gastric Cancer. *J. Surg. Oncol.* **2012**, *106*, 286–293. [CrossRef] [PubMed]

85. Oidovsambuu, O.; Nyamsuren, G.; Liu, S.; Göring, W.; Engel, W.; Adham, I.M. Adhesion Protein VSIG1 Is Required for the Proper Differentiation of Glandular Gastric Epithelia. *PLoS ONE* **2011**, *6*, e25908. [CrossRef]

86. COSMIC Catalogue of Somatic Mutations in Cancer. Available online: https://cancer.sanger.ac.uk/cosmic/mutation/overview?id=95736066#references (accessed on 15 February 2019).

87. Lehmann, U.; Celikkaya, G.; Hasemeier, B.; Länger, F.; Kreipe, H. Promoter Hypermethylation of the Death-Associated Protein Kinase Gene in Breast Cancer Is Associated with the Invasive Lobular Subtype. *Cancer Res.* **2002**, *62*, 6634–6638. [PubMed]

88. O'Reilly, J.-A.; Fitzgerald, J.; Fitzgerald, S.; Kenny, D.; Kay, E.W.; O'Kennedy, R.; Kijanka, G.S. Diagnostic Potential of Zinc Finger Protein-Specific Autoantibodies and Associated Linear B-Cell Epitopes in Colorectal Cancer. *PLoS ONE* **2015**, *10*, e0123469. [CrossRef]
89. Zhu, Z.; Teng, Z.; van Duijnhoven, F.J.B.; Dong, M.; Qian, Y.; Yu, H.; Yang, J.; Han, R.; Su, J.; Du, W.; et al. Interactions between RASA2, CADM1, HIF1AN Gene Polymorphisms and Body Fatness with Breast Cancer: A Population-Based Case-Control Study in China. *Oncotarget* **2017**, *8*, 98258–98269. [CrossRef]
90. Suber, T.L.; Nikolli, I.; O'Brien, M.E.; Londino, J.; Zhao, J.; Chen, K.; Mallampalli, R.K.; Zhao, Y. FBXO17 Promotes Cell Proliferation through Activation of Akt in Lung Adenocarcinoma Cells. *Respir. Res.* **2018**, *19*, 206. [CrossRef]
91. Cancer Dependency Map. Available online: https://score.depmap.sanger.ac.uk (accessed on 13 September 2019).
92. Zhao, X.; Wu, X.; Wang, H.; Yu, H.; Wang, J. USP53 Promotes Apoptosis and Inhibits Glycolysis in Lung Adenocarcinoma through FKBP51-AKT1 Signaling. *Mol. Carcinog.* **2020**, *59*, 1000–1011. [CrossRef]
93. Winter, S.F.; Lukes, L.; Hunter, K.W. Arid4b Is a Potential Breast Cancer Progression Modifier Gene. *Cancer Res.* **2010**, *70*, 2371. [CrossRef]

Article

Extracellular Vesicles from BMSCs Prevent Glucocorticoid-Induced BMECs Injury by Regulating Autophagy via the PI3K/Akt/mTOR Pathway

Jinhui Ma [1],[†], Mengran Shen [2],[†], Debo Yue [1], Weiguo Wang [1], Fuqiang Gao [1],[*] and Bailiang Wang [1],[*]

[1] Department of Orthopaedic Surgery, Center for Osteonecrosis and Joint Preserving & Reconstruction, China-Japan Friendship Hospital, Beijing 100029, China; majinhui@zryhyy.com.cn (J.M.); yuedebo@zryhyy.com.cn (D.Y.); wangweiguo@zryhyy.com.cn (W.W.)
[2] Department of Orthopaedic Surgery, Peking University China-Japan Friendship School of Clinical Medicine, Beijing 100029, China; 2111210577@stu.pku.edu.cn
[*] Correspondence: gaofuqiang@bjmu.edu.cn (F.G.); wangbailiang@zryhyy.com.cn (B.W.)
[†] These authors contributed equally to this work.

Abstract: Osteonecrosis of the femoral head (ONFH) is a common clinical disease with a high disability rate. Injury of bone microvascular endothelial cells (BMECs) caused by glucocorticoid administration is one of the important causes of ONFH, and there is currently a lack of effective clinical treatments. Extracellular vesicles derived from bone stem cells (BMSC-EVs) can prevent ONFH by promoting angiogenesis and can inhibit cell apoptosis by regulating autophagy via the PI3K/Akt/mTOR signaling pathway. The present study aimed to investigate the effect of extracellular vesicles derived from bone marrow stem cells (BMSC) on a glucocorticoid-induced injury of BMECs and possible mechanisms. We found that BMSC-EVs attenuated glucocorticoid-induced viability, angiogenesis capacity injury, and the apoptosis of BMECs. BMSC-EVs increased the LC3 level, but decreased p62 (an autophagy protein receptor) expression, suggesting that BMSC-Exos activated autophagy in glucocorticoid-treated BMECs. The protective effects of BMSC-EVs on the glucocorticoid-induced injury of BMECs was mimicked by a known stimulator of autophagy (rapamycin) and could be enhanced by co-treatment with an autophagy inhibitor (LY294002). BMSC-EVs also suppressed the PI3K/Akt/mTOR signaling pathway, which regulates cell autophagy, in glucocorticoid-treated BMECs. In conclusion, the results indicate that BMSC-EVs prevent the glucocorticoid-induced injury of BMECs by regulating autophagy via the PI3K/Akt/mTOR pathway.

Keywords: autophagy; extracellular vesicles; PI3K/Akt/mTOR pathway; bone marrow mesenchymal stem cells; bone microvascular endothelial cells

Citation: Ma, J.; Shen, M.; Yue, D.; Wang, W.; Gao, F.; Wang, B. Extracellular Vesicles from BMSCs Prevent Glucocorticoid-Induced BMECs Injury by Regulating Autophagy via the PI3K/Akt/mTOR Pathway. *Cells* **2022**, *11*, 2104. https://doi.org/10.3390/cells11132104

Academic Editors: Jean Christopher Chamcheu, Claudia Bürger, Shile Huang and Bruce A. Bunnell

Received: 30 March 2022
Accepted: 26 May 2022
Published: 3 July 2022
Corrected: 9 July 2024

1. Introduction

Osteonecrosis of the femoral head (ONFH) is the death of bone (including bone cells, bone marrow hematopoietic cells, and fat cells) due to various factors [1]. The pathogenesis of non-traumatic ONFH is still unclear, but may be attributed to altered lipid metabolism/fat emboli, cell and bone death, increased mechanical stress, elevated intracortical pressure, and bone remodeling imbalance [2,3]. These factors diminish femoral blood perfusion through common pathologic mechanisms, including vascular endothelial damage and microvascular thrombosis. Bone microvascular endothelial cells (BMECs) are highly active endocrine cells, which comprise a monolayer structure attached to the inner wall of bone and form bone microvessels. Reduced numbers and impaired function of endothelial progenitor cells have been associated with an increased risk of ONFH [4]. An injury to BMECs caused by glucocorticoid could result in local blood hypercoagulation, microvascular thrombosis, and vascular occlusion, leading to necrosis of the femoral head in the dominant area [5,6]. A large number of reactive oxygen species produced by the injured BMECs could reduce the synthesis of vasodilator substances and further aggravate the injury of BMECs, eventually leading to necrosis of bone cells and bone marrow [7]. Promoting angiogenesis and maintaining vascular permeability of the femoral head is of great significance for the prevention and treatment of ONFH. Therefore, the molecular mechanism of bone microcirculation disorder in the femoral head caused by glucocorticoid-induced damage of BMECs must be further studied.

In recent years, the influence of autophagy on endothelial cells has attracted great interest [8]. Autophagy is a highly conserved biological phenomenon characterized by the formation of double-membrane vesicles called autophagosomes, and the subsequent engulfment and delivery of various cellular components (proteins, organelles, and invading pathogens) to lysosomes for degradation and material recycling [9,10]. The phosphoinositide 3-kinase/protein kinase B/mammalian target of the rapamycin (PI3K/Akt/mTOR) signaling pathway is a prototypic survival pathway that plays a central role in diverse cellular functions, including proliferation, growth, survival, metabolism, and autophagy [11,12]. As a downstream effector of the PI3K/Akt pathway, mTOR activity is the key to autophagosome formation and maturation. The mTOR, a conserved serine/threonine kinase, can integrate various signaling pathways associated with growth factors, stress, and nutrients to promote cell survival and inhibit autophagy [13]. The study conducted by An Y et al. proved that autophagy determines the therapeutic effect of MSCs in cutaneous wound healing through the promotion of endothelial cell angiogenesis, and further revealed that autophagy enhanced the vascular endothelial growth factor secretion from MSCs to promote the angiogenesis of ECs by directly phosphorylating ERK1/2 [14]. In the study by Liao Y et al., Western blot and immunofluorescence staining results showed that the LC3-II/LC3-I ratio and the group of Beclin-1 (autophagy-related proteins) in dexamethasone-induced bone-marrow-derived endothelial progenitor cells (BM-EPCs) gradually decreased from the 12 h time point to the 24 h time point, reaching the lowest level at the 48 h time point; dexamethasone could inhibit autophagy levels in EPCs, and pravastatin could ameliorate ONFH by upregulating the autophagy activity in (BM-EPCs) [15]. Therefore, modulating autophagy in ECs may also be a potential target for the treatment of ONFH.

Recent studies have demonstrated that the autophagy in vascular ECs can be regulated by a range of biological factors and chemical compounds, and may have significant impacts on the fate of ECs. Meanwhile, a previous study suggested that EVs secreted by induced pluripotent stem-cell-derived mesenchymal stem cells significantly enhanced the proliferation, migration, and tube-forming capacities of ECs in vitro by activating the PI3K/Akt signaling pathway [16]. EVs are a class of membrane-bound vesicles with a diameter of 30 to 2000 nm, depending on their origin, that are involved in delivering functional biochemicals, including cytokines, proteins, lipids, and RNAs (mRNAs and miRNAs) into the target cell to stimulate a particular biological function. As determined by their biogenesis, the three main classes of EVs are exosomes, microvesicles, and apoptotic bodies [17–19]. The transplantation of EVs has been confirmed to exert similar therapeutic effects to direct stem cell transplantation in tissue repair. Some studies found that the transplantation of EVs secreted by human-induced pluripotent stem-cell-derived mesenchymal stem cells could promote angiogenesis of ischemic tissue in limb ischemia and skin defects that might be useful for other ischemic diseases, including ONFH [20,21]. Some studies have demonstrated the potential relationship of the PI3K/Akt pathway and stem cells in the biological activity of vascular endothelial cells [22]. It was reported that endothelial autophagy could be inhabited via the PI3K/Akt/mTOR signaling pathway, which could contribute to endothelial cell dysfunction. These results suggests that MSC-EVs can regulate ECs' autophagy via the PI3K/Akt/mTOR signaling pathway, affecting the biological function of ECs. Since exosomes are important functional products of MSCs, the biological function of ECs affected by EVs might be associated with endothelial autophagy via the PI3K/Akt/mTOR signaling pathway. Based on these studies, we hypothesize that the transplantation of MSC-EVs might prevent the progression of ONFH, and the PI3K/Akt/mTOR pathway might be involved in the autophagy of BMECs triggered by MSC-EVs.

2. Materials and Methods

2.1. BMECs Isolation, Identification and Culture

The human femoral head was obtained from patients undergoing total hip arthroplasty to generate BMECs after obtaining informed consent from patients. Cancellous bone of the femoral head was made into small bone fragments, then placed in a 50 mL centrifuge tube containing serum-free Dulbecco's Modified Eagle's Medium (DMEM; Gibco, Grand Island, NY, USA). The centrifuge tube was oscillated every 5 min. We then added 0.1% trypsin -0.1% EDTA (Procell, Wuhan, China), and digested it in a water bath at 37 °C for 5 min. After digestion, the liquid was filtered through a 200-mesh metal screen, centrifuged at $2000\times g$ for 10 min. The cell precipitates were re-suspended in a complete endothelial culture medium, then inoculated in 2% gelation-coated culture plates, followed by incubation at 37 °C with 5% CO_2 for 24 h. The nonadherent cells were washed off with DMEM and the adherent cells were incubated with a complete endothelial culture medium at 37 °C with 5% CO_2 until the cells grew to nearly 80% confluence. The magnetic beads coated with Ulex europaeus agglutinin I were added into the cell culture well, then we added 1 mL of 0.1% trypsin for digestion after the magnetic beads combined well under the microscope. The digested cell fluid was collected in a centrifuge tube, then the combined cells and uncombined cells were separated using a magnetic bead collector and collected. A fluorescence microscope (Olympus, Beijing, China) was used to detect the typical marker on the cell surface after immunofluorescence staining. The collected cells were inoculated into a 2% gelatin-coated culture flask and cultured with a complete endothelial cell culture medium (ingredients: 80% M199, 20% FBS, 2 mmol/L of glutamine, 100 U/mL of penicillin, and 100 μg/mL of endothelial cell growth factor) at 37 °C with 5% CO_2. The passage was carried out when the monolayer cells had grown to cover 80% of the culture flask bottom area, at which point we took the third generation of cells for the next step of the experiment.

2.2. Extracellular Vesicle Isolation and Identification

2.2.1. Generation of BMSCs from Bone Marrow of a Mouse

The mice were adopted in the generation of the MSCs. The mice femur and tibia were exposed under sterile conditions in supine fixation and rinsed twice with PBS (Procell, Wuhan, China). The bone marrow cavity was exposed and rinsed twice with 2 mL of MEMα (Procell, Wuhan, China), then the bone marrow was collected. The bone marrow was filtered through a 200-mesh cell screen and collected in a 15 mL centrifuge tube, centrifuged at 1200 rpm for 5 min. The cell precipitates were re-suspended with 4 mL of MEMα (Procell, Wuhan, China), then added into a 15 mL centrifuge tube with 4 mL of mouse bone marrow lymphocyte separation solution, centrifuged at 2000 rpm for 20 min. The intermediate white membrane cells were transferred into a 15 mL centrifuge tube, to which we added 10 mL of PBS to dilute, centrifuged at 1500 rpm for 5 min. The cell precipitates were re-suspended in a mouse MSCs complete culture medium (Procell, Wuhan, China), and the cells were inoculated with 2×10^6 cells/mL in a polylysine-pre-coated petri dish, then incubated at 37 °C in 5% CO_2 constant temperature for 3 days. The passage was carried out when the cells were fully grown, and the cell morphology was observed under a phase contrast microscope (Olympus). The second generation of cells was selected for the next step of the experiment.

2.2.2. Isolation and Identification of BMSC-EVs

EVs were isolated from MSC supernatants as previously described [23,24]. The second generation of BMSCs with good growth were collected, centrifuged at $3000\times g$ for 15 min to remove cellular debris, then the cell supernatant was collected after centrifugation. A total of 5 mL of cell supernatant was collected in a centrifuge tube, and exosome precipitation reagent (Rengen Bioscience, Liaoning, China) was added, mixed, and left standing at 4 °C for 30 min, then ultracentrifuged at $100,000\times g$ at 4 °C for 30 min. The supernatant was removed, and the EVs' precipitation was re-suspended with 100 μL of PBS. The re-suspended EVs were transferred into the purification column, centrifuged at $2000\times g$ at

4 °C for 5 min, and then collected. Transmission electron microscopy (JEOL, Tokyo, Japan) was used to observe the morphology of the EVs. Nanoparticle tracking analysis (NTA) was used to detect the size distribution and concentration of EVs using ZetaView (Particle Metrix, Meerbusch, Germany). The protein concentration of the EVs was measured using a bicinchoninic acid (BCA) protein assay kit (Beyotime, Shanghai, China). A Western blot analysis was performed to identify surface markers of MSC-EVs, including CD9, CD63, and CD81 [25]. The images were captured.

2.3. Cell Treatment

2.3.1. Establishment of Glucocorticoid-Induced BMECs Injury Model

The third generation of BMECs was selected and cultured. Cells in the logarithmic growth stage and in a good growth state were seeded into 24-well Matrigel (Corning, Corning, NY, USA) culture plates at a density of 5×10^3 cells per well overnight. The cells in the well were added into the medium containing a series of concentration-gradient hydrocortisone amounts (0 mg/mL, 0.03 mg/mL, 0.1 mg/mL, 0.3 mg/mL and 1 mg/mL); continued culturing for 12 h, 24 h, and 48 h at 37 °C with 5% CO_2; and then each well was incubated with 10 μL of CCK-8 solution for 4 h away from light before we measured the absorbance at 450 nm using a Thermo Varioskan LUX multimode microplate reader. The appropriate concentration of hydrocortisone will be used for the next step of the experiment.

2.3.2. Administration of EVs to Glucocorticoid-Induced BMECs Injury Model

The third generation BMECs was selected and randomly divided into five groups: the control group (no special treatment), model group (treated with 0.1 mg/mL of hydrocortisone), model + EVs group(0.1 mg/mL of hydrocortisone + 400 μg/mL of exos), model + rapamycin (MedChemExpress, Monmouth Junction, NJ, USA) group (0.1 mg/mL of hydrocortisone + 50 nM of rapamycin), and model + EVs + LY294002 (MedChemExpress, Monmouth Junction, NJ, USA) group (0.1 mg/mL hydrocortisone + 400 μg/mL EVs + 25 μM LY294002). Each group had three samples.

2.4. Cell Migration and Invasion Ability Analysis and Capillary Network Formation Assay

The third generation of BMECs was selected and washed with 3 mL of PBS, then digested with 0.25% trypsin, centrifuged at 1000 rpm for 5 min. The precipitates were washed to remove residual serum and re-suspended in a serum-free medium. The BMECs were preconditioned as aforementioned. The cell concentration was diluted to 3×10^5 cell/mL in each group. A total of 200 μL of cell suspension was plated into the upper chambers of a transwell plate (Corning), and 800μL of 10% FBS medium, placed in the lower chamber, was used as a chemoattractant. Then, the cells were cultured at 37 °C with 5% CO_2 for 24 h. The membranes were fixed with ethanol and stained with crystal violet (Beyotime), then mounted and observed under a light microscope (Olympus), and the number of cells was counted.

A tube formation assay was performed to investigate the endothelial cell network formation. Briefly, BMECs were seeded onto Matrigel-coated 24-well plates at a density of 1×10^5 cells per well. The BMECs were preconditioned as aforementioned and cultured at 37 °C with 5% CO_2 for 24 h. Images were captured, and the number of meshes and tubule lengths was quantified by the Image J software.

2.5. Cell Apoptosis Analysis

Annexin V-FITC/PI kits (Elabscience, Wuhan, China) were used to assess cell apoptosis. The BMECs were preconditioned as aforementioned, then removed from the medium and washed with PBS. The BMECs were harvested using 0.25% trypsinization and transferred to Eppendorf tubes, then re-suspended with 300 μL of binding buffer, to which 5 μL of annexin V and 5 μL of PI was added. These were incubated for 10 min at room temperature in the dark and analyzed by a flow cytometer.

2.6. Cell Viability Assay

Cell Counting Kit-8 was used to test cell viability. Approximately 5×10^3 BMECs were seeded in 96-well plates with 100 μL of medium in each well. The BMECs were preconditioned as aforementioned, then cultured at 37 °C with 5% CO_2 for 24 h. Then, each group was incubated with 10 μL of CCK-8 solution for 4 h away from light before the absorbance was measured at 450 nm by a Thermo Varioskan LUX multimode microplate reader.

2.7. Western Blot Analysis

The BMECs of each group were collected and the total protein was extracted by a total protein extraction kit (Beyotime). A BCA protein assay kit (Beyotime) was used to measure the protein concentration. A 60μg sample of each group was loaded onto 10% SDS-PAGE electrophoresis, then transferred to a polyvinylidene difluoride membrane. The membrane was blocked with TBST containing 5% skim milk, followed by incubation with the primary antibodies of LC3 (Abcam, Cambridge, UK, dilution 1:1000), PI3K (Proteintech Group, Wuhan, China, dilution 1:2000), P62 (Abcam, Cambridge, UK, dilution 1:1000), p-Akt (Abcam, Cambridge, UK, dilution 1:5000), Akt (Proteintech Group, Wuhan, China, dilution 1:2000), p-mTOR (Bioss, Beijing, China, dilution 1:1000), mTOR (Bioss, Beijing, China, dilution 1:1000), and GADPH (Hangzhou Goodhere Biotechnology, China, dilution 1:1000) at 4 °C overnight, then washed with TBST for 10 min and incubated with secondary antibodies for 2 h at room temperature.

Finally, the result was visualized by a chemiluminescence detection system (FluorChem M, ProteinSimple, San Jose, CA, USA).

2.8. Immunofluorescence Analysis

The BMECs were grown on glass coverslips in 24-well plates. After being treated by group, the BMECs were fixed with 3.7% paraformaldehyde at 37 °C for 30 min, and then made into frozen slices with a thickness of 5 μm after sucrose gradient dehydration. The BMECs were penetrated with 0.25% TritonX-100 at 37 °C for 30 min, and blocked BMECs with 10% goat serum in PBST (PBS + 0.1% Tween 20) for 30 min. The BMECs were incubated in the primary antibody of LC3 (Abcam; dilution 1:200) in a humidified chamber for 1 h at room temperature. The cells were then incubated with the secondary antibody for 1 h at room temperature in the dark after being washed three times for 5 min with PBS. Then, we decanted the secondary antibody solution and washed the cells three times for 5 min with PBS in the dark. Finally, we mounted the coverslip with a drop of mounting medium with 4′6-diamidino-2-phenylindole (DAPI). All imaging analyses were performed using fluorescence microscopy (Olympus).

2.9. Transmission Electron Microscopy

The BMECs were grown on glass coverslips in 24-well plates. After being treated by group, BMECs were washed with a phosphoric acid buffer, and then fixed with 1% osmium acid (Pelco) at 4 °C for 3 h. The BMECs were washed three times with the buffer, then dehydrated with ethanol, replaced by propylene oxide, and then polymerized in an oven at 70 °C. The samples were observed and photographed under a transmission electron microscope (JEM1230) after being sliced by the ultrathin slicer (EM UC6) and stained with uranium dioxy acetate (Spi-Chem) and lead citrate (Spi-Chem).

3. Statistical Analysis

All data are shown as mean ± standard deviation (SD). Differences between groups were assessed by one-way analysis of variance (ANOVA). Statistical analyses were performed using SPSS 18.0 software (SPSS, Inc., Chicago, IL, USA). p values < 0.05 were considered statistically significant.

4. Results

4.1. Isolation, Purification, and Culture of Microvascular Endothelial Cells

The BMECs were successfully isolated from a rabbit femoral head by using magnetic beads, and then being cultured. Immunofluorescence staining showed that the cells expressed typical marker molecules vWF and CD31 of endothelial cells (Figure 1A–D).

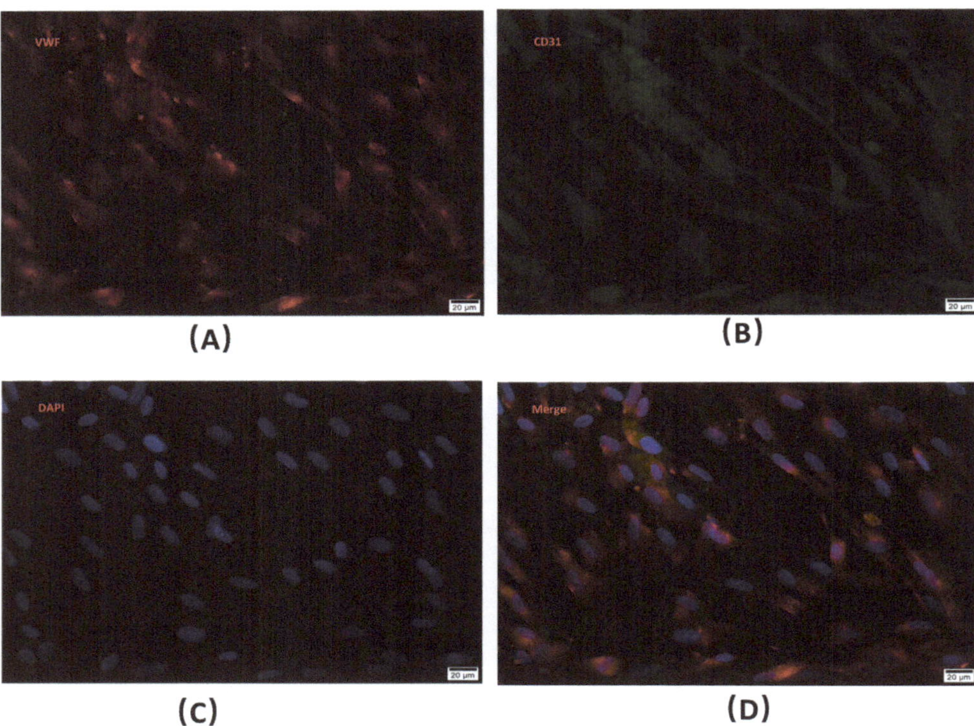

Figure 1. (**A–D**) Identification of microvascular endothelial cells. BMECs-specific markers were detected via immunofluorescence staining.

4.2. Characterization of BMSC-EVs

The BMSCs were successfully isolated and cultured from the mouse. The cells displayed a homogeneous fibroblastic-like morphology (Figure 2A). SEM, Western blotting and dynamic light scattering were used to characterize the purified particles derived from MSCs. SEM images showed that EVs exhibited spheroidal morphology (Figure 2B). The results of Western blotting confirmed the expression of CD9, CD63, and CD81 in EVs, which are surface markers exceptionally enriched in EVs (Figure 2C). The protein concentration of EVs was 4.63 mg/mL, tested by BCA. NTA revealed that the average diameter was 99.3 nm with a mean concentration of 1.2×10^9 particles/mL, and the number of exosomes with the diameter of 84.4 nm was the largest (Figure 2D).

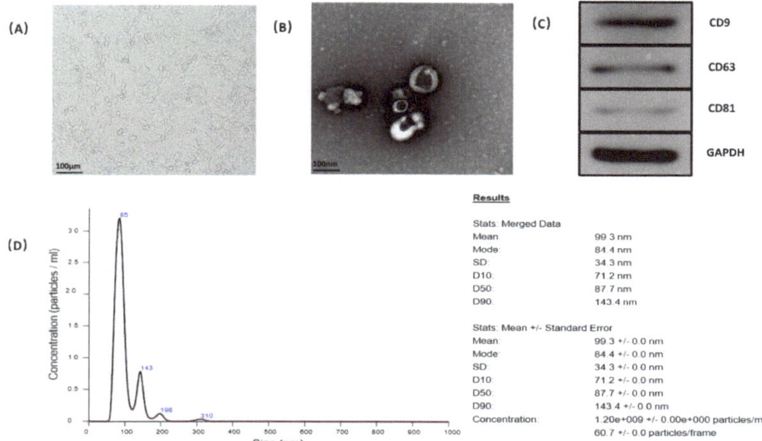

Figure 2. Characterization of BMSC-EVs. (**A**) The fibroblast-like morphology of BMSCs shown by microscope. (**B**) The morphology of BMSC-EVs shown by scanning electron microscopy. (**C**) Expression of CD9, CD63, and CD81 incorporation into BMSC-EVs shown by Western blotting. (**D**) Identification of size and concentration of BMSC by nanoparticle tracking analysis.

4.3. Appropriate Concentration of Glucocorticoid-Damaging BMECs

BMECs were treated with different concentrations of gradient hydrocortisone. Cell Counting Kit-8 was used to test the cell viability (Figure 3A–D). The indication of the viability of BMECs was a decreased tendency as the concentration of hydrocortisone increased and treating time was extended. The 0.1 mg/mL of hydrocortisone was selected in the subsequent experiments as the appropriate concentration (Figure 3E–H).

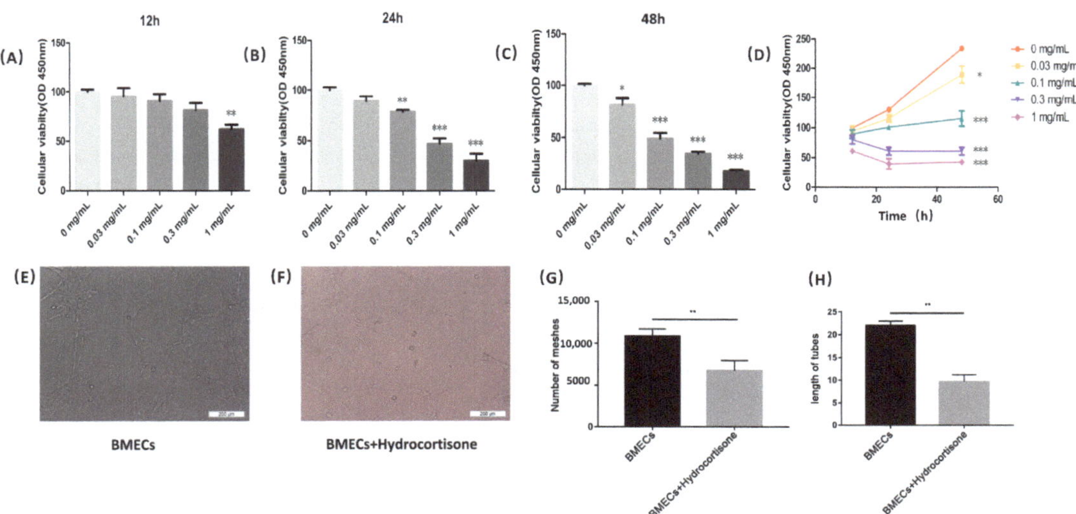

Figure 3. Hydrocortisone decreased BMECs' viability. (**A–D**) CCK-8 was used to measure viability of BMECs treated with different concentrations of hydrocortisone for different times. (**E–H**) When 0.1 mg/mL of hydrocortisone was used to treat BMECs, the number of meshes and tube length of BMECs were significantly reduced, indicating that BMECs injury model was successfully established. OD value; * $p < 0.05$, ** $p < 0.01$, *** $p < 0.005$. Each bar represents the mean ± SD of three independent experiments.

4.4. The Influence of BMSC-EVs on Glucocorticoid-Induced BMECs Injury

4.4.1. BMSC-Derived EVs Promote Migration, Invasion Capacity, and Angiogenesis of BMECs

In a transwell assay, hydrocortisone significantly decreased the migration and invasion capacity of BMECs compared to the control group. Compared with the model group, the inhibited effect was reversed after being treated with the BMSC-EVs. Simultaneously, the inhibition of migration and invasion capacity of BMECs was reversed with the administration of rapamycin. Compared with the Model + EVs group, the migration and invasion capacity of BMECs improved further after adding LY294002 (Figure 4A–D). In the tube formation assay, the model group showed a significant antiangiogenic manifestation compared with control group. BMSC-EVs, rapamycin and LY294002 reversed the inhibitory effect of angiogenesis and increased the loop formation ability of BMECs (Figure 4E). The number of meshes and lengths of tubes increased after adding BMSC-EVs, rapamycin, and LY294002 (Figure 4F). These results showed that BMSC-EVs can promote angiogenesis.

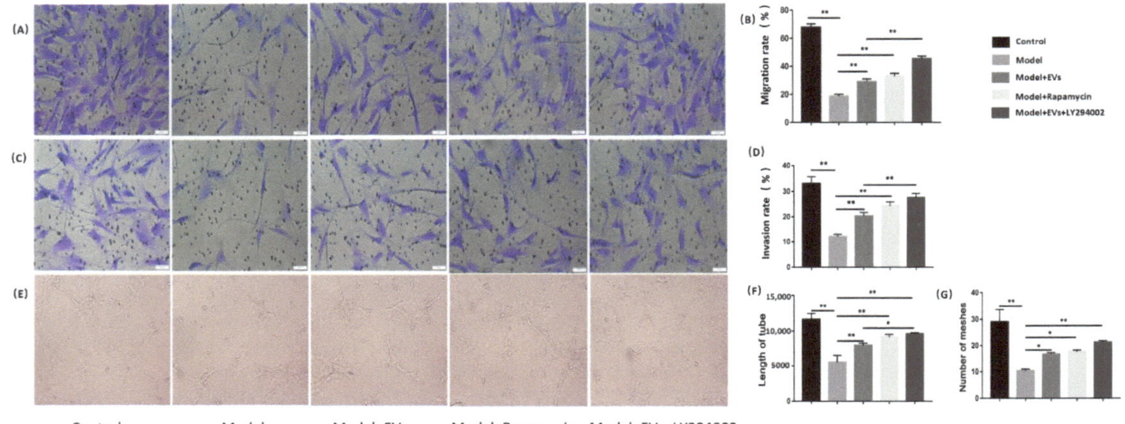

Figure 4. Angiogenesis was promoted by BMSC-EVs. (**A**) The migration capacity of BMECs was investigated by transwell assay in different groups. (**B**) Quantitative analysis of the migration rate of BMECs. (**C**) The invasion capacity of BMECs was investigated by transwell assay in different groups. (**D**) Quantitative analysis of the invasion rate of BMECs. (**E**) Tube formation assay for detecting the tube-forming ability of BMECs in the different groups. (**F,G**) Quantitative analysis of tube formation. The value of the total mesh area, total length was measured. Each bar represents the mean ± SD of three independent experiments. * $p < 0.05$, ** $p < 0.01$.

4.4.2. BMSC-Derived EVs Prevented Glucocorticoid-Induced Apoptosis of BMECs

Flow cytometry demonstrated that the percentage of apoptotic cells increased from 4.42% (control group) to 26.60% (model group) after treatment with hydrocortisone ($p < 0.01$). MSCs-EVs showed a protective effect on hydrocortisone-induced apoptosis of BMECs. When MSCs-EVs were added, the percentage of apoptotic cells decreased from 26.60% (model group) to 13.76% (model + exos group) ($p < 0.01$). When rapamycin was added, the percentage of apoptotic cells decreased from 26.60% (model group) to 14.34% (model + rapamycin group) ($p < 0.01$), and its effect was similar to the effect caused by BMSC-EVs. When the PI3K inhibitor LY294002 was added, the percentage of apoptotic cells further decreased from 13.76% (model + EVs group) to 7.36% (model + EVs + LY294002 group) (Figure 5A,B).

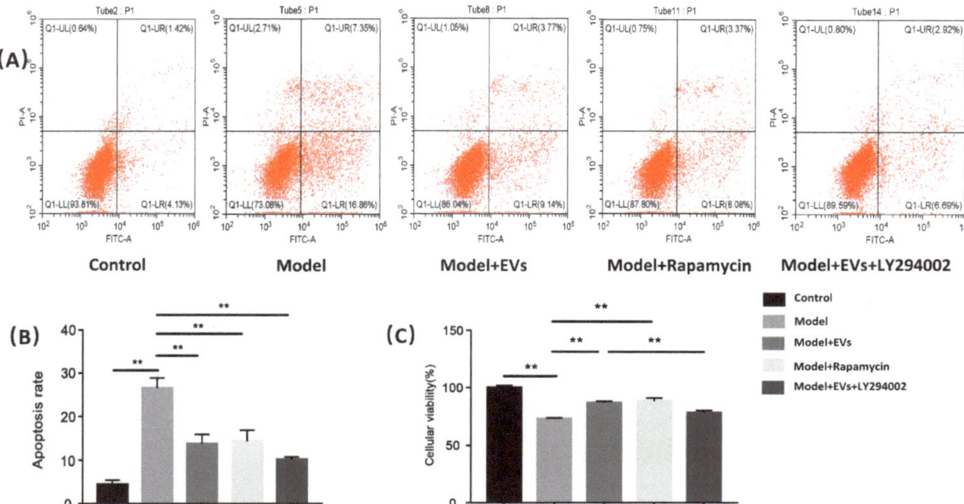

Figure 5. BMSC-EV-derived exosomes BMECs' viability and protected BMECs against glucocorticoid-induced apoptosis. (**A**) Apoptosis was quantified by flow cytometry after staining with annexin V-FITC/PI. (**B**) Percentage of apoptotic cells in different groups. (**C**) CCK-8 was used to measure viability of BMECs in different groups. Each bar represents the mean ± SD of three independent experiments. ** $p < 0.01$.

4.4.3. BMSC-Derived EVs Alleviate the Decreased Cell Viability of BMECs Induced by GCs

Cell Counting Kit-8 was used to test the cell viability. Compared with the control group, the cell viability significantly decreased after being treated with hydrocortisone. Compared with the model group, the cell viability significantly increased after being treated with BMSC-EVs, indicating that BMSC-EVs effectively alleviate glucocorticoid-induced BMEC injury. Meanwhile, the addition of rapamycin as an autophagy activator can effectively alleviate the decreased viability caused by hydrocortisone, suggesting rapamycin alleviated the injury of BMECs caused by hydrocortisone by activating autophagy. Compared with the model + EVs group, the treatment with LY294002 as the PI3K inhibitor effectively inhibited the alleviating effect of BMEC-EVs on glucocorticoid-induced injury, suggesting that BMSC-EVs alleviated hydrocortisone-induced injury mainly through the activation of autophagy (Figure 5C).

4.4.4. BMSC-EVs Regulated Autophagy of BMECs

LY294002 was used as an autophagy inhibitor, and rapamycin was used as an autophagy inducer to observe the effects of BMSC-derived exosomes on cell autophagy. Western blot analysis revealed that hydrocortisone significantly decreased the ratio of LC3-II/I, and increased the expression of P62 (an autophagy protein receptor) (Figure 6B–D). Accordingly, immunofluorescence analysis showed the number and intensity of punctate LC3 fluorescence decreased in the model group. Compared with the model group, the ratio of LC3-II/I was upregulated, and the expression of p62 was downregulated in the model + EVs group, and its effect was similar to the model + rapamycin group. When LY294002 was added, the ratio of LC3-II/I was further upregulated and the expression of p62 was further downregulated. Moreover, the number and intensity of punctate LC3 fluorescence increased in the model + EVs group, and it was more than the model + rapamycin group, but less than the model + EVs + LY294002 group (Figure 6A).

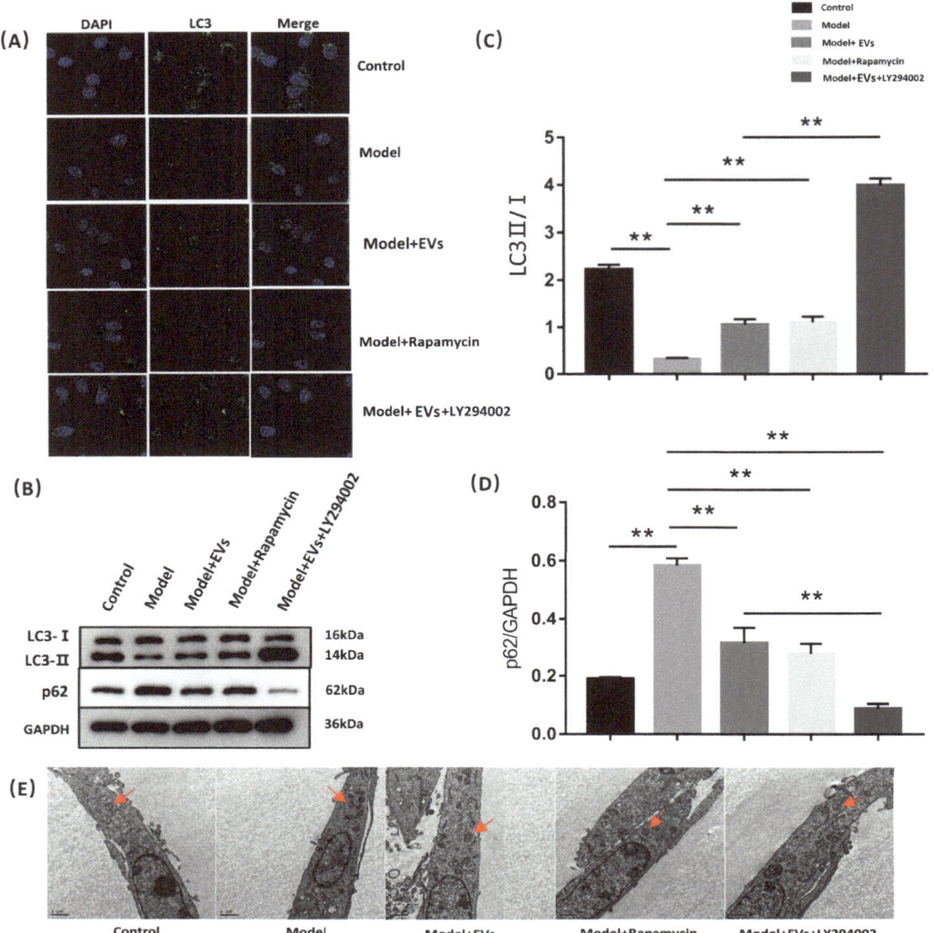

Figure 6. BMSC-EVs regulated autophagy of BMECs. (**A**) Immunohistochemical analysis of autophagic marker LC3. Scale bar = 10 μm. (**B**) Western blot analysis of LC3 and P62 in different groups. (**C,D**) Quantification of LC3 and P62 in different groups. (**E**) Morphological changes of autophagosomes were observed by TEM in different groups and the red arrow represents the autophagosome location. Each bar represents the mean ± SD of three independent experiments. ** $p < 0.01$. DAPI, 4′,6-diamidino-2-phenylindole; EVs, BMSC-derived extracellular vesicles; GAPDH, glyceraldehyde 3-phosphate dehydrogenase; mTOR, mammalian target of rapamycin.

Transmission electron microscopy was used to observe the morphology. Compared with the control group, hydrocortisone made the autophagosome fold and shrink and reduce its number. The morphology of the autophagosome in the model + EVs group was similar to that of the control group, while the change of the autophagosome in the model + EVs + LY294002 group was not significant compared with the model + EVs group (Figure 6E).

These results suggested that BMSC-EVs can promote autophagy of BMECs, and the inhibition of the PI3K signal further increased the level of autophagy.

4.4.5. BMSC-Derived EVs Prevent Glucocorticoid-Induced BMECs Injury by Regulating Autophagy via the PI3K/Akt/mTOR Pathway

To assess whether BMSC-EVs prevent glucocorticoid-induced BMEC injury by regulating autophagy via the PI3/Akt/mTOR pathway, LY294002 (an inhibitor of PI3K) and rapamycin (an inhibitor of mTOR) were used. As shown in Figure 7, hydrocortisone induced a significant upregulation of PI3K, p-mTOR/mTOR, and p-Akt/Akt expression, and this effect could be reversed by treatment with BMSC-EVs (Figure 7A–D). When rapamycin was added, the level of p-mTOR/mTOR and PI3K significantly downregulated, but the level of p-Akt/Akt slightly upregulated. Compared with the model + EVs group, the expression of p-mTOR/mTOR, p-Akt/Akt, and PI3K downregulated in the model + EVs + Y294002 group (Figure 7A–D).

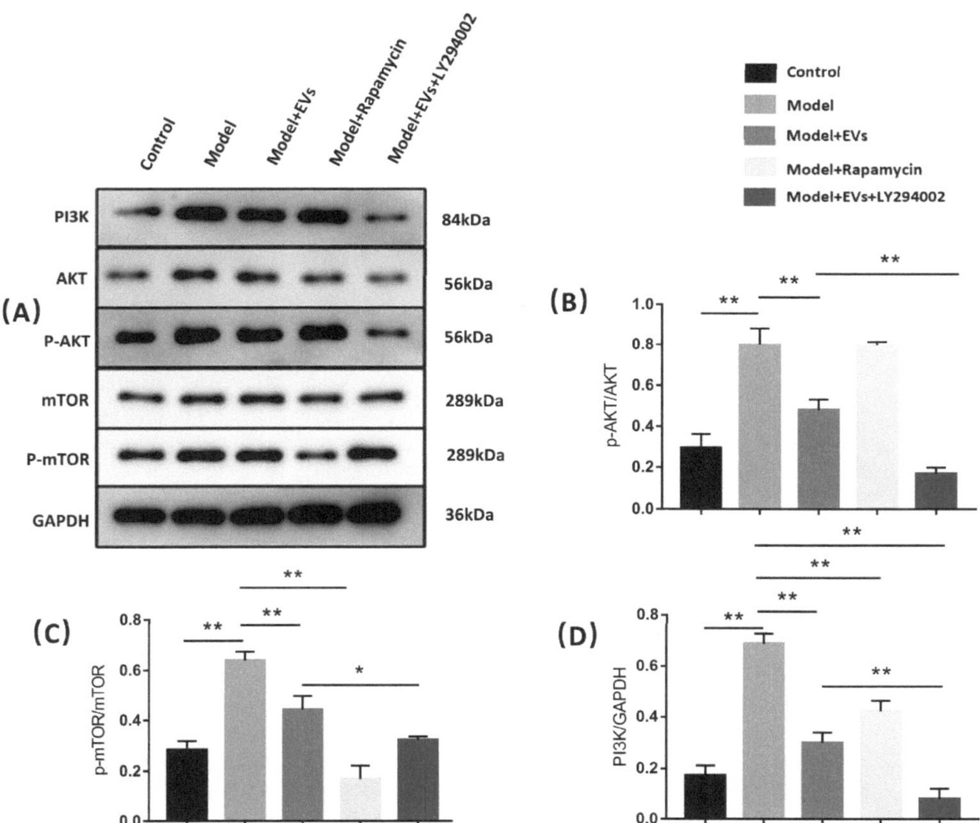

Figure 7. BMSC-EVs regulated the level of BMECs autophagy via PI3K/Akt/mTOR pathway. (**A**) Western blot analysis. (**B–D**) Quantification of p-Akt, Akt, p-mTOR, mTOR and PI3K in five groups. Each bar represents the mean ± SD of three independent experiments. * $p < 0.05$, ** $p < 0.01$. EVs, BMSC-derived extracellular vesicles; GAPDH, glyceraldehyde 3-phosphate dehydrogenase; mTOR, mammalian target of rapamycin.

5. Discussion

Long-term or short-term overuse of GCs can cause local ischemia and secondary osteonecrosis, which are considered core pathological mechanisms in ONFH. BMEC is an important part of bone microcirculation, and BMEC injury caused by GCs is an important mechanism of ONFH [26]. The mechanism of injury caused by GCs to BMECs is not clear.

Sbardella et al. found that dexamethasone inhibited trabecular meshwork cells by downregulating autophagy, a kind of secretory endothelial cell similar to BMECs, causing metabolic disorder and even apoptosis [27]. Under normal circumstances, autophagy is at a low level, and only moderate autophagy has a protective effect. Inappropriate autophagy will inhibit the normal function of autophagy, leading to apoptosis and injury to the cell [28,29]. Meanwhile, Liu et al. reported that extracellular vesicles secreted from human-induced pluripotent stem cells improved migration and the tube-formation ability of human umbilical vascular endothelial cells [16]. The in vitro results suggested BMSC-Exos and autophagy might be associated with the progression of steroid-induced ONFH.

Stem cell transplantation has been shown to improve local blood to ONFH by releasing various cytokines to promote angiogenesis [30]. However, the potential risk and problems are the major limitation for the possibility of clinical application. EVs derived from MSCs are studied extensively and demonstrate equal efficacy as MSCs [31]. Previous studies have identified that transplantation of exosomes derived from MSCs did not induce an immune reaction in vivo [20,32]. EVs secreted by stem cells have been demonstrated to have a similar effect to promote angiogenesis [22]. Liu et al. reported that MSC-derived exosomes inhibited H9C2 cell apoptosis induced by hypoxia and serum deprivation, and improved H9C2 cell viability by reducing excess autophagy activity via activation of the PI3K/Akt/mTOR pathway [33]. The extraction of EVs was performed by a combination of ultrafiltration and purification. The obtained EVs exhibited round-shaped morphology with an average diameter of 99.3 nm, and expressed typical EV surface markers including CD9, CD63, and CD81, with a mean concentration of 1.2×10^9 particles/mL.

The potential mechanism of BMSC-EVs preventing glucocorticoid-induced BMEC injury was also investigated in this study. The most important function of EVs is their role in communication from host cells to target cells. EVs transport a variety of proteins, lipids, RNA, and other substances to the site of injury, and play an important role in angiogenesis, anti-apoptosis, and anti-inflammatory responses [34]. We found that BMSC-EVs could alleviate the decreased cell viability induced by glucocorticoids, improve migration, invasion capacity, and tube formation, and prevent glucocorticoid-induced apoptosis of BMECs. We further found that BMSC-EVs upregulated the expression of LC3 and downregulated the expression of P62, and reversed the glucocorticoid-induced folding and contraction of autophagosome. This illustrated that BMSC-EVs inhibited glucocorticoid-induced BMEC apoptosis, improved cell viability, and promoted angiogenesis via regulation of autophagy activity. Interactions between autophagy and apoptotic components suggested complex crosstalk. mTOR, a serine/threonine kinase, can promote cell growth and inhibit autophagy via various pathways, and is also an important downstream target of Akt, which plays a critical role in regulating apoptosis [13,35]. Hsu et al. demonstrated that hyperphosphatemia-induced protective autophagy in endothelial cells through the inhibition of Akt/mTOR signaling and inhibited high-Pi-induced autophagy aggravates endothelial cell apoptosis [36]. Our results showed that GCs could activate the PI3K/Akt/mTOR pathway, inhibit BMEC viability, angiogenesis, and cause cell apoptosis. However, when the pathway was blocked by the PI3K inhibitor and mTOR inhibitor, the detrimental effect elicited by glucocorticoids was alleviated. When BMSC-EVs were added, the detrimental effect elicited by GCs was also alleviated. These results suggest that GCs can inhibit autophagy by activating the PI3K/Akt/mTOR pathway, causing a decrease in BMEC viability, migration, invasion and tube-forming capacity, and even apoptosis. BMSC-EVs can prevent glucocorticoid-induced BMEC injury by regulating autophagy via the PI3K/Akt/mTOR pathway. Our experiment also has some limitations: first, we did not use filters to isolate BMSC-EVs; second, we did not detect negative markers of EVs; third, we did not use AngioTool software in the tube formation assay. We will further improve our experimental method in future experiments.

6. Conclusions

We identified that GCs inhibit autophagy by activating the PI3K/Akt/mTOR pathway, causing a decrease in BMEC viability, angiogenesis capacity, and even apoptosis. BMSC-EVs can regulate autophagy via the PI3K/Akt/mTOR pathway of BMECs to enhance cell viability and angiogenesis capacity. Our findings suggest that BMSC-EVs may be a promising therapeutic approach in the treatment of ONFH.

Author Contributions: J.M. was responsible for the performance of the whole experiment and analyzing the experimental data. M.S. was responsible for writing the manuscript. F.G. and B.W. provided guidance for the design of the experiment. W.W. and D.Y. participated in revision of the manuscript. All authors have read and agreed to the published version of the manuscript.

Funding: This project was supported by grants from the Beijing Natural Science Foundation (7204301,7182146), Fundamental Research Funds for the Central Universities (3332021088), Elite Medical Professionals project of China-Japan Friendship Hospital (NO.ZRJY2021-TD01, NO.ZRJY2021-GG12), the National Natural Science Foundation of China (81672236, 81871830), the Ningxia Natural Science Foundation (2020AAC03337), and the Joint Project of BRC-BC (Biomedical Translational Engineering Research Center of BUCT-CJFH) (RZ2020-02).

Institutional Review Board Statement: The study was approved by the Animal Care Committee of China-Japan Friendship Hospital and was strictly conducted according to the animal experiment guidelines (Approval Code: Zryhyy12-20-01-7).

Informed Consent Statement: Informed consent was obtained from all subjects involved in the study.

Data Availability Statement: All data generated or analyzed during this study are included in this published article and are available from the corresponding author upon reasonable request.

Conflicts of Interest: The authors declare no conflict of interest.

References

1. Mont, M.A.; Cherian, J.J.; Sierra, R.J.; Jones, L.C.; Lieberman, J.R. Nontraumatic Osteonecrosis of the Femoral Head: Where Do We Stand Today? A Ten-Year Update. *J. Bone Jt. Surg. Am.* **2015**, *97*, 1604–1627. [CrossRef]
2. Zalavras, C.G.; Lieberman, J.R. Osteonecrosis of the femoral head: Evaluation and treatment. *J. Am. Acad. Orthop. Surg.* **2014**, *22*, 455–464. [CrossRef]
3. Zheng, Y.; Zheng, Z.; Zhang, K.; Zhu, P. Osteonecrosis in systemic lupus erythematosus: Systematic insight from the epidemiology, pathogenesis, diagnosis and management. *Autoimmun. Rev.* **2022**, *21*, 102992. [CrossRef]
4. Feng, Y.; Yang, S.H.; Xiao, B.J.; Xu, W.H.; Ye, S.N.; Xia, T.; Zheng, D.; Liu, X.Z.; Liao, Y.F. Decreased in the number and function of circulation endothelial progenitor cells in patients with avascular necrosis of the femoral head. *Bone* **2010**, *46*, 32–40. [CrossRef]
5. Zhao, D.Y.; Yu, Q.S.; Guo, W.S.; Cheng, L.M. [Effect of icariin on the proteomic expression profile of bone microvascular endothelial cells of human femoral head against steroids-induced lesion]. *Zhonghua Yi Xue Za Zhi* **2016**, *96*, 1026–1030. [CrossRef]
6. Starklint, H.; Lausten, G.S.; Arnoldi, C.C. Microvascular obstruction in avascular necrosis. Immunohistochemistry of 14 femoral heads. *Acta Orthop. Scand.* **1995**, *66*, 9–12. [CrossRef]
7. Yang, Y.; Lou, J.; Li, Z.; Sun, W.; Wang, B.; Jia, Y. [Effect of glucocorticoid on production of reactive oxygen species in bone microvascular endothelial cells]. *Zhongguo Xiu Fu Chong Jian Wai Ke Za Zhi* **2011**, *25*, 533–537.
8. Jiang, F. Autophagy in vascular endothelial cells. *Clin. Exp. Pharmacol. Physiol.* **2016**, *43*, 1021–1028. [CrossRef]
9. Boya, P.; Reggiori, F.; Codogno, P. Emerging regulation and functions of autophagy. *Nat. Cell Biol.* **2013**, *15*, 713–720. [CrossRef]
10. Levine, B.; Mizushima, N.; Virgin, H.W. Autophagy in immunity and inflammation. *Nature* **2011**, *469*, 323–335. [CrossRef]
11. Peng, D.J.; Wang, J.; Zhou, J.Y.; Wu, G.S. Role of the Akt/mTOR survival pathway in cisplatin resistance in ovarian cancer cells. *Biochem. Biophys. Res. Commun.* **2010**, *394*, 600–605. [CrossRef]
12. Rubinsztein, D.C.; Codogno, P.; Levine, B. Autophagy modulation as a potential therapeutic target for diverse diseases. *Nat. Rev. Drug Discov.* **2012**, *11*, 709–730. [CrossRef]
13. Saxton, R.A.; Sabatini, D.M. mTOR Signaling in Growth, Metabolism, and Disease. *Cell* **2017**, *169*, 361–371. [CrossRef]
14. An, Y.; Liu, W.J.; Xue, P.; Ma, Y.; Zhang, L.Q.; Zhu, B.; Qi, M.; Li, L.Y.; Zhang, Y.J.; Wang, Q.T.; et al. Autophagy promotes MSC-mediated vascularization in cutaneous wound healing via regulation of VEGF secretion. *Cell Death Dis.* **2018**, *9*, 58. [CrossRef]
15. Liao, Y.; Zhang, P.; Yuan, B.; Li, L.; Bao, S. Pravastatin Protects Against Avascular Necrosis of Femoral Head via Autophagy. *Front. Physiol.* **2018**, *9*, 307. [CrossRef]

16. Liu, X.; Li, Q.; Niu, X.; Hu, B.; Chen, S.; Song, W.; Ding, J.; Zhang, C.; Wang, Y. Exosomes Secreted from Human-Induced Pluripotent Stem Cell-Derived Mesenchymal Stem Cells Prevent Osteonecrosis of the Femoral Head by Promoting Angiogenesis. *Int. J. Biol. Sci.* **2017**, *13*, 232–244. [CrossRef]
17. Katsuda, T.; Kosaka, N.; Takeshita, F.; Ochiya, T. The therapeutic potential of mesenchymal stem cell-derived extracellular vesicles. *Proteomics* **2013**, *13*, 1637–1653. [CrossRef]
18. El Andaloussi, S.; Mäger, I.; Breakefield, X.O.; Wood, M.J. Extracellular vesicles: Biology and emerging therapeutic opportunities. *Nat. Rev. Drug Discov.* **2013**, *12*, 347–357. [CrossRef]
19. Chaput, N.; Théry, C. Exosomes: Immune properties and potential clinical implementations. *Semin. Immunopathol.* **2011**, *33*, 419–440. [CrossRef]
20. Zhang, J.; Guan, J.; Niu, X.; Hu, G.; Guo, S.; Li, Q.; Xie, Z.; Zhang, C.; Wang, Y. Exosomes released from human induced pluripotent stem cells-derived MSCs facilitate cutaneous wound healing by promoting collagen synthesis and angiogenesis. *J. Transl. Med.* **2015**, *13*, 49. [CrossRef]
21. Hu, G.W.; Li, Q.; Niu, X.; Hu, B.; Liu, J.; Zhou, S.M.; Guo, S.C.; Lang, H.L.; Zhang, C.Q.; Wang, Y.; et al. Exosomes secreted by human-induced pluripotent stem cell-derived mesenchymal stem cells attenuate limb ischemia by promoting angiogenesis in mice. *Stem Cell Res. Ther.* **2015**, *6*, 10. [CrossRef] [PubMed]
22. Hu, Y.; Tao, R.; Chen, L.; Xiong, Y.; Xue, H.; Hu, L.; Yan, C.; Xie, X.; Lin, Z.; Panayi, A.C.; et al. Exosomes derived from pioglitazone-pretreated MSCs accelerate diabetic wound healing through enhancing angiogenesis. *J. Nanobiotechnol.* **2021**, *19*, 150. [CrossRef] [PubMed]
23. Théry, C.; Amigorena, S.; Raposo, G.; Clayton, A. Isolation and characterization of exosomes from cell culture supernatants and biological fluids. *Curr. Protoc. Cell Biol.* **2006**, *30*, 3–22. [CrossRef] [PubMed]
24. Witwer, K.W.; Buzás, E.I.; Bemis, L.T.; Bora, A.; Lässer, C.; Lötvall, J.; Nolte-'t Hoen, E.N.; Piper, M.G.; Sivaraman, S.; Skog, J.; et al. Standardization of sample collection, isolation and analysis methods in extracellular vesicle research. *J. Extracell. Vesicles* **2013**, *2*, 20360. [CrossRef]
25. Théry, C.; Zitvogel, L.; Amigorena, S. Exosomes: Composition, biogenesis and function. *Nat. Rev. Immunol.* **2002**, *2*, 569–579. [CrossRef]
26. Séguin, C.; Kassis, J.; Busque, L.; Bestawros, A.; Theodoropoulos, J.; Alonso, M.L.; Harvey, E.J. Non-traumatic necrosis of bone (osteonecrosis) is associated with endothelial cell activation but not thrombophilia. *Rheumatology* **2008**, *47*, 1151–1155. [CrossRef]
27. Sbardella, D.; Tundo, G.R.; Coletta, M.; Manni, G.; Oddone, F. Dexamethasone Downregulates Autophagy through Accelerated Turn-Over of the Ulk-1 Complex in a Trabecular Meshwork Cells Strain: Insights on Steroid-Induced Glaucoma Pathogenesis. *Int. J. Mol. Sci.* **2021**, *22*, 5891. [CrossRef]
28. Li, Z.Y.; Wu, Y.F.; Xu, X.C.; Zhou, J.S.; Wang, Y.; Shen, H.H.; Chen, Z.H. Autophagy as a double-edged sword in pulmonary epithelial injury: A review and perspective. *Am. J. Physiol. Lung Cell. Mol. Physiol.* **2017**, *313*, L207–L217. [CrossRef]
29. Chai, P.; Ni, H.; Zhang, H.; Fan, X. The Evolving Functions of Autophagy in Ocular Health: A Double-edged Sword. *Int. J. Biol. Sci.* **2016**, *12*, 1332–1340. [CrossRef]
30. Maruyama, M.; Moeinzadeh, S.; Guzman, R.A.; Zhang, N.; Storaci, H.W.; Utsunomiya, T.; Lui, E.; Huang, E.E.; Rhee, C.; Gao, Q.; et al. The efficacy of lapine preconditioned or genetically modified IL4 over-expressing bone marrow-derived mesenchymal stromal cells in corticosteroid-associated osteonecrosis of the femoral head in rabbits. *Biomaterials* **2021**, *275*, 120972. [CrossRef]
31. Kiang, J.G. Chapter Six—Mesenchymal stem cells and exosomes in tissue regeneration and remodeling: Characterization and therapy. In *Tissue Barriers in Disease, Injury and Regeneration*; Gorbunov, N.V., Ed.; Elsevier: Amsterdam, The Netherlands, 2021; pp. 159–185.
32. Wen, D.; Peng, Y.; Liu, D.; Weizmann, Y.; Mahato, R.I. Mesenchymal stem cell and derived exosome as small RNA carrier and Immunomodulator to improve islet transplantation. *J. Control. Release* **2016**, *238*, 166–175. [CrossRef] [PubMed]
33. Liu, H.; Sun, X.; Gong, X.; Wang, G. Human umbilical cord mesenchymal stem cells derived exosomes exert antiapoptosis effect via activating PI3K/Akt/mTOR pathway on H9C2 cells. *J. Cell. Biochem.* **2019**, *120*, 14455–14464. [CrossRef] [PubMed]
34. Lazar, E.; Benedek, T.; Korodi, S.; Rat, N.; Lo, J.; Benedek, I. Stem cell-derived exosomes—An emerging tool for myocardial regeneration. *World J. Stem Cells* **2018**, *10*, 106–115. [CrossRef]
35. Sussman, M.A.; Völkers, M.; Fischer, K.; Bailey, B.; Cottage, C.T.; Din, S.; Gude, N.; Avitabile, D.; Alvarez, R.; Sundararaman, B.; et al. Myocardial AKT: The omnipresent nexus. *Physiol. Rev.* **2011**, *91*, 1023–1070. [CrossRef] [PubMed]
36. Hsu, Y.J.; Hsu, S.C.; Huang, S.M.; Lee, H.S.; Lin, S.H.; Tsai, C.S.; Shih, C.C.; Lin, C.Y. Hyperphosphatemia induces protective autophagy in endothelial cells through the inhibition of Akt/mTOR signaling. *J. Vasc. Surg.* **2015**, *62*, 210–221.e2. [CrossRef]

Review

Insight into the Role of the PI3K/Akt Pathway in Ischemic Injury and Post-Infarct Left Ventricular Remodeling in Normal and Diabetic Heart

Bartosz Walkowski [1], Marcin Kleibert [1,*,†], Miłosz Majka [1,*,†] and Małgorzata Wojciechowska [1,2]

[1] Laboratory of Centre for Preclinical Research, Department of Experimental and Clinical Physiology, Medical University of Warsaw, Banacha 1b, 02-097 Warsaw, Poland; walkowski.bartosz@wp.pl (B.W.); malgorzata.wojciechowska2@wum.edu.pl (M.W.)

[2] Invasive Cardiology Unit, Independent Public Specialist Western Hospital John Paul II, Daleka 11, 05-825 Grodzisk Mazowiecki, Poland

* Correspondence: marcin.kleibert@gmail.com (M.K.); miloszmajka98@gmail.com (M.M.)

† These authors contributed equally to this work.

Abstract: Despite the significant decline in mortality, cardiovascular diseases are still the leading cause of death worldwide. Among them, myocardial infarction (MI) seems to be the most important. A further decline in the death rate may be achieved by the introduction of molecularly targeted drugs. It seems that the components of the PI3K/Akt signaling pathway are good candidates for this. The PI3K/Akt pathway plays a key role in the regulation of the growth and survival of cells, such as cardiomyocytes. In addition, it has been shown that the activation of the PI3K/Akt pathway results in the alleviation of the negative post-infarct changes in the myocardium and is impaired in the state of diabetes. In this article, the role of this pathway was described in each step of ischemia and subsequent left ventricular remodeling. In addition, we point out the most promising substances which need more investigation before introduction into clinical practice. Moreover, we present the impact of diabetes and widely used cardiac and antidiabetic drugs on the PI3K/Akt pathway and discuss the molecular mechanism of its effects on myocardial ischemia and left ventricular remodeling.

Keywords: PI3K/Akt pathway; myocardial infarction; heart failure; left ventricular remodeling; diabetes; apoptosis; necroptosis; ferroptosis; pyroptosis; fibrosis

Citation: Walkowski, B.; Kleibert, M.; Majka, M.; Wojciechowska, M. Insight into the Role of the PI3K/Akt Pathway in Ischemic Injury and Post-Infarct Left Ventricular Remodeling in Normal and Diabetic Heart. *Cells* **2022**, *11*, 1553. https://doi.org/10.3390/cells11091553

Academic Editors: Jean Christopher Chamcheu, Claudia Bürger and Shile Huang

Received: 31 March 2022
Accepted: 3 May 2022
Published: 5 May 2022

Publisher's Note: MDPI stays neutral with regard to jurisdictional claims in published maps and institutional affiliations.

1. Introduction

Cardiovascular diseases (CVD) are the leading cause of death globally, and myocardial infarction (MI) is of fundamental importance [1]. MI is caused by the restriction of blood flow through the coronary arteries, which results in an imbalance between myocardial demand and the blood supply. It results in the activation of many intracellular pathways which promotes the activation of programmed and unprogrammed cell death, leading to the impairment of cardiac functions [2,3]. The most common cause of MI is a rupture of an atherosclerotic plaque. Numerous risk factors such as smoking, a high level of low-density cholesterol, hypertension, diabetes mellitus, and a lack of physical activity stimulate atherosclerotic plaque progression. The complexity of atherosclerosis progression and vulnerability of plaques were well-described by Libby et al. and Anderson et al. [4,5].

The 1-year mortality rate among patients after MI is still high [6]. However, improvements in the acute treatment of MI have increased the number of surviving patients, who are at higher risk of recurrent infarction and development of post-infarction heart failure (HF) [7–9].

At the same time, cardiovascular diseases are associated with growing public and private expenditure on healthcare worldwide [10]. Therefore, a new therapeutic approach is needed to reduce the cost of CVD treatment. It seems that targeted modification of molecular pathways may improve the outcome for patients.

Post-infarction HF is mostly caused by adverse remodeling of the left ventricle (LV). This is a series of molecular, cellular, and interstitial changes following MI [11–13]. It involves the inflammation, fibrosis, and hypertrophy of the myocardium, and changes in the vascular bed and the conducive system of the heart [14,15]. Altered remodeling results in electrophysiological disorders and ventricular dysfunction and leads to heart failure, which significantly impacts a patient's prognosis [16–19]. One of the main risk factors of post-MI HF development is diabetes mellitus (DM), which impairs cell communication, involving many molecular pathways including PI3K/Akt [20].

The phosphatidylinositol 3-kinase (PI3K)/protein kinase B(Akt) pathway is one of the most significant intracellular signal transduction pathways. Even though it was relatively recently discovered, it has caused a revolution mainly in personalized oncology. PI3K was identified by Lewis Cantley's group in 1988 [21], while Akt kinase was identified by Stephen Staal in 1987 [22]. It has become a subject of great interest in the scientific community and the focus of numerous basic and clinical studies.

The proper activity of PI3K and Akt appears to be essential for the development, functioning, and survival of the organism. The lack of their expression leads to a blockade of cell division and embryonic lethality [23,24]. In response to extracellular signals, the PI3K/Akt pathway controls cell metabolism, growth, proliferation, and the stress response. It plays a key role in the proper functioning of most human organs and is associated with the development of diseases [25–27]. In the myocardium, Akt is a central signal transductor, and the PI3K/Akt pathway is responsible for proper metabolism and cell response. Indeed, it plays a role in cardiovascular diseases, including chronic (CCS) and acute coronary syndromes (ACS) or HF. In this review, we summarize the currently known role of the PI3K/Akt pathway in the pathophysiology of myocardial infarction and the post-infarction remodeling of the left ventricle, and we indicate the potential targets of therapy. In addition, we discuss the clinical importance of the impact of diabetes and commonly used drugs on its activity in the heart.

2. The PI3K/Akt Pathway and Its Impact on the Peri-Infarct Processes

The main molecules involved in the PI3K/Akt pathway are presented in the figure below (Figure 1a) This pathway consists of tyrosine kinase receptors (RTKs), phosphatidylinositol 3-kinase (PI3K), phosphatidylinositol-4,5-bisphosphate (PIP2), phosphatidylinositol-3,4,5-bisphosphate (PIP3), and Akt (Figure 1a).

2.1. Components and Mechanisms of PI3K/Akt Activation

Receptor tyrosine kinases (RTKs) are surface receptors composed of three functional domains: an extracellular ligand-binding domain, a transmembrane domain and an intracellular tyrosine kinase domain [28]. After bounding a ligand (growth factors (GFs), cytokines, hormones), RTK forms a dimer, which activates the intracellular domain and the mutual autophosphorylation of each monomer [29]. PI3K kinase consists of two domains: catalytic P110 and regulatory P85, which can be activated by RTK or indirectly by adapter molecules (including the insulin receptor substrate IRS, or the GTP-binding protein RAS) [29,30]. Activated PI3K phosphorylates the hydroxyl group in the third position of the inositol ring of phosphatidylinositol [31]. Phosphatidylinositol-4,5-bisphosphate (PIP2) and phosphatidylinositol-3,4,5-trisphosphate (PIP3) are small phospholipid components of cell membranes and important signaling transductors. In the PI3K/Akt pathway, the 3-position PIP3 phosphate group can bind to both phosphoinositide-dependent kinase-1 (PDK1) and Akt (also known as PKB (protein kinase B)) and recruit the Akt protein in the cell membrane, enabling PDK1 to access the PKB [32]. Akt/PKB is a highly conserved serine/threonine kinase. It plays a key role in many cellular processes and serves a dominant role in the signal transduction of the entire PI3K pathway. The amino acid structure of Akt, from the N-terminus to the C-terminus, consists of three recognizable domains: a pleckstrin homology (PH) domain, the central catalytic domain, and the carboxy-terminal regulatory domain. The PH domain acts as a mediator in membrane translocation following Akt activation, whereas the catalytic domain binds ATP. Complete activation of Akt requires phosphorylation by PDK1 at the T308 site [33] and by the mammalian target of rapamycin complex 2 (mTORC2) [34,35] or DNA-dependent protein kinase (DNA-PK) [36] at the S473 site. Activated Akt is transported from the cell membrane to other regions of the cell to phosphorylate its substrates, and consequently, either to suppress or enhance their activity. Therefore, Akt mediates diverse important cellular processes, including cell growth and proliferation, cell survival, and gene expression, as detailed in later sections of this article [37]. Its three isoforms are distinguished, namely Akt1, Akt2, and Akt3 (also known as PKBa, PKBb, and PKBc, respectively) [38,39]. Their expression varies between tissues and disease states [40,41].

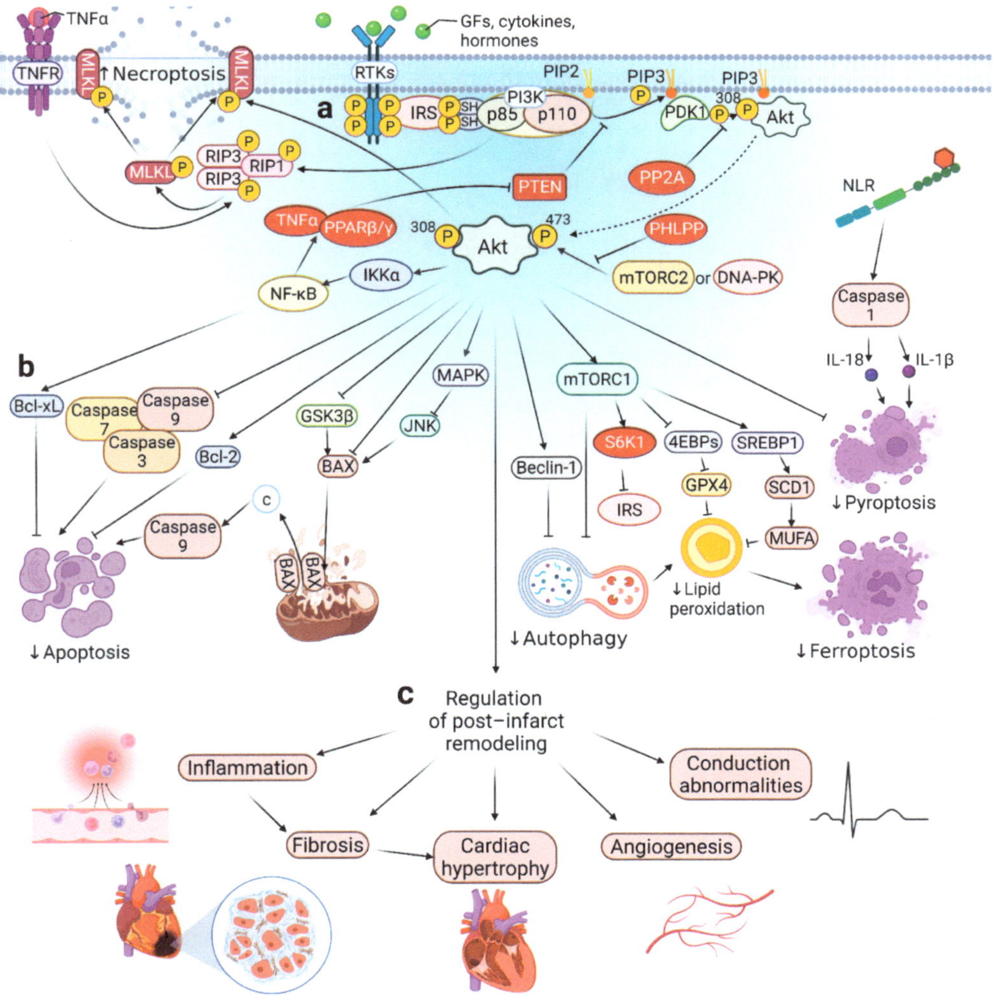

Figure 1. Schematic of components and mechanisms of PI3K/Akt activation and regulation (**a**). PI3K/Akt pathway regulates cell death during MI (**b**). PI3K/Akt signaling is a key component in the

regulation of numerous processes in the peri-infarct period and post-infarction remodeling of the left ventricle (c). Akt—protein kinase B, Bax—Bcl-2-associated X protein, Bcl-xL—B-cell lymphoma-extra large, Bcl-2—B-cell lymphoma 2, c—cytochrome c, DNA-PK—DNA-dependent protein kinase, GFs—growth factors, GPX4—glutathione peroxidase 4, GSK3β—glycogen synthase kinase 3 β, IKKα—IκB kinase α, IL-1β, IL-18—interleukins 1β and 18, IRS—insulin receptor substrate, JNK—c-Jun NH_2-terminal kinase, MAPK—mitogen-activated protein kinase, MUFA—monounsaturated fatty acids, MLKL—mixed lineage kinase domain-like protein, mTORC1, mTORC2—mammalian target of rapamycin complex 1 and 2, NF-κB—nuclear factor κ-light-chain-enhancer of activated B cells, NLR—NOD-like receptor, P—phosphate, PDK1—phosphoinositide-dependent kinase-1, PHLPP—pleckstrin homology domain leucine-rich repeat protein phosphatase, PIP2—phosphatidylinositol-4,5-bisphosphate, PIP3—phosphatidylinositol-3,4,5-trisphosphate, PI3K—phosphoinositide 3-kinase, PPARβ/γ—peroxisome proliferator-activated receptor β/γ, PP2A—protein phosphatase 2, PTEN—phosphatase and tensin homolog, RIP1, RIP3—receptor-interacting protein 1 and 3, RTK—receptor tyrosine kinase, SCD1—stearoyl-CoA desaturase-1, SREBP1—sterol regulatory element-binding protein-1, S6K1—ribosomal S6 kinase-1, TNFα—tumor necrosis factor α, TNFR—tumor necrosis factor receptor, 4EBPs—eukaryotic translation initiation factor 4E-binding protein 1.

2.2. Regulation of the PI3K/Akt Signaling

Due to the importance of this pathway, its activity must be carefully regulated. Down-regulation of the PI3K/Akt pathway may be achieved via two mechanisms. Firstly, phosphatase and tensin homolog (PTEN) may specifically dephosphorylate the 3-phosphate of the inositol ring in PIP3 to PIP2 and therefore reduce the concentration of PIP3 required for PI3K activation [42–44]. Inactivation of PTEN results in the constitutive activation of Akt and the mammalian target of rapamycin complex 1 (mTORC1), which is an evolutionarily conserved regulator of translation and ribosome biogenesis. Therefore, it leads to the disturbance of the size of the cells and growth regulation [45]. Another mechanism is dephosphorylation of Akt either at Thr308 by protein phosphatase 2A (PP2A) or at Ser473 by pleckstrin homology domain leucine-rich repeat protein phosphatase (PHLPP) [46]. The PI3K/Akt pathway itself also has feedback mechanisms. Akt phosphorylates IκB kinase α (IKKα) at the Thr23 site, which in turn phosphorylates nuclear factor κ-light-chain-enhancer of activated B cells (NF-κB). Activation of NF-κB regulates peroxisome proliferator-activated receptor delta (PPARβ/δ) agonists and tumor necrosis factor α (TNFα), which in turn repress PTEN expression as positive feedback. In addition, a negative feedback loop is initiated by mTORC1 and ribosomal S6 kinase-1 (S6K1) activation, which phosphorylates the insulin receptor substrate (IRS-1) and prevents its binding to RTKs, which results in the suppression of PI3K.

2.3. Impact of PI3K/Akt on Peri-Infarct Processes

The activity of the described pathway has an impact on many processes associated with the peri-infarct period, such as myocardial fibrosis, and other components of post-infarct left ventricular remodeling (Figure 1b,c). All of them will be described one-by-one in the context of PI3K/Akt. In addition, the influence of some comorbidities and commonly administered drugs in all of these processes is discussed.

3. Activity of the PI3K/Akt Pathway in Myocardial Infarction

Both the physical obstruction of a coronary vessel and the redistribution of blood flow reduce the oxygen supply and are responsible for irreversible damage of the cardiomyocytes [47]. Necrosis has been recognized as the main pathway of cardiomyocyte death during MI. However, the role of programmed types of cell death in ischemic and post-infarct heart injury has gained attention over the past two decades. Their activity determines the size of the infarcted area, and consequently, impairs the functions of the myocardium [48]. Inhibition of these processes, both pharmacologically and genetically, could improve cardiac functions. Moreover, the recently common division of cell death into programmed apoptosis and unregulated necrosis is slowly being turned into the field of study of MI. Recent studies also indicate necroptosis, ferroptosis, pyroptosis, and parthanatos as programmed types of cell death, also involved in MI pathogenesis [49–51]. All these processes ultimately result in the loss of the integrity of the cell membrane and the inflammatory response triggered by the release of the cytoplasmic content into the environment. The PI3K/Akt pathway appears to be directly involved in many of these processes (Figure 1b). For this reason, its proper regulation may contribute to reducing the damage caused by MI. The sections below describe these impacts in detail.

3.1. Necrosis

Necrosis is the unprogrammed type of cell death that is responsible for infarct size to the greatest extent. The beginning of necrosis is the imbalance between myocardial demand for oxygen and blood supply during MI, which causes the activation of anaerobic metabolism. This leads to the intracellular accumulation of H^+, which arises from anaerobic glycolysis. In addition, the lack of ATP causes the dysfunction of the Na^+/K^+ ion pump. The increased level of H^+ causes the need for their removal by the Na^+/H^+ exchanger. In response to elevated levels of Na^+ (associated with the Na^+/K^+ ion pump failure and the Na^+/H^+ exchanger) inside of the cell, the Na^+/Ca^{2+} exchanger operates in reverse mode and causes an increase in Ca^{2+} concentration. Elevated levels of all of these ions cause the swelling of the cell and promote the rupture of the cell membrane and necrosis. In one basic study, it was shown that pretreatment with HDL can protect the cardiomyocytes against oxygen and glucose deprivation-dependant necrosis. In addition, the authors brought about the inhibition of the PI3K/Akt pathway which abolished the protective effect of HDL [52]. Moreover, the PI3K/Akt pathway is associated with the generation of reactive oxygen species (ROS) which enhance necrosis [53–55]. There is evidence from numerous studies on cancer cells that disturbed signaling within this pathway contributes to increased levels of ROS. This may occur both directly through the modulation of mitochondrial bioenergetics and the activation of NADPH oxidases (NOXs), and indirectly, where ROS is produced as a metabolic by-product [54,56–59]. Necrosis is involved in most of the myocardial post-infarct changes. Due to cell membrane rapture, it activates and promotes inflammation in the ischemic heart, which is the inducer of other processes such as fibrosis, hypertrophy, and hemodynamic dysfunction, which are responsible for the adverse remodeling of the left ventricle.

3.2. Programmed Type of Cell Death

3.2.1. Apoptosis

Cardiomyocyte apoptosis can be induced by the intrinsic pathway in response to DNA damage, increased ROS and cytosolic calcium levels, or by the extrinsic pathway as a consequence of activation of sarcolemmal death receptors (FAS or TNFα) [60]. During MI, the PI3K/Akt pathway activity downregulates the expression of numerous proapoptotic molecules, which are briefly described in this section [61–63]. Their increased expression is associated with a higher activity of apoptosis and an increase in the number of cells dying as a result of this type of cell death. For instance, Akt phosphorylates cysteine-aspartic proteases (caspases) such as caspase-3, caspase-7, and caspase-9 (at Ser196). This prevents a caspase cascade, leading to cell death [64].

Moreover, the PI3K/Akt pathway upregulates the expression of the anti-apoptotic molecule B-cell lymphoma 2 (BCL-2), which blocks the process of the formation of the mitochondrial pore by which cytochrome c is released. Consequently, once PI3K/Akt signaling is active, the release of cytochrome c, which induces apoptosis by the intrinsic pathway, is reduced [65,66]. Moreover, Kim et al. showed in a cell culture that Akt kinase can decrease the activity of apoptosis signal-regulating kinase 1 (ASK1) [67]. ASK1 is a mitogen-activated protein kinase (MAPK) and can be activated by multiple cytotoxic stressors [68]. Akt phosphorylates ASK1 at Ser83 and thus inhibits it. This consequently suppresses the activity of c-Jun N-terminal kinase (JNK) and activating transcription factor 2 (ATF-2) in intact cells, which finally results in the inhibition of apoptosis induced by ASK1 [67].

In addition to the abovementioned, Akt also phosphorylates molecules such as BCL-2 associated agonist of cell death (BAD) and glycogen synthase kinase 3 beta (GSK3β), which results in the decreased activity of apoptotic pathways [69]. Blume-Jenes et al. [70] revealed in cell line studies that PI3K/Akt can also regulate cell death of its downstream factors. For example, it phosphorylates Ser136 residues of the BAD protein, a pro-apoptotic protein of the BCL-2 family, thereby depriving the apoptotic complex of its function. This causes translocation from the mitochondrial membrane to the cytosol and consequently negatively regulates the pro-apoptotic activity of Bax. Interestingly, Akt can also phosphorylate BAD at other sites, such as Ser99, Ser75, and Ser118 [71–73]. Phosphorylation of GSK3β on a highly conserved N-terminal regulatory site at Ser9 contributes to both myocardial necrosis (by mitochondrial permeability transition pore (mPTP) opening) and apoptosis (via various mechanisms including phosphorylation of Bax and destabilization of pro-survival beta-catenin) [74,75]. Bax can also be phosphorylated at residue S184 directly by Akt, which inhibits its conformational change for mitochondrial membrane distribution [76,77].

Statins can reduce apoptosis through the PI3K/Akt pathway. It was shown in animals that pitavastatin encapsulated in nanoparticles can activate this pathway and significantly reduce the TUNEL-positive cells in comparison with pitavastatin alone and a placebo. The protective effect of this lipid-lowering drug was diminished by a PI3K/Akt pathway inhibitor (wortmannin) [78]. In addition, it was shown that this pathway can be important in some pleiotropic effects of statins, such as improvement of endothelial function (defined as increased NO production), increases in proliferation and migration, and reduction in apoptosis of cardiac microvascular endothelial cells [79]. Additionally, angiotensin-converting enzyme inhibitors (ACEIs) may have a positive impact on cardiomyocytes. One of the ACEIs (benazepril) reduced apoptosis activity in an in vitro model of doxorubicin cardiotoxicity (H9c2 cell line). This positive effect was mediated by the activation of the PI3K/Akt pathway. It was shown that benazepril treatment restored the phosphorylation of Akt reduced by the doxorubicin therapy. In addition, the administration of an Akt inhibitor diminished the cardioprotective effect of this ACEI. [80]. Due to the fact that ACEIs are commonly used drugs among all patients after MI, an experiment which assesses the role of this group of drugs in apoptosis regulation in the context of the PI3K/Akt pathway is needed. The reduction in apoptosis activity and the improvement of the hemodynamic function after MI was also reported as an effect of eplerenone and bisoprolol [81,82]. Furthermore, the combination of drugs may have an even better impact on cardiomyocyte survival than therapy with a single substance. It was shown in animals that a combination of rosuvastatin with carvedilol significantly reduced the increase in biomarkers of MI (such as troponin and CK-MB). The authors observed the simultaneous activation of the PI3K/Akt pro-survival pathway [83].

Moreover, Liu et al. revealed that PI3K/Akt/nuclear factor signaling erythroid 2-related factor 2 (Nrf2), by upregulating the expression of heme oxygenase-1 (HO-1), can protect H9c2 cardiomyocytes from ischemia-reperfusion injury (IRI)-induced apoptosis [84]. They administered hydroxysafflor yellow A (HSYA), a chemical compound with a previously noticed protective effect [85,86], during reperfusion. This resulted in the increased expression and activity of HO-1, Akt phosphorylation, translocation of nuclear factor Nrf2, and, consequently, a decrease in apoptosis. Inhibition of PI3K by LY294002 abolished these positive effects.

In addition, Feng et al. showed in their study on mice that ischemia induces the expression of PTEN and the levels of this PI3K/Akt inhibitor remain elevated in the post-infarct period [65]. They selectively inhibited PTEN with bisperoxovanadium 5-hydroxipyridine-2-carboxylic acid (BPV(HOpic)) (1 mg/kg) in mice with MI induced by the ligation of the left anterior descending artery (LAD), and observed that BPV treatment promoted angiogenesis and reduced cardiomyocyte apoptosis, resulting in reduced infarct size. Taken together, such improved cardiac function is a result of PI3K/Akt pathway activation, which leads to a reduction in the number of cardiomyocytes undergoing apoptosis in the infarcted tissue compared with the myocardium where this pathway is inhibited [61,62,65]. Additionally, Wang et al. showed on a swine model that both mRNA and the protein level of PTEN increase after ischemia. This process was attenuated by atorvastatin treatment, which confirms that this type of lipid-lowering drug can have a positive impact on the survival of cardiomyocytes [87].

3.2.2. Necroptosis

Necroptosis is associated with both physiological and pathological processes, including embryonal development, inflammatory reaction, and IRI [50,88]. In addition, it is one of the key pathways in the loss of functional cardiomyocytes during MI [89,90]. The main stimulus of necroptosis is TNFα, which once combined with its receptor (TNFR) leads to the activation of the pathway consisting of receptor-interacting serine/threonine-protein 1 (RIP1) and RIP3 and the mixed kinase domain-like protein (MLKL) pathway [91,92]. PI3K/Akt signaling appears to have a key role in promoting this process by phosphorylating and oligomerizing RIP1, RIP3, and MLKL in response to TNFα stimulation [93,94]. Inhibition of both PI3K (its catalytic subunit p110α) and Akt in mouse fibrosarcoma L929 led to the inhibition of necroptosis by the suppression of the above mediators [95].

Tuuminen et al. (2016) showed that simvastatin may affect the necroptosis activity. Pretreatment of both rat donors and recipients of the transplanted heart with simvastatin can reduce RIP kinase-1 and kinase-3 activity in comparison with the placebo. In addition, it reduced expression of caspase-3 and caspase-9 involved in apoptosis [96]. Therefore, inhibition of PI3K may prevent necroptosis and increase cell viability and heart function in MI patients.

3.2.3. Ferroptosis

Ferroptosis is a relatively recently discovered iron-dependent regulated type of cell death with growing importance in MI pathogenesis, especially in its early and middle stages [97–100]. The PI3K/Akt pathway has been shown to play a significant role in its regulation [101].

For instance, Sun et al. [102] demonstrated in a cardiomyocyte (H9c2) cell culture that PI3K/Akt activation has a cardioprotective effect by reducing oxidative stress and inhibiting ferroptosis induced by doxorubicin (DOX) and lapatinib (LAP) [103,104]. These substances inhibit both PI3K and Akt phosphorylation and alter the mitochondrial membrane potential, reduce ATP, and increase the level of cytochrome C. Activation of the PI3K/Akt pathway using 30 μM 740Y-P resulted in a reversal of the deleterious effects of DOX and LAP and increased survival of the H9c2 cardiomyocytes [105,106]. Moreover, Yi et al. indicated that PI3K/Akt signaling inhibits ferroptosis in cancer cells via a pathway consisting of mTOR complex 1 (mTORC1), sterol regulatory element-binding proteins 1 (SREBP1) and stearoyl-coenzyme A desaturase 1 (SCD1), where SCD1 finally leads to the production of monounsaturated fatty acids [107]. Interestingly, they revealed that lipogenesis appeared to protect cells from oxidative stress and ferroptotic death.

Studies on the impact of the PI3K/Akt pathway on ferroptosis and cardiomyocyte viability are limited [107,108]. Recently, Jiang et al. published results of a bioinformatic analysis that revealed that the expression of 17 ferroptosis-related genes can be associated with the presence and development of ischemic and idiopathic cardiomyopathy. These genes were mainly involved in the regulation of apoptosis, cellular response to FGF stimulus, and response to some unspecified drugs by the MAPK and PI3K/Akt pathways. Moreover, the authors proposed some drugs and substances which can potentially be used in treatment [109]. This study provides useful information for further preclinical studies evaluating the role of PI3K/Akt signaling in the regulation of ferroptosis during MI and post-infarction HF.

3.2.4. Pyroptosis

Pyroptosis is a death pathway that begins with the activation of one of the NOD-like receptors (NLRs), which then leads to the activation of caspase-1 that finally activates pro-inflammatory cytokines, including interleukin-1β (IL-1β) and interleukin-18 (IL-18) [110]. As levels of these molecules have been found to increase during MI, inducing inflammation in the myocardium, regulation of this pathway may result in a degree of damage to the cardiomyocytes during infarction [111]. Overactivation of this highly inflammatory pathway negatively affects the infarct area and impairs myocardial contractility [112–114]. Although this type of cell death is most induced after infection with intracellular pathogens, its important effect has been noted during IRI.

Recently, Gio et al. [111] indicated that the radioprotective 105 kD protein (RP105)/ PI3K/Akt pathway is directly involved in the regulation of pyroptosis. After the administration of piperine to rats, they observed the inhibition of miR-383, which led to the activation of RP105/PI3K/Akt signaling. A consequence of the observed increased activity of the PI3K/Akt pathway was a decrease in pyroptosis occurring in the myocardium, and thus a greater survival of cardiomyocytes [115–117].

In addition, Do Carmo et al. [113] observed that administration of VX-785, a clinically available, highly selective caspase-1 inhibitor, reduced the infarct size in rats during analysis performed on Langendorff-perfused rat hearts. Moreover, the administration of a PI3K inhibitor—wortmannin—abolished all protective effects. This is more proof of the importance of the PI3K/Akt pathway in the protection of cardiomyocytes against pyroptosis.

Although pyroptosis is a recently known and still little-understood pathway of cell death, the proven influence of PI3K/Akt in its course offers further potential therapeutic targets in alleviating MI-induced death of the cardiomyocytes. However, further investigation is needed.

3.3. Ischemic Conditioning

The development of a novel strategy to limit infarction is of great clinical importance because the main predictor of a patient's prognosis is infarct size. One of the special interests is ischemic conditioning (IC)—a leading paradigm of cardioprotection [90]. IC means induction of short periods of myocardial ischemia and reperfusion before the onset of MI (preconditioning) or during reperfusion (postconditioning). Applying these cycles at a remote site is referred to as remote ischemic conditioning (RIC). In contrast to ischemic preconditioning, whose beneficial effects may be considered for patients with pre-infarction angina, ischemic postconditioning can be applied for patients undergoing PCI [90]. Apart from mechanical interventions, certain groups of drugs, such as volatile anesthetics and G protein-coupled receptor (GPCR) agonists, can also initiate pharmacological IC [118]. Moreover, exercise and pre-infarction angina can also promote transduction signals of IC, which can reduce infarct size and improve left ventricular function [119].

PI3K/Akt signaling is considered to be the main pro-survival kinase cascade mediating the IC-induced protective effect by forming a RISK pathway parallel to MEK1-ERK1/2 [90,120]. PI3K activity is required during both the pre-ischemic trigger phase and the post-ischemic mediator phase of IC to reduce infarct size and mortality [121,122]. In addition, IC increases the levels of Akt phosphorylation during both phases of IC. It plays a pivotal role as a mediator of IC and its pharmacological inhibition abolishes the infarct size either during the trigger phase or at reperfusion [123,124]. However, Barsukevich et al. showed that the role of the PI3K and RISK pathways may depend on the time of applied ischemia in the case of postconditioning. Immediate IC (10 s of reperfusion) increases phosphorylation of PI3K-AKT and ERK1/2, while early or delayed IC (applied after 10 min or 30 min of reperfusion) had no effect on phosphorylation in the rat model [125]. Nevertheless, postconditioning applied for 10 s, or 10, 30, 45, or 60 min after the onset of reperfusion retains the cardioprotective function, which is possibly dependent on a different mechanism. Another study showed that remote postconditioning increases phosphorylation of Akt and ERK1/2 in the early stages of reperfusion and then it gradually decreases [126].

Moreover, inhibition of PTEN may be important for increasing the activity of the PI3K/Akt signal in the course of postconditioning [127]. Furthermore, intermittent hypoxia increases the activity of the PI3K/Akt signal and enhances Ser473 Akt phosphorylation in the mouse model. It was associated with an increased capillary network in the myocardium and improved cardiac function, as well as reduced infarct size [128]. Administration of wortmannin (a PI3K inhibitor) reduces the level of Akt phosphorylation and its beneficial effects. This suggests that PI3K/Akt plays an important role in long-term preconditioning. An increasing number of studies indicates that extracellular vesicles (EVs) may play an essential role in mediating IC. EVs obtained from IC rats ameliorate IRI via activating the PI3K/Akt pathway. They decrease apoptosis and reduce the infarct size by increasing phosphorylation of PI3K and Akt [129]. Application of a PI3K inhibitor reverses these effects. Moreover, a study undertaken by Lessen et al. showed that EVs from the plasma of healthy human volunteers after RIC may not only provide cardioprotection via increasing mTOR expression, but also demonstrate accumulation in the damaged myocardium compared with sham-operated hearts [130]. Therefore, EVs may provide new therapeutic options for alleviating myocardial IRI via the PI3K/Akt pathway. All these observations highlight the importance of the PI3K/Akt pathway during MI and confirm its protective role.

4. The Role of the PI3K/Akt Pathway in Post-Infarction Left Ventricular Remodeling

Post-infarction cardiac remodeling is a maladaptive, complex, and multifactorial process of regional and global structural and functional changes in the myocardium, and presents as a common complication of acute MI [16,131,132]. In most studies, the authors assessed changes which involved the left ventricle. Thus, in the sections below, we refer to remodeling of this chamber of the heart. This ultimately results in cardiac hypertrophy (CH) and heart failure (HF) [13,133]. Such modifications occur initially as a consequence of the loss of a viable myocardium and abrupt increase in loading conditions [134]. Then, the course of change depends on a plethora of determinants, including the size and location of the necrosis, the timing and efficacy of reperfusion, the exuberant inflammatory response, the increased wall stress in the border zone, and the remote myocardium, neurohormonal activation, and dysregulation of transcription [135–140]. All of these factors affect each other and create a vicious circle that leads to the gradual deterioration of heart function. There are numerous known causes that influence the course of this phenomenon after MI, and the most important clinically include arterial hypertension (HT), obesity, diabetes mellitus (DM), and ischemic heart disease, among others. The PI3K/Akt pathway plays a crucial role in post-infarction left ventricular remodeling (Figures 1c and 2). Its activity provides cardioprotection and promotes the repair and healing of myocardial cells after MI [141]. Chen et al. showed that both PI3K and Akt activity were increased after MI in both human and mouse hearts. Moreover, the systemic as well as cardiac cell-specific inhibition resulted in greater cardiovascular risks and an increase in mortality of the mice. These results were associated with enhanced myocardial apoptosis and inflammation, reduced angiogenesis, and adaptive hypertrophy [142]. In the study undertaken by Feng et al., activation of the PI3K/Akt pathway by the PTEN inhibitor improved cardiac function 14 days after MI in mice. The reduced cardiomyocyte apoptosis promoted angiogenesis and activated the PI3K/Akt/vascular endothelial growth factor (VEGF) signaling pathway which resulted in significantly increased left ventricular fraction (LVEF), + dp/dtmax, and pressure–volume loops in the LV, as well as decreased left ventricular end-diastolic pressure (LVEDP). The contribution of the PI3K pathway to each of the stages of post-infarct remodeling and the impact of diabetes, is presented below.

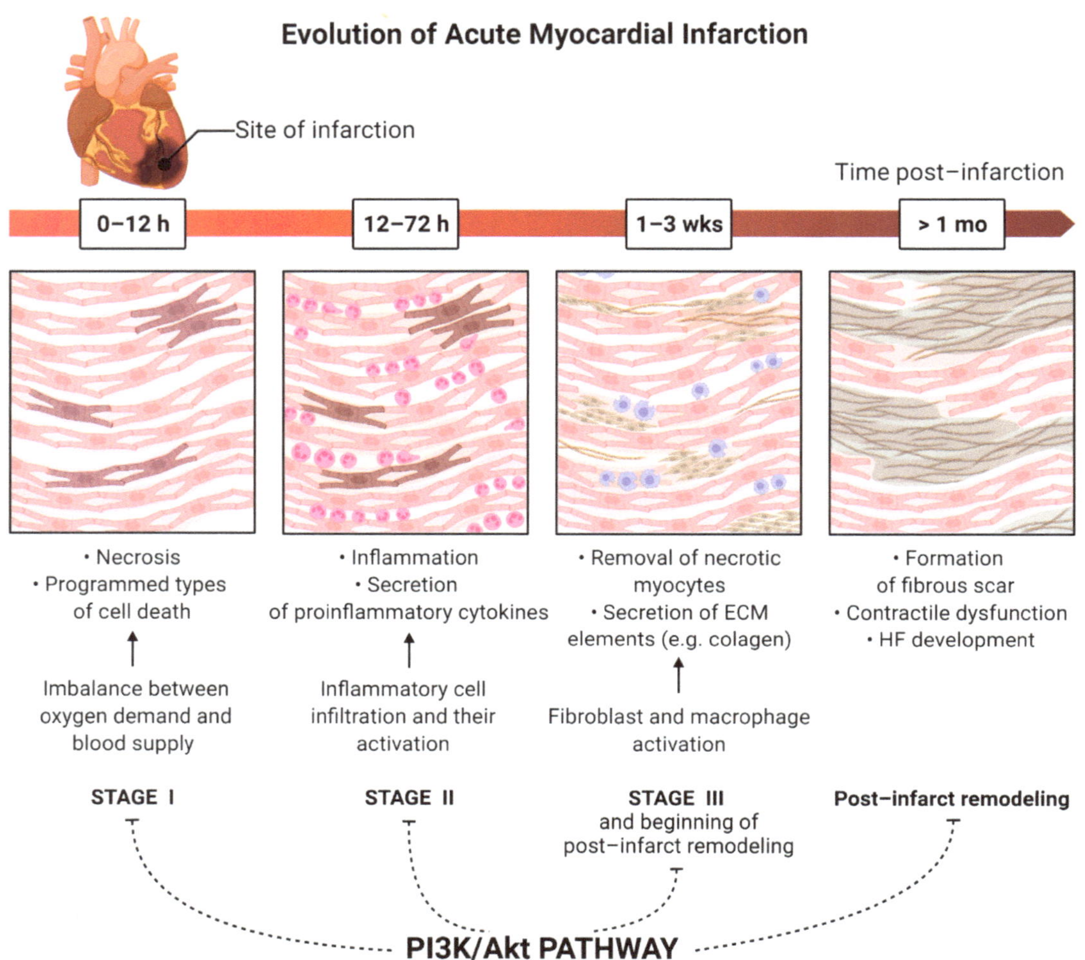

Figure 2. The processes regulated by the PI3K/Akt pathway which are involved in the histologic evolution of acute myocardial infarction. Stage I—histologic changes and processes directly induced by ischemia, stage II—beginning of inflammatory process, stage III and post-infarct left ventricular remodeling—irreversible histologic changes in the myocardium such as collagen secretion. ECM—extracellular matrix, HF—heart failure.

4.1. Inflammation

Injury of the cardiac cells and the extracellular matrix due to acute ischemia provides a strong systemic inflammatory response associated with increased production of pro-inflammatory cytokines [143]. These cytokines released during MI affect both the area of necrosis and the surrounding tissues and determine the course of post-MI left ventricular remodeling [144]. Both excessive and prolonged inflammatory reaction during MI are associated with an unfavorable prognosis for the patient. This process promotes myocardial damage, remodeling, and dysfunction of the left ventricle. As a large body of existing evidence from basic research and clinical trials indicates, this overactivity contributes to conditions such as chronic cardiac dilatation, left ventricular systolic dysfunction (LVSD), and HF [145–148]. There are multiple pathways involved in the course of post-MI inflammatory response, and PI3K/Akt signaling is one of the key ones [149]

C-reactive protein (CRP) is the most reactive serum protein of acute-phase inflammation and one of the most important prognostic biomarkers of atherosclerosis and cardiovascular disease (CVD) [150,151]. It affects the cell cycle and the inflammatory process of the cardiomyocytes [152,153]. CRP concentration increases significantly in the first hours of MI [131,154], and its level is a clinically important predictor of the course of left ventricular remodeling and patient prognosis [155,156]. Importantly, apart from inflammation, CRP is also a mediator of inflammation with prothrombotic and proapoptotic properties [157,158], which further worsens the prognosis of patients. Boras et al. noted that CRP in combination with a Notch-3 activator promotes angiogenesis in bovine aortic endothelial cells via the PI3K/Akt pathway [159]. Notably, pharmacological blockade of the PI3K/Akt survival pathway by LY294002 completely inhibited angiogenesis induced by CRP/Notch-3. Similarly, Chen et al. also noticed that CRP, through Akt activation, stimulates angiogenesis [160]. They revealed that the mechanism for this is the increase in the expression of vascular endothelial growth factor-A (VEGF-A) in activating hypoxia-induced factor-1α (HIF-1α) in adipose-derived stem cells (ADSCs).

In contrast, Tanigaki et al. in their studies on mice (C57BL/6) showed that CRP attenuates insulin-induced Akt phosphorylation in endothelial cells [161]. Insulin promotes the protective cardiovascular endothelial functions by activating Akt, which then phosphorylates endothelial NO synthase (eNOS) in Ser1179, stimulating the production of cardioprotective NO [162]. CRP seems to impair this signaling and therefore increases the risk of CVD. Lee et al. noted that the reduction in the activity of the PI3K/Akt pathway mediated by CRP is related, inter alia, to the increase in the protein and mRNA levels of PTEN induced by CRP [163]. To sum up, the CRP protein and the PI3K/Akt pathway are undoubtedly strongly linked, and together play a significant role in regulating the inflammatory process. However, as the results of the direction of this interaction are inconclusive and there is insufficient research on these mechanisms during MI, there is a need for further research in this field.

In addition to CRP, the PI3K/Akt pathway also regulates the inflammation via other mechanisms. Parajuli et al. demonstrated that PTEN regulates the expression of important pro-inflammatory cytokines through the PI3K/Akt/IL-10/TNF-α signaling pathway and thus plays a significant role in post-MI remodeling of the myocardium in mice [164]. They indicated that inactivation of PTEN results in the induction of the inflammatory process after MI by reducing the expression of TNF-α and matrix metalloproteinase-2 (MMP-2), and by increasing the production of IL-10. In turn, the overexpression of PTEN can be observed during MI, which inhibits the PI3K/Akt pathway and thus causes the opposite effect on IL-10, TNF-α, and MMP-2 expression, and in addition increases leukocyte infiltration in the myocardium and increases cardiomyocyte mortality. Yuanji Ma et al. showed that the PI3K/Akt/Nrf2 pathway plays a role in attenuating inflammatory cells [165]. They showed that the stimulation of Akt phosphorylation with atorvastatin during angiotensin 2-induced oxidative stress activated Nrf2 in bone marrow-derived dendritic cells (BMDCs). As a result, it inhibited their maturation, which consequently promoted antioxidant and anti-inflammatory responses. Moreover, the use of the PI3K inhibitor LY2994002 abolished these positive effects [84,166,167].

It seems that the PI3K/Akt pathway may also play a role in other mechanisms of the inflammatory response. For example, the latest reports suggest that it variously mediates the signaling of interleukin-1 (IL-1), an important pro-inflammatory cytokine [168–170]. However, as these studies are sparse and have been performed on tissues other than the myocardium, it is not possible to draw direct conclusions. In recent years, clinical trials with IL-1 inhibitors (e.g., with canakinumab and anakinra) have been promising, and suggest a reduction in the incidence of cardiovascular complications among post-MI patients [171–173]. Therefore, more research is needed to better understand the mechanisms of these interactions in the context of PI3K/Akt.

The drugs used after MI onset can modulate the inflammation process. Among numerous positive effects of statin therapy, they can inhibit the expression of pro-inflammatory cytokines, such as TNF-α, by direct activation of nuclear factor (NF)-κBα [174]. Moreover, the downexpression of the PI3K/Akt pathway can enhance the immune system activation [174,175]. Thus, regulation of the PI3K/Akt pathway can be used to reduce the inflammation and infiltration of the myocardium by immune cells, consequently improving post-MI left ventricular remodeling [176].

Taken together, the results of experimental and clinical trials on anti-inflammatory strategies targeting the PI3K/Akt pathway in MI patients are promising regarding mitigating the effects of infarction. For this reason, there is undoubtedly a need for further research, both on existing and potentially new therapies.

4.2. Autophagy

Autophagy is an evolutionarily conserved cellular process of degradation of both unnecessary and dysfunctional components of the cell, including mitochondria and long-lived macromolecules. Moreover, it ensures the availability of energy substrates and reduces oxidative stress that would otherwise promote cell death [177,178]. Therefore, it is an important regulator of cardiac homeostasis and function. Autophagy is a lysosome-dependent process that utilizes double-membrane structures (autophagosomes) as a form of intracellular transport, and enables the cell to degrade and recycle its components [179,180]. Autophagy preserves the structure and function of the cardiomyocytes under baseline conditions, and during infarction and may be induced by various factors associated with MI, such as nutrient deprivation, hypoxia, ROS, damaged organelles, and protein aggregates [181–183]. It could be of high importance in the peri-infarct zone. There are multiple significant mechanisms of autophagy regulation in the cardiomyocytes, the disruption of which impairs autophagy activation and exacerbates myocardial injury [183–185]. The PI3K/Akt pathway plays a significant role in some of them, as we describe in this section.

One of the key pathways is the PI3K/Akt/mTOR signaling pathway. mTORC1 is an important serine/threonine kinase and is an upstream repressor of autophagy. Kim et al. observed that mTORC1 in vitro inhibits autophagosome formation by phosphorylation of unc-51-like autophagy activating kinase 1 (Ulk1) at Ser757 [186]. In addition, mTORC1 also acts at the gene level. As Martina et al. noticed, it has been observed to inhibit the transport of nuclear transcription factor EB (TFEB), a regulator of autophagy and lysosomal biogenesis, and thus downregulates the transcription of specific autophagy-related (Atg) genes [187]. Under physiological conditions, autophagy is activated in cells under stress. In contrast, Sciarretta et al. revealed that Akt phosphorylates diverse substrates during MI, indirectly causing forced activation of mTORC1 in the myocardium [184,188]. This sequence of events leads to the inhibition of autophagy and fiercely intensifies ischemic injury [184]. Interestingly, besides Akt, mTORC1 can also be regulated via other cellular signaling pathways such as MAPK/ERK, JNK, and Wnt signaling [189–191].

Moreover, PI3K can also inhibit autophagy directly by phosphorylating Beclin-1, a core component of Beclin 1-PI3KC3 complex, a lipid–kinase complex playing a crucial role in autophagosome nucleation [192]. Phosphorylation of Beclin-1 enhances its interaction with BCL-2, and this Beclin 1–BCL-2 interaction not only inhibits autophagy but also dissociates BCL-2 from Bax, thereby activating Bax and stimulating apoptosis. Matsui et al. revealed that autophagy via Beclin-1 is particularly activated during reperfusion by the mass-produced ROS during that period [185]. In addition, in mice (C57BL/6J) studies, they showed that mice with systemic heterozygous Beclin 1 deletion display significantly reduced autophagy and ischemic injury. Interestingly, Beclin 1 can both positively and negatively regulate autophagy, hence the optimum concentration range of this molecule is needed for autophagy to function efficiently [193].

Via both the abovementioned pathways, PI3K/Akt signaling negatively modulates autophagy and therefore has a protective effect on the heart. Yan et al. also observed that in the pig heart, MI-induced autophagy could be a homeostatic mechanism which inhibits apoptosis and hence limits its destructive effects [182]. Taken together, stimulation of autophagy may have a protective effect on the heart [194,195], which may be another potential target of the therapy.

In addition, by regulation of autophagy and apoptosis, the activity of the PI3K/Akt/mTOR pathway contributes to myocardial remodeling following MI [196]. It may suppress the progression of HF and cardiac hypertrophy (CH). Pharmacological inhibition of autophagy (e.g., with rapamycin) exacerbates cardiac dysfunction and dilation in the chronic phase of MI [197,198]. Nevertheless, the activity of autophagy seems not to be sufficient to control the quality of proteins and organelles during MI. Consequently, misfolded proteins accumulate, and mitochondrial dysfunction develops in post-MI hearts. Regulation of key proteins in this pathway has been shown to reduce infarct size [199]. These observations suggest that autophagy may potentially contribute to the heart's healing process by mechanisms such as activation of the repairing process, angiogenesis, or promotion of cardiac regeneration. Thus, modulation of autophagy could be used to alleviate HF and CH after MI.

4.3. Fibrosis

Myocardial fibrosis is a repair process which involves multiple modifications in the interstitial myocardial collagen network, which results in the impairment of the cardiac structure and function. The replacement of the necrotic myocardium by connective tissue is the main mechanism of fibrosis during post-infarction left ventricular remodeling [200]. As cardiomyocytes are displaced with fibrous tissue, many detrimental effects occur. These include excitation–contraction coupling and the systolic and diastolic function of the heart, which can lead to HF [201]. The PI3K/Akt pathway can regulate fibrosis in several ways. Cardiomyocyte apoptosis initiates the fibrotic response, and this process may involve both immune modulation and paracrine signaling [202]. Therefore, the PI3K/Akt pathway performs a protective effect in post-infarction fibrosis by inhibiting the death of cardiomyocytes [203].

Activation of the PI3K/Akt pathway may significantly increase cardiac dysfunction when its inhibition has the reverse effect, and increases fibrosis in the infarcted area [66]. Indeed, it has been proved that the administration to animals of one of the PI3K/Akt inhibitors such as wortmannin or LY294002 reduces fibrosis and post-infarct remodeling [121]. It seems that Akt agonists or PTEN inhibitors may show a positive effect on post-infarct remodeling and improve a patient's outcome, but all the conducted studies have focused on the oncologic utility of this group of substances [65,204]. It seems that they may be used as a part of antifibrotic therapy after MI, but the assessment of their safety and tolerance is needed [204].

In contrast, Zhao et al. in their research noted that the active PI3K/Akt pathway may promote myocardial fibrosis [205]. The authors showed that the activation of this pathway by long noncoding RNA (LncRNA) myocardial infarction-associated transcript (MIAT) promotes the expression of inflammatory factors in the myocardium. Conversely, when LncRNA MIAT was silenced, the levels of vascular endothelial growth factor (VEGF), Akt, and PI3K levels were significantly downregulated. In addition, as this silencing resulted in a reduction in collagen expression, regulation of these interactions could potentially improve cardiac repair and consequently ameliorate HF.

One of the initial processes involved in myocardial remodeling is the activation of fibroblasts. It was shown that the response of cardiomyocytes to fibroblast growth factor (FGF) can be impaired after MI. The expression of dysregulated proteins can promote activation of fibroblasts and cause the accumulation of collagen types I and III, which are the major fibrins of the myocardial collagen matrix. This post-infarct remodeling of the interstitium, mainly of the left ventricle, is a major cause of cardiac hypertrophy and can be responsible for the development of HF after MI. It seems that PI3K/Akt activity may be involved in these processes [109]. Some drugs which are prescribed for patients after MI have an antifibrotic effect. ACEIs and mineralocorticoid receptor antagonists (MRAs) are well-known drugs that present this effect. However, there are no data showing that the

PI3K/Akt pathway is directly involved in this process. However, it has been shown that the stimulation of β-receptors can stimulate neonatal rat cardiac fibroblasts to engage in protein synthesis with a simultaneous increase in PI3K activity. This process may be involved in collagen production after MI, so it seems that the administration of β-adrenolytics can reduce the activity of fibroblasts and improve cardiac function among patients not only due to the positive-inotropic and anti-arrhythmic effect of these group of drugs [206].

4.4. Cardiac Hypertrophy

Hypertrophy of the myocardium occurs as a response to stress stimuli such as infarction and is an important component of left ventricular remodeling after MI. Such structural changes are associated with increased cardiomyocyte apoptosis and fibrosis and result in contractile dysfunction, dilatation, and consequently, the development of HF [207,208]. There are numerous triggers of cardiac hypertrophy, such as mechanical stress and humoral stimulation, which can lead to multiple metabolic responses [209]. The PI3K/Akt signaling pathway and its interaction with its downstream effectors are altered and play a key role in regulating this process [210].

Short-term activation of Akt promotes physiological hypertrophy, and despite leading to mild myocardial enlargement, it has a cardioprotective effect [211] by attenuating damage to ischemia in endothelial cells [212]. Nevertheless, long-term Akt activation induces pathological hypertrophy and HF [213]. In this case, PI3K/Akt increases angiogenesis, which was mentioned in the previous section, but these blood vessels are, however, unorganized and reminiscent of tumor vasculature [212]. In addition, Zhao et al. [205] in their research on human cardiac fibroblasts (HCF) and rats (C57BL/6) have noticed that the activation of the PI3K/Akt pathway by myocardial infarction-associated transcript (MIAT) promotes myocardial fibrosis and the expression of various inflammatory factors such as IL-1β, IL-6, and TNF-α mRNA, and numerous proteins that collectively contribute to the occurrence of HF. Muting their MIAT reduced the incidence of HF. Meng et al. [214], on the other hand, found that the PI3K/Akt pathway in mice (C57BL/6J) is involved in the development of hypertrophy by means of yet another molecule, the aforementioned BCL-2, as it maintains cardiomyocyte survival.

All the mentioned reports suggest that PI3K/Akt signaling plays a significant role in post-infarction myocardial hypertrophy, and that adequate regulation of this pathway could potentially be an important therapeutic target to reduce the incidence of HF. However, as these studies were performed only on an animal model, the effectiveness of this regulation needs to be evaluated in clinical trials.

4.5. Angiogenesis

As a constant supply of blood to cardiomyocytes is of fundamental importance for their proper functioning, an adequate vascularization system in the myocardium is essential. Angiogenesis is a complex process consisting of endothelial proliferation, cell migration, and eventually the formation of blood vessels. It has been noticed that PI3K/Akt plays a pivotal role in activating angiogenesis and in forming collateral circulation via, among others, regulating the transcription and expression of proangiogenic cytokines such as vascular VEGF, angiopoietin-1 (Ang-1), and bFGF (which in addition reverse-stimulates Akt phosphorylation) [215,216]. As these growth factors stimulate the formation of blood vessels, they improve the supply of oxygen and nutrients to the heart. Consequently, this activation has the potential to both rescue the myocardium at early stages after AMI and prevent subsequent ischemic-related heart failure [217].

Moreover, overexpression of PTEN (PI3K/Akt inhibitor) attenuates these positive effects and, for example, inhibits angiogenesis and contributes to thrombosis by inducing endothelial dysfunction [218,219].

Research shows that the downregulation of PTEN could promote angiogenesis through increasing the expression of VEGF as well as reinforcing the signal transduction of the VEGF-binding cell [220,221]. In addition, such PTEN inhibition results in an increase in the expression of the cluster of differentiation 31 (CD31), also known as platelet endothelial cell adhesion molecule (PECAM-1). This molecule participates in the creation of adhesive interactions between endothelial cells and adhesion receptors. This makes it possible to mitigate the impairing effect of MI on capillary density [65].

As angiogenesis has such a cardioprotective potential in the early stages after MI, therapies focused on the PI3K/Akt pathway have become a novel treatment strategy for patients after infarction [217,222].

4.6. Conduction Disturbances

Post-infarction severe arrhythmias, especially ventricular arrhythmias (VA), are the leading cause of death in MI patients. The most common arrythmias observed in patients with MI are ventricular tachycardia (VT, 17–21%), advanced and complete heart block (23–35%), ventricular fibrillation (VF, 24–29%), and atrial fibrillation (AF, 11–20%) [223]. Such arrhythmias are the result of sympathetic remodeling, a phenomenon of disturbance of the spatial distribution and density of the myocardial sympathetic innervation caused by denervation in the infarcted zone and hyperinnervation in the infarcted border zone [224,225]. Such a sympathetic remodeling may cause electrophysiological disturbances and increased heterogeneity of noradrenergic transmission, and consequently increase the risk of dangerous arrhythmias [226]. The key cytokine involved in sympathetic remodeling following MI is the nerve growth factor (NGF), part of the NGF/TrKA/PI3K/Akt pathway that regulates neuronal plasticity and survival [227,228]. PI3K/Akt signaling is essential for NGF and other neurotrophic activities [26]. Li et al. showed that the appropriate regulation of the PI3K/Akt pathway by inhibiting its excessive activity reduces the abnormal excitability, automatism, and conductivity of residual myocytes, which result in the initiation of arrhythmia after myocardial infarction in rats [229,230]. Taken together, targeting the NGF/TrKA/PI3K/Akt pathway has the potential to regulate factors associated with sympathetic remodeling and thus potentially reduce the incidence of post-MI arrhythmias. [230]. However, these positive effects were only noticed in preclinical trials, and further investigation is required in this field.

5. Diabetes

Experimental studies on both in vitro and animal models of DM indicates the higher susceptibility of cardiomyocytes to IRI. Infarct size in animal models seems to be enlarged, however some results are contradictory [231]. Moreover, DM is associated with significantly higher mortality one year after ACS compared with non-DM patients, and this partially results from more severe damage of the myocardium [232]. Clinical studies show that the median infarct size is larger in diabetic patients and reperfusion is impaired [233,234]. Moreover, we observe the altered expression of the PI3K/Akt pathway in the diabetic heart. Preclinical studies show that this may play a crucial role in cardiac susceptibility to IRI and impaired cardioprotection in diabetic patients [235].

The cardiomyocytes of the adult rat are exposed to increased intracellular oxidative stress and apoptosis in hyperglycemic and free fatty acid conditions. Treatment with ghrelin decreases the apoptosis rate by 25–44% and is associated with increased activation of the PI3K/Akt pathway. This promotes phosphorylation of both Akt and ERK1/2 and the nuclear translocation of NFκB, which promotes anti-apoptotic BCL-2 and Bcl-xL expression [236]. The antiapoptotic effect of ghrelin is reversed by treatment with wortmannin (a PI3K inhibitor). Therefore, hyperglycemia impairs the antiapoptotic effect of PI3K/Akt in cardiomyocytes and its promotion can protect cells. Activation of PI3K/Akt is also diminished in the hearts of Sprague Dawley rats with streptozotocin (STZ)-induced diabetes. It decreases the expression of p308-Akt, p473-Akt, p136-BAD, Bcl-2, and the Bax/Bcl-2 ratio when compared with the non-diabetic group. Treatment with luteolin increases the expression of fibroblast growth factor receptor 2 (FGFR2) and leukemia inhibitory factor (LIF), which results in the activation of the PI3K/Akt pathway. It is associated with reduced cardiomyocyte apoptosis, diminished infarct size, and preserved cardiac function after IRI [237]. Chen et al. showed that Sprague Dawley rats with DM induced by a high-fat diet combined with low-dose STZ have a decreased level of Akt phosphorylation, and it is not enhanced by IRI. In opposition, activity of the PI3K/Akt pathway significantly increases in non-DM rats, which is connected with smaller infarct size and a lower apoptosis rate when compared with the DM group after IRI. Inhibition of JNK activity can partially restore the function of the PI3K/Akt pathway, which results in decreased apoptosis and improved cardiac function in DM rats [238]. Using a similar model, An et al. showed that apelin treatment can increase PI3K activity in DM rats and this leads to increased expression of its downstream effector, impairing the Notch1/Hes1 pathway. This results in decreased expression of iNOS and an enhanced eNOS level, alleviated cardiomyocyte apoptosis and lower infarct size after IRI, and improved cardiac function 6 weeks after IRI [239]. The impact of the PI3K/Akt signal on the Notch1/Hes1 pathway and its beneficial effect on the diabetic myocardium after IRI was also presented by other researchers [240].

The impaired cardioprotective effect of the PI3K/Akt pathway in the diabetic heart also results in non-efficient ischemic conditioning. Tsnag et al. showed that hearts of diabetic Goto–Kakizaki (GK) rats require more cycles of preconditioning in order to experience a significant reduction in the infarct size than non-diabetic Wistar rats [241]. Even though one cycle of IC induces phosphorylation of Akt (Ser 473) compared with the control GK hearts (24.7 ± 5.0 arbitrary units (AU) vs. 6.2 ± 0.9 AU, respectively; $p < 0.05$) it is not related with a reduction in the infarct size. However, three cycles of IC induce higher levels of p-Akt (42.4 ± 3.2 AU; $p < 0.01$) and results in decreased infarct size compared to the control hearts (20.8 ± 2.6 vs. $46.6 \pm 5.2\%$; $p < 0.01$). Inhibition of PI3K signaling with LY294002 reduced the p-Akt level (20.8 ± 2.1 AU; $p < 0.01$) and abrogated the infarct size ($44.9 \pm 6.4\%$; $p < 0.01$). In non-diabetic Wistar rats, one cycle was sufficient to reduce the infarct size and significantly increase Akt phosphorylation (72.4 ± 7.5 and 74.2 ± 9.1 vs. 37.1 ± 5.7 AU, respectively; $p < 0.01$). Therefore, the threshold for cardioprotection provided by activity of the PI3K/Akt pathway is elevated in diabetic myocardium. DM inhibits preconditioning and postconditioning, affecting PI3K/Akt on many levels [122,242]. For example, activation of STAT3 is strongly reduced in STZ-induced diabetic rats, which results in the inhibition of PI3K/Akt activity and decreased cardioprotection [243]. This effect may be caused by high glucose-dependent oxidative stress, which abrogates remifentanil preconditioning in STZ-induced diabetic rats [244]. DM decreases phosphorylation of Akt on ser473 and STAT3 on Tyr705 and ser727, without influencing total Akt and total STAT3 expression at baseline. Moreover, it can be restored by antioxidant treatment with N-acetyl cysteine (NAC), by improving Akt and STAT3 activation, and by facilitating the cross talk between the PI3K/Akt and JAK2/STAT3 signaling pathways, resulting in attenuated myocardial IRI and post-MI LV dysfunction. Furthermore, sevoflurane postconditioning can be diminished by the inactivation of T-LAK cell-originated protein kinase (TOPK). This protein kinase can inactivate PTEN, thus inducing PI3K/Akt activity. Alteration of this pathway was observed in STZ-induced diabetic mice and cardiac cells exposed to high glucose in vitro [245]. Moreover, selective inhibition of PTEN may preserve the cardioprotective effect of ischemic postconditioning in STZ-induced diabetic rats [246]. Other studies indicate that impairment of PI3K/Akt signaling results in decreased activity of its downstream effector, GSK-3β, leading to ineffective cardioprotection [247,248].

Diabetes mellitus is also a strong predictor of post-MI HF development. Among patients with first anterior MI, DM was associated with a greater risk of cardiovascular death or rehospitalization for HF [249]. Other authors have also observed an increase in the left atrial volume index (LAVi) as well as left atrial enlargement at 20-month follow-up [250]. These studies indicate the long-term elevation in LV diastolic pressure post-MI among diabetic patients, which may be caused by adverse remodeling and might mediate the risk. DM is also an independent predictor of the higher risk of rehospitalization of HF among patients who underwent PCI ((HR) 1.576, $p = 0.010$) [251]. In a study by Akashi et al., DM was associated with a higher risk of major adverse cardiovascular events (MACEs) ((HR) 2.79, $p = 0.017$) and HF hospitalization ((HR) 3.62, $p = 0.023$) after STEMI, and left ventricular remodeling at the baseline was an independent risk factor [252]. Moreover, Yang et al. showed that insulin resistance is also a strong predictor of post-MI HF. They observed greater LV dilation after STEMI among patients with impaired fasting glucose (IFG), impaired glucose tolerance (IGT), and high HOMA-IR levels, who underwent primary percutaneous coronary intervention and were followed up for 12 months [253]. The mechanisms underlying the higher risk of HF after MI in diabetic patients remain unknown. However, we can observe similar changes in animal studies, and the PI3K/Akt pathway may be an important mediator.

Multiple studies show aggravated left ventricular remodeling and HF after MI in experimental diabetes. Hyperglycemia increases the apoptosis rate of cardiomyocytes and induces cardiac fibrosis, leading to left ventricular enlargement and dysfunction [254,255]. Moreover, Backlund et al. showed that the number of apoptotic cells in the border zone of infarction as well as in the non-infarction site is significantly higher among diabetic rats at 12 weeks after MI. Moreover, Thakker et al. observed an altered inflammatory response and healing process leading to adverse cardiac remodeling in diet-induced obesity and insulin resistance after myocardial IRI [256].

Modification of the PI3K/Akt expression may play an important role in lower survival rate, worse LV function, and aggravated myocardial fibrosis. Vahtola et al. showed that pronounced cardiomyocyte hypertrophy, increased fibrosis, and cardiomyocyte apoptosis in diabetic rats 12 weeks after MI were associated with decreased Akt activation and increased nuclear localization of FOXO3a (forkhead box O3) [257]. These changes resulted in the increased expression of PTEN and the inhibition of the cardioprotective properties of the PI3K/Akt pathway. Moreover, Akt played a key role in mediating the cardiac remodeling induced by EGFR stimulation in diabetic mice. This results in ROS generation and cardiac damage which can be decreased by EGFR inhibitors [258].

It is not only the diabetic state and hyperglycemia that can impact the activity of the PI3K/Akt pathway and influence the infarction size and post-infarction left ventricular remodeling, but also common antidiabetic drugs can do so.

Metformin has cardioprotective effects which are mediated by many pathways including PI3K/Akt. Bhamra et al. showed that rats receiving metformin had reduced infarct size in comparison with the placebo group ($35 +/- 2.7\%$ vs. $62 +/- 3.0\%$, $p < 0.05$). This protective effect was also observed in the animal model of DM ($43 +/- 4.7\%$ metformin vs. $60 +/- 3.8\%$ control, $p < 0.05$) [259]. The authors observed that it was accompanied by a significant increase in Akt phosphorylation. Furthermore, the addition of the PI3K inhibitor abolished the metformin-induced Akt phosphorylation and the protective effect [259]. Moreover, metformin can be effective in the treatment of diabetic cardiomyopathy. In an animal study, administration of metformin significantly reduced the expression of caspase-1 and nucleotide-binding domain, leucine-rich-containing family and pyrin domain-containing-3 (NLRP3) inflammasome, which resulted in the inhibition of pyroptosis as a consequence. Moreover, this antidiabetic drug can inhibit the expression of mTOR and reduce the activity of autophagy [260]. All this evidence shows that metformin has a cardioprotective effect.

Insulin activates PI3K and Akt which provide cardioprotection via the subsequent activation of many eNOS, mTOR, and reduction in ROS production. Administration of insulin stimulates eNOS to produce NO, one of the most important vasoprotective substances. In addition, it was observed that wortmannin (a PI3K inhibitor) can abolish this positive effect [261]. Another basic study showed that insulin can reduce apoptosis and necrosis by suppressing ROS production, which was also mediated by the PI3K/Akt pathway [262]. Moreover, insulin can activate mTOR kinase, which may inhibit inflammation and fibrosis, responsible for adverse left ventricular remodeling and hemodynamic dysfunction [263]. It seems that activation of the PI3K/Akt pathway is one of the main targets of insulin and the mechanism of the cardioprotective effect of this antidiabetic drug.

SGLT2 inhibitors are primary antidiabetic drugs which caused a revolution in the treatment of HF. The administration of these drugs significantly reduces mortality among patients with HF, even without diabetes [264]. The mechanism of action is still not fully understood. It seems that some of the positive effects can be mediated by the PI3K/Akt pathway. Hasan et al. showed that SGLT2 inhibitors can attenuate oxidative stress induced by isoprenaline by reducing inflammation [265]. They revealed that administration of canagliflozin can reduce fibrosis of the Long–Evans rat heart. It was associated with diminished reduction in Akt phosphorylation by the SGLT2 inhibitor. The authors simultaneously noted numerous changes in the expression of other proteins such as AMPK and eNOS. The significance of the PI3K/Akt pathway in cardioprotection should be confirmed.

Additionally, the GLP-1 mimetics can attenuate adverse left ventricular remodeling. Robinson et al. showed that the infusion of exendin-4 for 4 weeks after MI can improve survival and protect against cardiac dysfunction [266]. A broad spectrum of its activities decreases cardiomyocyte apoptosis, attenuates fibrosis, and inhibits myocardial inflammation by the modulation of the Akt/GSK3β pathway, among others.

To summarize, restoring the correct function of the PI3K/Akt pathway in the diabetic heart is a potential target for diminishing infarct size and improving cardiac function after IRI. The PI3K/Akt pathway may also play a significant role, and its dysregulation results in adverse remodeling, leading to cardiac dysfunction. Pharmacological treatment may improve the outcome for patients after MI, affecting the cardioprotection mediated by this pathway. This could improve the outcome and reduce post-MI mortality in diabetic patients.

6. MicroRNA

MicroRNAs are a short non-coding RNAs of about 22 nucleotides that can mediate gene silencing by guiding Argonaute (AGO) protein to target sites in the 3′ untranslated region of the mRNAs. MiRNAs can be secreted out of a cell and mediate cell communication in a paracrine and endocrine manner. They can be transported to target cells by binding to proteins, including AGO, or within extracellular vesicles t(EVs) [267]. Moreover, miRNA levels in circulation differ in various pathological states, including myocardial infarction and diabetes. This can be associated with the increased vulnerability of the heart to ischemic injury or the adverse post-MI remodeling of the left ventricle [268]. Below, we present miRNAs that can regulate cardiac cells' metabolism and target the PI3K/Akt pathway activity (Table 1), and describe the most interesting examples.

Expression of miR-21 is significantly increased in hearts subjected to IRI after 2 and 7 days of reperfusion compered to sham-treated hearts [269]. MiR-21 regulates activity of the PI3K/Akt pathway by targeting the 3′-UTR of a PTEN mRNA. Its inhibition results in elevated phosphorylation of Akt, which causes induction of matrix metalloprotease-2 (MMP-2) in cardiac fibroblasts in vitro [269]. MMP-2 is involved in the remodeling of the extracellular matrix and may lead to the impaired ventricular contraction [270]. Moreover, mir-21-induced activity of Akt signaling decreases the Bax/Bcl-2 ratio and expression of caspase-3, which can inhibit apoptosis of human H9C2 cardiomyocytes and lead to a reduction in the infarct size in the SD rats IRI model [271,272]. Furthermore, miR-21 induces the Akt/mTOR signaling pathway, which inhibits autophagic activity and alleviates apoptosis of H9C2 cells [273].

Interestingly, inhibition of PTEN by miR-21 mediates the cardioprotective effect of ischemic postconditioning and results in alleviated myocardial apoptosis, decreased infarct size and improved left ventricle function in mice [274]. Loss of miR-21 reverses all these beneficial effects and can also reverse the cardioprotective effect of isoflurane preconditioning, which is associated with increased phosphorylation of Akt and eNOS [275]. Moreover, mir-21 can be also transported inside EVs and therefore be a potential MI regenerative therapy target [276]. Interestingly, dysregulation of miR-21-5p within EVs from heart failure patients causes loss of their beneficial effects and impairs their ability to promote both angiogenesis and cardiomyocyte proliferation. In contrast, they exacerbate left ventricle remodeling and decrease LVEF in mice 3 weeks after intramyocardial injection [277]. Restoration of miR-21 expression in the EVs of HF patients rescues their cardioprotective effect. Therefore, miR-21 plays an important role in regulation of the PI3K/Akt signal in cardiomyocytes and may impact the heart's ischemic injury and remodeling.

MiR-126 plays an important role in angiogenesis and vascular homeostasis. It targets the 3′-UTR of Phosphoinositide-3-Kinase Regulatory Subunit 2 (PIK3R2), which is a gene encoding regulatory subunit p85β. Its inhibition results in up-regulated activity of PI3K and Akt signals. Song et al. reported that expression of miR-126 may be triggered by HIF-1α in endothelial cells of post-MI hearts, which can affect hypoxia-induced tube formation and angiogenic signaling. Adaptive exercise training seven days after MI induction can improve myocardial angiogenesis and cardiac function, increasing LVSP, +dp/dt max, and decreasing LVEDP compared with the MI SD rat group. This effect is associated with increased expression of HIF-1α and miR-126, and reduced expression of PIK3R2 and Sprouty-related protein 1 (SPRED1). Moreover, all these results can be attenuated by HIF-1a inhibitor. Further study on human umbilical vein endothelial cells (HUVECs) suggests that miR-126 may be involved in tube formation under hypoxia through the PI3K/AKT/eNOS pathway [278]. PIK3R2 is also a negative regulator of the VEGF pathway, and its inhibition by miR-126 promotes the function of endothelial progenitor cells (EPC) under hypoxic conditions. Overexpression of miR-126 enhances viability, migration and tube-forming ability of EPCs, which can play important role in vascular integrity and repair after MI [279].

Both obesity and diabetes downregulate miR-126 expression level (Table 1). Gomes et al. showed that this is associated with decreased capillary density in skeletal muscles of obese Zucker rats. Interestingly, exercise training restores miR-126 expression, and promotes angiogenesis by decreasing PI3KR2 activity and inducing VEGF and eNOS. It results in enhanced skeletal muscle capillary rarefaction [280]. Furthermore, miR-126 is downregulated in the EPCs from diabetic patients and impair their function by targeting SPRED1 and impairing Ras/ERK/VEGF and PI3K/Akt/eNOS signal pathways [281]. Moreover, downregulation of miR-126 may result in increased apoptosis of vascular endothelial cells and elevated myocardial susceptibility to IRI [282,283]. Therefore, miR-126 can play an important role in regulation of PI3K/Akt signal transduction during myocardial infarction and remodeling. It is a potential post-MI prognostic factor and target for miRNA-based therapeutic interventions for DM complications. Interestingly, the circulating level of miR-126 is downregulated in diabetic patients, what can be associated with the impaired cardioprotective function of the PI3K/Akt pathway. Furthermore, the above effects on angiogenesis may be also mediated by EVs-derived miR-126, and exosomes from miR-126-overexpressing mesenchymal stem cells (MSCs) may be a promising therapeutic strategy to promote angiogenesis and wound repair [284].

MiR-145 can also play a cardioprotective role, attenuating both myocardial ischemia–reperfusion injury and post-infarct remodeling via targeting the PI3K/Akt pathway. It inhibits cardiac cells' apoptosis, ROS activity and increases cells' viability under hypoxic conditions in vitro. These effects are positively related to the activation of PI3K/Akt and expression of SGK1 (serum- and glucocorticoid-regulated kinase (1)—a downstream effector of PI3K associated with cell survival [285]. Moreover, miR-145 promotes autophagy and reduces the infarct size in AMI models in vivo via targeting FRS2 (fibroblast growth factor receptor substrate (2) and activating the PI3K/Akt/mTOR pathway [286,287]. Additionally, miR-145 can inhibit post-infarct remodeling. It reduces both LV systolic and diastolic dimensions and improves ejection fraction and +dP/dt in rabbit MI model [286]. Moreover, it attenuates post-infarct fibrosis and prolonged action potential duration via activation of Akt/CREB cascades, which leads to reduced response to β-adrenergic stimuli [288]. On the other hand, miR-145 has been shown to attenuate fibrosis and improve cardiac function via suppressing the AKT/GSK-3β/β-catenin signaling pathway in fibroblasts [289]. Nevertheless, miR-145 seems to have beneficial effect on cardiac repair, and its augmentation can be an attractive target for preventing both cardiac IRI and remodeling after MI. A study on 246 patients with first STEMI who underwent successful PCI showed that circulating miR-145 on day 5 post-MI is independently associated with MACE (HR 7.174, 95% CI 4.208–12.229 [$p < 0.0001$] and cardiac death (HR 5.628, 95% CI 1.990–15.911 [$p = 0.0012$]) during the one-year follow-up period [290]. Moreover, miR-145 levels peak at day 1 post-AMI, which negatively correlates with the EF (Spearman $\rho = 0.65$, $p < 0.0001$) [291].

Table 1. A list of microRNAs which can impair PI3K/Akt signal activity and potential mechanisms of their impact on ischemia–reperfusion injury and post-infarction remodeling.

miRNA	DM2	Ref.	The Potential Regulatory Mechanism
miR-19	-	-	MiR-19a protects H9C2 cardiomyocytes against H/R-induced apoptosis by inhibiting PTEN [292]; MiR-19b promotes NRCFs proliferation and migration by targeting PTEN [293].
miR-21	up *	[294–297]	Targets PTEN expression and promotes adverse ventricular remodeling by induction of MMP-2 in cardiac fibroblasts [269], alleviates cardiomyocytes apoptosis and reduces infarct size through decreasing Bax/Bcl-2 ratio and caspase-3 expression [271], decreases cardiomyocytes autophagy [273].
miR-34	up	[294]	Inhibition of miR-34a attenuates MI-induced LV remodeling in mice and induces Akt phosphorylation [298]; activates PI3K/AKT pathways via up-regulating ZEB1 in cardiomyocytes and attenuates hypoxia-induced injury [299]; protects H9C2 cardiomyocytes from high-glucose-induced injury [300].
miR-122	up	[295,301]	Aggravates oxygen–glucose deprivation and reperfusion apoptosis of H9C2 cardiomyocytes inhibiting AKT/GSK-3β/β-catenin and AKT/mTOR pathway signaling [302,303];
miR-126	down	[295,304–308]	Targets PIK3R2 and SPRED1 expression, resulting in elevated activity of the PI3K/Akt signal and improved angiogenesis, left ventricle function after MI, and alleviated apoptosis of both endothelial cells and cardiomyocytes [278–284]
miR-130	up	[295,309,310]	Attenuates LV dysfunction and remodeling after MI targeting PTEN and increasing activity of Akt [311]
miR-145	down	[312]	Inhibits cardiac cells apoptosis and ROC activity by enhancing the PI3K/Akt and SGK1 activity [285], promotes autophagy and reduces myocardial infarct size targeting FRS2 and inducing PI3K/Akt/mTOR activity [286,287], attenuates fibrosis via activation of the Akt/CREB and suppression of the AKT/GSK-3β/β-catenin pathways [288,289].
miR-155	up	[313,314]	Targets IKKi expression decreasing its cardioprotective role in activating Akt and NF-κB, independent of the PI3K, and enhances cardiac hypertrophy [315], inhibits the AKT/CREB pathway signal and impairs the left ventricle function [316].
miR-223	up	[310,317,318]	Inhibits angiogenic function of CMECs by decreasing the PI3K/Akt signal activity [319], regulates cardiac hypertrophy by modulating the p-Akt activation [320], and mediates cardiac fibrosis after MI targeting RASA1 expression, which promotes MEK1/2, ERK1/2 and AKT phosphorylation [321].
miR-320	up	[295,306]	Increases vulnerability of cardiomyocytes to hypoxia/reoxygenation injury targeting expression of Akt3 [322].
miR-375	up	[323,324]	Exacerbates inflammation and cardiomyocyte apoptosis, decreases angiogenesis and impairs the LV function after MI by a reduction in PDK-1 expression, which results in decreased Akt Thr-308 phosphorylation [325,326].

(DM—diabetes mellitus, *—indicates a contradictory finding where the miRNA was found to be downregulated in at least one study). Akt—protein kinase B, Bax—Bcl-2-associated X protein, Bcl-2—B-cell lymphoma 2, CMECs—cardiac microvascular endothelial cells, CREB—cAMP response element-binding protein, ERK1/2—extracellular signal-regulated kinase1/2, FRS2—fibroblast growth factor receptor substrate 2, GSK-3β—glycogen synthase kinase 3 β, LV—left ventricle, MEK1/2—mitogen-activated protein kinase kinase1/2, MI—myocardial infarction, MMP-2—metalloproteinase 2, mTOR—mammalian target of rapamycin, NF-κB—nuclear factor κ-light-chain-enhancer of activated B cells, NRCFs—neonatal rat cardiac fibroblasts, IKKi—inducible IkB kinase, H/R—hypoxia/reoxigenation, p-Akt—phosphorylated Akt, PDK-1—phosphoinositide-dependent kinase-1, PI3K—phosphoinositide 3-kinase, PIK3R2—phosphoinositide 3-kinase regulatory subunit 2, PTEN—phosphatase and tensin homolog, RASA1—Ras p21 protein activator 1, ROC—Ras of complex proteins, SGK1—serum and glucocorticoid-regulated kinase 1, SPRED1—Sprouty-related EVH1 domain containing 1, Thr-308—threonine 308, ZEB1—zinc finger E-box binding homeobox 1.

MiRNAs may play and important role in the modulation of PI3K/Akt activity in the heart. They may serve as potential biomarkers and prognostic factors of the infarct size and adverse post-MI remodeling. Moreover, miRNAs can mediate the effect of DM on the impaired PI3K/Akt signaling in the heart.

7. Further Perspectives

The PI3K/Akt pathway plays an important role in the survival and function of cardiomyocytes. It seems that this pathway may be an ideal target to protect the myocardium against most of the post-infarct changes described above. In addition, PI3K and other components of the described pathway are among the essential links in the inflammatory response after ischemia. Substances activating this pathway (e.g., SC79) are at an advanced stage of development in basic research, but still the results are not conclusive [327–329]. It seems that these drugs can be useful in patients after MI and may reduce the risk of post-infarct HF. However, there are no sufficient in vivo analyses or clinical trials to test the safety and clinical efficacy of these drugs. Moreover, this therapy may potentially promote carcinogenesis or stimulate the growth of pre-existing undetected tumors, and this has to be checked in clinical trials [330]. Due to this, there is still a long way before activators of the PI3K/Akt pathway can be introduced into clinical practice.

Besides the activators of this pathway, substances which inhibit (e.g., LY294002) its activation are under investigation for oncology [44]. Taking into consideration all of the facts mentioned above, it seems that this drug can have a potential cardiotoxic effect. It can increase the risk of MI and post-infarct HF, but this has to be checked.

8. Conclusions

The PI3K/Akt pathway regulates the metabolism of cardiac cells on multiple levels, and plays an important role in the pathophysiology of myocardial infarction and the post-infarction remodeling of the left ventricle. The activation of this pathway is responsible for the cardioprotective effect of many drugs such as statins and ACEI, prescribed for patients with high cardiovascular risk or after MI. Moreover, altered activity of the PI3K/Akt pathway during diabetes may be responsible for the greater risk of death or post-MI HF development in this group of patients. The results of thebasic studies show that the modulation of this pathway by selective inhibitors or miRNA can influence both ischemic damage and post-infarction LV remodeling. Akt seems to be a key component of this signal transduction and is responsible for most of the positive effects of its activation in cardiomyocytes. There is still a long way to go before introducing drugs based on the activation or inhibition of this pathway, but this may bring significant clinical benefits.

Author Contributions: Conceptualization—M.K. and M.M.; resources—B.W., M.K. and M.M.; writing—original draft preparation—B.W., M.K. and M.M.; writing—review and editing—M.K., M.M. and M.W.; visualization—B.W., M.M. and M.K.; supervision—M.W.; funding acquisition—M.K., M.M. and M.W. All authors have read and agreed to the published version of the manuscript.

Funding: This research was funded by the Publishing Fund of the Medical University of Warsaw, grant number: 1MA/N/2022.

Institutional Review Board Statement: Not applicable.

Informed Consent Statement: Not applicable.

Data Availability Statement: Not applicable.

Conflicts of Interest: The authors declare no conflict of interest.

References

1. Saleh, M.; Ambrose, J.A. Understanding myocardial infarction. *F1000Research* **2018**, *7*, 1378. [CrossRef] [PubMed]
2. Boateng, S.; Sanborn, T. Acute myocardial infarction. *Dis. Mon.* **2013**, *59*, 83–96. [CrossRef] [PubMed]
3. Reddy, K.; Khaliq, A.; Henning, R.J. Recent advances in the diagnosis and treatment of acute myocardial infarction. *World J. Cardiol.* **2015**, *7*, 243–276. [CrossRef] [PubMed]
4. Libby, P.; Buring, J.E.; Badimon, L.; Hansson, G.K.; Deanfield, J.; Bittencourt, M.S.; Tokgozoglu, L.; Lewis, E.F. Atherosclerosis. *Nat. Rev. Dis. Primers* **2019**, *5*, 56. [CrossRef]
5. Anderson, J.L.; Morrow, D.A. Acute Myocardial Infarction. *N. Engl. J. Med.* **2017**, *376*, 2053–2064. [CrossRef]
6. Alabas, O.A.; Jernberg, T.; Pujades-Rodriguez, M.; Rutherford, M.J.; West, R.M.; Hall, M.; Timmis, A.; Lindahl, B.; Fox, K.A.A.; Hemingway, H.; et al. Statistics on mortality following acute myocardial infarction in 842 897 Europeans. *Cardiovasc. Res.* **2020**, *116*, 149–157. [CrossRef]
7. Lu, L.; Liu, M.; Sun, R.; Zheng, Y.; Zhang, P. Myocardial Infarction: Symptoms and Treatments. *Cell Biochem. Biophys.* **2015**, *72*, 865–867. [CrossRef]
8. Schumacher, B.; Pecher, P.; von Specht, B.U.; Stegmann, T. Induction of neoangiogenesis in ischemic myocardium by human growth factors: First clinical results of a new treatment of coronary heart disease. *Circulation* **1998**, *97*, 645–650. [CrossRef]
9. Ezekowitz, J.A.; Kaul, P.; Bakal, J.A.; Armstrong, P.W.; Welsh, R.C.; McAlister, F.A. Declining in-hospital mortality and increasing heart failure incidence in elderly patients with first myocardial infarction. *J. Am. Coll. Cardiol.* **2009**, *53*, 13–20. [CrossRef]
10. Mela, A.; Rdzanek, E.; Poniatowski, L.A.; Jaroszynski, J.; Furtak-Niczyporuk, M.; Galazka-Sobotka, M.; Olejniczak, D.; Niewada, M.; Staniszewska, A. Economic Costs of Cardiovascular Diseases in Poland Estimates for 2015–2017 Years. *Front. Pharmacol.* **2020**, *11*, 1231. [CrossRef]
11. Kologrivova, I.; Shtatolkina, M.; Suslova, T.; Ryabov, V. Cells of the Immune System in Cardiac Remodeling: Main Players in Resolution of Inflammation and Repair after Myocardial Infarction. *Front. Immunol.* **2021**, *12*, 664457. [CrossRef] [PubMed]
12. Gibb, A.A.; Hill, B.G. Metabolic Coordination of Physiological and Pathological Cardiac Remodeling. *Circ. Res.* **2018**, *123*, 107–128. [CrossRef] [PubMed]
13. Bellis, A.; Di Gioia, G.; Mauro, C.; Mancusi, C.; Barbato, E.; Izzo, R.; Trimarco, B.; Morisco, C. Reducing Cardiac Injury during ST-Elevation Myocardial Infarction: A Reasoned Approach to a Multitarget Therapeutic Strategy. *J. Clin. Med.* **2021**, *10*, 2968. [CrossRef]
14. Ibanez, B.; James, S.; Agewall, S.; Antunes, M.J.; Bucciarelli-Ducci, C.; Bueno, H.; Caforio, A.L.P.; Crea, F.; Goudevenos, J.A.; Halvorsen, S.; et al. 2017 ESC Guidelines for the management of acute myocardial infarction in patients presenting with ST-segment elevation: The Task Force for the management of acute myocardial infarction in patients presenting with ST-segment elevation of the European Society of Cardiology (ESC). *Eur. Heart J.* **2018**, *39*, 119–177. [CrossRef] [PubMed]
15. Collet, J.P.; Thiele, H.; Barbato, E.; Barthelemy, O.; Bauersachs, J.; Bhatt, D.L.; Dendale, P.; Dorobantu, M.; Edvardsen, T.; Folliguet, T.; et al. 2020 ESC Guidelines for the management of acute coronary syndromes in patients presenting without persistent ST-segment elevation. *Eur. Heart J.* **2021**, *42*, 1289–1367. [CrossRef] [PubMed]
16. Cohn, J.N.; Ferrari, R.; Sharpe, N. Cardiac remodeling—Concepts and clinical implications: A consensus paper from an international forum on cardiac remodeling. *J. Am. Coll. Cardiol.* **2000**, *35*, 569–582. [CrossRef]
17. Anand, I.S.; Florea, V.G.; Solomon, S.D.; Konstam, M.A.; Udelson, J.E. Noninvasive assessment of left ventricular remodeling: Concepts, techniques, and implications for clinical trials. *J. Card Fail.* **2002**, *8*, S452–S464. [CrossRef]
18. Han, B.; Trew, M.L.; Zgierski-Johnston, C.M. Cardiac Conduction Velocity, Remodeling and Arrhythmogenesis. *Cells* **2021**, *10*, 2923. [CrossRef]
19. Bahit, M.C.; Kochar, A.; Granger, C.B. Post-Myocardial Infarction Heart Failure. *JACC Heart Fail.* **2018**, *6*, 179–186. [CrossRef]
20. Jenča, D.; Melenovský, V.; Stehlik, J.; Staněk, V.; Kettner, J.; Kautzner, J.; Adámková, V.; Wohlfahrt, P. Heart failure after myocardial infarction: Incidence and predictors. *ESC Heart Fail.* **2021**, *8*, 222–237. [CrossRef]
21. Whitman, M.; Downes, C.P.; Keeler, M.; Keller, T.; Cantley, L. Type I phosphatidylinositol kinase makes a novel inositol phospholipid, phosphatidylinositol-3-phosphate. *Nature* **1988**, *332*, 644–646. [CrossRef] [PubMed]
22. Staal, S.P. Molecular cloning of the akt oncogene and its human homologues AKT1 and AKT2: Amplification of AKT1 in a primary human gastric adenocarcinoma. *Proc. Natl. Acad. Sci. USA* **1987**, *84*, 5034–5037. [CrossRef] [PubMed]
23. Bi, L.; Okabe, I.; Bernard, D.J.; Nussbaum, R.L. Early embryonic lethality in mice deficient in the p110beta catalytic subunit of PI 3-kinase. *Mamm. Genome* **2002**, *13*, 169–172. [CrossRef] [PubMed]
24. Marques, M.; Kumar, A.; Cortes, I.; Gonzalez-Garcia, A.; Hernandez, C.; Moreno-Ortiz, M.C.; Carrera, A.C. Phosphoinositide 3-kinases p110alpha and p110beta regulate cell cycle entry, exhibiting distinct activation kinetics in G1 phase. *Mol. Cell. Biol.* **2008**, *28*, 2803–2814. [CrossRef]

25. Huang, X.; Liu, G.; Guo, J.; Su, Z. The PI3K/AKT pathway in obesity and type 2 diabetes. *Int. J. Biol. Sci.* **2018**, *14*, 1483–1496. [CrossRef] [PubMed]

26. Rai, S.N.; Dilnashin, H.; Birla, H.; Singh, S.S.; Zahra, W.; Rathore, A.S.; Singh, B.K.; Singh, S.P. The Role of PI3K/Akt and ERK in Neurodegenerative Disorders. *Neurotox. Res.* **2019**, *35*, 775–795. [CrossRef]

27. Zeng, B.; Liu, L.; Liao, X.; Zhang, C.; Ruan, H. Thyroid hormone protects cardiomyocytes from H(2)O(2)-induced oxidative stress via the PI3K-AKT signaling pathway. *Exp. Cell Res.* **2019**, *380*, 205–215. [CrossRef]

28. Fruman, D.A.; Chiu, H.; Hopkins, B.D.; Bagrodia, S.; Cantley, L.C.; Abraham, R.T. The PI3K Pathway in Human Disease. *Cell* **2017**, *170*, 605–635. [CrossRef]

29. Guo, H.; German, P.; Bai, S.; Barnes, S.; Guo, W.; Qi, X.; Lou, H.; Liang, J.; Jonasch, E.; Mills, G.B.; et al. The PI3K/AKT Pathway and Renal Cell Carcinoma. *J. Genet. Genom.* **2015**, *42*, 343–353. [CrossRef]

30. Vanhaesebroeck, B.; Guillermet-Guibert, J.; Graupera, M.; Bilanges, B. The emerging mechanisms of isoform-specific PI3K signalling. *Nat. Rev. Mol. Cell Biol.* **2010**, *11*, 329–341. [CrossRef]

31. Cantley, L.C. The phosphoinositide 3-kinase pathway. *Science* **2002**, *296*, 1655–1657. [CrossRef] [PubMed]

32. Leslie, N.R.; Biondi, R.M.; Alessi, D.R. Phosphoinositide-regulated kinases and phosphoinositide phosphatases. *Chem. Rev.* **2001**, *101*, 2365–2380. [CrossRef] [PubMed]

33. Alessi, D.R.; James, S.R.; Downes, C.P.; Holmes, A.B.; Gaffney, P.R.J.; Reese, C.B.; Cohen, P. Characterization of a 3-phosphoinositide-dependent protein kinase which phosphorylates and activates protein kinase Bα. *Curr. Biol.* **1997**, *7*, 261–269. [CrossRef]

34. Sarbassov, D.D.; Guertin, D.A.; Ali, S.M.; Sabatini, D.M. Phosphorylation and regulation of Akt/PKB by the rictor-mTOR complex. *Science* **2005**, *307*, 1098–1101. [CrossRef]

35. Sciarretta, S.; Zhai, P.; Maejima, Y.; del Re, D.P.; Nagarajan, N.; Yee, D.; Liu, T.; Magnuson, M.A.; Volpe, M.; Frati, G.; et al. mTORC2 regulates cardiac response to stress by inhibiting MST1. *Cell Rep.* **2015**, *11*, 125–136. [CrossRef]

36. Feng, J.; Park, J.; Cron, P.; Hess, D.; Hemmings, B.A. Identification of a PKB/Akt hydrophobic motif Ser-473 kinase as DNA-dependent protein kinase. *J. Biol. Chem.* **2004**, *279*, 41189–41196. [CrossRef]

37. Brazil, D.P.; Hemmings, B.A. Ten years of protein kinase B signalling: A hard Akt to follow. *Trends Biochem. Sci.* **2001**, *26*, 657–664. [CrossRef]

38. Bellacosa, A.; Testa, J.R.; Moore, R.; Larue, L. A portrait of AKT kinases: Human cancer and animal models depict a family with strong individualities. *Cancer Biol. Ther.* **2004**, *3*, 268–275. [CrossRef]

39. Toulany, M.; Maier, J.; Iida, M.; Rebholz, S.; Holler, M.; Grottke, A.; Juker, M.; Wheeler, D.L.; Rothbauer, U.; Rodemann, H.P. Akt1 and Akt3 but not Akt2 through interaction with DNA-PKcs stimulate proliferation and post-irradiation cell survival of K-RAS-mutated cancer cells. *Cell Death Discov.* **2017**, *3*, 17072. [CrossRef]

40. Shen, Y.H.; Zhang, L.; Ren, P.; Nguyen, M.T.; Zou, S.; Wu, D.; Wang, X.L.; Coselli, J.S.; LeMaire, S.A. AKT2 confers protection against aortic aneurysms and dissections. *Circ. Res.* **2013**, *112*, 618–632. [CrossRef]

41. Lee, M.Y.; Luciano, A.K.; Ackah, E.; Rodriguez-Vita, J.; Bancroft, T.A.; Eichmann, A.; Simons, M.; Kyriakides, T.R.; Morales-Ruiz, M.; Sessa, W.C. Endothelial Akt1 mediates angiogenesis by phosphorylating multiple angiogenic substrates. *Proc. Natl. Acad. Sci. USA* **2014**, *111*, 12865–12870. [CrossRef] [PubMed]

42. Sansal, I.; Sellers, W.R. The biology and clinical relevance of the PTEN tumor suppressor pathway. *J. Clin. Oncol.* **2004**, *22*, 2954–2963. [CrossRef] [PubMed]

43. Lee, Y.R.; Chen, M.; Pandolfi, P.P. The functions and regulation of the PTEN tumour suppressor: New modes and prospects. *Nat. Rev. Mol. Cell Biol.* **2018**, *19*, 547–562. [CrossRef]

44. Yang, J.; Nie, J.; Ma, X.; Wei, Y.; Peng, Y.; Wei, X. Targeting PI3K in cancer: Mechanisms and advances in clinical trials. *Mol. Cancer* **2019**, *18*, 26. [CrossRef]

45. Grunwald, V.; DeGraffenried, L.; Russel, D.; Friedrichs, W.E.; Ray, R.B.; Hidalgo, M. Inhibitors of mTOR reverse doxorubicin resistance conferred by PTEN status in prostate cancer cells. *Cancer Res.* **2002**, *62*, 6141–6145. [PubMed]

46. Carracedo, A.; Pandolfi, P.P. The PTEN-PI3K pathway: Of feedbacks and cross-talks. *Oncogene* **2008**, *27*, 5527–5541. [CrossRef] [PubMed]

47. Krijnen, P.A.; Nijmeijer, R.; Meijer, C.J.; Visser, C.A.; Hack, C.E.; Niessen, H.W. Apoptosis in myocardial ischaemia and infarction. *J. Clin. Pathol.* **2002**, *55*, 801–811. [CrossRef]

48. Chiong, M.; Wang, Z.V.; Pedrozo, Z.; Cao, D.J.; Troncoso, R.; Ibacache, M.; Criollo, A.; Nemchenko, A.; Hill, J.A.; Lavandero, S. Cardiomyocyte death: Mechanisms and translational implications. *Cell Death Dis.* **2011**, *2*, e244. [CrossRef]

49. Xie, Y.; Hou, W.; Song, X.; Yu, Y.; Huang, J.; Sun, X.; Kang, R.; Tang, D. Ferroptosis: Process and function. *Cell Death Differ.* **2016**, *23*, 369–379. [CrossRef]

50. Vanden Berghe, T.; Linkermann, A.; Jouan-Lanhouet, S.; Walczak, H.; Vandenabeele, P. Regulated necrosis: The expanding network of non-apoptotic cell death pathways. *Nat. Rev. Mol. Cell Biol.* **2014**, *15*, 135–147. [CrossRef]

51. Conrad, M.; Angeli, J.P.; Vandenabeele, P.; Stockwell, B.R. Regulated necrosis: Disease relevance and therapeutic opportunities. *Nat. Rev. Drug Discov.* **2016**, *15*, 348–366. [CrossRef] [PubMed]

52. Durham, K.K.; Chathely, K.M.; Trigatti, B.L. High-density lipoprotein protects cardiomyocytes against necrosis induced by oxygen and glucose deprivation through SR-B1, PI3K, and AKT1 and 2. *Biochem. J.* **2018**, *475*, 1253–1265. [CrossRef] [PubMed]

53. Murphy, M.P. How mitochondria produce reactive oxygen species. *Biochem. J.* **2009**, *417*, 1–13. [CrossRef] [PubMed]

54. Chatterjee, S.; Browning, E.A.; Hong, N.; DeBolt, K.; Sorokina, E.M.; Liu, W.; Birnbaum, M.J.; Fisher, A.B. Membrane depolarization is the trigger for PI3K/Akt activation and leads to the generation of ROS. *Am. J. Physiol. Heart Circ. Physiol.* **2012**, *302*, H105–H114. [CrossRef]

55. Festjens, N.; Vanden Berghe, T.; Vandenabeele, P. Necrosis, a well-orchestrated form of cell demise: Signalling cascades, important mediators and concomitant immune response. *Biochim. Biophys. Acta* **2006**, *1757*, 1371–1387. [CrossRef]

56. Bijur, G.N.; Jope, R.S. Rapid accumulation of Akt in mitochondria following phosphatidylinositol 3-kinase activation. *J. Neurochem.* **2003**, *87*, 1427–1435. [CrossRef]

57. Chen, Q.; Vazquez, E.J.; Moghaddas, S.; Hoppel, C.L.; Lesnefsky, E.J. Production of reactive oxygen species by mitochondria: Central role of complex III. *J. Biol. Chem.* **2003**, *278*, 36027–36031. [CrossRef]

58. Kukreja, R.C.; Kontos, H.A.; Hess, M.L.; Ellis, E.F. PGH synthase and lipoxygenase generate superoxide in the presence of NADH or NADPH. *Circ. Res.* **1986**, *59*, 612–619. [CrossRef]

59. Mailloux, R.J.; Gardiner, D.; O'Brien, M. 2-Oxoglutarate dehydrogenase is a more significant source of $O_2{}^-/H_2O_2$ than pyruvate dehydrogenase in cardiac and liver tissue. *Free Radic. Biol. Med.* **2016**, *97*, 501–512. [CrossRef]

60. Elmore, S. Apoptosis: A review of programmed cell death. *Toxicol. Pathol.* **2007**, *35*, 495–516. [CrossRef]

61. Cao, W.; Xie, Y.H.; Li, X.Q.; Zhang, X.K.; Chen, Y.T.; Kang, R.; Chen, X.; Miao, S.; Wang, S.W. Burn-induced apoptosis of cardiomyocytes is survivin dependent and regulated by PI3K/Akt, p38 MAPK and ERK pathways. *Basic Res. Cardiol.* **2011**, *106*, 1207–1220. [CrossRef] [PubMed]

62. Tang, X.L.; Liu, J.X.; Dong, W.; Li, P.; Li, L.; Lin, C.R.; Zheng, Y.Q.; Cong, W.H.; Hou, J.C. Cardioprotective effect of protocatechuic acid on myocardial ischemia/reperfusion injury. *J. Pharmacol. Sci.* **2014**, *125*, 176–183. [CrossRef] [PubMed]

63. Duronio, V. The life of a cell: Apoptosis regulation by the PI3K/PKB pathway. *Biochem. J.* **2008**, *415*, 333–344. [CrossRef] [PubMed]

64. Manning, B.D.; Cantley, L.C. AKT/PKB signaling: Navigating downstream. *Cell* **2007**, *129*, 1261–1274. [CrossRef] [PubMed]

65. Feng, Q.; Li, X.; Qin, X.; Yu, C.; Jin, Y.; Qian, X. PTEN inhibitor improves vascular remodeling and cardiac function after myocardial infarction through PI3k/Akt/VEGF signaling pathway. *Mol. Med.* **2020**, *26*, 111. [CrossRef] [PubMed]

66. Ruan, Y.; Jin, Q.; Zeng, J.; Ren, F.; Xie, Z.; Ji, K.; Wu, L.; Wu, J.; Li, L. Grape Seed Proanthocyanidin Extract Ameliorates Cardiac Remodelling after Myocardial Infarction Through PI3K/AKT Pathway in Mice. *Front. Pharmacol.* **2020**, *11*, 585984. [CrossRef]

67. Kim, A.H.; Khursigara, G.; Sun, X.; Franke, T.F.; Chao, M.V. Akt phosphorylates and negatively regulates apoptosis signal-regulating kinase 1. *Mol. Cell. Biol.* **2001**, *21*, 893–901. [CrossRef]

68. Takeda, K.; Noguchi, T.; Naguro, I.; Ichijo, H. Apoptosis signal-regulating kinase 1 in stress and immune response. *Annu. Rev. Pharmacol. Toxicol.* **2008**, *48*, 199–225. [CrossRef]

69. Linseman, D.A.; Butts, B.D.; Precht, T.A.; Phelps, R.A.; Le, S.S.; Laessig, T.A.; Bouchard, R.J.; Florez-McClure, M.L.; Heidenreich, K.A. Glycogen synthase kinase-3beta phosphorylates Bax and promotes its mitochondrial localization during neuronal apoptosis. *J. Neurosci.* **2004**, *24*, 9993–10002. [CrossRef]

70. Blume-Jensen, P.; Janknecht, R.; Hunter, T. The Kit receptor promotes cell survival via activation of PI 3-kinase and subsequent Akt-mediated phosphorylation of Bad on Ser136. *Curr. Biol.* **1998**, *8*, 779–785. [CrossRef]

71. Polzien, L.; Baljuls, A.; Rennefahrt, U.E.E.; Fischer, A.; Schmitz, W.; Zahedi, R.P.; Sickmann, A.; Metz, R.; Albert, S.; Benz, R.; et al. Identification of novel in vivo phosphorylation sites of the human proapoptotic protein BAD: Pore-forming activity of BAD is regulated by phosphorylation. *J. Biol. Chem.* **2009**, *284*, 28004–28020. [CrossRef] [PubMed]

72. Putcha, G.V.; Le, S.; Frank, S.; Besirli, C.G.; Clark, K.; Chu, B.; Alix, S.; Youle, R.J.; LaMarche, A.; Maroney, A.C.; et al. JNK-Mediated BIM Phosphorylation Potentiates BAX-Dependent Apoptosis. *Neuron* **2003**, *38*, 899–914. [CrossRef]

73. Kotrasova, V.; Keresztesova, B.; Ondrovicova, G.; Bauer, J.A.; Havalova, H.; Pevala, V.; Kutejova, E.; Kunova, N. Mitochondrial Kinases and the Role of Mitochondrial Protein Phosphorylation in Health and Disease. *Life* **2021**, *11*, 82. [CrossRef] [PubMed]

74. Bergmann, M.W.; Rechner, C.; Freund, C.; Baurand, A.; El Jamali, A.; Dietz, R. Statins inhibit reoxygenation-induced cardiomy-ocyte apoptosis: Role for glycogen synthase kinase 3beta and transcription factor beta-catenin. *J. Mol. Cell. Cardiol.* **2004**, *37*, 681–690. [CrossRef] [PubMed]

75. Zhai, P.; Sciarretta, S.; Galeotti, J.; Volpe, M.; Sadoshima, J. Differential roles of GSK-3beta during myocardial ischemia and ischemia/reperfusion. *Circ. Res.* **2011**, *109*, 502–511. [CrossRef] [PubMed]

76. Li, R.; Ding, C.; Zhang, J.; Xie, M.; Park, D.; Ding, Y.; Chen, G.; Zhang, G.; Gilbert-Ross, M.; Zhou, W.; et al. Modulation of Bax and mTOR for Cancer Therapeutics. *Cancer Res.* **2017**, *77*, 3001–3012. [CrossRef]

77. Kale, J.; Kutuk, O.; Brito, G.C.; Andrews, T.S.; Leber, B.; Letai, A.; Andrews, D.W. Phosphorylation switches Bax from promoting to inhibiting apoptosis thereby increasing drug resistance. *EMBO Rep.* **2018**, *19*, e45235. [CrossRef]

78. Nagaoka, K.; Matoba, T.; Mao, Y.; Nakano, Y.; Ikeda, G.; Egusa, S.; Tokutome, M.; Nagahama, R.; Nakano, K.; Sunagawa, K.; et al. A New Therapeutic Modality for Acute Myocardial Infarction: Nanoparticle-Mediated Delivery of Pitavastatin Induces Cardio-protection from Ischemia-Reperfusion Injury via Activation of PI3K/Akt Pathway and Anti-Inflammation in a Rat Model. *PLoS ONE* **2015**, *10*, e0132451. [CrossRef]

79. Pan, Q.; Xie, X.; Guo, Y.; Wang, H. Simvastatin promotes cardiac microvascular endothelial cells proliferation, migration and survival by phosphorylation of p70 S6K and FoxO3a. *Cell Biol. Int.* **2014**, *38*, 599–609. [CrossRef]

80. Zhan, L.; Wang, X.; Zhang, Y.; Zhu, G.; Ding, Y.; Chen, X.; Jiang, W.; Wu, S. Benazepril hydrochloride protects against doxorubicin cardiotoxicity by regulating the PI3K/Akt pathway. *Exp. Ther. Med.* **2021**, *22*, 1082. [CrossRef]

81. Mahajan, U.B.; Patil, P.D.; Chandrayan, G.; Patil, C.R.; Agrawal, Y.O.; Ojha, S.; Goyal, S.N. Eplerenone pretreatment protects the myocardium against ischaemia/reperfusion injury through the phosphatidylinositol 3-kinase/Akt-dependent pathway in diabetic rats. *Mol. Cell. Biochem.* **2018**, *446*, 91–103. [CrossRef] [PubMed]

82. Wang, J.; Liu, J.; Xie, L.; Cai, X.; Ma, X.; Gong, J. Bisoprolol, a β(1) antagonist, protects myocardial cells from ischemia-reperfusion injury via PI3K/AKT/GSK3β pathway. *Fundam. Clin. Pharmacol.* **2020**, *34*, 708–720. [CrossRef] [PubMed]

83. Baraka, S.A.; Tolba, M.F.; Elsherbini, D.A.; El-Naga, R.N.; Awad, A.S.; El-Demerdash, E. Rosuvastatin and low-dose carvedilol combination protects against isoprenaline-induced myocardial infarction in rats: Role of PI3K/Akt/Nrf2/HO-1 signalling. *Clin. Exp. Pharmacol. Physiol.* **2021**, *48*, 1358–1370. [CrossRef] [PubMed]

84. Liu, S.X.; Zhang, Y.; Wang, Y.F.; Li, X.C.; Xiang, M.X.; Bian, C.; Chen, P. Upregulation of heme oxygenase-1 expression by hydroxysafflor yellow A conferring protection from anoxia/reoxygenation-induced apoptosis in H9c2 cardiomyocytes. *Int. J. Cardiol.* **2012**, *160*, 95–101. [CrossRef]

85. Liu, Y.N.; Zhou, Z.M.; Chen, P. Evidence that hydroxysafflor yellow A protects the heart against ischaemia-reperfusion injury by inhibiting mitochondrial permeability transition pore opening. *Clin. Exp. Pharmacol. Physiol.* **2008**, *35*, 211–216. [CrossRef]

86. Ji, D.B.; Zhu, M.C.; Zhu, B.; Zhu, Y.Z.; Li, C.L.; Ye, J.; Zhu, H.B. Hydroxysafflor yellow A enhances survival of vascular endothelial cells under hypoxia via upregulation of the HIF-1 alpha-VEGF pathway and regulation of Bcl-2/Bax. *J. Cardiovasc. Pharmacol.* **2008**, *52*, 191–202. [CrossRef]

87. Wang, J.; Chen, H.; Zhou, Y.; Su, Q.; Liu, T.; Wang, X.T.; Li, L. Atorvastatin Inhibits Myocardial Apoptosis in a Swine Model of Coronary Microembolization by Regulating PTEN/PI3K/Akt Signaling Pathway. *Cell. Physiol. Biochem.* **2016**, *38*, 207–219. [CrossRef]

88. Galluzzi, L.; Vanden Berghe, T.; Vanlangenakker, N.; Buettner, S.; Eisenberg, T.; Vandenabeele, P.; Madeo, F.; Kroemer, G. Programmed necrosis from molecules to health and disease. *Int. Rev. Cell Mol. Biol.* **2011**, *289*, 1–35. [CrossRef]

89. Piamsiri, C.; Maneechote, C.; Siri-Angkul, N.; Chattipakorn, S.C.; Chattipakorn, N. Targeting necroptosis as therapeutic potential in chronic myocardial infarction. *J. Biomed. Sci.* **2021**, *28*, 25. [CrossRef]

90. Heusch, G. Myocardial ischaemia-reperfusion injury and cardioprotection in perspective. *Nat. Rev. Cardiol.* **2020**, *17*, 773–789. [CrossRef]

91. Van Herreweghe, F.; Festjens, N.; Declercq, W.; Vandenabeele, P. Tumor necrosis factor-mediated cell death: To break or to burst, that's the question. *Cell. Mol. Life Sci.* **2010**, *67*, 1567–1579. [CrossRef] [PubMed]

92. Zhang, D.W.; Zheng, M.; Zhao, J.; Li, Y.Y.; Huang, Z.; Li, Z.; Han, J. Multiple death pathways in TNF-treated fibroblasts: RIP3- and RIP1-dependent and independent routes. *Cell Res.* **2011**, *21*, 368–371. [CrossRef] [PubMed]

93. Cai, Z.; Jitkaew, S.; Zhao, J.; Chiang, H.C.; Choksi, S.; Liu, J.; Ward, Y.; Wu, L.G.; Liu, Z.G. Plasma membrane translocation of trimerized MLKL protein is required for TNF-induced necroptosis. *Nat. Cell. Biol.* **2014**, *16*, 55–65. [CrossRef] [PubMed]

94. Sun, L.; Wang, H.; Wang, Z.; He, S.; Chen, S.; Liao, D.; Wang, L.; Yan, J.; Liu, W.; Lei, X.; et al. Mixed lineage kinase domain-like protein mediates necrosis signaling downstream of RIP3 kinase. *Cell* **2012**, *148*, 213–227. [CrossRef]

95. Hu, S.; Chang, X.; Zhu, H.; Wang, D.; Chen, G. PI3K mediates tumor necrosis factor induced-necroptosis through initiating RIP1-RIP3-MLKL signaling pathway activation. *Cytokine* **2020**, *129*, 155046. [CrossRef]

96. Tuuminen, R.; Holmstrom, E.; Raissadati, A.; Saharinen, P.; Rouvinen, E.; Krebs, R.; Lemstrom, K.B. Simvastatin pretreatment reduces caspase-9 and RIPK1 protein activity in rat cardiac allograft ischemia-reperfusion. *Transpl. Immunol.* **2016**, *37*, 40–45. [CrossRef]

97. Dixon, S.J.; Lemberg, K.M.; Lamprecht, M.R.; Skouta, R.; Zaitsev, E.M.; Gleason, C.E.; Patel, D.N.; Bauer, A.J.; Cantley, A.M.; Yang, W.S.; et al. Ferroptosis: An iron-dependent form of nonapoptotic cell death. *Cell* **2012**, *149*, 1060–1072. [CrossRef]

98. Hu, H.; Chen, Y.; Jing, L.; Zhai, C.; Shen, L. The Link Between Ferroptosis and Cardiovascular Diseases: A Novel Target for Treatment. *Front. Cardiovasc. Med.* **2021**, *8*, 710963. [CrossRef]

99. Wu, X.; Li, Y.; Zhang, S.; Zhou, X. Ferroptosis as a novel therapeutic target for cardiovascular disease. *Theranostics* **2021**, *11*, 3052–3059. [CrossRef]

100. Zhao, W.K.; Zhou, Y.; Xu, T.T.; Wu, Q. Ferroptosis: Opportunities and Challenges in Myocardial Ischemia-Reperfusion Injury. *Oxid Med. Cell Longev.* **2021**, *2021*, 9929687. [CrossRef]

101. Zhao, S.; Li, P.; Wu, W.; Wang, Q.; Qian, B.; Li, X.; Shen, M. Roles of ferroptosis in urologic malignancies. *Cancer Cell Int.* **2021**, *21*, 676. [CrossRef] [PubMed]

102. Sun, L.; Wang, H.; Xu, D.; Yu, S.; Zhang, L.; Li, X. Lapatinib induces mitochondrial dysfunction to enhance oxidative stress and ferroptosis in doxorubicin-induced cardiomyocytes via inhibition of PI3K/AKT signaling pathway. *Bioengineered* **2022**, *13*, 48–60. [CrossRef] [PubMed]

103. D'Amato, V.; Raimondo, L.; Formisano, L.; Giuliano, M.; De Placido, S.; Rosa, R.; Bianco, R. Mechanisms of lapatinib resistance in HER2-driven breast cancer. *Cancer Treat. Rev.* **2015**, *41*, 877–883. [CrossRef] [PubMed]

104. Wang, H.; Li, F.; Du, C.; Wang, H.; Mahato, R.I.; Huang, Y. Doxorubicin and lapatinib combination nanomedicine for treating resistant breast cancer. *Mol. Pharm.* **2014**, *11*, 2600–2611. [CrossRef]

105. Mei, S.; Hong, L.; Cai, X.; Xiao, B.; Zhang, P.; Shao, L. Oxidative stress injury in doxorubicin-induced cardiotoxicity. *Toxicol. Lett.* **2019**, *307*, 41–48. [CrossRef]

106. Segredo, M.P.; Salvadori, D.M.; Rocha, N.S.; Moretto, F.C.; Correa, C.R.; Camargo, E.A.; de Almeida, D.C.; Reis, R.A.; Freire, C.M.; Braz, M.G.; et al. Oxidative stress on cardiotoxicity after treatment with single and multiple doses of doxorubicin. *Hum. Exp. Toxicol.* **2014**, *33*, 748–760. [CrossRef]

107. Yi, J.; Zhu, J.; Wu, J.; Thompson, C.B.; Jiang, X. Oncogenic activation of PI3K-AKT-mTOR signaling suppresses ferroptosis via SREBP-mediated lipogenesis. *Proc. Natl. Acad. Sci. USA* **2020**, *117*, 31189–31197. [CrossRef]

108. Li, G.; Yang, J.; Zhao, G.; Shen, Z.; Yang, K.; Tian, L.; Zhou, Q.; Chen, Y.; Huang, Y. Dysregulation of ferroptosis may involve in the development of non-small-cell lung cancer in Xuanwei area. *J. Cell. Mol. Med.* **2021**, *25*, 2872–2884. [CrossRef]

109. Jiang, Y.; Chen, L.; Chao, Z.; Chen, T.; Zhou, Y. Ferroptosis Related Genes in Ischemic and Idiopathic Cardiomyopathy: Screening for Potential Pharmacological Targets. *Front. Cell Dev. Biol.* **2022**, *10*, 817819. [CrossRef]

110. Yu, P.; Zhang, X.; Liu, N.; Tang, L.; Peng, C.; Chen, X. Pyroptosis: Mechanisms and diseases. *Signal Transduct. Target. Ther.* **2021**, *6*, 128. [CrossRef]

111. Guo, X.; Hu, S.; Liu, J.J.; Huang, L.; Zhong, P.; Fan, Z.X.; Ye, P.; Chen, M.H. Piperine protects against pyroptosis in myocardial ischaemia/reperfusion injury by regulating the miR-383/RP105/AKT signalling pathway. *J. Cell. Mol. Med.* **2021**, *25*, 244–258. [CrossRef] [PubMed]

112. Liu, W.; Shen, J.; Li, Y.; Wu, J.; Luo, X.; Yu, Y.; Zhang, Y.; Gu, L.; Zhang, X.; Jiang, C.; et al. Pyroptosis inhibition improves the symptom of acute myocardial infarction. *Cell Death Dis.* **2021**, *12*, 852. [CrossRef] [PubMed]

113. Do Carmo, H.; Arjun, S.; Petrucci, O.; Yellon, D.M.; Davidson, S.M. The Caspase 1 Inhibitor VX-765 Protects the Isolated Rat Heart via the RISK Pathway. *Cardiovasc. Drugs Ther.* **2018**, *32*, 165–168. [CrossRef] [PubMed]

114. Reed, G.W.; Rossi, J.E.; Cannon, C.P. Acute myocardial infarction. *Lancet* **2017**, *389*, 197–210. [CrossRef]

115. Guo, X.; Jiang, H.; Chen, J. RP105-PI3K-Akt axis: A potential therapeutic approach for ameliorating myocardial ischemia/reperfusion injury. *Int. J. Cardiol.* **2016**, *206*, 95–96. [CrossRef]

116. Qin, Q.; Cui, L.; Zhou, Z.; Zhang, Z.; Wang, Y.; Zhou, C. Inhibition of microRNA-141-3p Reduces Hypoxia-Induced Apoptosis in H9c2 Rat Cardiomyocytes by Activating the RP105-Dependent PI3K/AKT Signaling Pathway. *Med. Sci. Monit.* **2019**, *25*, 7016–7025. [CrossRef]

117. Guo, X.; Li, X.-Y.; Hu, S.; Wu, G.; Chen, Z.; Liu, J.-J.; Ye, P.; Chen, M.-H. SH2B1 protects cardiomyocytes from ischemia/reperfusion injury via the activation of the PI3K/AKT pathway. *Int. Immunopharmacol.* **2020**, *83*, 105910. [CrossRef]

118. Yellon, D.M.; Downey, J.M. Preconditioning the myocardium: From cellular physiology to clinical cardiology. *Physiol. Rev.* **2003**, *83*, 1113–1151. [CrossRef]

119. Rezkalla, S.H.; Kloner, R.A. Ischemic preconditioning and preinfarction angina in the clinical arena. *Nat. Clin. Pract. Cardiovasc. Med.* **2004**, *1*, 96–102. [CrossRef]

120. Hausenloy, D.J.; Yellon, D.M. Preconditioning and postconditioning: United at reperfusion. *Pharmacol. Ther.* **2007**, *116*, 173–191. [CrossRef]

121. Su, F.; Zhao, L.; Zhang, S.; Wang, J.; Chen, N.; Gong, Q.; Tang, J.; Wang, H.; Yao, J.; Wang, Q.; et al. Cardioprotection by PI3K-mediated signaling is required for anti-arrhythmia and myocardial repair in response to ischemic preconditioning in infarcted pig hearts. *Lab. Investig.* **2015**, *95*, 860–871. [CrossRef] [PubMed]

122. Rossello, X.; Riquelme, J.A.; Davidson, S.M.; Yellon, D.M. Role of PI3K in myocardial ischaemic preconditioning: Mapping pro-survival cascades at the trigger phase and at reperfusion. *J. Cell. Mol. Med.* **2017**, *22*, 926–935. [CrossRef]

123. Cai, Z.P.; Parajuli, N.; Zheng, X.; Becker, L. Remote ischemic preconditioning confers late protection against myocardial ischemia-reperfusion injury in mice by upregulating interleukin-10. *Basic Res. Cardiol.* **2012**, *107*, 277. [CrossRef] [PubMed]

124. Yang, X.; Cohen, M.V.; Downey, J.M. Mechanism of cardioprotection by early ischemic preconditioning. *Cardiovasc. Drugs Ther.* **2010**, *24*, 225–234. [CrossRef] [PubMed]

125. Barsukevich, V.; Basalay, M.; Sanchez, J.; Mrochek, A.; Whittle, J.; Ackland, G.L.; Gourine, A.V.; Gourine, A. Distinct cardioprotective mechanisms of immediate, early and delayed ischaemic postconditioning. *Basic Res. Cardiol.* **2015**, *110*, 452. [CrossRef]

126. Wang, X.; Wang, J.; Tu, T.; Iyan, Z.; Mungun, D.; Yang, Z.; Guo, Y. Remote Ischemic Postconditioning Protects against Myocardial Ischemia-Reperfusion Injury by Inhibition of the RAGE-HMGB1 Pathway. *Biomed. Res. Int.* **2018**, *2018*, 4565630. [CrossRef] [PubMed]

127. Li, C.M.; Shen, S.W.; Wang, T.; Zhang, X.H. Myocardial ischemic post-conditioning attenuates ischemia reperfusion injury via PTEN/Akt signal pathway. *Int. J. Clin. Exp. Med.* **2015**, *8*, 15801–15807.

128. Milano, G.; Abruzzo, P.M.; Bolotta, A.; Marini, M.; Terraneo, L.; Ravara, B.; Gorza, L.; Vitadello, M.; Burattini, S.; Curzi, D.; et al. Impact of the phosphatidylinositide 3-kinase signaling pathway on the cardioprotection induced by intermittent hypoxia. *PLoS ONE* **2013**, *8*, e76659. [CrossRef]

129. Zhang, J.; Zhang, X. Ischaemic preconditioning-induced serum exosomes protect against myocardial ischaemia/reperfusion injury in rats by activating the PI3K/AKT signalling pathway. *Cell Biochem. Funct.* **2020**, *39*, 287–295. [CrossRef]

130. Lassen, T.R.; Just, J.; Hjortbak, M.V.; Jespersen, N.R.; Stenz, K.T.; Gu, T.; Yan, Y.; Su, J.; Hansen, J.; Bæk, R.; et al. Cardioprotection by remote ischemic conditioning is transferable by plasma and mediated by extracellular vesicles. *Basic Res. Cardiol.* **2021**, *116*, 16. [CrossRef]

131. Swiatkiewicz, I.; Kozinski, M.; Magielski, P.; Fabiszak, T.; Sukiennik, A.; Navarese, E.P.; Odrowaz-Sypniewska, G.; Kubica, J. Value of C-reactive protein in predicting left ventricular remodelling in patients with a first ST-segment elevation myocardial infarction. *Mediat. Inflamm.* **2012**, *2012*, 250867. [CrossRef] [PubMed]

132. Bolognese, L.; Neskovic, A.N.; Parodi, G.; Cerisano, G.; Buonamici, P.; Santoro, G.M.; Antoniucci, D. Left ventricular remodeling after primary coronary angioplasty: Patterns of left ventricular dilation and long-term prognostic implications. *Circulation* **2002**, *106*, 2351–2357. [CrossRef] [PubMed]

133. Bhatt, A.S.; Ambrosy, A.P.; Velazquez, E.J. Adverse Remodeling and Reverse Remodeling After Myocardial Infarction. *Curr. Cardiol. Rep.* **2017**, *19*, 71. [CrossRef] [PubMed]

134. Pfeffer, M.A.; Braunwald, E. Ventricular remodeling after myocardial infarction. Experimental observations and clinical implications. *Circulation* **1990**, *81*, 1161–1172. [CrossRef]

135. Carrick, D.; Haig, C.; Rauhalammi, S.; Ahmed, N.; Mordi, I.; McEntegart, M.; Petrie, M.C.; Eteiba, H.; Lindsay, M.; Watkins, S.; et al. Pathophysiology of LV Remodeling in Survivors of STEMI: Inflammation, Remote Myocardium, and Prognosis. *JACC Cardiovasc. Imaging* **2015**, *8*, 779–789. [CrossRef]

136. Dutka, M.; Bobinski, R.; Korbecki, J. The relevance of microRNA in post-infarction left ventricular remodelling and heart failure. *Heart Fail. Rev.* **2019**, *24*, 575–586. [CrossRef]

137. Stone, G.W.; Selker, H.P.; Thiele, H.; Patel, M.R.; Udelson, J.E.; Ohman, E.M.; Maehara, A.; Eitel, I.; Granger, C.B.; Jenkins, P.L.; et al. Relationship Between Infarct Size and Outcomes Following Primary PCI: Patient-Level Analysis From 10 Randomized Trials. *J. Am. Coll. Cardiol.* **2016**, *67*, 1674–1683. [CrossRef]

138. Thomas, T.P.; Grisanti, L.A. The Dynamic Interplay Between Cardiac Inflammation and Fibrosis. *Front. Physiol.* **2020**, *11*, 529075. [CrossRef]

139. Savoye, C.; Equine, O.; Tricot, O.; Nugue, O.; Segrestin, B.; Sautiere, K.; Elkohen, M.; Pretorian, E.M.; Taghipour, K.; Philias, A.; et al. Left ventricular remodeling after anterior wall acute myocardial infarction in modern clinical practice (from the REmodelage VEntriculaire [REVE] study group). *Am. J. Cardiol.* **2006**, *98*, 1144–1149. [CrossRef]

140. Funaro, S.; La Torre, G.; Madonna, M.; Galiuto, L.; Scara, A.; Labbadia, A.; Canali, E.; Mattatelli, A.; Fedele, F.; Alessandrini, F.; et al. Incidence, determinants, and prognostic value of reverse left ventricular remodelling after primary percutaneous coronary intervention: Results of the Acute Myocardial Infarction Contrast Imaging (AMICI) multicenter study. *Eur. Heart J.* **2009**, *30*, 566–575. [CrossRef]

141. Meng, H.; Zhang, Y.; An, S.T.; Chen, Y. Annexin A3 gene silencing promotes myocardial cell repair through activation of the PI3K/Akt signaling pathway in rats with acute myocardial infarction. *J. Cell. Physiol.* **2019**, *234*, 10535–10546. [CrossRef] [PubMed]

142. Chen, X.; Zhabyeyev, P.; Azad, A.K.; Vanhaesebroeck, B.; Grueter, C.E.; Murray, A.G.; Kassiri, Z.; Oudit, G.Y. Pharmacological and cell-specific genetic PI3Kalpha inhibition worsens cardiac remodeling after myocardial infarction. *J. Mol. Cell. Cardiol.* **2021**, *157*, 17–30. [CrossRef] [PubMed]

143. Frangogiannis, N.G. Regulation of the inflammatory response in cardiac repair. *Circ. Res.* **2012**, *110*, 159–173. [CrossRef] [PubMed]

144. Nian, M.; Lee, P.; Khaper, N.; Liu, P. Inflammatory cytokines and postmyocardial infarction remodeling. *Circ. Res.* **2004**, *94*, 1543–1553. [CrossRef]

145. Takahashi, T.; Anzai, T.; Kaneko, H.; Mano, Y.; Anzai, A.; Nagai, T.; Kohno, T.; Maekawa, Y.; Yoshikawa, T.; Fukuda, K.; et al. Increased C-reactive protein expression exacerbates left ventricular dysfunction and remodeling after myocardial infarction. *Am. J. Physiol. Heart Circ. Physiol.* **2010**, *299*, H1795–H1804. [CrossRef]

146. Suleiman, M.; Khatib, R.; Agmon, Y.; Mahamid, R.; Boulos, M.; Kapeliovich, M.; Levy, Y.; Beyar, R.; Markiewicz, W.; Hammerman, H.; et al. Early inflammation and risk of long-term development of heart failure and mortality in survivors of acute myocardial infarction predictive role of C-reactive protein. *J. Am. Coll. Cardiol.* **2006**, *47*, 962–968. [CrossRef]

147. Swiatkiewicz, I.; Magielski, P.; Kubica, J.; Zadourian, A.; DeMaria, A.N.; Taub, P.R. Enhanced Inflammation is a Marker for Risk of Post-Infarct Ventricular Dysfunction and Heart Failure. *Int. J. Mol. Sci.* **2020**, *21*, 807. [CrossRef]

148. Swiatkiewicz, I.; Magielski, P.; Kubica, J. C-Reactive Protein as a Risk Marker for Post-Infarct Heart Failure over a Multi-Year Period. *Int. J. Mol. Sci.* **2021**, *22*, 3169. [CrossRef]

149. Westman, P.C.; Lipinski, M.J.; Luger, D.; Waksman, R.; Bonow, R.O.; Wu, E.; Epstein, S.E. Inflammation as a Driver of Adverse Left Ventricular Remodeling After Acute Myocardial Infarction. *J. Am. Coll. Cardiol.* **2016**, *67*, 2050–2060. [CrossRef]

150. Sproston, N.R.; Ashworth, J.J. Role of C-Reactive Protein at Sites of Inflammation and Infection. *Front. Immunol.* **2018**, *9*, 754. [CrossRef]

151. Fu, Y.; Wu, Y.; Liu, E. C-reactive protein and cardiovascular disease: From animal studies to the clinic (Review). *Exp. Ther. Med.* **2020**, *20*, 1211–1219. [CrossRef] [PubMed]

152. Bisoendial, R.J.; Boekholdt, S.M.; Vergeer, M.; Stroes, E.S.; Kastelein, J.J. C-reactive protein is a mediator of cardiovascular disease. *Eur. Heart J.* **2010**, *31*, 2087–2091. [CrossRef] [PubMed]

153. Choi, J.W.; Lee, K.H.; Kim, S.H.; Jin, T.; Lee, B.S.; Oh, J.; Won, H.Y.; Kim, S.Y.; Kang, S.M.; Chung, J.H. C-reactive protein induces p53-mediated cell cycle arrest in H9c2 cardiac myocytes. *Biochem. Biophys. Res. Commun.* **2011**, *410*, 525–530. [CrossRef] [PubMed]

154. Orn, S.; Manhenke, C.; Ueland, T.; Damas, J.K.; Mollnes, T.E.; Edvardsen, T.; Aukrust, P.; Dickstein, K. C-reactive protein, infarct size, microvascular obstruction, and left-ventricular remodelling following acute myocardial infarction. *Eur. Heart J.* **2009**, *30*, 1180–1186. [CrossRef]

155. Mather, A.N.; Fairbairn, T.A.; Artis, N.J.; Greenwood, J.P.; Plein, S. Relationship of cardiac biomarkers and reversible and irreversible myocardial injury following acute myocardial infarction as determined by cardiovascular magnetic resonance. *Int. J. Cardiol.* **2013**, *166*, 458–464. [CrossRef]

156. Fertin, M.; Hennache, B.; Hamon, M.; Ennezat, P.V.; Biausque, F.; Elkohen, M.; Nugue, O.; Tricot, O.; Lamblin, N.; Pinet, F.; et al. Usefulness of serial assessment of B-type natriuretic peptide, troponin I, and C-reactive protein to predict left ventricular remodeling after acute myocardial infarction (from the REVE-2 study). *Am. J. Cardiol.* **2010**, *106*, 1410–1416. [CrossRef]

157. Pepys, M.B.; Hirschfield, G.M.; Tennent, G.A.; Gallimore, J.R.; Kahan, M.C.; Bellotti, V.; Hawkins, P.N.; Myers, R.M.; Smith, M.D.; Polara, A.; et al. Targeting C-reactive protein for the treatment of cardiovascular disease. *Nature* **2006**, *440*, 1217–1221. [CrossRef]

158. Abbate, A.; Biondi-Zoccai, G.G.L.; Bussani, R.; Dobrina, A.; Camilot, D.; Feroce, F.; Rossiello, R.; Baldi, F.; Silvestri, F.; Biasucci, L.M.; et al. Increased myocardial apoptosis in patients with unfavorable left ventricular remodeling and early symptomatic post-infarction heart failure. *J. Am. Coll. Cardiol.* **2003**, *41*, 753–760. [CrossRef]

159. Boras, E.; Slevin, M.; Alexander, M.Y.; Aljohi, A.; Gilmore, W.; Ashworth, J.; Krupinski, J.; Potempa, L.A.; Al Abdulkareem, I.; Elobeid, A.; et al. Monomeric C-reactive protein and Notch-3 co-operatively increase angiogenesis through PI3K signalling pathway. *Cytokine* **2014**, *69*, 165–179. [CrossRef]

160. Chen, J.; Gu, Z.; Wu, M.; Yang, Y.; Zhang, J.; Ou, J.; Zuo, Z.; Wang, J.; Chen, Y. C-reactive protein can upregulate VEGF expression to promote ADSC-induced angiogenesis by activating HIF-1alpha via CD64/PI3k/Akt and MAPK/ERK signaling pathways. *Stem Cell Res. Ther.* **2016**, *7*, 114. [CrossRef]

161. Tanigaki, K.; Mineo, C.; Yuhanna, I.S.; Chambliss, K.L.; Quon, M.J.; Bonvini, E.; Shaul, P.W. C-reactive protein inhibits insulin activation of endothelial nitric oxide synthase via the immunoreceptor tyrosine-based inhibition motif of FcgammaRIIB and SHIP-1. *Circ. Res.* **2009**, *104*, 1275–1282. [CrossRef] [PubMed]

162. Muniyappa, R.; Montagnani, M.; Koh, K.K.; Quon, M.J. Cardiovascular actions of insulin. *Endocr. Rev.* **2007**, *28*, 463–491. [CrossRef] [PubMed]

163. Lee, B.S.; Kim, S.H.; Oh, J.; Jin, T.; Choi, E.Y.; Park, S.; Lee, S.H.; Chung, J.H.; Kang, S.M. C-reactive protein inhibits survivin expression via Akt/mTOR pathway downregulation by PTEN expression in cardiac myocytes. *PLoS ONE* **2014**, *9*, e98113. [CrossRef] [PubMed]

164. Parajuli, N.; Yuan, Y.; Zheng, X.; Bedja, D.; Cai, Z.P. Phosphatase PTEN is critically involved in post-myocardial infarction remodeling through the Akt/interleukin-10 signaling pathway. *Basic Res. Cardiol.* **2012**, *107*, 248. [CrossRef] [PubMed]

165. Ma, Y.; Chen, Z.; Zou, Y.; Ge, J. Atorvastatin represses the angiotensin 2-induced oxidative stress and inflammatory response in dendritic cells via the PI3K/Akt/Nrf 2 pathway. *Oxid. Med. Cell. Longev.* **2014**, *2014*, 148798. [CrossRef]

166. Liu, Y.; Zhang, S.; Xue, J.; Wei, Z.; Ao, P.; Shen, B.; Ding, L. CGRP Reduces Apoptosis of DRG Cells Induced by High-Glucose Oxidative Stress Injury through PI3K/AKT Induction of Heme Oxygenase-1 and Nrf-2 Expression. *Oxid. Med. Cell. Longev.* **2019**, *2019*, 2053149. [CrossRef] [PubMed]

167. Zhao, S.M.; Gao, H.L.; Wang, Y.L.; Xu, Q.; Guo, C.Y. Attenuation of High Glucose-Induced Rat Cardiomyocyte Apoptosis by Exendin-4 via Intervention of HO-1/Nrf-2 and the PI3K/AKT Signaling Pathway. *Chin. J. Physiol.* **2017**, *60*, 89–96. [CrossRef]

168. Xiao, Z.; Peng, J.; Gan, N.; Arafat, A.; Yin, F. Interleukin-1beta Plays a Pivotal Role via the PI3K/Akt/mTOR Signaling Pathway in the Chronicity of Mesial Temporal Lobe Epilepsy. *Neuroimmunomodulation* **2016**, *23*, 332–344. [CrossRef]

169. Li, B.; Smith, T.J. PI3K/AKT pathway mediates induction of IL-1RA by TSH in fibrocytes: Modulation by PTEN. *J. Clin. Endocrinol. Metab.* **2014**, *99*, 3363–3372. [CrossRef]

170. Teshima, S.; Nakanishi, H.; Nishizawa, M.; Kitagawa, K.; Kaibori, M.; Yamada, M.; Habara, K.; Kwon, A.H.; Kamiyama, Y.; Ito, S.; et al. Up-regulation of IL-1 receptor through PI3K/Akt is essential for the induction of iNOS gene expression in hepatocytes. *J. Hepatol.* **2004**, *40*, 616–623. [CrossRef]

171. Abbate, A.; Toldo, S.; Marchetti, C.; Kron, J.; Van Tassell, B.W.; Dinarello, C.A. Interleukin-1 and the Inflammasome as Therapeutic Targets in Cardiovascular Disease. *Circ. Res.* **2020**, *126*, 1260–1280. [CrossRef] [PubMed]

172. Abbate, A.; Van Tassell, B.W.; Biondi-Zoccai, G.; Kontos, M.C.; Grizzard, J.D.; Spillman, D.W.; Oddi, C.; Roberts, C.S.; Melchior, R.D.; Mueller, G.H.; et al. Effects of interleukin-1 blockade with anakinra on adverse cardiac remodeling and heart failure after acute myocardial infarction [from the Virginia Commonwealth University-Anakinra Remodeling Trial (2) (VCU-ART2) pilot study]. *Am. J. Cardiol.* **2013**, *111*, 1394–1400. [CrossRef] [PubMed]

173. Everett, B.M.; Cornel, J.H.; Lainscak, M.; Anker, S.D.; Abbate, A.; Thuren, T.; Libby, P.; Glynn, R.J.; Ridker, P.M. Anti-Inflammatory Therapy With Canakinumab for the Prevention of Hospitalization for Heart Failure. *Circulation* **2019**, *139*, 1289–1299. [CrossRef]

174. Choi, H.W.; Shin, P.G.; Lee, J.H.; Choi, W.S.; Kang, M.J.; Kong, W.S.; Oh, M.J.; Seo, Y.B.; Kim, G.D. Anti-inflammatory effect of lovastatin is mediated via the modulation of NF-kappaB and inhibition of HDAC1 and the PI3K/Akt/mTOR pathway in RAW264.7 macrophages. *Int. J. Mol. Med.* **2018**, *41*, 1103–1109. [CrossRef]

175. Xie, S.; Chen, M.; Yan, B.; He, X.; Chen, X.; Li, D. Identification of a role for the PI3K/AKT/mTOR signaling pathway in innate immune cells. *PLoS ONE* **2014**, *9*, e94496. [CrossRef] [PubMed]

176. Chartoumpekis, D.; Ziros, P.G.; Psyrogiannis, A.; Kyriazopoulou, V.; Papavassiliou, A.G.; Habeos, I.G. Simvastatin lowers reactive oxygen species level by Nrf2 activation via PI3K/Akt pathway. *Biochem. Biophys. Res. Commun.* **2010**, *396*, 463–466. [CrossRef] [PubMed]

177. Mizushima, N.; Komatsu, M. Autophagy: Renovation of cells and tissues. *Cell* **2011**, *147*, 728–741. [CrossRef]

178. Choi, A.M.; Ryter, S.W.; Levine, B. Autophagy in human health and disease. *N. Engl. J. Med.* **2013**, *368*, 651–662. [CrossRef]

179. Levine, B.; Kroemer, G. Autophagy in the pathogenesis of disease. *Cell* **2008**, *132*, 27–42. [CrossRef]
180. Kobayashi, S. Choose Delicately and Reuse Adequately: The Newly Revealed Process of Autophagy. *Biol. Pharm. Bull.* **2015**, *38*, 1098–1103. [CrossRef]
181. Marambio, P.; Toro, B.; Sanhueza, C.; Troncoso, R.; Parra, V.; Verdejo, H.; Garcia, L.; Quiroga, C.; Munafo, D.; Diaz-Elizondo, J.; et al. Glucose deprivation causes oxidative stress and stimulates aggresome formation and autophagy in cultured cardiac myocytes. *Biochim. Biophys. Acta* **2010**, *1802*, 509–518. [CrossRef] [PubMed]
182. Yan, L.; Vatner, D.E.; Kim, S.J.; Ge, H.; Masurekar, M.; Massover, W.H.; Yang, G.; Matsui, Y.; Sadoshima, J.; Vatner, S.F. Autophagy in chronically ischemic myocardium. *Proc. Natl. Acad. Sci. USA* **2005**, *102*, 13807–13812. [CrossRef] [PubMed]
183. Sciarretta, S.; Maejima, Y.; Zablocki, D.; Sadoshima, J. The Role of Autophagy in the Heart. *Annu. Rev. Physiol.* **2018**, *80*, 1–26. [CrossRef] [PubMed]
184. Sciarretta, S.; Zhai, P.; Shao, D.; Maejima, Y.; Robbins, J.; Volpe, M.; Condorelli, G.; Sadoshima, J. Rheb is a critical regulator of autophagy during myocardial ischemia: Pathophysiological implications in obesity and metabolic syndrome. *Circulation* **2012**, *125*, 1134–1146. [CrossRef] [PubMed]
185. Matsui, Y.; Takagi, H.; Qu, X.; Abdellatif, M.; Sakoda, H.; Asano, T.; Levine, B.; Sadoshima, J. Distinct roles of autophagy in the heart during ischemia and reperfusion: Roles of AMP-activated protein kinase and Beclin 1 in mediating autophagy. *Circ. Res.* **2007**, *100*, 914–922. [CrossRef]
186. Kim, J.; Kundu, M.; Viollet, B.; Guan, K.L. AMPK and mTOR regulate autophagy through direct phosphorylation of Ulk1. *Nat. Cell Biol.* **2011**, *13*, 132–141. [CrossRef]
187. Martina, J.A.; Chen, Y.; Gucek, M.; Puertollano, R. MTORC1 functions as a transcriptional regulator of autophagy by preventing nuclear transport of TFEB. *Autophagy* **2012**, *8*, 903–914. [CrossRef]
188. Sciarretta, S.; Volpe, M.; Sadoshima, J. Mammalian target of rapamycin signaling in cardiac physiology and disease. *Circ. Res.* **2014**, *114*, 549–564. [CrossRef]
189. Mendoza, M.C.; Er, E.E.; Blenis, J. The Ras-ERK and PI3K-mTOR pathways: Cross-talk and compensation. *Trends Biochem. Sci.* **2011**, *36*, 320–328. [CrossRef]
190. Kwak, D.; Choi, S.; Jeong, H.; Jang, J.H.; Lee, Y.; Jeon, H.; Lee, M.N.; Noh, J.; Cho, K.; Yoo, J.S.; et al. Osmotic stress regulates mammalian target of rapamycin (mTOR) complex 1 via c-Jun N-terminal Kinase (JNK)-mediated Raptor protein phosphorylation. *J. Biol. Chem.* **2012**, *287*, 18398–18407. [CrossRef]
191. Majid, S.; Saini, S.; Dahiya, R. Wnt signaling pathways in urological cancers: Past decades and still growing. *Mol. Cancer* **2012**, *11*, 7. [CrossRef] [PubMed]
192. Xie, Y.; Kang, R.; Tang, D. Role of the Beclin 1 Network in the Cross-Regulation Between Autophagy and Apoptosis. In *Autophagy: Cancer, Other Pathologies, Inflammation, Immunity, Infection, and Aging*; Academic Press: Cambridge, MA, USA, 2016; pp. 75–88.
193. Ma, X.; Liu, H.; Foyil, S.R.; Godar, R.J.; Weinheimer, C.J.; Hill, J.A.; Diwan, A. Impaired autophagosome clearance contributes to cardiomyocyte death in ischemia/reperfusion injury. *Circulation* **2012**, *125*, 3170–3181. [CrossRef] [PubMed]
194. Bhattacharya, D.; Mukhopadhyay, M.; Bhattacharyya, M.; Karmakar, P. Is autophagy associated with diabetes mellitus and its complications? A review. *EXCLI J.* **2018**, *17*, 709–720. [CrossRef] [PubMed]
195. Chen, Y.; Hua, Y.; Li, X.; Arslan, I.M.; Zhang, W.; Meng, G. Distinct Types of Cell Death and the Implication in Diabetic Cardiomyopathy. *Front. Pharmacol.* **2020**, *11*, 42. [CrossRef]
196. Buss, S.J.; Muenz, S.; Riffel, J.H.; Malekar, P.; Hagenmueller, M.; Weiss, C.S.; Bea, F.; Bekeredjian, R.; Schinke-Braun, M.; Izumo, S.; et al. Beneficial effects of Mammalian target of rapamycin inhibition on left ventricular remodeling after myocardial infarction. *J. Am. Coll. Cardiol.* **2009**, *54*, 2435–2446. [CrossRef]
197. Zhang, D.; Contu, R.; Latronico, M.V.; Zhang, J.; Rizzi, R.; Catalucci, D.; Miyamoto, S.; Huang, K.; Ceci, M.; Gu, Y.; et al. MTORC1 regulates cardiac function and myocyte survival through 4E-BP1 inhibition in mice. *J. Clin. Investig.* **2010**, *120*, 2805–2816. [CrossRef]
198. Shioi, T.; McMullen, J.R.; Tarnavski, O.; Converso, K.; Sherwood, M.C.; Manning, W.J.; Izumo, S. Rapamycin attenuates load-induced cardiac hypertrophy in mice. *Circulation* **2003**, *107*, 1664–1670. [CrossRef]
199. Maejima, Y.; Kyoi, S.; Zhai, P.; Liu, T.; Li, H.; Ivessa, A.; Sciarretta, S.; Del Re, D.P.; Zablocki, D.K.; Hsu, C.P.; et al. Mst1 inhibits autophagy by promoting the interaction between Beclin1 and Bcl-2. *Nat. Med.* **2013**, *19*, 1478–1488. [CrossRef]
200. Krenning, G.; Zeisberg, E.M.; Kalluri, R. The origin of fibroblasts and mechanism of cardiac fibrosis. *J. Cell. Physiol.* **2010**, *225*, 631–637. [CrossRef]
201. Gyongyosi, M.; Winkler, J.; Ramos, I.; Do, Q.T.; Firat, H.; McDonald, K.; Gonzalez, A.; Thum, T.; Diez, J.; Jaisser, F.; et al. Myocardial fibrosis: Biomedical research from bench to bedside. *Eur. J. Heart Fail.* **2017**, *19*, 177–191. [CrossRef]
202. Yu, P.; Ma, S.; Dai, X.; Cao, F. Elabela alleviates myocardial ischemia reperfusion-induced apoptosis, fibrosis and mitochondrial dysfunction through PI3K/AKT signaling. *Am. J. Transl. Res.* **2020**, *12*, 4467–4477. [PubMed]
203. Wang, L.; Tian, X.; Cao, Y.; Ma, X.; Shang, L.; Li, H.; Zhang, X.; Deng, F.; Li, S.; Guo, T.; et al. Cardiac Shock Wave Therapy Improves Ventricular Function by Relieving Fibrosis Through PI3K/Akt Signaling Pathway: Evidence From a Rat Model of Post-infarction Heart Failure. *Front. Cardiovasc. Med.* **2021**, *8*, 693875. [CrossRef] [PubMed]
204. Qin, W.; Cao, L.; Massey, I.Y. Role of PI3K/Akt signaling pathway in cardiac fibrosis. *Mol. Cell. Biochem.* **2021**, *476*, 4045–4059. [CrossRef] [PubMed]

205. Zhao, X.; Ren, Y.; Ren, H.; Wu, Y.; Liu, X.; Chen, H.; Ying, C. The mechanism of myocardial fibrosis is ameliorated by myocardial infarction-associated transcript through the PI3K/Akt signaling pathway to relieve heart failure. *J. Int. Med. Res.* **2021**, *49*, 3000605211031433. [CrossRef]

206. Colombo, F.; Noel, J.; Mayers, P.; Mercier, I.; Calderone, A. beta-Adrenergic stimulation of rat cardiac fibroblasts promotes protein synthesis via the activation of phosphatidylinositol 3-kinase. *J. Mol. Cell. Cardiol.* **2001**, *33*, 1091–1106. [CrossRef]

207. Mann, D.L.; Bogaev, R.; Buckberg, G.D. Cardiac remodelling and myocardial recovery: Lost in translation? *Eur. J. Heart Fail.* **2010**, *12*, 789–796. [CrossRef]

208. Shah, A.M.; Solomon, S.D. A unified view of ventricular remodelling. *Eur. J. Heart Fail.* **2010**, *12*, 779–781. [CrossRef]

209. Shimizu, I.; Minamino, T. Physiological and pathological cardiac hypertrophy. *J. Mol. Cell. Cardiol.* **2016**, *97*, 245–262. [CrossRef]

210. Chaanine, A.H.; Hajjar, R.J. AKT signalling in the failing heart. *Eur. J. Heart Fail.* **2011**, *13*, 825–829. [CrossRef]

211. Shiojima, I.; Yefremashvili, M.; Luo, Z.; Kureishi, Y.; Takahashi, A.; Tao, J.; Rosenzweig, A.; Kahn, C.R.; Abel, E.D.; Walsh, K. Akt signaling mediates postnatal heart growth in response to insulin and nutritional status. *J. Biol. Chem.* **2002**, *277*, 37670–37677. [CrossRef]

212. Phung, T.L.; Ziv, K.; Dabydeen, D.; Eyiah-Mensah, G.; Riveros, M.; Perruzzi, C.; Sun, J.; Monahan-Earley, R.A.; Shiojima, I.; Nagy, J.A.; et al. Pathological angiogenesis is induced by sustained Akt signaling and inhibited by rapamycin. *Cancer Cell* **2006**, *10*, 159–170. [CrossRef] [PubMed]

213. Kemi, O.J.; Ceci, M.; Wisloff, U.; Grimaldi, S.; Gallo, P.; Smith, G.L.; Condorelli, G.; Ellingsen, O. Activation or inactivation of cardiac Akt/mTOR signaling diverges physiological from pathological hypertrophy. *J. Cell. Physiol.* **2008**, *214*, 316–321. [CrossRef] [PubMed]

214. Meng, X.; Cui, J.; He, G. Bcl-2 Is Involved in Cardiac Hypertrophy through PI3K-Akt Pathway. *Biomed. Res. Int.* **2021**, *2021*, 6615502. [CrossRef] [PubMed]

215. Zhang, S.; Zhao, L.; Shen, L.; Xu, D.; Huang, B.; Wang, Q.; Lin, J.; Zou, Y.; Ge, J. Comparison of various niches for endothelial progenitor cell therapy on ischemic myocardial repair: Coexistence of host collateralization and Akt-mediated angiogenesis produces a superior microenvironment. *Arterioscler. Thromb. Vasc. Biol.* **2012**, *32*, 910–923. [CrossRef]

216. Ma, J.; Sawai, H.; Ochi, N.; Matsuo, Y.; Xu, D.; Yasuda, A.; Takahashi, H.; Wakasugi, T.; Takeyama, H. PTEN regulates angiogenesis through PI3K/Akt/VEGF signaling pathway in human pancreatic cancer cells. *Mol. Cell. Biochem.* **2009**, *331*, 161–171. [CrossRef]

217. Cochain, C.; Channon, K.M.; Silvestre, J.S. Angiogenesis in the infarcted myocardium. *Antioxid. Redox. Signal.* **2013**, *18*, 1100–1113. [CrossRef]

218. Cheng, S.; Zhang, X.; Feng, Q.; Chen, J.; Shen, L.; Yu, P.; Yang, L.; Chen, D.; Zhang, H.; Sun, W.; et al. Astragaloside IV exerts angiogenesis and cardioprotection after myocardial infarction via regulating PTEN/PI3K/Akt signaling pathway. *Life Sci.* **2019**, *227*, 82–93. [CrossRef]

219. Kuo, H.M.; Lin, C.Y.; Lam, H.C.; Lin, P.R.; Chan, H.H.; Tseng, J.C.; Sun, C.K.; Hsu, T.F.; Wu, C.C.; Yang, C.Y.; et al. PTEN overexpression attenuates angiogenic processes of endothelial cells by blockade of endothelin-1/endothelin B receptor signaling. *Atherosclerosis* **2012**, *221*, 341–349. [CrossRef]

220. Jiang, B.H.; Zheng, J.Z.; Aoki, M.; Vogt, P.K. Phosphatidylinositol 3-kinase signaling mediates angiogenesis and expression of vascular endothelial growth factor in endothelial cells. *Proc. Natl. Acad. Sci. USA* **2000**, *97*, 1749–1753. [CrossRef]

221. Skinner, H.D.; Zheng, J.Z.; Fang, J.; Agani, F.; Jiang, B.H. Vascular endothelial growth factor transcriptional activation is mediated by hypoxia-inducible factor 1alpha, HDM2, and p70S6K1 in response to phosphatidylinositol 3-kinase/AKT signaling. *J. Biol. Chem.* **2004**, *279*, 45643–45651. [CrossRef]

222. Robich, M.P.; Chu, L.M.; Oyamada, S.; Sodha, N.R.; Sellke, F.W. Myocardial therapeutic angiogenesis: A review of the state of development and future obstacles. *Expert Rev. Cardiovasc. Ther.* **2011**, *9*, 1469–1479. [CrossRef] [PubMed]

223. Gorenek, B.; Blomstrom Lundqvist, C.; Brugada Terradellas, J.; Camm, A.J.; Hindricks, G.; Huber, K.; Kirchhof, P.; Kuck, K.H.; Kudaiberdieva, G.; Lin, T.; et al. Cardiac arrhythmias in acute coronary syndromes: Position paper from the joint EHRA, ACCA, and EAPCI task force. *EuroIntervention* **2015**, *10*, 1095–1108. [CrossRef] [PubMed]

224. Yu, T.S.; Ge, L.Z.; Cao, J.M. Research Advances in Sympathetic Remodeling after Myocardial Infarction and Its Significance in Forensic Science. *Fa Yi Xue Za Zhi* **2019**, *35*, 68–73. [CrossRef] [PubMed]

225. Gardner, R.T.; Ripplinger, C.M.; Myles, R.C.; Habecker, B.A. Molecular Mechanisms of Sympathetic Remodeling and Arrhythmias. *Circ. Arrhythm. Electrophysiol.* **2016**, *9*, e001359. [CrossRef] [PubMed]

226. Cao, J.M.; Fishbein, M.C.; Han, J.B.; Lai, W.W.; Lai, A.C.; Wu, T.J.; Czer, L.; Wolf, P.L.; Denton, T.A.; Shintaku, I.P.; et al. Relationship between regional cardiac hyperinnervation and ventricular arrhythmia. *Circulation* **2000**, *101*, 1960–1969. [CrossRef] [PubMed]

227. Allen, S.J.; Watson, J.J.; Shoemark, D.K.; Barua, N.U.; Patel, N.K. GDNF, NGF and BDNF as therapeutic options for neurodegeneration. *Pharmacol. Ther.* **2013**, *138*, 155–175. [CrossRef]

228. Wei, K.; Liu, L.; Xie, F.; Hao, X.; Luo, J.; Min, S. Nerve growth factor protects the ischemic heart via attenuation of the endoplasmic reticulum stress induced apoptosis by activation of phosphatidylinositol 3-kinase. *Int. J. Med. Sci.* **2015**, *12*, 83–91. [CrossRef]

229. Li, C.Y.; Li, Y.G. Cardiac Sympathetic Nerve Sprouting and Susceptibility to Ventricular Arrhythmias after Myocardial Infarction. *Cardiol. Res. Pract.* **2015**, *2015*, 698368. [CrossRef]

230. Li, S.S.; Kang, N.; Li, X.L.; Yuan, J.; Ling, R.; Li, P.; Li, J.L. LianXia Formula Granule Attenuates Cardiac Sympathetic Remodeling in Rats with Myocardial Infarction via the NGF/TrKA/PI3K/AKT Signaling Pathway. *Evid.-Based Complement. Altern. Med.* **2021**, *2021*, 5536406. [CrossRef]

231. Miki, T.; Itoh, T.; Sunaga, D.; Miura, T. Effects of diabetes on myocardial infarct size and cardioprotection by preconditioning and postconditioning. *Cardiovasc. Diabetol.* **2012**, *11*, 67. [CrossRef]

232. Donahoe, S.M.; Stewart, G.C.; McCabe, C.H.; Mohanavelu, S.; Murphy, S.A.; Cannon, C.P.; Antman, E.M. Diabetes and mortality following acute coronary syndromes. *JAMA* **2007**, *298*, 765–775. [CrossRef] [PubMed]

233. Alegria, J.R.; Miller, T.D.; Gibbons, R.J.; Yi, Q.L.; Yusuf, S.; Collaborative Organization of RheothRx Evaluation (CORE) Trial Investigators. Infarct size, ejection fraction, and mortality in diabetic patients with acute myocardial infarction treated with thrombolytic therapy. *Am. Heart J.* **2007**, *154*, 743–750. [CrossRef] [PubMed]

234. Marso, S.P.; Miller, T.; Rutherford, B.D.; Gibbons, R.J.; Qureshi, M.; Kalynych, A.; Turco, M.; Schultheiss, H.P.; Mehran, R.; Krucoff, M.W.; et al. Comparison of myocardial reperfusion in patients undergoing percutaneous coronary intervention in ST-segment elevation acute myocardial infarction with versus without diabetes mellitus (from the EMERALD Trial). *Am. J. Cardiol.* **2007**, *100*, 206–210. [CrossRef] [PubMed]

235. Mocanu, M.M.; Yellon, D.M. PTEN, the Achilles' heel of myocardial ischaemia/reperfusion injury? *Br. J. Pharmacol.* **2007**, *150*, 833–838. [CrossRef]

236. Kui, L.; Weiwei, Z.; Ling, L.; Daikun, H.; Guoming, Z.; Linuo, Z.; Renming, H. Ghrelin inhibits apoptosis induced by high glucose and sodium palmitate in adult rat cardiomyocytes through the PI3K-Akt signaling pathway. *Regul. Pept.* **2009**, *155*, 62–69. [CrossRef]

237. Sun, D.; Huang, J.; Zhang, Z.; Gao, H.; Li, J.; Shen, M.; Cao, F.; Wang, H. Luteolin limits infarct size and improves cardiac function after myocardium ischemia/reperfusion injury in diabetic rats. *PLoS ONE* **2012**, *7*, e33491. [CrossRef]

238. Chen, Q.; Xu, T.; Li, D.; Pan, D.; Wu, P.; Luo, Y.; Ma, Y.; Liu, Y. JNK/PI3K/Akt signaling pathway is involved in myocardial ischemia/reperfusion injury in diabetic rats: Effects of salvianolic acid A intervention. *Am. J. Transl. Res.* **2016**, *8*, 2534–2548.

239. An, S.; Wang, X.; Shi, H.; Zhang, X.; Meng, H.; Li, W.; Chen, D.; Ge, J. Apelin protects against ischemia-reperfusion injury in diabetic myocardium via inhibiting apoptosis and oxidative stress through PI3K and p38-MAPK signaling pathways. *Aging* **2020**, *12*, 25120–25137. [CrossRef]

240. Yu, L.; Li, Z.; Dong, X.; Xue, X.; Liu, Y.; Xu, S.; Zhang, J.; Han, J.; Yang, Y.; Wang, H. Polydatin Protects Diabetic Heart against Ischemia-Reperfusion Injury via Notch1/Hes1-Mediated Activation of Pten/Akt Signaling. *Oxid. Med. Cell. Longev.* **2018**, *2018*, 2750695. [CrossRef]

241. Tsang, A.; Hausenloy, D.J.; Mocanu, M.M.; Carr, R.D.; Yellon, D.M. Preconditioning the diabetic heart: The importance of Akt phosphorylation. *Diabetes* **2005**, *54*, 2360–2364. [CrossRef]

242. Yin, X.; Zheng, Y.; Zhai, X.; Zhao, X.; Cai, L. Diabetic inhibition of preconditioning- and postconditioning-mediated myocardial protection against ischemia/reperfusion injury. *Exp. Diabetes Res.* **2012**, *2012*, 198048. [CrossRef] [PubMed]

243. Drenger, B.; Ostrovsky, I.A.; Barak, M.; Nechemia-Arbely, Y.; Ziv, E.; Axelrod, J.H. Diabetes blockade of sevoflurane postconditioning is not restored by insulin in the rat heart: Phosphorylated signal transducer and activator of transcription 3- and phosphatidylinositol 3-kinase-mediated inhibition. *Anesthesiology* **2011**, *114*, 1364–1372. [CrossRef] [PubMed]

244. Lei, S.; Su, W.; Xia, Z.Y.; Wang, Y.; Zhou, L.; Qiao, S.; Zhao, B.; Xia, Z.; Irwin, M.G. Hyperglycemia-Induced Oxidative Stress Abrogates Remifentanil Preconditioning-Mediated Cardioprotection in Diabetic Rats by Impairing Caveolin-3-Modulated PI3K/Akt and JAK2/STAT3 Signaling. *Oxid. Med. Cell. Longev.* **2019**, *2019*, 9836302. [CrossRef] [PubMed]

245. Gao, S.; Wang, R.; Dong, S.; Wu, J.; Perek, B.; Xia, Z.; Yao, S.; Wang, T. Inactivation of TOPK Caused by Hyperglycemia Blocks Diabetic Heart Sensitivity to Sevoflurane Postconditioning by Impairing the PTEN/PI3K/Akt Signaling. *Oxid. Med. Cell. Longev.* **2021**, *2021*, 6657529. [CrossRef]

246. Xue, R.; Lei, S.; Xia, Z.Y.; Wu, Y.; Meng, Q.; Zhan, L.; Su, W.; Liu, H.; Xu, J.; Liu, Z.; et al. Selective inhibition of PTEN preserves ischaemic post-conditioning cardioprotection in STZ-induced Type 1 diabetic rats: Role of the PI3K/Akt and JAK2/STAT3 pathways. *Clin. Sci.* **2016**, *130*, 377–392. [CrossRef]

247. Cheng, X.; Hu, J.; Wang, Y.; Ye, H.; Li, X.; Gao, Q.; Li, Z. Effects of Dexmedetomidine Postconditioning on Myocardial Ischemia/Reperfusion Injury in Diabetic Rats: Role of the PI3K/Akt-Dependent Signaling Pathway. *J. Diabetes Res.* **2018**, *2018*, 3071959. [CrossRef]

248. Tai, W.; Shi, E.; Yan, L.; Jiang, X.; Ma, H.; Ai, C. Diabetes abolishes the cardioprotection induced by sevoflurane postconditioning in the rat heart in vivo: Roles of glycogen synthase kinase-3β and its upstream pathways. *J. Surg. Res.* **2012**, *178*, 96–104. [CrossRef]

249. Lamblin, N.; Fertin, M.; de Groote, P.; Bauters, C. Cardiac remodeling and heart failure after a first anterior myocardial infarction in patients with diabetes mellitus. *J. Cardiovasc. Med.* **2012**, *13*, 353–359. [CrossRef]

250. Shah, A.M.; Hung, C.L.; Shin, S.H.; Skali, H.; Verma, A.; Ghali, J.K.; Kober, L.; Velazquez, E.J.; Rouleau, J.L.; McMurray, J.J.; et al. Cardiac structure and function, remodeling, and clinical outcomes among patients with diabetes after myocardial infarction complicated by left ventricular systolic dysfunction, heart failure, or both. *Am. Heart J.* **2011**, *162*, 685–691. [CrossRef]

251. Nakatani, D.; Sakata, Y.; Mizuno, H.; Shimizu, M.; Suna, S.; Usami, M.; Ito, H.; Yasumura, Y.; Hirayama, A.; Takeda, H.; et al. Impact of Diabetes Mellitus on Rehospitalization for Heart Failure Among Survivors of Acute Myocardial Infarction in the Percutaneous Coronary Intervention Era. *Circ. J.* **2009**, *73*, 662–666. [CrossRef]

252. Akashi, N.; Tsukui, T.; Yamamoto, K.; Seguchi, M.; Taniguchi, Y.; Sakakura, K.; Wada, H.; Momomura, S.-I.; Fujita, H. Comparison of clinical outcomes and left ventricular remodeling after ST-elevation myocardial infarction between patients with and without diabetes mellitus. *Heart Vessel.* **2021**, *36*, 1445–1456. [CrossRef] [PubMed]

253. Yang, C.D.; Shen, Y.; Lu, L.; Ding, F.H.; Yang, Z.K.; Zhang, R.Y.; Shen, W.F.; Jin, W.; Wang, X.Q. Insulin resistance and dysglycemia are associated with left ventricular remodeling after myocardial infarction in non-diabetic patients. *Cardiovasc. Diabetol.* **2019**, *18*, 6657529. [CrossRef] [PubMed]
254. Shiomi, T.; Tsutsui, H.; Ikeuchi, M.; Matsusaka, H.; Hayashidani, S.; Suematsu, N.; Wen, J.; Kubota, T.; Takeshita, A. Streptozotocin-induced hyperglycemia exacerbates left ventricular remodeling and failure after experimental myocardial infarction. *J. Am. Coll. Cardiol.* **2003**, *42*, 165–172. [CrossRef]
255. Backlund, T.; Palojoki, E.; Saraste, A.; Eriksson, A.; Finckenberg, P.; Kyto, V.; Lakkisto, P.; Mervaala, E.; Voipio-Pulkki, L.M.; Laine, M.; et al. Sustained cardiomyocyte apoptosis and left ventricular remodelling after myocardial infarction in experimental diabetes. *Diabetologia* **2004**, *47*, 325–330. [CrossRef]
256. Thakker, G.D.; Frangogiannis, N.G.; Bujak, M.; Zymek, P.; Gaubatz, J.W.; Reddy, A.K.; Taffet, G.; Michael, L.H.; Entman, M.L.; Ballantyne, C.M. Effects of diet-induced obesity on inflammation and remodeling after myocardial infarction. *Am. J. Physiol. Heart Circ. Physiol.* **2006**, *291*, H2504–H2514. [CrossRef]
257. Vahtola, E.; Louhelainen, M.; Forstén, H.; Merasto, S.; Raivio, J.; Kaheinen, P.; Kytö, V.; Tikkanen, I.; Levijoki, J.; Mervaala, E. Sirtuin1-p53, forkhead box O3a, p38 and post-infarct cardiac remodeling in the spontaneously diabetic Goto-Kakizaki rat. *Cardiovasc. Diabetol.* **2010**, *9*, 5. [CrossRef]
258. Liang, D.; Zhong, P.; Hu, J.; Lin, F.; Qian, Y.; Xu, Z.; Wang, J.; Zeng, C.; Li, X.; Liang, G. EGFR inhibition protects cardiac damage and remodeling through attenuating oxidative stress in STZ-induced diabetic mouse model. *J. Mol. Cell. Cardiol.* **2015**, *82*, 63–74. [CrossRef]
259. Bhamra, G.S.; Hausenloy, D.J.; Davidson, S.M.; Carr, R.D.; Paiva, M.; Wynne, A.M.; Mocanu, M.M.; Yellon, D.M. Metformin protects the ischemic heart by the Akt-mediated inhibition of mitochondrial permeability transition pore opening. *Basic Res. Cardiol.* **2008**, *103*, 274–284. [CrossRef]
260. Yang, F.; Qin, Y.; Wang, Y.; Meng, S.; Xian, H.; Che, H.; Lv, J.; Li, Y.; Yu, Y.; Bai, Y.; et al. Metformin Inhibits the NLRP3 Inflammasome via AMPK/mTOR-dependent Effects in Diabetic Cardiomyopathy. *Int. J. Biol. Sci.* **2019**, *15*, 1010–1019. [CrossRef]
261. Ji, L.; Fu, F.; Zhang, L.; Liu, W.; Cai, X.; Zhang, L.; Zheng, Q.; Zhang, H.; Gao, F. Insulin attenuates myocardial ischemia/reperfusion injury via reducing oxidative/nitrative stress. *Am. J. Physiol. Endocrinol. Metab.* **2010**, *298*, E871–E880. [CrossRef]
262. Teshima, Y.; Takahashi, N.; Thuc, L.C.; Nishio, S.; Nagano-Torigoe, Y.; Miyazaki, H.; Ezaki, K.; Yufu, K.; Hara, M.; Nakagawa, M.; et al. High-glucose condition reduces cardioprotective effects of insulin against mechanical stress-induced cell injury. *Life Sci.* **2010**, *87*, 154–161. [CrossRef] [PubMed]
263. Aoyagi, T.; Kusakari, Y.; Xiao, C.Y.; Inouye, B.T.; Takahashi, M.; Scherrer-Crosbie, M.; Rosenzweig, A.; Hara, K.; Matsui, T. Cardiac mTOR protects the heart against ischemia-reperfusion injury. *Am. J. Physiol. Heart Circ. Physiol.* **2012**, *303*, H75–H85. [CrossRef] [PubMed]
264. Cardoso, R.; Graffunder, F.P.; Ternes, C.M.P.; Fernandes, A.; Rocha, A.V.; Fernandes, G.; Bhatt, D.L. SGLT2 inhibitors decrease cardiovascular death and heart failure hospitalizations in patients with heart failure: A systematic review and meta-analysis. *EClinicalMedicine* **2021**, *36*, 100933. [CrossRef] [PubMed]
265. Hasan, R.; Lasker, S.; Hasan, A.; Zerin, F.; Zamila, M.; Chowdhury, F.I.; Nayan, S.I.; Rahman, M.M.; Khan, F.; Subhan, N.; et al. Canagliflozin attenuates isoprenaline-induced cardiac oxidative stress by stimulating multiple antioxidant and anti-inflammatory signaling pathways. *Sci. Rep.* **2020**, *10*, 14459. [CrossRef] [PubMed]
266. Robinson, E.; Cassidy, R.S.; Tate, M.; Zhao, Y.; Lockhart, S.; Calderwood, D.; Church, R.; Mcgahon, M.K.; Brazil, D.P.; Mcdermott, B.J.; et al. Exendin-4 protects against post-myocardial infarction remodelling via specific actions on inflammation and the extracellular matrix. *Basic Res. Cardiol.* **2015**, *110*, 20. [CrossRef]
267. O'Brien, J.; Hayder, H.; Zayed, Y.; Peng, C. Overview of MicroRNA Biogenesis, Mechanisms of Actions, and Circulation. *Front. Endocrinol.* **2018**, *9*, 402. [CrossRef]
268. Majka, M.; Kleibert, M.; Wojciechowska, M. Impact of the Main Cardiovascular Risk Factors on Plasma Extracellular Vesicles and Their Influence on the Heart's Vulnerability to Ischemia-Reperfusion Injury. *Cells* **2021**, *10*, 3331. [CrossRef]
269. Roy, S.; Khanna, S.; Hussain, S.-R.A.; Biswas, S.; Azad, A.; Rink, C.; Gnyawali, S.; Shilo, S.; Nuovo, G.J.; Sen, C.K. MicroRNA expression in response to murine myocardial infarction: miR-21 regulates fibroblast metalloprotease-2 via phosphatase and tensin homologue. *Cardiovasc. Res.* **2009**, *82*, 21–29. [CrossRef]
270. Wang, G.; Bergman, M.; Nguyen, A.; Turcato, S.; Swigart, P.; Rodrigo, M.; Simpson, P.; Karliner, J.; Lovett, D.; Baker, A. Cardiac transgenic matrix metalloproteinase-2 expression directly induces impaired contractility. *Cardiovasc. Res.* **2006**, *69*, 688–696. [CrossRef]
271. Yang, Q.; Yang, K.; Li, A.Y. Trimetazidine protects against hypoxia-reperfusion-induced cardiomyocyte apoptosis by increasing microRNA-21 expression. *Int. J. Clin. Exp. Pathol.* **2015**, *8*, 3735–3741.
272. Ma, N.; Bai, J.; Zhang, W.; Luo, H.; Zhang, X.; Liu, D.; Qiao, C. Trimetazidine protects against cardiac ischemia/reperfusion injury via effects on cardiac miRNA-21 expression, Akt and the Bcl-2/Bax pathway. *Mol. Med. Rep.* **2016**, *14*, 4216–4222. [CrossRef] [PubMed]
273. Huang, Z.; Wu, S.; Kong, F.; Cai, X.; Ye, B.; Shan, P.; Huang, W. MicroRNA-21 protects against cardiac hypoxia/reoxygenation injury by inhibiting excessive autophagy in H9c2 cells via the Akt/mTOR pathway. *J. Cell. Mol. Med.* **2017**, *21*, 467–474. [CrossRef] [PubMed]

274. Tu, Y.; Wan, L.; Fan, Y.; Wang, K.; Bu, L.; Huang, T.; Cheng, Z.; Shen, B. Ischemic Postconditioning-Mediated miRNA-21 Protects against Cardiac ischemia/reperfusion Injury via PTEN/Akt Pathway. *PLoS ONE* **2013**, *8*, e75872. [CrossRef]
275. Qiao, S.; Olson, J.M.; Paterson, M.; Yan, Y.; Zaja, I.; Liu, Y.; Riess, M.L.; Kersten, J.R.; Liang, M.; Warltier, D.C.; et al. MicroRNA-21 Mediates Isoflurane-induced Cardioprotection against Ischemia–Reperfusion Injury via Akt/Nitric Oxide Synthase/Mitochondrial Permeability Transition Pore Pathway. *Anesthesiology* **2015**, *123*, 786–798. [CrossRef] [PubMed]
276. Mayourian, J.; Ceholski, D.K.; Gorski, P.A.; Mathiyalagan, P.; Murphy, J.F.; Salazar, S.I.; Stillitano, F.; Hare, J.M.; Sahoo, S.; Hajjar, R.J.; et al. Exosomal microRNA-21-5p Mediates Mesenchymal Stem Cell Paracrine Effects on Human Cardiac Tissue Contractility. *Circ. Res.* **2018**, *122*, 933–944. [CrossRef] [PubMed]
277. Qiao, L.; Hu, S.; Liu, S.; Zhang, H.; Ma, H.; Huang, K.; Li, Z.; Su, T.; Vandergriff, A.; Tang, J.; et al. microRNA-21-5p dysregulation in exosomes derived from heart failure patients impairs regenerative potential. *J. Clin. Investig.* **2019**, *129*, 2237–2250. [CrossRef] [PubMed]
278. Song, W.; Liang, Q.; Cai, M.; Tian, Z. HIF-1α-induced up-regulation of microRNA-126 contributes to the effectiveness of exercise training on myocardial angiogenesis in myocardial infarction rats. *J. Cell. Mol. Med.* **2020**, *24*, 12970–12979. [CrossRef] [PubMed]
279. Zhang, Y.; Xu, Y.; Zhou, K.; Kao, G.; Xiao, J. MicroRNA-126 and VEGF enhance the function of endothelial progenitor cells in acute myocardial infarction. *Exp. Ther. Med.* **2022**, *23*, 142. [CrossRef]
280. Gomes, J.L.; Fernandes, T.; Soci, U.P.; Silveira, A.C.; Barretti, D.L.; Negrão, C.E.; Oliveira, E.M. Obesity Downregulates MicroRNA-126 Inducing Capillary Rarefaction in Skeletal Muscle: Effects of Aerobic Exercise Training. *Oxid. Med. Cell. Longev.* **2017**, *2017*, 2415246. [CrossRef]
281. Meng, S.; Cao, J.T.; Zhang, B.; Zhou, Q.; Shen, C.X.; Wang, C.Q. Downregulation of microRNA-126 in endothelial progenitor cells from diabetes patients, impairs their functional properties, via target gene Spred-1. *J. Mol. Cell. Cardiol.* **2012**, *53*, 64–72. [CrossRef]
282. Chen, L.; Wang, J.; Wang, B.; Yang, J.; Gong, Z.; Zhao, X.; Zhang, C.; Du, K. MiR-126 inhibits vascular endothelial cell apoptosis through targeting PI3K/Akt signaling. *Ann. Hematol.* **2016**, *95*, 365–374. [CrossRef] [PubMed]
283. Li, S.H.; Wang, M.S.; Ke, W.L.; Wang, M.R. Naringenin alleviates myocardial ischemia reperfusion injury by enhancing the myocardial miR-126-PI3K/AKT axis in streptozotocin-induced diabetic rats. *Exp. Ther. Med.* **2021**, *22*, 810. [CrossRef] [PubMed]
284. Zhang, L.; Ouyang, P.; He, G.; Wang, X.; Song, D.; Yang, Y.; He, X. Exosomes from microRNA-126 overexpressing mesenchymal stem cells promote angiogenesis by targeting the PIK3R2-mediated PI3K/Akt signalling pathway. *J. Cell. Mol. Med.* **2021**, *25*, 2148–2162. [CrossRef] [PubMed]
285. Sun, N.; Meng, F.; Xue, N.; Pang, G.; Wang, Q.; Ma, H. Inducible miR-145 expression by HIF-1α protects cardiomyocytes against apoptosis via regulating SGK1 in simulated myocardial infarction hypoxic microenvironment. *Cardiol. J.* **2013**, *25*, 268–278. [CrossRef] [PubMed]
286. Higashi, K.; Yamada, Y.; Minatoguchi, S.; Baba, S.; Iwasa, M.; Kanamori, H.; Kawasaki, M.; Nishigaki, K.; Takemura, G.; Kumazaki, M.; et al. MicroRNA-145 repairs infarcted myocardium by accelerating cardiomyocyte autophagy. *Am. J. Physiol. Heart Circ. Physiol.* **2015**, *309*, H1813–H1826. [CrossRef]
287. Yan, L.; Guo, N.; Cao, Y.; Zeng, S.; Wang, J.; Lv, F.; Wang, Y.; Cao, X. miRNA145 inhibits myocardial infarctioninduced apoptosis through autophagy via Akt3/mTOR signaling pathway in vitro and in vivo. *Int. J. Mol. Med.* **2018**, *42*, 1537–1547. [CrossRef]
288. Liu, Z.; Tao, B.; Fan, S.; Cui, S.; Pu, Y.; Qiu, L.; Xia, H.; Xu, L. Over-expression of microRNA-145 drives alterations in β-adrenergic signaling and attenuates cardiac remodeling in heart failure post myocardial infarction. *Aging* **2020**, *12*, 11603–11622. [CrossRef]
289. Cui, S.; Liu, Z.; Tao, B.; Fan, S.; Pu, Y.; Meng, X.; Li, D.; Xia, H.; Xu, L. miR-145 attenuates cardiac fibrosis through the AKT/GSK-3beta/beta-catenin signaling pathway by directly targeting SOX9 in fibroblasts. *J. Cell. Biochem.* **2021**, *122*, 209–221. [CrossRef]
290. Dong, Y.M.; Liu, X.X.; Wei, G.Q.; Da, Y.N.; Cha, L.; Ma, C.S. Prediction of long-term outcome after acute myocardial infarction using circulating miR-145. *Scand. J. Clin. Lab. Investig.* **2015**, *75*, 85–91. [CrossRef]
291. Zhang, M.; Cheng, Y.J.; Sara, J.D.; Liu, L.J.; Liu, L.P.; Zhao, X.; Gao, H. Circulating MicroRNA-145 is Associated with Acute Myocardial Infarction and Heart Failure. *Chin. Med. J.* **2017**, *130*, 51–56. [CrossRef]
292. Sun, G.; Lu, Y.; Li, Y.; Mao, J.; Zhang, J.; Jin, Y.; Li, Y.; Sun, Y.; Liu, L.; Li, L. miR-19a protects cardiomyocytes from hypoxia/reoxygenation-induced apoptosis via PTEN/PI3K/p-Akt pathway. *Biosci. Rep.* **2017**, *37*, BSR20170899. [CrossRef] [PubMed]
293. Zhong, C.; Wang, K.; Liu, Y.; Lv, D.; Zheng, B.; Zhou, Q.; Sun, Q.; Chen, P.; Ding, S.; Xu, Y.; et al. miR-19b controls cardiac fibroblast proliferation and migration. *J. Cell. Mol. Med.* **2016**, *20*, 1191–1197. [CrossRef] [PubMed]
294. Dehaini, H.; Awada, H.; El-Yazbi, A.; Zouein, F.A.; Issa, K.; Eid, A.A.; Ibrahim, M.; Badran, A.; Baydoun, E.; Pintus, G.; et al. MicroRNAs as Potential Pharmaco-targets in Ischemia-Reperfusion Injury Compounded by Diabetes. *Cells* **2019**, *8*, 152. [CrossRef] [PubMed]
295. Kim, H.; Bae, Y.-U.; Lee, H.; Kim, H.; Jeon, J.S.; Noh, H.; Han, D.C.; Byun, D.W.; Kim, S.H.; Park, H.K.; et al. Effect of diabetes on exosomal miRNA profile in patients with obesity. *BMJ Open Diabetes Res. Care* **2020**, *8*, e001403. [CrossRef]
296. Chien, H.-Y.; Lee, T.-P.; Chen, C.-Y.; Chiu, Y.-H.; Lin, Y.-C.; Lee, L.-S.; Li, W.-C. Circulating microRNA as a diagnostic marker in populations with type 2 diabetes mellitus and diabetic complications. *J. Chin. Med. Assoc.* **2015**, *78*, 204–211. [CrossRef]
297. Sekar, D.; Venugopal, B.; Sekar, P.; Ramalingam, K. Role of microRNA 21 in diabetes and associated/related diseases. *Gene* **2016**, *582*, 14–18. [CrossRef]

298. Bernardo, B.C.; Gao, X.-M.; Winbanks, C.E.; Boey, E.J.H.; Tham, Y.K.; Kiriazis, H.; Gregorevic, P.; Obad, S.; Kauppinen, S.; Du, X.-J.; et al. Therapeutic inhibition of the miR-34 family attenuates pathological cardiac remodeling and improves heart function. *Proc. Natl. Acad. Sci. USA* **2012**, *109*, 17615–17620. [CrossRef]

299. Shi, K.; Sun, H.; Zhang, H.; Xie, D.; Yu, B. miR-34a-5p aggravates hypoxia-induced apoptosis by targeting ZEB1 in cardiomyocytes. *Biol. Chem.* **2019**, *400*, 227–236. [CrossRef]

300. Zhu, Y.; Qian, X.; Li, J.; Lin, X.; Luo, J.; Huang, J.; Jin, Z. Astragaloside-IV protects H9C2(2-1) cardiomyocytes from high glucose-induced injury via miR-34a-mediated autophagy pathway. *Artif. Cells Nanomed. Biotechnol.* **2019**, *47*, 4172–4181. [CrossRef]

301. Wang, R.; Hong, J.; Cao, Y.; Shi, J.; Gu, W.; Ning, G.; Zhang, Y.; Wang, W. Elevated circulating microRNA-122 is associated with obesity and insulin resistance in young adults. *Eur. J. Endocrinol.* **2015**, *172*, 291–300. [CrossRef]

302. Gong, L.; Chang, H.; Xu, H. LncRNA MALAT1 knockdown alleviates oxygen-glucose deprivation and reperfusion induced cardiomyocyte apoptotic death by regulating miR-122. *Exp. Mol. Pathol.* **2019**, *111*, 104325. [CrossRef] [PubMed]

303. Zhang, Z.; Li, H.; Cui, Z.; Zhou, Z.; Chen, S.; Ma, J.; Hou, L.; Pan, X.; Li, Q. Long non-coding RNA UCA1 relieves cardiomyocytes H9c2 injury aroused by oxygen-glucose deprivation via declining miR-122. *Artif. Cells Nanomed. Biotechnol.* **2019**, *47*, 3492–3499. [CrossRef] [PubMed]

304. Jansen, F.; Wang, H.; Przybilla, D.; Franklin, B.S.; Dolf, A.; Pfeifer, P.; Schmitz, T.; Flender, A.; Endl, E.; Nickenig, G.; et al. Vascular endothelial microparticles-incorporated microRNAs are altered in patients with diabetes mellitus. *Cardiovasc. Diabetol.* **2016**, *15*, 49. [CrossRef] [PubMed]

305. Wu, K.; Yang, Y.; Zhong, Y.; Ammar, H.M.; Zhang, P.; Guo, R.; Liu, H.; Cheng, C.; Koroscil, T.M.; Chen, Y.; et al. The effects of microvesicles on endothelial progenitor cells are compromised in type 2 diabetic patients via downregulation of the miR-126/VEGFR2 pathway. *Am. J. Physiol. Endocrinol. Metab.* **2016**, *310*, E828–E837. [CrossRef] [PubMed]

306. Villard, A.; Marchand, L. Diagnostic Value of Cell-free Circulating Micrornas for Obesity and Type 2 Diabetes: A Meta-analysis. *J. Mol. Biomark. Diagn.* **2015**, *6*, 251. [CrossRef]

307. Ortega, F.J.; Mercader, J.M.; Moreno-Navarrete, J.M.; Rovira, O.; Guerra, E.; Esteve, E.; Xifra, G.; Martinez, C.; Ricart, W.; Rieusset, J.; et al. Profiling of circulating microRNAs reveals common microRNAs linked to type 2 diabetes that change with insulin sensitization. *Diabetes Care* **2014**, *37*, 1375–1383. [CrossRef]

308. Zampetaki, A.; Kiechl, S.; Drozdov, I.; Willeit, P.; Mayr, U.; Prokopi, M.; Mayr, A.; Weger, S.; Oberhollenzer, F.; Bonora, E.; et al. Plasma MicroRNA Profiling Reveals Loss of Endothelial MiR-126 and Other MicroRNAs in Type 2 Diabetes. *Circ. Res.* **2010**, *107*, 810–817. [CrossRef]

309. Gan, L.; Xie, D.; Liu, J.; Lau, W.B.; Christopher, T.A.; Lopez, B.; Zhang, L.; Gao, E.; Koch, W.; Ma, X.L.; et al. Small Extracellular Microvesicles Mediated Pathological Communications between Dysfunctional Adipocytes and Cardiomyocytes as a Novel Mechanisms Exacerbating Ischemia/Reperfusion Injury in Diabetic Mice. *Circulation* **2020**, *141*, 968–983. [CrossRef]

310. Liang, Y.Z.; Li, J.J.H.; Xiao, H.B.; He, Y.; Zhang, L.; Yan, Y.X. Identification of stress-related microRNA biomarkers in type 2 diabetes mellitus: A systematic review and meta-analysis. *J. Diabetes* **2020**, *12*, 633–644. [CrossRef]

311. Lu, C.; Wang, X.; Ha, T.; Hu, Y.; Liu, L.; Zhang, X.; Yu, H.; Miao, J.; Kao, R.; Kalbfleisch, J.; et al. Attenuation of cardiac dysfunction and remodeling of myocardial infarction by microRNA-130a are mediated by suppression of PTEN and activation of PI3K dependent signaling. *J. Mol. Cell. Cardiol.* **2015**, *89*, 87–97. [CrossRef]

312. Shahrokhi, S.Z.; Saeidi, L.; Sadatamini, M.; Jafarzadeh, M.; Rahimipour, A.; Kazerouni, F. Can miR-145-5p be used as a marker in diabetic patients? *Arch. Physiol. Biochem.* **2020**. [CrossRef] [PubMed]

313. Deng, Z.B.; Poliakov, A.; Hardy, R.W.; Clements, R.; Liu, C.; Liu, Y.; Wang, J.; Xiang, X.; Zhang, S.; Zhuang, X.; et al. Adipose tissue exosome-like vesicles mediate activation of macrophage-induced insulin resistance. *Diabetes* **2009**, *58*, 2498–2505. [CrossRef] [PubMed]

314. Li, S.; Wei, J.; Zhang, C.; Li, X.; Meng, W.; Mo, X.; Zhang, Q.; Liu, Q.; Ren, K.; Du, R.; et al. Cell-Derived Microparticles in Patients with Type 2 Diabetes Mellitus: A Systematic Review and Meta-Analysis. *Cell. Physiol. Biochem.* **2016**, *39*, 2439–2450. [CrossRef] [PubMed]

315. Yuan, Y.; Wang, J.; Chen, Q.; Wu, Q.; Deng, W.; Zhou, H.; Shen, D. Long non-coding RNA cytoskeleton regulator RNA (CYTOR) modulates pathological cardiac hypertrophy through miR-155-mediated IKKi signaling. *Biochim. Biophys. Acta Mol. Basis Dis.* **2019**, *1865*, 1421–1427. [CrossRef] [PubMed]

316. Gao, L.; Li, T.; Li, S.; Song, Z.; Chang, Y.; Yuan, L. Schisandrin A protects against isoproterenol-induced chronic heart failure via miR-155. *Mol. Med. Rep.* **2021**, *25*, 24. [CrossRef]

317. Sanchez-Ceinos, J.; Rangel-Zuniga, O.A.; Clemente-Postigo, M.; Podadera-Herreros, A.; Camargo, A.; Alcala-Diaz, J.F.; Guzman-Ruiz, R.; Lopez-Miranda, J.; Malagon, M.M. miR-223-3p as a potential biomarker and player for adipose tissue dysfunction preceding type 2 diabetes onset. *Mol. Ther. Nucleic Acids* **2021**, *23*, 1035–1052. [CrossRef]

318. Yan, L.-N.; Zhang, X.; Xu, F.; Fan, Y.-Y.; Ge, B.; Guo, H.; Li, Z.-L. Four-microRNA signature for detection of type 2 diabetes. *World J. Clin. Cases* **2020**, *8*, 1923–1931. [CrossRef]

319. Dai, G.-H.; Liu, N.; Zhu, J.-W.; Yao, J.; Yang, C.; Ma, P.-Z.; Song, X.-B. Qi-Shen-Yi-Qi Dripping Pills Promote Angiogenesis of Ischemic Cardiac Microvascular Endothelial Cells by Regulating MicroRNA-223-3p Expression. *Evid.-Based Complement. Altern. Med.* **2016**, *2016*, 5057328. [CrossRef]

320. Yang, L.; Li, Y.; Wang, X.; Mu, X.; Qin, D.; Huang, W.; Alshahrani, S.; Nieman, M.; Peng, J.; Essandoh, K.; et al. Overexpression of miR-223 Tips the Balance of Pro- and Anti-hypertrophic Signaling Cascades toward Physiologic Cardiac Hypertrophy. *J. Biol. Chem.* **2016**, *291*, 15700–15713. [CrossRef]

321. Liu, X.; Xu, Y.; Deng, Y.; Li, H. MicroRNA-223 Regulates Cardiac Fibrosis After Myocardial Infarction by Targeting RASA1. *Cell. Physiol. Biochem.* **2018**, *46*, 1439–1454. [CrossRef]

322. Cao, L.; Chai, S. miR-320-3p is involved in morphine pre-conditioning to protect rat cardiomyocytes from ischemia/reperfusion injury through targeting Akt3. *Mol. Med. Rep.* **2020**, *22*, 1480–1488. [CrossRef] [PubMed]

323. Sun, K.; Chang, X.; Yin, L.; Li, J.; Zhou, T.; Zhang, C.; Chen, X. Expression and DNA methylation status of microRNA-375 in patients with type 2 diabetes mellitus. *Mol. Med. Rep.* **2014**, *9*, 967–972. [CrossRef] [PubMed]

324. Wu, X.; Li, Y.; Man, B.; Li, D. Assessing MicroRNA-375 Levels in Type 2 Diabetes Mellitus (T2DM) Patients and Their First-Degree Relatives with T2DM. *Diabetes Metab. Syndr. Obes. Targets Ther.* **2021**, *14*, 1445–1451. [CrossRef] [PubMed]

325. Garikipati, V.N.S.; Verma, S.K.; Jolardarashi, D.; Cheng, Z.; Ibetti, J.; Cimini, M.; Tang, Y.; Khan, M.; Yue, Y.; Benedict, C.; et al. Therapeutic inhibition of miR-375 attenuates post-myocardial infarction inflammatory response and left ventricular dysfunction via PDK-1-AKT signalling axis. *Cardiovasc. Res.* **2017**, *113*, 938–949. [CrossRef]

326. Yan, M.; Liu, Q.; Jiang, Y.; Wang, B.; Ji, Y.; Liu, H.; Xie, Y. Long Noncoding RNA LNC_000898 Alleviates Cardiomyocyte Apoptosis and Promotes Cardiac Repair After Myocardial Infarction Through Modulating the miR-375/PDK1 Axis. *J. Cardiovasc. Pharmacol.* **2020**, *76*, 77–85. [CrossRef]

327. Zheng, K.; Zhang, Q.; Lin, G.; Li, Y.; Sheng, Z.; Wang, J.; Chen, L.; Lu, H.H. Activation of Akt by SC79 protects myocardiocytes from oxygen and glucose deprivation (OGD)/re-oxygenation. *Oncotarget* **2017**, *8*, 14978–14987. [CrossRef]

328. Moreira, J.B.; Wohlwend, M.; Alves, M.N.; Wisloff, U.; Bye, A. A small molecule activator of AKT does not reduce ischemic injury of the rat heart. *J. Transl. Med.* **2015**, *13*, 76. [CrossRef]

329. Matsui, T.; Tao, J.; del Monte, F.; Lee, K.H.; Li, L.; Picard, M.; Force, T.L.; Franke, T.F.; Hajjar, R.J.; Rosenzweig, A. Akt activation preserves cardiac function and prevents injury after transient cardiac ischemia in vivo. *Circulation* **2001**, *104*, 330–335. [CrossRef]

330. Peng, Y.; Wang, Y.; Zhou, C.; Mei, W.; Zeng, C. PI3K/Akt/mTOR Pathway and Its Role in Cancer Therapeutics: Are We Making Headway? *Front. Oncol.* **2022**, *12*, 819128. [CrossRef]

Article

PI3K/AKT/mTOR Signaling Pathway Is Required for JCPyV Infection in Primary Astrocytes

Michael P. Wilczek [1], Francesca J. Armstrong [1], Colleen L. Mayberry [1], Benjamin L. King [1,2] and Melissa S. Maginnis [1,2,*]

[1] Department of Molecular and Biomedical Sciences, University of Maine, Orono, ME 04469, USA; michael.wilczek@maine.edu (M.P.W.); francesca.armstrong@maine.edu (F.J.A.); colleen.mayberry@maine.edu (C.L.M.); benjamin.l.king@maine.edu (B.L.K.)
[2] Graduate School in Biomedical Sciences and Engineering, University of Maine, Orono, ME 04469, USA
* Correspondence: melissa.maginnis@maine.edu

Abstract: Astrocytes are a main target of JC polyomavirus (JCPyV) in the central nervous system (CNS), where the destruction of these cells, along with oligodendrocytes, leads to the fatal disease progressive multifocal leukoencephalopathy (PML). There is no cure currently available for PML, so it is essential to discover antivirals for this aggressive disease. Additionally, the lack of a tractable in vivo models for studying JCPyV infection makes primary cells an accurate alternative for elucidating mechanisms of viral infection in the CNS. This research to better understand the signaling pathways activated in response to JCPyV infection reveals and establishes the importance of the PI3K/AKT/mTOR signaling pathway in JCPyV infection in primary human astrocytes compared to transformed cell lines. Using RNA sequencing and chemical inhibitors to target PI3K, AKT, and mTOR, we have demonstrated the importance of this signaling pathway in JCPyV infection of primary astrocytes not observed in transformed cells. Collectively, these findings illuminate the potential for repurposing drugs that are involved with inhibition of the PI3K/AKT/mTOR signaling pathway and cancer treatment as potential therapeutics for PML, caused by this neuroinvasive virus.

Keywords: JC polyomavirus; PML; astrocytes; SVGA cells; primary cells; PI3K; AKT; mTOR; rapamycin; wortmannin

check for **updates**

Citation: Wilczek, M.P.; Armstrong, F.J.; Mayberry, C.L.; King, B.L.; Maginnis, M.S. PI3K/AKT/mTOR Signaling Pathway Is Required for JCPyV Infection in Primary Astrocytes. *Cells* **2021**, *10*, 3218. https://doi.org/10.3390/cells10113218

Academic Editors: Jean Christopher Chamcheu, Claudia Bürger and Shile Huang

Received: 8 October 2021
Accepted: 12 November 2021
Published: 18 November 2021

Publisher's Note: MDPI stays neutral with regard to jurisdictional claims in published maps and institutional affiliations.

1. Introduction

JC polyomavirus (JCPyV) is a human-specific pathogen and is the causative agent of a fatal disease in the central nervous system (CNS) known as progressive multifocal leukoencephalopathy (PML) [1–4]. The virus is present in ~60 to 80% of the adult population, where initial infection is thought to occur in tonsillar tissue [5–8], allowing for secondary infections in circulating B cells, bone marrow, and kidneys [7]. In nearly all infected individuals, JCPyV is characterized as a persistent, asymptomatic infection, and the virus is periodically shed in the urine [5–7,9–11]. However, during immunosuppression, JCPyV can reactivate and spread to the CNS [9,12,13], causing the fatal, demyelinating disease PML [1–4]. Due to the immunosuppressive state associated with HIV infection, a large proportion of individuals diagnosed with PML are infected with HIV; however, due to more effective treatments related to HIV/AIDS, PML incidence is decreasing in this population [14–16]. Unfortunately, new risk groups are emerging, with a higher proportion of PML cases diagnosed in patients with hematological malignancies and in patients taking immunomodulatory therapies for immune-mediated diseases [16]. Specifically, individuals with multiple sclerosis (MS) undergoing natalizumab treatment are significantly at risk for PML development [16–19]. Current treatment for PML is focused on removal of immunosuppressive therapies or treatment of the underlying immunosuppression [19–22]. Additional treatment options, including adoptive T cell transfer and checkpoint inhibitors, prolong life expectancy; however, these treatments are still relatively new, only address the underlying immunosuppression and can result in severe morbidity [19,20,23–25].

In the CNS, astrocytes are a main target during JCPyV infection [26]. The lytic destruction of these cells, along with JCPyV infection of oligodendrocytes, leads to PML [1–4]. Once a cell becomes infected, JCPyV hijacks the host cell machinery to produce viral proteins, including Large T Antigen (T Ag), to create a favorable environment for JCPyV DNA replication; this occurs through the binding of T Ag to retinoblastoma (Rb) and sequestering p53 [27,28]. As transformation ensues and viral DNA replication progresses, the late gene expression of JCPyV occurs [28,29]. Here, viral proteins (VP) 1, 2, and 3 are transcribed and translated to encapsidate the viral DNA [30], before subsequent release of the newly formed virions into the surrounding environment.

Initial studies were performed to better understand JCPyV infection in astrocytes, and the infectious cycle was compared to infection in SVG-A cells (SVGAs) [31]. SVGAs are a mixed population of glial cells, mostly comprised of astrocytes, immortalized with SV40 T Ag to support robust levels of JCPyV infection [32–34]. Our previous research demonstrated that the infectious cycle was delayed in primary normal human astrocytes (NHAs) [31]. This delay was most likely the consequence of SV40 T Ag, as the immortalization of NHAs with this protein (referred to as NHA-Ts) resulted in levels of infection comparable to that observed in SVGAs [31]. In addition, differences in JCPyV infection between cell types was attributed to variation in cyclin expression [31]. Cyclins are commonly used as markers for the cell cycle [26,35–37], and it was demonstrated that JCPyV infection was able to drive the cell into S phase by the accumulation of cyclin E in the nucleus of NHAs [31]. The S phase is needed for successful DNA viral replication [28]. However, as the infectious cycle ensued, cyclin B1, a marker of the G_2/M phase, accumulated in the cytoplasm of NHAs, allowing for productive viral infection; this accumulation of cyclin B1 was not observed during infection of immortalized cells at the same time points [31].

SV40 T Ag is known to dysregulate the cell cycle and activate cellular pathways that can potentially confound mechanisms during JCPyV infection [28,38]. One of these pathways is the AKT signaling pathway, important in cell growth and survival [39–45]. Previous research has demonstrated that SV40 T Ag can inhibit apoptosis in *ts*13 cells by activating the AKT signaling pathway and can directly phosphorylate AKT in U2OS cells [46,47]. In relation to this, our understanding of the AKT signaling pathway during JCPyV infection is limited. Currently, it has been shown that a component of the AKT pathway, specifically phosphoinositide 3-kinase γ (PI3Kγ), is involved in JCPyV infection in cells immortalized with SV40 T Ag, such as SVGAs; the authors hypothesize that this pathway was activated upon stimulation of G protein-coupled receptors (GPCRs) [48]. However, PI3K signaling is complex and is activated by various mechanisms, including Ras, a component of the mitogen-activated protein kinase (MAPK) pathway, and has numerous isoforms defined by their catalytic and regulatory subunits [49]. As a result of SV40 T Ag activating AKT, and being able to influence other pathways, it is unclear if the previously identified PI3K/AKT signaling pathway was due to the immortalization of cells by SV40 T Ag or by JCPyV infection. Additionally, genes downstream of AKT, such as mechanistic target of rapamycin (mTOR), have also been implicated in JCPyV infection [50]. Overall, these experiments demonstrated the importance of PI3Kγ and mTOR [48,50], but due to the consequences of immortalization, additional research should elucidate these proteins in a primary cell line.

There have been many advances in understanding the PI3K/AKT/mTOR signaling pathway, as this pathway is frequently altered in human cancer [51]. As a result, there have been numerous chemical inhibitors developed targeting the PI3K/AKT/mTOR signaling pathway, some of which are in clinical trials or are already FDA approved [52]. This includes MK2206, an AKT inhibitor [53], and rapamycin, also known clinically as sirolimus (the generic drug name), an mTOR inhibitor, that has been FDA-approved for decades and used to prevent organ transplant rejection and to protect coronary stents [54–56]. Additionally, a clinical study has demonstrated that mTOR inhibitors, such as rapamycin, can cross the blood–brain barrier (BBB), providing a premise for the treatment of neurological disorders [57,58].

Along with the diverse drugs targeting this pathway and their ability to traverse the BBB, they have also been implicated in JCPyV infection, contributing to a framework for future PML therapeutics. MK2206 has been demonstrated to reduce JCPyV DNA replication in an oligodendrocyte cell line [59]; however, mTOR inhibitors, such as rapamycin, increased expression of JCPyV T Ag in an immortalized kidney cell line—human embryonic kidney (HEK) 293 cells [50]. Due to these recent findings, the lack of an effective treatment for PML, and observed differences in JCPyV infection in immortalized cells, components of the PI3K/AKT/mTOR signaling pathway were analyzed through RNA sequencing (RNA-seq) analysis, and chemical inhibitors that target this signaling pathway were examined for their capacity to reduce JCPyV infection. These studies were conducted in primary human astrocytes (i.e., NHAs) and compared to cell types immortalized with SV40 T Ag, SVGAs, and NHA-Ts, to determine whether transformation of cells with T Ag yielded differing outcomes in activation of the PI3K/AKT/mTOR pathway and the JCPyV infectious cycle. This study also further characterizes the mechanisms of JCPyV infection of NHAs and describes how additional cellular pathways are possibly intertwined, such as pathways that are required to transform the cell, to support viral infection.

2. Materials and Methods

2.1. Cells and Viruses

The maintenance of primary normal human astrocytes (i.e., NHAs) has been previously described [31]. In brief, NHAs were purchased from Lonza Walkersville Inc. (Walkersville, MD, USA), where they were isolated from a 19-week-gestation female with no detected levels of HIV, hepatitis B virus (HBV), or hepatitis C virus (HCV). They were cultured according to the manufacturer's guidelines in astrocyte growth medium and supplemented with SingleQuots supplements (Lonza Walkersville Inc. (Walkersville, MD, USA)) and 1% penicillin-streptomycin (P-S) (Corning, Corning, NY, USA). All experiments were performed at low passages (P2 to P10). SVGAs were graciously provided by the Atwood Laboratory (Brown University). They were cultured in complete minimum essential medium (MEM) (Corning, Corning, NY, USA), with 10% fetal bovine serum (FBS) (Atlanta Biologicals, Bio-Techne, Flowery Branch, GA, USA), 1% P-S, and 0.1% Plasmocin prophylactic (InvivoGen, San Diego, CA, USA). All cell types were grown in a humidified incubator at 37 °C with 5% CO_2. The generation of NHA-Ts has been previously described [31]; they were cultured similarly to SVGAs, however, 16% FBS was used. The expression of SV40 T Ag in NHA-Ts was consistently monitored using epifluorescence microscopy staining for SV40 T Ag (Table 1 below). All cell types were grown in a humidified incubator at 37 °C with 5% CO_2. JCPyV strains were provided by the Atwood Laboratory (Brown University, Providence, RI, USA). The generation and production of the lysate viral strains of Mad-1/SVEΔ have been described previously [60].

Table 1. Antibodies used in immunofluorescence and ICW assays.

Protein	1° Antibody (Dilution, Source)	2° Antibody (Dilution, Manufacturer)
JCPyV T Ag	PAB962 (1:5, hybridoma, Tevethia Lab, Penn State University, State College, PA, USA)	anti-mouse Alexa Fluor 594 (1:1000, Thermo Fisher Scientific (Waltham, MA, USA))
JCPyV VP1	Ab34756 (1:1000, Abcam)	
pERK (P-p44/42 MAPK at T202/Y204)	9101S (1:750, CST)	
pAKT (S473)	4060S (1:400, CST)	anti-rabbit IRDye 800CW (1:10,000, LI-COR, Lincoln, NE, USA)
pmTOR (S2448)	44-1125G (1:1000, Invitrogen, Waltham, MA, USA)	
CST, Cell Signaling Technology; ICW, In-Cell Western assay		

2.2. JCPyV Infection

All cell types were seeded in 96-well plates with ~10,400 cells/well, for 70% confluency at the time of infection. Cells were infected at 37 °C for 1 h with 42 µL/well of MEM containing 10% FBS, 1% P-S, and 0.1% Plasmocin prophylactic, across all cell types. The multiplicities of infection (MOIs) are indicated in the figure legend. Following the 1 h incubation, cells were fed with 100 µL/well of medium and incubated at 37 °C for the duration of the infection. Cells were fixed at timepoints indicated in the legends and either stained for indirect immunofluorescence to determine % infection or stained for ICW to measure protein expression.

2.3. Chemical Inhibitor Treatments

All chemical inhibitors were reconstituted in DMSO. Inhibitors that targeted the PI3K/AKT/mTOR signaling pathway were used to pretreat cells for 30 min, while U0126 pretreatment was for 1 h. The concentrations of chemical inhibitors are indicated in the figure legends. All chemical inhibitors were diluted in MEM containing 10% FBS, 1% P-S, and 0.1% Plasmocin prophylactic, prior to JCPyV infection. Following the 1 h viral incubation, all cell types were fed with 100 µL/well of media containing the chemical inhibitor or the DMSO control. For experiments that quantified protein expression, all chemical inhibitors were diluted in incomplete MEM (0% FBS), and cells were treated for 24 h and then fixed in 4% PFA for subsequent analysis. U0126 was purchased from Cell Signaling Technology (Danvers, MA, USA), (#9903S), wortmannin was purchased from Sigma (St. Louis, MO, USA) (#W1628), rapamycin was purchased from Frontier Scientific (Logan, UT, USA (#JK948477), and MK2206 and PP242 was purchased from Selleckchem (Houston, TX, USA) (#S1078 and #S2218). The concentrations of all the chemical inhibitors utilized did not affect cell viability as measured by 4,5-dimethylthiazol-2-yl)-5-(3A-carboxymethoxyphenyl)-2-(4-sulfophenyl)-2H-tetrazolium (MTS) assay (Promega, Madison, WI, USA) (data not shown).

2.4. Indirect Immunofluorescence Staining and Quantitation of JCPyV Infection

Following JCPyV infection, all cell types were stained for T Ag or VP1 (Table 1) at RT. When quantifying for JCPyV T Ag, NHAs, SVGAs and NHA-Ts were fixed at 48 hpi, and when quantifying for both viral proteins, cells were fixed at 72 hpi and stained for both JCPyV T Ag and VP1. Cells were fixed with 4% PFA for 10 min and washed with 1 X phosphate-buffered saline (PBS) with 0.01% Tween. NHAs, SVGAs, and NHA-Ts were then permeabilized using PBS-0.5% Triton X-100 for 15 min and blocked with PBS with 0.01% Tween and 10% goat serum for 45 min. Cells were then stained for T Ag or both viral proteins using the antibodies listed in Table 1 at RT for 1 h. Following the 1° antibody incubation, all cell types were washed three times in PBS-0.01% Tween and counterstained with an anti-mouse Alexa Fluor 594 2° antibody at RT for 1 h. Cells were then washed with PBS-0.01% Tween and the nuclei were stained using DAPI (4′,6-diamidino-2-phenylindole) at RT for 5 min. Finally, cells were washed with PBS-0.01% Tween and stored in PBS-0.01% Tween.

To quantify infectivity and nuclear expression of viral protein, T Ag or VP1 over the total number of DAPI-positive cells (percent infection), a Nikon Eclipse Ti epifluorescence microscope (Micro Video Instruments, Inc., Avon, MA, USA) equipped with a 20× objective was used. T Ag- or VP1-expressing cells were counted manually; however, the DAPI-positive cells were determined using a binary algorithm in Nikon NIS-Elements Basic Research software (Micro Video Instruments, Inc.) (version 4.50.00, 64 bit). This algorithm separated cells based on three variables: intensity, diameter, and circularity, resulting in an accurate measurement of the total number of cells in each field of view [31,61–64].

2.5. ICW Assay to Measure Protein Expression Using LI-COR Software

Protein expression measuring phosphorylated ERK, phosphorylated AKT, and phosphorylated mTOR were performed at 24 h post-treatment using an ICW assay [64,65]. Cells were plated in a 96-well plate at ~100% confluency and treated with the various chemical inhibitors for 24 h. Following treatment of the inhibitors, cells were fixed in 4% PFA and washed with PBS-0.01% Tween. Cells were then permeabilized with $1\times$ PBS-0.5% Triton X-100 at RT for 15 min and blocked with Tris-buffered saline (TBS) Odyssey buffer (LI-COR) at RT for 1 h. All cell types were stained with the respective 1° antibody (Table 1) in TBS Odyssey blocking buffer at 4 °C for ~16 h while rocking. The following day, the 1° antibody was removed, and the cells were washed twice with PBS-0.01% Tween. NHAs, SVGAs, and NHA-Ts were then counterstained with 2° antibody, as indicated in Table 1 and CellTag (1:500, LI-COR) for 1 h. Finally, cells were washed with PBS-0.01% Tween and the wells were aspirated. Prior to scanning, the bottom of the plate was cleaned with 70% ethanol and the lid was removed. Plates were weighted with a silicone mat (LI-COR) and imaged using the LI-COR Odyssey CLx infrared imaging system (LI-COR) to detect both the 700 and 800 nm intensities. The imaging settings for the LI-COR were as follows: medium quality, 42 μm resolution, with a 3.0 mm focus offset; when the scan was complete, the 700 and 800 nm channels were aligned and the ICW analysis was performed in Image Studio (version 5.2) (LI-COR). Protein quantification was determined in two steps. First, background fluorescence from the 800 nm channel (wells that only received 2° antibody) were subtracted to the 800 nm channel, in which the protein of interest was being quantified. Next, the ratio was determined using this new value (protein of interest), normalized to the 700 nm channel (CellTag) or the overall number of cells in each well [61,65,66].

2.6. RNA-seq and Pathway Analysis

RNA-seq data is available under the accession number: GSE183322. The read counts were generated as previously described [66] and analyzed using RStudio (version 1.2.1335, Boston, MA, USA) and R/edgeR (version 3.30.3) http://bioconductor.org/packages/release/bioc/html/edgeR.html) (accessed on 1 June 2021) [67]. Genes expressed were mapped to the PI3K/AKT signaling pathway from KEGG (https://www.kegg.jp) (accessed on 1 June 2021) [68–70] by obtaining the gene symbols and then converting them to Ensembl Gene IDs using EnsemblBioMart (Ensembl version 103, https://www.ensembl.org) (accessed on 1 June 2021) [71]. The Ensembl Gene IDs were matched to the RNA-seq data to acquire the genes expressed in the PI3K/AKT signaling pathway. The list of expressed PI3K/AKT genes was then merged with the results of the R/edgeR analysis [66] and subdivided based on an unadjusted p value of less than 0.10. Venn diagrams showing the overlap between the differentially expressed genes for each cell type and timepoint were created using the R package, eulerr (version 6.1.0, https://cran.r-project.org/web/packages/eulerr/) (accessed on 1 June 2021) [72].

2.7. Statistical Analysis and Graphing in RStudio

A two-sample Student's t test assuming unequal variances was used to compare the mean values for at least triplicate samples when the data was normally distributed. Non-normal distribution of data was determined by both the Shapiro–Wilk's test and a quantile-quantile (Q-Q plot) in R. Statistical analyses were performed using the Wilcoxon signed rank test, to compare the median values for two populations, or the Kruskal–Wallis test to compare the median values of more than two populations. If the Kruskal–Wallis test determined a significant result with at least two groups, then the pairwise Wilcoxon rank sum test, along with the Bonferroni adjustment, was used to determine the pairs of groups that were statistically different. The Student's t test was determined in Microsoft Excel (version 2110, Microsoft, Redmond, WA, USA), and all other statistical tests were performed using R. Statistical analyses for the RNA-seq data was performed using R/edge R (version 3.30.3, http://bioconductor.org/packages/release/bioc/html/edgeR.html) (accessed on 1 June 2021) [67].

3. Results

3.1. U0126, a Common MEK Inhibitor, Does Not Reduce JCPyV Infection in Primary Astrocytes

In addition to the PI3K/AKT/mTOR signaling pathway implicated in JCPyV infection [48,50,59], the mitogen-activated protein kinase, extracellular signal-regulated kinase (MAPK/ERK) pathway is required for JCPyV infection [61,73,74]. This pathway is temporally regulated, specifically being phosphorylated upon JCPyV infection, and infection of immortalized cells is significantly reduced when cells are treated with ERK siRNAs or inhibitors [61,66,73,74]. The MAPK/ERK pathway overlaps with the PI3K/AKT/mTOR pathway and both pathways have important roles in cell survival and differentiation [39,75]. To determine if inhibiting the phosphorylation of ERK decreases JCPyV in primary astrocytes, a well-known MEK inhibitor, U0126, was tested for its effects on JCPyV infection. NHAs, SVGAs, and NHA-Ts were treated with U0126 at 10 μM or the DMSO vehicle control (vehicle control), and subsequently infected with JCPyV (Figure 1A). U0126 did not decrease infection in NHAs at 48 hpi compared to SVGAs and NHA-Ts (Figure 1A, top). As a negative control, all cell types were infected with SV40, which does not require MEK for infection [76,77], and U0126 did not influence SV40 infection in any cell type (Figure 1A, bottom). To confirm the inhibitory effect of U0126 in NHAs, SVGAs, and NHA-Ts, ERK phosphorylation following treatment of the MEK inhibitor for 1 h was evaluated by In-Cell Western (ICW; Figure 1B). U0126 nearly abolished ERK phosphorylation in all three cell types (Figure 1C). PMA, an ERK activator, was used as a control to further determine the specificity of the MEK inhibitor. All cell types were treated with U0126 for 1 h, treated with PMA for 5 min, and ERK phosphorylation was nearly reduced to comparable levels to cells treated with U0126 alone (Figure 1C). Together, these data suggest that although elevated levels of ERK phosphorylation promote JCPyV infection [66], inhibition of ERK phosphorylation does not decrease JCPyV infection in primary astrocytes when compared to cells immortalized with SV40 T Ag. Thus, these data suggest that the mechanisms of MAPK cell signaling activation utilized in JCPyV infection of NHAs differs from those in SVGAs and NHA-Ts.

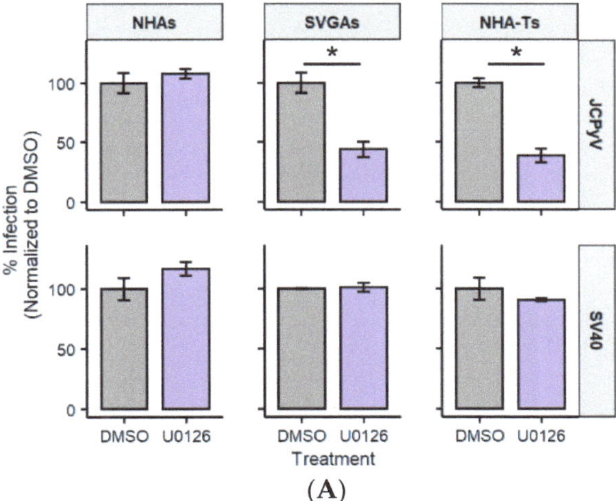

Figure 1. *Cont.*

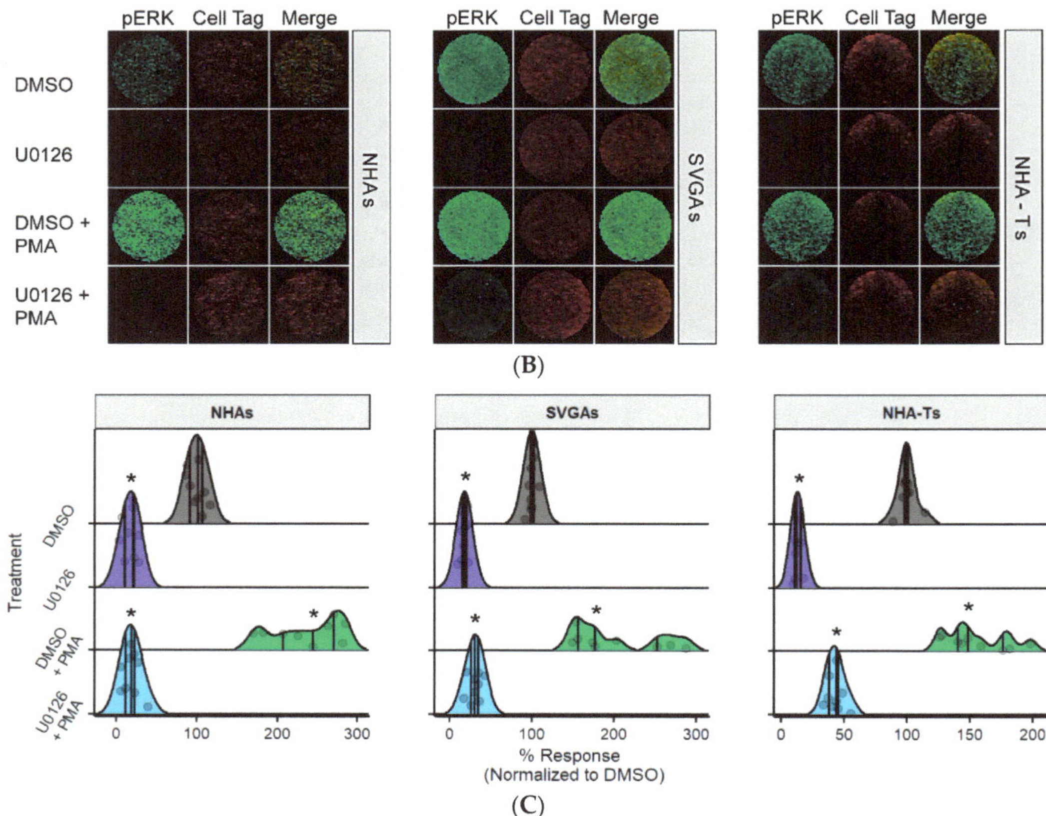

Figure 1. MEK inhibitor U0126 does not decrease JCPyV infection in NHAs. (**A**) NHAs, SVGAs, and NHA-Ts were pretreated with the MEK inhibitor U0126 at 10 µM or DMSO at an equivalent volume control and then infected with JCPyV (MOI = 2.0 FFU/cell; **A**, top) or SV40 (MOI = 2.0 FFU/cell; **A**, bottom) at 37 °C for 1 h. Cells were incubated in media containing DMSO or U0126 for 48 h and then fixed and stained by indirect immunofluorescence. Percent infection was determined by counting the number of JCPyV T Ag- or SV40 VP1-positive nuclei divided by the number of DAPI-positive nuclei for five ×20 fields of view for triplicate samples. Data is representative of three individual experiments. Error bars indicate SD. Student's *t* test was used to determine statistical significance comparing DMSO to U0126, with each cell type and viral protein. *, $p < 0.01$. (**B**) NHAs, SVGAs, and NHA-Ts were pretreated with DMSO or U0126 for 1 h in addition to PMA (40 nM) treatment of the indicated wells for the final 5 min at 37 °C. Cells were fixed and stained for pERK (green) or CellTag (red). (**C**) Percentage of pERK for each treatment was quantitated by In-Cell Western signal intensity values per [(pERK)/Cell Tag × 100% = % response] within each ICW analysis with LI-COR software (representative image shown in **B**). Density ridgeline plots represent the distribution of samples (individual points) with the lower quartile, median and upper quartile denoted as black lines in each distribution. Colored points represent individual points for each treatment (3 replicates, performed in triplicate). A Kruskal–Wallis test, along with the Bonferroni adjustment, was used to compare treatments in each cell type; however, only significance with the DMSO treatment versus other treatments are illustrated. *, $p < 0.01$. Data are representative of three independent experiments performed in triplicate.

3.2. AKT Phosphorylation Is Moderately Increased in NHAs during U0126 Treatment Compared to SVGAs and NHA-Ts

The ERK signaling cascade can interact with the AKT signaling pathway [78] and they work together to induce cellular transformation and survival [39,75]. Research has demonstrated that treatment of cells with MEK inhibitors, such as U0126, results in increased AKT activity in various cell types [79–81]. To determine if U0126 influences AKT phosphorylation, NHAs, SVGAs, and NHA-Ts were treated with U0126 and AKT phosphorylation was measured (Figure 2). Cells were serum-starved and treated with U0126 (10 μM) or the vehicle control for 24 h, and both ERK and AKT phosphorylation were measured by ICW (Figure 2A). U0126 increased AKT phosphorylation in NHAs, albeit by ~20%, while AKT phosphorylation was not altered in SVGAs and NHA-Ts (Figure 2B). This experiment demonstrated that the crosstalk between the AKT signaling pathway and the ERK signaling cascade occurs in primary astrocytes. Additionally, it provides a premise for explaining why U0126 did not decrease JCPyV infection in NHAs; U0126 inhibits ERK phosphorylation, suggesting that the virus may also hijack the AKT signaling pathway to support viral replication.

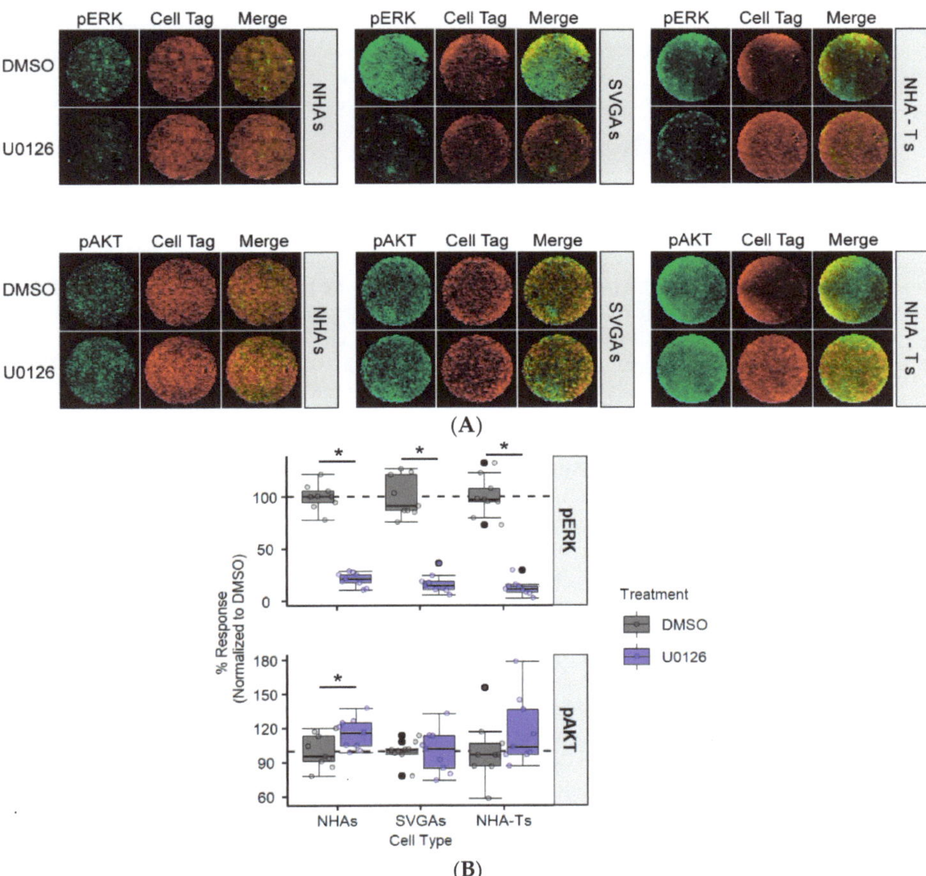

Figure 2. U0126 increases AKT phosphorylation in NHAs. (**A**) NHAs, SVGAs, and NHA-Ts were treated with DMSO or U0126 (10 μM) at 37 °C for 24 h. Cells were fixed and stained for pERK or pAKT (green) or CellTag (red). Entire wells of a 96-well plate are shown (**B**) Percentage of pERK and pAKT for each cell type was quantitated by ICW signal intensity values per [(pERK or pAKT)/Cell Tag × 100% = % response] within each ICW analysis with LI-COR software. Box and whisker plots represent the distribution of samples (individual points), with the lower quartile, median and upper quartile denoted as black lines. Colored points represent individual points for each cell type (3 replicates, performed in triplicate), and outliers are represented by black circles. The dashed line indicates the normalized DMSO % response of pERK and pAKT. A Wilcoxon rank sum exact test was used to compare treatments in each cell type. Data are representative of three independent experiments performed in triplicate. *, $p < 0.05$.

3.3. PI3K/AKT Signaling Pathway Genes Are Upregulated during JCPyV Infection in NHAs

Previous research has demonstrated that the PI3K/AKT signaling pathway, specifically the AKT and PI3Kγ steps, is required for JCPyV infection in SVGAs and an oligodendrocyte cell line [48,59,82]. To determine if this pathway is implicated in JCPyV infection in NHAs, RNA-seq data was analyzed [66] over the course of JCPyV infection in NHAs and SVGAs. Differentially-expressed genes (unadjusted p value < 0.10) were mapped to the PI3K/AKT signaling pathway from the Kyoto Encyclopedia of Genes and Genomes (KEGG) database [68–70]. The table of genes with their p values and log-fold changes (FCs) can be found in the Supplementary Material (Table S1). At 24 hpi, 25 genes that mapped to the pathway were downregulated in SVGAs but upregulated in NHAs (Figure 3, top). At 48 hpi, only one gene was upregulated in SVGAs compared to 23 genes upregulated in NHAs (Figure 3, middle). At 96 hpi, approximately 51% of the genes that mapped to the pathway were upregulated in NHAs versus ~38% in SVGAs (Figure 3, bottom). These data illustrate that compared to SVGAs, JCPyV infection in NHAs activated more genes in the PI3K-AKT signaling pathway, providing additional evidence for the involvement of this pathway during infection in primary astrocytes.

3.4. AKT Is Differentially Expressed and Required for JCPyV Infection in NHAs

To further validate the previous results, the expression of the three isoforms of AKT (*AKT1*, *AKT2* and *AKT3*) were analyzed (Figure 4A). Both *ATK1* and *ATK2* were significantly downregulated in SVGAs at 24 and 96 hpi (unadjusted p value < 0.05) and upregulated in NHAs (unadjusted p value < 0.1; Figure 4A). Due to the differential gene expression observed, NHAs, SVGAs, and NHA-Ts were pretreated with the AKT inhibitor, MK2206, or the vehicle control at increasing concentrations, and subsequently infected with JCPyV (Figure 4B). A dose-dependent decrease in viral infection was observed in NHAs; however, a dose-dependent increase was measured in SVGAs and NHA-Ts (Figure 4B). Considering the significant differences in JCPyV infection among cell types, MK2206 was validated for its inhibitory effects on AKT phosphorylation by ICW (Figure 4C). Cells were serum-starved and treated with the highest concentration of MK2206 for 24 h. The chemical inhibitor significantly reduced AKT phosphorylation in all three cell types compared to the vehicle control, demonstrating that the inhibitor is working effectively at the concentrations used (Figure 4D). This data revealed that JCPyV requires AKT during infection in NHAs and the utilization of this pathway is perhaps confounded by the immortalized properties of SVGAs and NHA-Ts.

3.5. PI3K Is Required for JCPyV Infection in NHAs

Upstream of AKT, PI3K was targeted using a chemical inhibitor to understand if additional proteins in the PI3K/AKT pathway are involved during JCPyV infection in NHAs. Cells were pretreated with the PI3K inhibitor, wortmannin, or the vehicle control at increasing concentrations, and subsequently infected with JCPyV (Figure 5A). Comparable to AKT inhibition, wortmannin significantly reduced JCPyV infection by 50% in NHAs, while increasing infection in SVGAs and not influencing infection in NHA-Ts (Figure 5A). The chemical inhibitor was validated for its inhibition of the PI3K/AKT signaling pathway by measuring AKT phosphorylation using ICW (Figure 5B). Cells were serum-starved and treated with the highest concentration of wortmannin. Across all three cell types, wortmannin significantly reduced AKT phosphorylation (Figure 5C). Altogether, these data illustrate that treatment of cells with the PI3K inhibitor equivalently impaired the PI3K/AKT signaling pathway, and a decrease in JCPyV infection during PIK3K inhibition was only observed in NHAs, verifying the results demonstrated with MK2206 treatment (Figure 4).

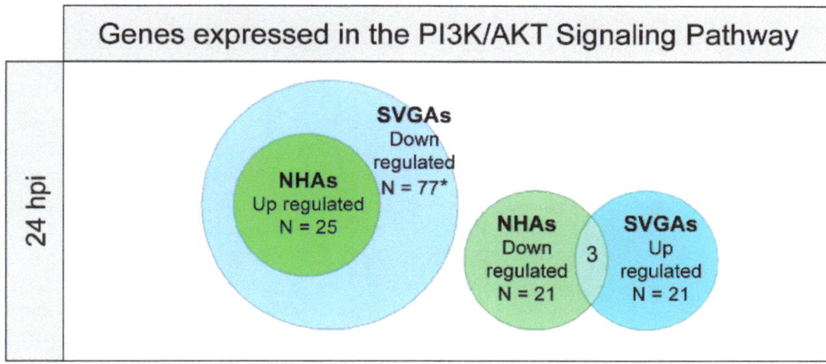

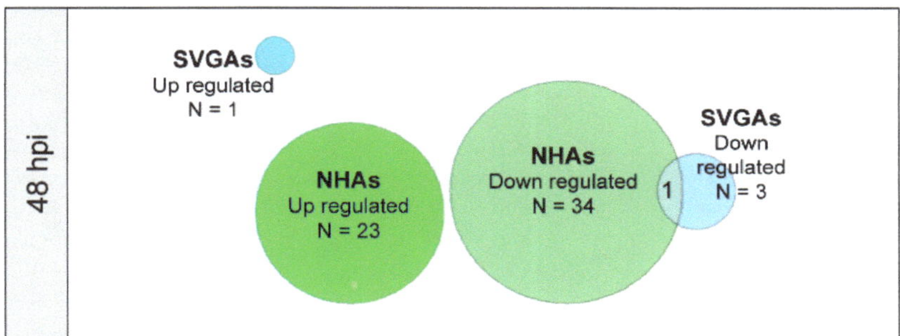

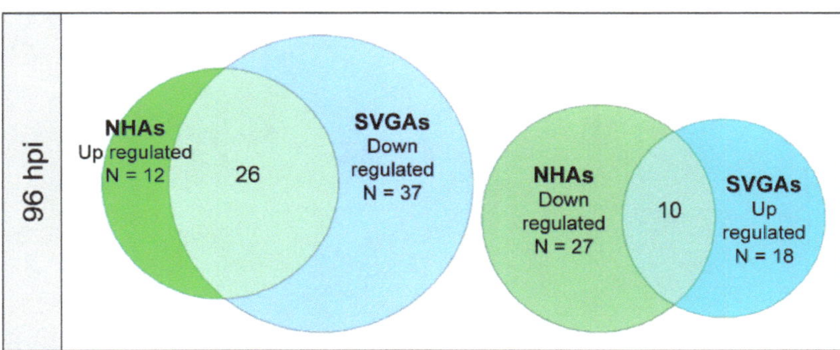

Figure 3. Genes in the PI3K/AKT pathway were differentially expressed during JCPyV infection in NHAs compared to viral infection in SVGAs. NHAs and SVGAs were either infected with JCPyV (MOI = 0.1 FFU/cell) or mock-infected with a vehicle control, and the transcriptomic profile was determined at 24, 48, and 96 h using RNA-seq. Genes of the PI3K/AKT pathway were determined from the KEGG database with an unadjusted p value < 0.10. Upregulated genes (Log FC > 0) and downregulated genes (Log FC < 0) are represented as Venn diagrams for each time point and cell type. The size of each circle is proportionate to the number of genes that apply to the above criteria. The green colors are representative of NHAs, while the blue colors are representative of SVGAs. Brighter shades represent upregulated genes and darker shades represent downregulated genes. (FC, fold-change; * 77 genes in total are down regulated in SVGAs, but 52 genes are exclusively downregulated in SVGAs.).

3.6. mTOR Inhibition Significantly Reduces JCPyV Infection in NHAs

A downstream target of the PI3K/AKT signaling pathway is mTOR [83], and previous research has determined that mTOR inhibition can increase JCPyV infection in an immortalized kidney cell line [50]. Moreover, our previous research has also demonstrated that JCPyV infection is delayed in NHAs, compared to SVGAs and NHA-Ts [31]. To determine if mTOR is required for JCPyV infection and if mTOR inhibition increases late viral protein expression in NHAs, all cell types were treated with either rapamycin, mTOR inhibitor PP242, or the vehicle control, and were subsequently infected with JCPyV, and infectivity was measured by assessing both early (T Ag) and late (VP1) viral protein production (Figure 6). Rapamycin and PP242 significantly reduced JCPyV infection in NHAs as quantified by both T Ag and VP1 production (Figure 6A,B). Furthermore, rapamycin significantly increased JCPyV infection in SVGAs and NHA-Ts (Figure 6A). However, PP242 did not influence viral infection in either SVGAs or NHA-Ts (Figure 6B). To validate the effectiveness of the chemical inhibitors in NHAs, SVGAs, and NHA-Ts, mTOR phosphorylation was measured by ICW, following serum starvation and treatment with each inhibitor for 24 h (Figure 6C). Rapamycin and PP242 significantly reduced mTOR phosphorylation, relatively, in each cell type; however, PP242 reduced phosphorylation to more appreciable levels when compared to rapamycin (Figure 6D). Overall, these findings demonstrate that mTOR phosphorylation, a downstream target of the PI3K/AKT signaling pathway, is required for JCPyV infection in NHAs. Furthermore, it substantiates previous findings that JCPyV uses alterative signaling pathways in primary astrocytes—viral mechanisms that are not observed in cell types that are immortalized with SV40 T Ag.

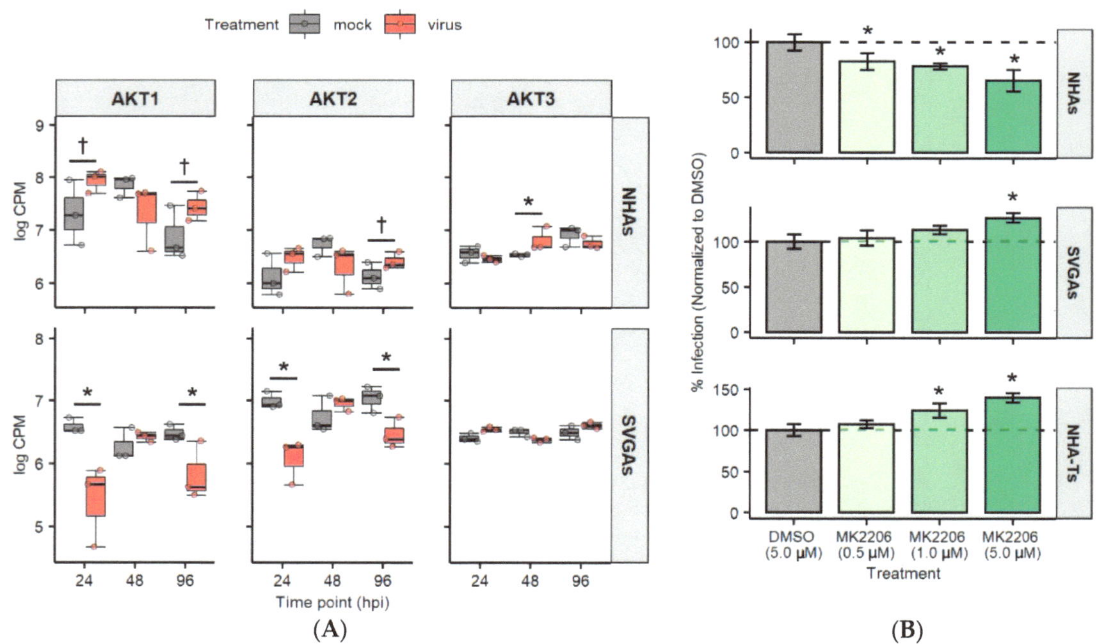

Figure 4. *Cont.*

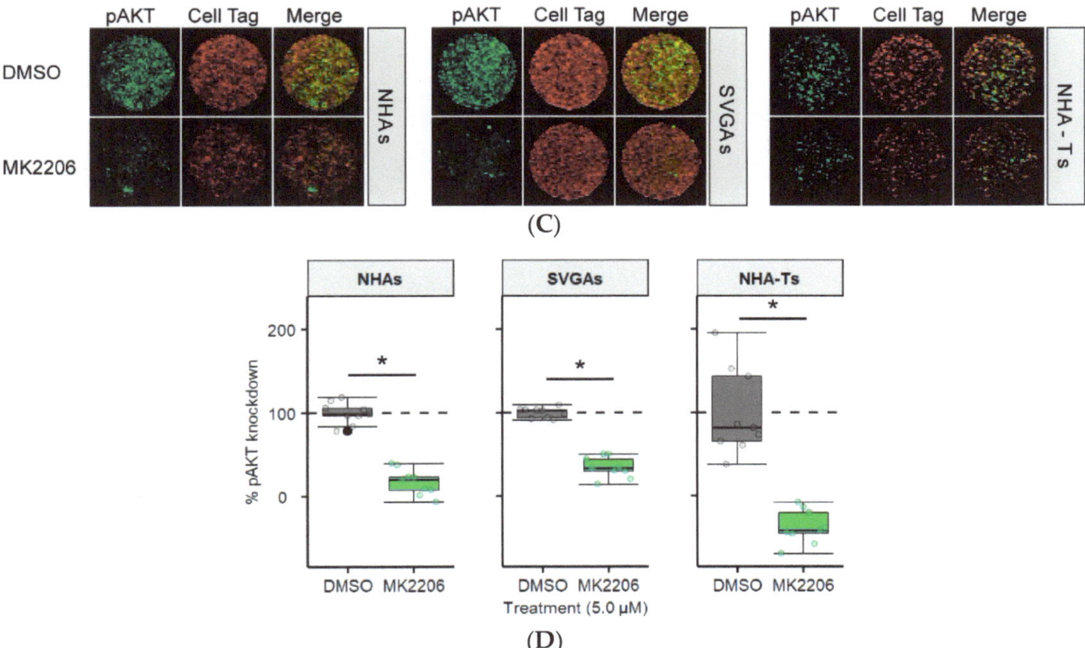

Figure 4. AKT is required for JCPyV infection in NHAs. (**A**) The log CPM was determined for the three isoforms of AKT (*AKT1*, *AKT2*, and *AKT3*) in mock-infected and JCPyV-infected cells at each time point from the RNA-seq data. Colored points represent individual points for each timepoint (3 replicates). †, $p < 0.10$; *, $p < 0.05$. (**B**) NHAs, SVGAs, and NHA-Ts were pretreated with indicated concentrations of AKT inhibitor MK2206 or DMSO at an equivalent volume control for 30 min and then infected with JCPyV (MOI = 1.0 FFU/cell) at 37 °C for 1 h. Cells were incubated in media containing DMSO or MK2206 for 48 h and then fixed and stained by indirect immunofluorescence. Infectivity was determined by counting the number of JCPyV T Ag-positive nuclei divided by the number of DAPI-positive nuclei for five ×20 fields of view for triplicate samples (% infection). Data is representative of three individual experiments. Error bars indicate SD. Student's *t* test was used to determine statistical significance, comparing DMSO to MK2206 for each cell type. (**C**) NHAs, SVGAs, and NHA-Ts were treated with DMSO or MK2206 (5.0 µM) at 37 °C for 24 h. Cells were fixed and stained for pAKT (green) or CellTag (red). (**D**) Percentage of pAKT for each cell type was quantitated by ICW signal intensity values per [(pAKT)/Cell Tag × 100% = % response] within each ICW analysis using LI-COR software (representative image shown in **C**). Box and whisker plots represent the distribution of samples (individual points) with the lower quartile, median, and upper quartile denoted as black lines. The dashed line indicates the normalized DMSO % response of pAKT. Colored points represent individual points for each treatment (3 replicates, performed in triplicate), and outliers are represented by black circles. A Wilcoxon rank sum exact test was used to compare treatments in each cell type. Data are representative of three independent experiments performed in triplicate. * $p < 0.05$.

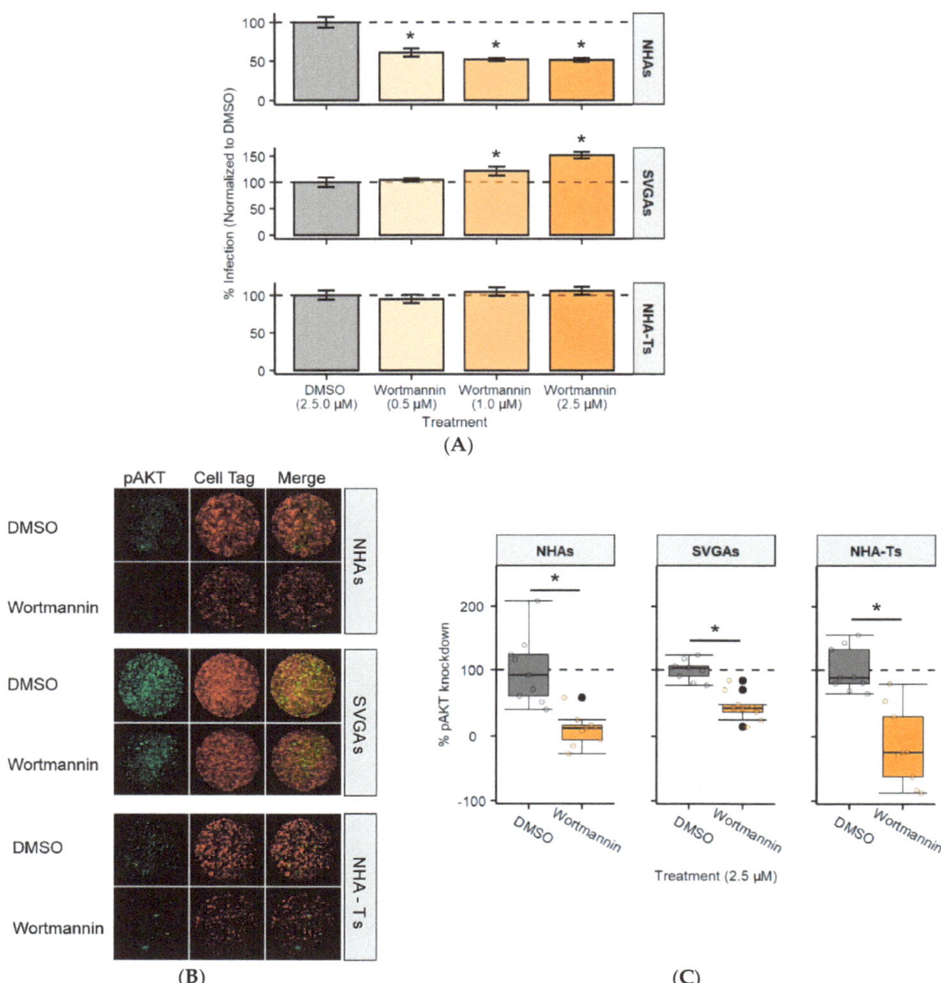

Figure 5. PI3K inhibitor wortmannin reduces JCPyV infection in NHAs. (**A**) NHAs, SVGAs, and NHA-Ts were pretreated with indicated concentrations of a PI3K inhibitor, wortmannin, or DMSO at an equivalent volume control and then infected with JCPyV (MOI = 1.0 FFU/cell) at 37 °C for 1 h. Cells were incubated in media containing DMSO or wortmannin for 48 h and then fixed and stained by indirect immunofluorescence. Infectivity was determined by counting the number of JCPyV T Ag-positive nuclei divided by the number of DAPI-positive nuclei for five ×20 fields of view for triplicate samples (% infection). Data is representative of three individual experiments. Error bars indicate SD. Student's *t* test was used to determine statistical significance comparing DMSO to wortmannin for each cell type. (**B**) NHAs, SVGAs, and NHA-Ts were treated with DMSO or wortmannin (2.5 μM) at 37 °C for 24 h. Cells were fixed and stained for pAKT (green) or CellTag (red). Entire wells of a 96-well plate are shown. (**C**) Percentage of pAKT for each cell type was quantitated by ICW signal intensity values per [(pAKT)/Cell Tag × 100% = % response] within each ICW analysis using LI-COR software (representative image shown in B). Box and whisker plots represent the distribution of samples (individual points) with the lower quartile, median and upper quartile denoted as black lines. The dashed line indicates the normalized DMSO % response of pAKT. Colored points represent individual points for each treatment (3 replicates, performed in triplicate), and outliers are represented by black circles. A Wilcoxon rank sum exact test was used to compare treatments in each cell type. Data are representative of three independent experiments performed in triplicate. *, $p < 0.05$.

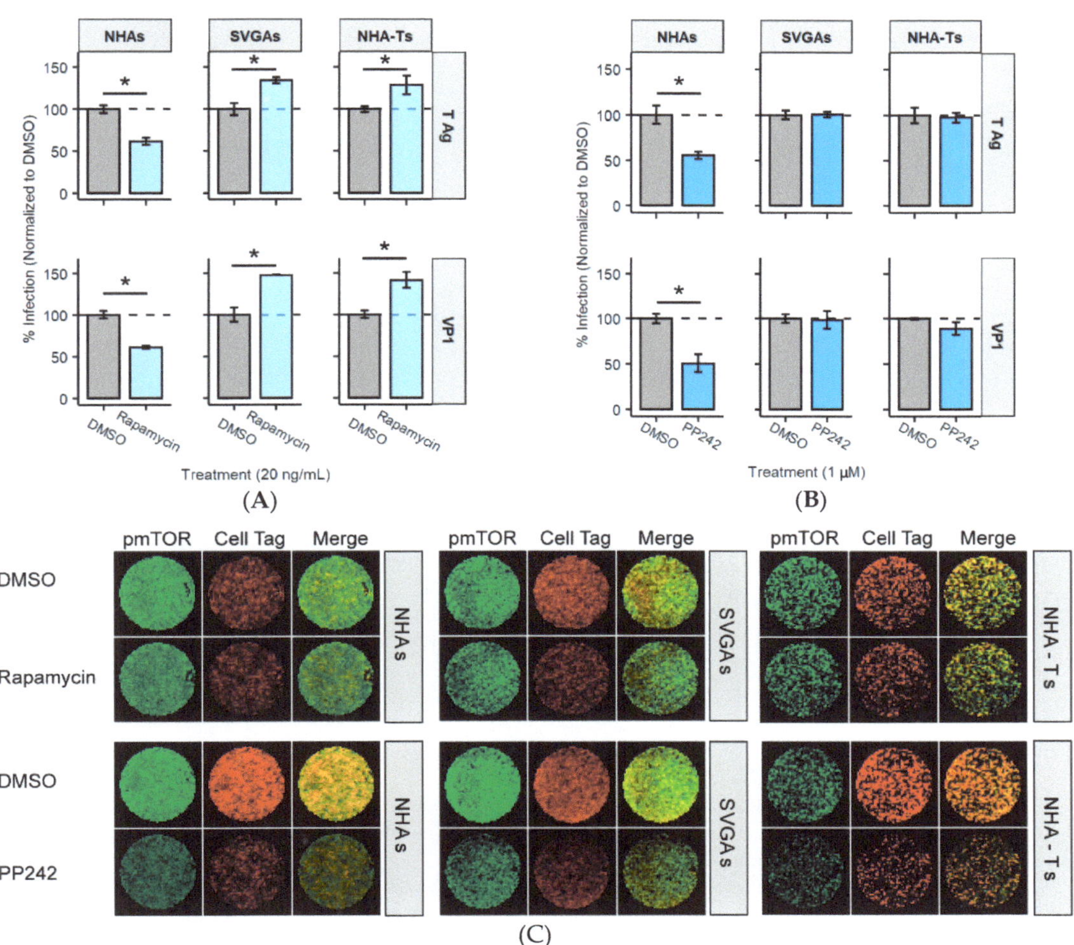

Figure 6. *Cont.*

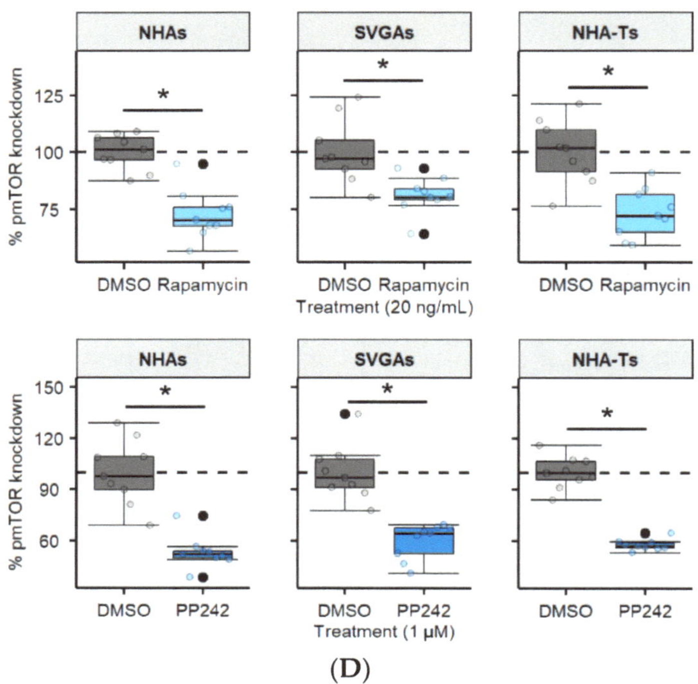

Figure 6. Inhibitors of mTOR, a target of the PI3K/AKT pathway, reduce JCPyV infection in NHAs. (**A,B**). NHAs, SVGAs, and NHA-Ts were pretreated with (**A**) rapamycin at 20 ng/mL, (**B**) PP242 at 1 µM, or the DMSO control at an equivalent volume control and then infected with JCPyV (MOI = 1.0 FFU/cell) at 37 °C for 1 h. Cells were incubated in media containing DMSO or mTOR inhibitors for 72 h and then fixed and stained by indirect immunofluorescence. Infectivity was determined by counting the number of JCPyV T Ag- or VP1- positive nuclei divided by the number of DAPI-positive nuclei for five ×20 fields of view for triplicate samples (% infection). Data is representative of three individual experiments performed in triplicate. Error bars indicate SD. Student's *t* test was used to determine statistical significance comparing DMSO to either mTOR inhibitor, with each cell type and viral protein. (**C**) All cell types were treated with DMSO or rapamycin at 20 ng/mL (top) or PP242 at 1 µM (bottom) at 37 °C for 24 h. Cells were fixed and stained for pAKT (green) or CellTag (red). Entire wells of a 96-well plate are shown. (**D**) Percentage of pmTOR for each cell type was quantitated by ICW signal intensity values per [(pmTOR)/Cell Tag × 100% = % response] within each ICW analysis with LI-COR software (representative image shown in **C**). Box and whisker plots represent the distribution of samples (individual points) with the lower quartile, median, and upper quartile denoted as black lines. The dashed line indicates the normalized DMSO % response of pmTOR. Colored points represent individual points for each treatment (3 replicates, performed in triplicate), and outliers are represented by black circles. A Wilcoxon rank sum exact test was used to compare treatments in each cell type. Data are representative of three independent experiments performed in triplicate. *, $p < 0.05$.

4. Discussion

Astrocytes are the main targets of JCPyV infection in the CNS, where the destruction of these cells, along with oligodendrocytes, leads to PML [1–4,26]. With no cure for this aggressive and ultimately fatal disease, more research is needed to reveal potential therapeutic targets. Cellular pathways, like the PI3K/AKT/mTOR signaling pathway, is an attractive candidate because there are numerous established drugs that inhibit this pathway for cancer treatment [51,52]. Additionally, there is evidence to suggest that these drugs can be repurposed to target coronaviruses, such as SARS-CoV-2 [84–87], the causative agent of COVID-19 [88]. In this study, we investigated the role of various chemical inhibitors of the PI3K/AKT/mTOR signaling pathway on JCPyV infection in primary astrocytes compared to immortalized cells. Our results demonstrate that viral infection is significantly reduced in primary astrocytes compared to cells transformed with SV40 T Ag, highlighting both the importance of this signaling pathway, and the need to use either ex vivo approaches or to validate findings in primary cells to further understand how JCPyV infection may occur in the human host.

JCPyV requires the MAPK/ERK pathway to successfully infect immortalized cells [61,66,73,74]. However, when a well-studied MEK inhibitor, U0126, did not reduce JCPyV infection in primary astrocytes, yet reduced JCPyV infection in cells immortalized with SV40 T Ag (Figure 1), additional studies were performed to identify alternative pathways that regulate JCPyV infection. Research has shown that the PI3K/AKT/mTOR signaling pathway intersects with the MAPK/ERK pathway [79–81]. Our findings also support this, as treatment of NHAs with U0126 led to an increase in phosphorylated AKT, while U0126 treatment did not exhibit any difference in AKT phosphorylation in SVGAs and NHA-Ts (Figure 2).

The PI3K/AKT/mTOR signaling pathway has been previously implicated in other polyomavirus infections as well [48,50,59,89–92]. Earlier research has demonstrated that the BK polyomavirus (BKPyV), murine polyomavirus, and JCPyV influence the PI3K/AKT signaling pathway by modulating the cellular phosphatase, protein phosphatase 2A (PP2A) [89–91]. Comparable to JCPyV, BKPyV also establishes an asymptomatic infection in the kidney, yet during immunosuppression BKPyV can cause nephropathy and hemorrhagic cystitis [93]. Treatment of cells with sirolimus (i.e., rapamycin) significantly reduces BKPyV infection; however, the concentration of the chemical inhibitor used in those studies was significantly different between immortalized and primary cells [89,92]. Hirsch et al. used a primary kidney cell line to determine the outcome of BKPyV infection during rapamycin treatment. They confirmed that using rapamycin at least a magnitude lower in concentration significantly reduces BKPyV infection in a primary kidney cell line, specifically during the first 24 h of infection (i.e., before viral genome replication) [92]. The authors concluded that these differences in concentration of the drug were the result of the transformed phenotype of the other cell line, causing it to require significantly higher concentrations of the inhibitor to reduce BKPyV infection. Unfortunately, understanding viral infection is challenging when using transformed cell lines, as transformation alters metabolic pathways and signal transduction within those pathways [94,95].

Our findings substantiate the consequences of immortalization in investigations of JCPyV infection with chemical inhibitors that target the PI3K/AKT/mTOR signaling pathway. Wortmannin, MK2206, rapamycin, and PP242—chemical inhibitors that target different steps of PI3K/AKT/mTOR signaling pathway—resulted in a significant reduction of JCPyV infection in NHAs that was not observed in SVGAs or NHA-Ts. These results, specifically with MK2206, are consistent with other published research reporting the impacts of JCPyV infection on an oligodendrocyte cell line treated with inhibitors of this pathway. MK2206 treatment of a glioma-derived stem cell line with oligodendrocyte precursor phenotypes (G144 cells) reduced JCPyV DNA replication [59]. It is important to note, that even though these cells are established through glioblastoma samples, G144 cells specifically display features that resemble normal fetal neural stem cells [96]. Together, these data suggest the importance of MK2206 as a potential antiviral for PML, as oligodendrocytes and astrocytes are the main cell types impacted by disease [1–4,26].

Our research also corresponds with findings demonstrated with BKPyV infection and mTOR inhibition with respect to both polyomavirus infection and cell-type dependent differences. Rapamycin significantly reduced JCPyV infection in primary astrocytes (Figure 6A), and similarly, BKPyV infection was reduced in a primary kidney cell line with rapamycin treatment [92]. Likewise, using both rapamycin and PP242—a secondary mTOR inhibitor—did not decrease JCPyV infection in SVGA and NHA-T immortalized cells (Figure 6A,B). Similarly, rapamycin and other mTOR inhibitors did not decrease but rather enhanced JCPyV replication in HEK293A cells [50]. HEK293A cells are transformed with adenovirus type 5 DNA [97], but not with SV40 T Ag; yet, immortalization through adenovirus leads to adenovirus oncogene E1A interactions with Rb and p53, disrupting important checkpoints in cell cycle and growth, similar to SV40 T Ag transformation [98–102]. Additionally, cells transformed with viral oncogenes can also influence PP2A, known to regulate numerous cellular pathways, such as the MAPK and PI3K signaling pathways—which are also important during JCPyV infection [91,103,104]. Together, these findings demonstrate both the requirement for the PI3K/AKT/mTOR signaling pathway and the importance of using primary cell lines to characterize polyomavirus infection.

Furthermore, rapamycin and PP242 are both mTOR inhibitors, yet the mode of inhibition is slightly different. mTOR forms two complexes in mammalian cells—mTOR complex 1 (mTORC1) and mTORC2—and activation of these complexes results in different functions for the cell [105]. The formation and activation of mTORC1 results in protein translation, cell growth, and autophagy, while mTORC2 results in survival, migration, and cytoskeletal organization [105]. Rapamycin has been demonstrated to inhibit mTORC1 more so than mTORC2—particularly in vitro [106]—while PP242 results in greater inhibition of both mTORC1 and mTORC2 [107]. These modes of mTOR inhibition may explain the differences observed in immortalized cells during JCPyV infection; however, it does substantiate the importance of mTOR during infection of primary astrocytes (Figure 6). Additionally, if the concentration of PP242 was increased, the results of JCPyV infection were similar. In fact, higher concentrations resulted in cytotoxic effects (data not shown), yet each chemical inhibitor was tested for cellular viability compared to the vehicle control in each cell type, and concentrations well-tolerated by cells were used in all the assays performed. Additionally, the inhibitory effect, measuring phosphorylation of AKT or mTOR, was similar across cell types—demonstrating that cell-type differences were not from cytotoxic effects or from inequitable impacts of AKT or mTOR phosphorylation from the chemical treatments.

Research has demonstrated that enhanced JCPyV infection from mTOR inhibition is perhaps due to the Skp-Cullin Fbox (SCF) E3 ligase, S-phase kinase-associated protein 2 (Skp2) [50,108]; the expression of this protein is highly variable in immortalized versus primary cells [109]. Skp2 is important in regulating the cell cycle, accumulates in the cell during the transition to the G1/S phase, remains highly expressed during S phase [110] and can also interact with polyomavirus Large T Ag [50,108]. Previous studies have concluded that the interaction of Skp2 and Large T Ag of numerous polyomaviruses (PyVs), including murine PyV, JCPyV, and BKPyV, was reduced with treatment with mTOR inhibitors, which resulted in an increase of Large T Ag expression [50]. Skp2 is highly expressed in glioma cell lines compared to normal astrocyte cell lines [109]; this could explain the differences observed between NHAs versus SVGAs and NHA-Ts. First, JCPyV Large T Ag expression is significantly lower in NHAs compared to SVGAs and NHA-Ts [31], and as a result, Skp2 may not be involved to the same extent in JCPyV infection of NHAs; thus, this interaction between Skp2 and Large T Ag is not sensitive to mTOR inhibition. However, future research should elucidate the mechanisms of viral protein production and the PI3K/AKT/mTOR signaling pathway.

Lastly, PI3K expression has been recently demonstrated to decrease JCPyV infection in SVGAs [48]. A reason for the differences in JCPyV infection between the findings reported here and by Clark et. al could be the PI3K isoform that was targeted, as well as the technique used. The authors determined that PI3K, specifically PI3Kγ, facilitates JCPyV infection in SVGAs through genetic knockdown approaches [48]. It is known that JCPyV facilitates entry into the cell through the utilization of a GPCR—the serotonin 5-hydroxytryptamine (5-HT$_2$) receptor [62,63,111]—which upon activation couples with PI3Kγ [48]. The authors speculated that knockdown of PI3Kγ disrupted early events of GPCR signaling, and as a result, disrupted possible virus capsid disassembly or trafficking to the endoplasmic reticulum or nucleus [48]. Wortmannin is one of the most well-characterized PI3K inhibitors and has been shown to interact strongly in vitro with PI3K, thus inhibiting numerous isoforms in the PI3K family [112,113]. However, wortmannin has also been demonstrated to have off-target effects, inhibiting other serine/threonine kinases of the PI3K family such as mTOR [114]. The differences that we have observed could be the consequence of wortmannin targeting other PI3Ks in the pathway and thus having similar results to the other inhibitors used in this research, specifically in SVGAs. Furthermore, JCPyV entry into primary astrocytes has not been extensively studied, and thus, we do not yet know whether GPCRs are utilized during entry, thereby activating PI3Kγ to the same extent as it is with viral entry of SVGAs. Additionally, recent work has demonstrated that JCPyV can use extracellular vesicles to infect SVGAs and astrocytes independently of the sialic acid attachment receptor through clathrin-dependent and independent mechanisms [115]. Future studies should define the role of PI3Ks in the viral infection of primary astrocytes using more targeted approaches such as siRNA, as well as in primary kidney cells. Nonetheless, wortmannin, MK2206, rapamycin, and PP242 significantly reduced JCPyV infection in primary astrocytes, while also inhibiting the phosphorylation of AKT and mTOR in all three cell types (Figure 7). Furthermore, RNA-seq analysis revealed numerous genes within the PI3K/AKT/mTOR pathway that were upregulated during JCPyV infection of NHAs, but downregulated in SVGAs, providing more evidence of the requirement of this pathway in primary astrocytes (Figures 3 and 4A). The genes β3 integrin (*ITGB3*) and interleukin-6 receptor (*IL-6R*) had a 3.4- and 2-fold increase, respectively, at 96 hpi in NHAs during JCPyV infection (Table S1). ITGB3 is a regulator of the PI3K/AKT/mTOR pathway [116] and has roles in cancer [116,117] and extracellular vesicles that induce cell signaling [118,119]. Interestingly, research has demonstrated that ITGB3 can regulate expression of matrix metalloproteinase 2 (MMP2) [120], a protein that is critical in the inflammatory response and in demyelinating diseases, such as MS [121–123]. Overall, this data illustrates how JCPyV activates the PI3K/AKT/mTOR pathway, yet this may have other implications in the viral-induced demyelinating disease PML. More research should determine if astrocytes induce an inflammatory response from activation of the PI3K/AKT/mTOR pathway.

In summary, this research has revealed and outlined the requirement of the PI3K/AKT/mTOR signaling pathway in JCPyV infection of primary human astrocytes. Using various chemical inhibitors, we have characterized how JCPyV uses this pathway to support viral infection in primary cells, and importantly, how immortalized characteristics may alter signaling events that, in turn, confound the requirement of this pathway in JCPyV infection of transformed cell lines. Overall, these findings will aid in the discovery of therapeutics to treat or slow the progression of PML, as no effective treatments are available for this fatal disease.

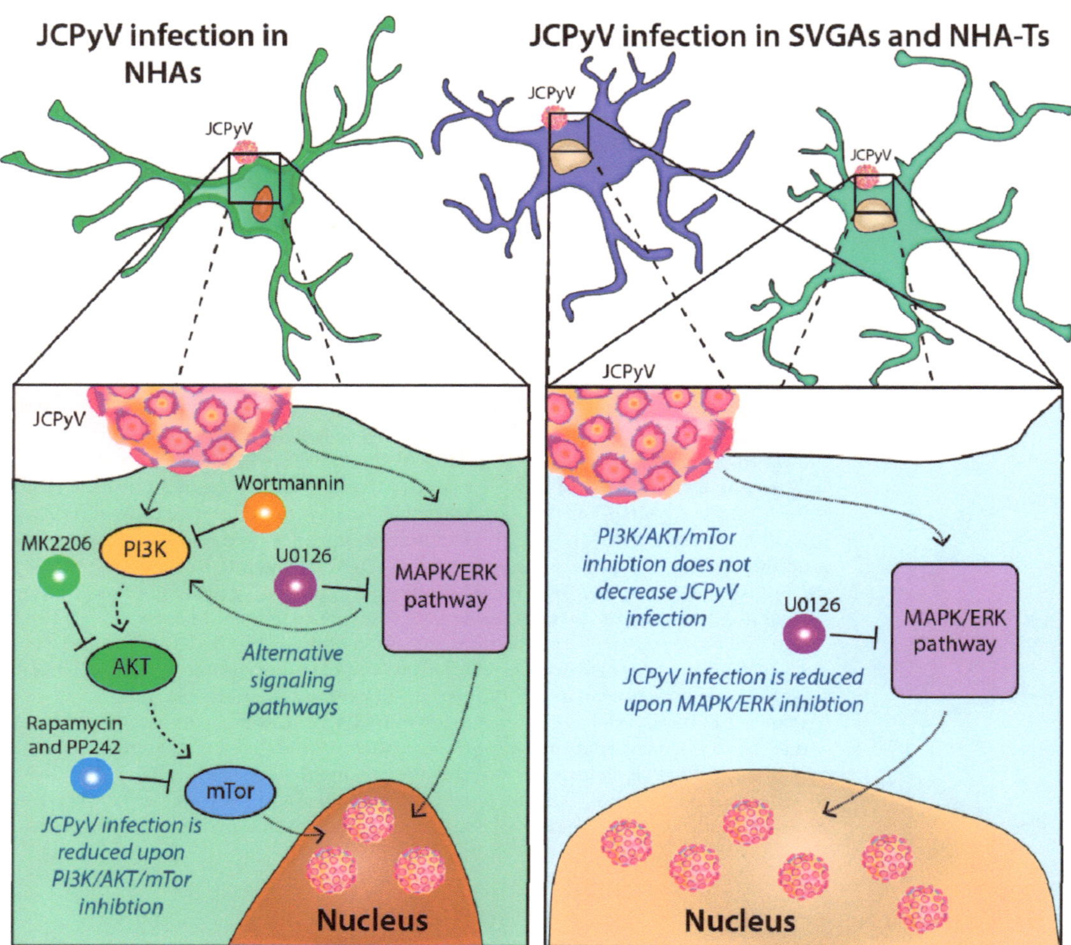

Figure 7. Chemical inhibitors that reduce JCPyV infection in NHAs, SVGAs, and NHA-Ts. Previous research has demonstrated the importance of the MAPK/ERK pathway in NHAs; however, JCPyV can also use the PI3K/AKT/mTOR signaling pathway for infection (**left**). Using chemical inhibitors targeting PI3K, AKT, and mTOR significantly reduced JCPyV infection in primary astrocytes (**left**). However, in immortalized cells, SVGAs and NHA-Ts, these chemical inhibitors did not reduce JCPyV infection and treatment of cells with U0126—an MEK inhibitor of the MAPK/ERK pathway—which significantly reduced JCPyV infection (**right**). This demonstrates that during the viral infection of immortalized cells, JCPyV is more dependent on the MAPK/ERK pathway and may not use other pathways, such as the PI3K/AKT/mTOR signaling pathway, to establish successful infection, compared to JCPyV infection in primary astrocytes.

Supplementary Materials: The following are available online at https://www.mdpi.com/article/10.3390/cells10113218/s1, Table S1: Log fold change of genes within the PI3K/AKT signaling pathway during JCPyV infection of NHAs and SVGAs.

Author Contributions: M.P.W., M.S.M. and B.L.K. conceived the study. M.P.W., M.S.M. and B.L.K. designed the experiments. M.P.W., F.J.A. and C.L.M. performed the experiments. M.P.W., M.S.M. and B.L.K. evaluated the data. M.P.W. and B.L.K. performed data analysis, and M.P.W. prepared the figures. M.P.W. and M.S.M. wrote the manuscript. All authors have read and agreed to the published version of the manuscript.

Funding: This research was supported by the National Institute of Allergy and Infectious Diseases grant number R15AI144686 (M.S.M.) of the National Institutes of Health and by the Maine IDeA Network of Biomedical Research Excellence (INBRE) through the National Institute of General Medical Sciences grant number P20GM103423 (M.S.M., B.L.K). Additionally, this work was also financially supported in part by funding from the University of Maine: Biomedical Sciences Accelerator Fund Faculty Award from the College of Natural Sciences, Forestry, and Agriculture (M.S.M., B.L.K.), Frederick H. Radke Undergraduate Research Fellowships (F.J.A.) from the Department of Molecular and Biomedical Sciences, and multiple grants from the University of Maine Graduate Student Government (M.P.W.).

Institutional Review Board Statement: Not applicable.

Informed Consent Statement: Not applicable.

Data Availability Statement: The RNA-seq data [66] analyzed in this study is deposited in the National Center for Biotechnology Information (NCBI) Gene Expression Omnibus (GEO) database, under accession number GSE183322.

Acknowledgments: We thank all past and current members of the Maginnis laboratory for your thorough feedback and critical discussion. Additionally, we thank the Atwood laboratory for cells, virus, and support.

Conflicts of Interest: The authors declare no conflict of interest.

References

1. Hirsch, H.H.; Kardas, P.; Kranz, D.; Leboeuf, C. The Human JC Polyomavirus (JCPyV): Virological Background and Clinical Implications. *APMIS* **2013**, *121*, 685–727. [CrossRef] [PubMed]
2. Padgett, B.L.; Walker, D.L.; ZuRhein, G.M.; Eckroade, R.J.; Dessel, B.H. Cultivation of Papova-like Virus from Human Brain with Progressive Multifocal Leucoencephalopathy. *Lancet* **1971**, *1*, 1257–1260. [CrossRef]
3. Silverman, L.; Rubinstein, L.J. Electron Microscopic Observations on a Case of Progressive Multifocal Leukoencephalopathy. *Acta Neuropathol.* **1965**, *5*, 215–224. [CrossRef] [PubMed]
4. Zurhein, G.; Chou, S.M. Particles Resembling Papova Viruses in Human Cerebral Demyelinating Disease. *Science* **1965**, *148*, 1477–1479. [CrossRef] [PubMed]
5. Kean, J.M.; Rao, S.; Wang, M.; Garcea, R.L. Seroepidemiology of Human Polyomaviruses. *PLoS Pathog.* **2009**, *5*, e1000363. [CrossRef]
6. Egli, A.; Infanti, L.; Dumoulin, A.; Buser, A.; Samaridis, J.; Stebler, C.; Gosert, R.; Hirsch, H.H. Prevalence of Polyomavirus BK and JC Infection and Replication in 400 Healthy Blood Donors. *J. Infect. Dis.* **2009**, *199*, 837–846. [CrossRef] [PubMed]
7. Monaco, M.C.; Atwood, W.J.; Gravell, M.; Tornatore, C.S.; Major, E.O. JC Virus Infection of Hematopoietic Progenitor Cells, Primary B Lymphocytes, and Tonsillar Stromal Cells: Implications for Viral Latency. *J. Virol.* **1996**, *70*, 7004–7012. [CrossRef]
8. Monaco, M.C.G.; Jensen, P.N.; Hou, J.; Durham, L.C.; Major, E.O. Detection of JC Virus DNA in Human Tonsil Tissue: Evidence for Site of Initial Viral Infection. *J. Virol.* **1998**, *72*, 9918–9923. [CrossRef]
9. Dubois, V.; Dutronc, H.; Lafon, M.E.; Poinsot, V.; Pellegrin, J.L.; Ragnaud, J.M.; Ferrer, A.M.; Fleury, H.J. Latency and Reactivation of JC Virus in Peripheral Blood of Human Immunodeficiency Virus Type 1-Infected Patients. *J. Clin. Microbiol.* **1997**, *35*, 2288–2292. [CrossRef] [PubMed]
10. Chapagain, M.L.; Nerurkar, V.R. Human Polyomavirus JC (JCV) Infection of Human B Lymphocytes: A Possible Mechanism for JCV Transmigration across the Blood-Brain Barrier. *J. Infect. Dis.* **2010**, *202*, 184–191. [CrossRef]
11. White, M.K.; Khalili, K. Pathogenesis of Progressive Multifocal Leukoencephalopathy—Revisited. *J. Infect. Dis.* **2011**, *203*, 578–586. [CrossRef] [PubMed]
12. Ferrante, P.; Caldarelli-Stefano, R.; Omodeo-Zorini, E.; Vago, L.; Boldorini, R.; Costanzi, G. PCR Detection of JC Virus DNA in Brain Tissue from Patients with and without Progressive Multifocal Leukoencephalopathy. *J. Med. Virol.* **1995**, *47*, 219–225. [CrossRef]
13. Gorelik, L.; Reid, C.; Testa, M.; Brickelmaier, M.; Bossolasco, S.; Pazzi, A.; Bestetti, A.; Carmillo, P.; Wilson, E.; McAuliffe, M.; et al. Progressive Multifocal Leukoencephalopathy (PML) Development Is Associated with Mutations in JC Virus Capsid Protein VP1 That Change Its Receptor Specificity. *J. Infect. Dis.* **2011**, *204*, 103–114. [CrossRef] [PubMed]
14. Khanna, N.; Elzi, L.; Mueller, N.J.; Garzoni, C.; Cavassini, M.; Fux, C.A.; Vernazza, P.; Bernasconi, E.; Battegay, M.; Hirsch, H.H.; et al. Incidence and Outcome of Progressive Multifocal Leukoencephalopathy over 20 Years of the Swiss HIV Cohort Study. *Clin. Infect. Dis.* **2009**, *48*, 1459–1466. [CrossRef]
15. Anand, P.; Hotan, G.C.; Vogel, A.; Venna, N.; Mateen, F.J. Progressive Multifocal Leukoencephalopathy: A 25-Year Retrospective Cohort Study. *Neurol. Neuroimmunol. Neuroinflamm.* **2019**, *6*, e618. [CrossRef]
16. Cortese, I.; Reich, D.S.; Nath, A. Progressive Multifocal Leukoencephalopathy and the Spectrum of JC Virus-Related Disease. *Nat. Rev. Neurol.* **2021**, *17*, 37–51. [CrossRef]

17. Carson, K.R.; Evens, A.M.; Richey, E.A.; Habermann, T.M.; Focosi, D.; Seymour, J.F.; Laubach, J.; Bawn, S.D.; Gordon, L.I.; Winter, J.N.; et al. Progressive Multifocal Leukoencephalopathy after Rituximab Therapy in HIV-Negative Patients: A Report of 57 Cases from the Research on Adverse Drug Events and Reports Project. *Blood* **2009**, *113*, 4834–4840. [CrossRef] [PubMed]
18. Bloomgren, G.; Richman, S.; Hotermans, C.; Subramanyam, M.; Goelz, S.; Natarajan, A.; Lee, S.; Plavina, T.; Scanlon, J.V.; Sandrock, A.; et al. Risk of Natalizumab-Associated Progressive Multifocal Leukoencephalopathy. *N. Engl. J. Med.* **2012**, *366*, 1870–1880. [CrossRef]
19. Pavlovic, D.; Patera, A.C.; Nyberg, F.; Gerber, M.; Liu, M.; Leukeoncephalopathy, C.P.M. Progressive Multifocal Leukoencephalopathy: Current Treatment Options and Future Perspectives. *Ther. Adv. Neurol. Diso.* **2015**, *8*, 255–273. [CrossRef]
20. Tan, I.L.; Koralnik, I.J.; Rumbaugh, J.A.; Burger, P.C.; King-Rennie, A.; McArthur, J.C. Progressive Multifocal Leukoencephalopathy in a Patient without Immunodeficiency. *Neurology* **2011**, *77*, 297–299. [CrossRef]
21. Vermersch, P.; Kappos, L.; Gold, R.; Foley, J.F.; Olsson, T.; Cadavid, D.; Bozic, C.; Richman, S. Clinical Outcomes of Natalizumab-Associated Progressive Multifocal Leukoencephalopathy(Podcast). *Neurology* **2011**, *76*, 1697–1704. [CrossRef]
22. Prosperini, L.; de Rossi, N.; Scarpazza, C.; Moiola, L.; Cosottini, M.; Gerevini, S.; Capra, R.; Italian PML Study Group. Study Natalizumab-Related Progressive Multifocal Leukoencephalopathy in Multiple Sclerosis: Findings from an Italian Independent Registry. *PLoS ONE* **2016**, *11*, e0168376. [CrossRef] [PubMed]
23. Balduzzi, A.; Lucchini, G.; Hirsch, H.H.; Basso, S.; Cioni, M.; Rovelli, A.; Zincone, A.; Grimaldi, M.; Corti, P.; Bonanomi, S.; et al. Polyomavirus JC-Targeted T-Cell Therapy for Progressive Multiple Leukoencephalopathy in a Hematopoietic Cell Transplantation Recipient. *Bone Marrow Transpl.* **2011**, *46*, 987–992. [CrossRef] [PubMed]
24. Muftuoglu, M.; Olson, A.; Marin, D.; Ahmed, S.; Mulanovich, V.; Tummala, S.; Chi, T.L.; Ferrajoli, A.; Kaur, I.; Li, L.; et al. Allogeneic BK Virus–Specific T Cells for Progressive Multifocal Leukoencephalopathy. *N. Engl. J. Med.* **2018**, *379*, 1443–1451. [CrossRef]
25. Cortese, I.; Muranski, P.; Enose-Akahata, Y.; Ha, S.-K.; Smith, B.; Monaco, M.; Ryschkewitsch, C.; Major, E.O.; Ohayon, J.; Schindler, M.K.; et al. Pembrolizumab Treatment for Progressive Multifocal Leukoencephalopathy. *N. Engl. J. Med.* **2019**, *380*, 1597–1605. [CrossRef]
26. Kondo, Y.; Windrem, M.S.; Zou, L.; Chandler-Militello, D.; Schanz, S.J.; Auvergne, R.M.; Betstadt, S.J.; Harrington, A.R.; Johnson, M.; Kazarov, A.; et al. Human Glial Chimeric Mice Reveal Astrocytic Dependence of JC Virus Infection. *J. Clin. Investig.* **2014**, *124*, 5323–5336. [CrossRef] [PubMed]
27. Dyson, N.; Bernards, R.; Friend, S.H.; Gooding, L.R.; Hassell, J.A.; Major, E.O.; Pipas, J.M.; Vandyke, T.; Harlow, E. Large T Antigens of Many Polyomaviruses Are Able to Form Complexes with the Retinoblastoma Protein. *J. Virol.* **1990**, *64*, 1353–1356. [CrossRef]
28. Valle, L.D.; Gordon, J.; Assimakopoulou, M.; Enam, S.; Geddes, J.F.; Varakis, J.N.; Katsetos, C.D.; Croul, S.; Khalili, K. Detection of JC Virus DNA Sequences and Expression of the Viral Regulatory Protein T-Antigen in Tumors of the Central Nervous System. *Cancer Res.* **2001**, *61*, 4287–4293.
29. Dickmanns, A.; Zeitvogel, A.; Simmersbach, F.; Weber, R.; Arthur, A.K.; Dehde, S.; Wildeman, A.G.; Fanning, E. The Kinetics of Simian Virus 40-Induced Progression of Quiescent Cells into S Phase Depend on Four Independent Functions of Large T Antigen. *J. Virol.* **1994**, *68*, 5496–5508. [CrossRef]
30. Ferenczy, M.W.; Marshall, L.J.; Nelson, C.D.; Atwood, W.J.; Nath, A.; Khalili, K.; Major, E.O. Molecular Biology, Epidemiology, and Pathogenesis of Progressive Multifocal Leukoencephalopathy, the JC Virus-Induced Demyelinating Disease of the Human Brain. *Clin. Microbiol. Rev.* **2012**, *25*, 471–506. [CrossRef]
31. Wilczek, M.P.; DuShane, J.K.; Armstrong, F.J.; Maginnis, M.S. JC Polyomavirus Infection Reveals Delayed Progression of the Infectious Cycle in Normal Human Astrocytes. *J. Virol.* **2019**, *94*, e01331-19. [CrossRef]
32. Lynch, K.J.; Frisque, R.J. Factors Contributing to the Restricted DNA Replicating Activity of JC Virus. *Virology* **1991**, *180*, 306–317. [CrossRef]
33. Sock, E.; Wegner, M.; Fortunato, E.A.; Grummt, F. Large T-Antigen and Sequences within the Regulatory Region of JC Virus Both Contribute to the Features of JC Virus DNA Replication. *Virology* **1993**, *197*, 537–548. [CrossRef]
34. Major, E.O.; Miller, A.E.; Mourrain, P.; Traub, R.G.; de Widt, E.; Sever, J. Establishment of a Line of Human Fetal Glial Cells That Supports JC Virus Multiplication. *Proc. Natl. Acad. Sci. USA* **1985**, *82*, 1257–1261. [CrossRef]
35. Ariza, A.; Mate, J.L.; Isamat, M.; Calatrava, A.; Fernandez-Vasalo, A.; Navas-Palacios, J.J. Overexpression of Ki-67 and Cyclins A and B1 in JC Virus-Infected Cells of Progressive Multifocal Leukoencephalopathy. *J. Neuropathol. Exp. Neurol.* **1998**, *57*, 226–230. [CrossRef] [PubMed]
36. Sanchez, V.; McElroy, A.K.; Spector, D.H. Mechanisms Governing Maintenance of Cdk1/Cyclin B1 Kinase Activity in Cells Infected with Human Cytomegalovirus. *J. Virol.* **2003**, *77*, 13214–13224. [CrossRef] [PubMed]
37. Marshall, A.; Rushbrook, S.; Davies, S.E.; Morris, L.S.; Scott, I.S.; Vowler, S.L.; Coleman, N.; Alexander, G. Relation between Hepatocyte G1 Arrest, Impaired Hepatic Regeneration, and Fibrosis in Chronic Hepatitis C Virus Infection. *Gastroenterology* **2005**, *128*, 33–42. [CrossRef] [PubMed]
38. Ahuja, D.; Sáenz-Robles, M.T.; Pipas, J.M. SV40 Large T Antigen Targets Multiple Cellular Pathways to Elicit Cellular Transformation. *Oncogene* **2005**, *24*, 7729–7745. [CrossRef] [PubMed]

39. Chang, F.; Lee, J.T.; Navolanic, P.M.; Steelman, L.S.; Shelton, J.G.; Blalock, W.L.; Franklin, R.A.; McCubrey, J.A. Involvement of PI3K/Akt Pathway in Cell Cycle Progression, Apoptosis, and Neoplastic Transformation: A Target for Cancer Chemotherapy. *Leukemia* **2003**, *17*, 590–603. [CrossRef]

40. Choudhury, G.G.; Karamitsos, C.; Hernandez, J.; Gentilini, A.; Bardgette, J.; Abboud, H.E. PI-3-Kinase and MAPK Regulate Mesangial Cell Proliferation and Migration in Response to PDGF. *Am. J. Physiol. Renal.* **1997**, *273*, F931–F938. [CrossRef]

41. Diehl, J.A.; Cheng, M.; Roussel, M.F.; Sherr, C.J. Glycogen Synthase Kinase-3β Regulates Cyclin D1 Proteolysis and Subcellular Localization. *Gene. Dev.* **1998**, *12*, 3499–3511. [CrossRef]

42. Gille, H.; Downward, J. Multiple Ras Effector Pathways Contribute to G1Cell Cycle Progression. *J. Biol. Chem.* **1999**, *274*, 22033–22040. [CrossRef]

43. Medema, R.H.; Kops, G.J.P.L.; Bos, J.L.; Burgering, B.M.T. AFX-like Forkhead Transcription Factors Mediate Cell-Cycle Regulation by Ras and PKB through P27kip1. *Nature* **2000**, *404*, 782–787. [CrossRef] [PubMed]

44. Muise-Helmericks, R.C.; Grimes, H.L.; Bellacosa, A.; Malstrom, S.E.; Tsichlis, P.N.; Rosen, N. Cyclin D Expression Is Controlled Post-Transcriptionally via a Phosphatidylinositol 3-Kinase/Akt-Dependent Pathway*. *J. Biol. Chem.* **1998**, *273*, 29864–29872. [CrossRef] [PubMed]

45. Choudhury, G.G. Akt Serine Threonine Kinase Regulates Platelet-Derived Growth Factor-Induced DNA Synthesis in Glomerular Mesangial Cells: Regulation of c-Fos and P27^{Kip1} Gene Expression. *J. Biol. Chem.* **2001**, *276*, 35636–35643. [CrossRef]

46. Yu, Y.; Alwine, J.C. Human Cytomegalovirus Major Immediate-Early Proteins and Simian Virus 40 Large T Antigen Can Inhibit Apoptosis through Activation of the Phosphatidylinositide 3′-OH Kinase Pathway and the Cellular Kinase Akt. *J. Virol.* **2002**, *76*, 3731–3738. [CrossRef]

47. Yu, Y.; Alwine, J.C. Interaction between Simian Virus 40 Large T Antigen and Insulin Receptor Substrate 1 Is Disrupted by the K1 Mutation, Resulting in the Loss of Large T Antigen-Mediated Phosphorylation of Akt. *J. Virol.* **2008**, *82*, 4521–4526. [CrossRef]

48. Clark, P.; Gee, G.V.; Albright, B.S.; Assetta, B.; Han, Y.; Atwood, W.J.; DiMaio, D. Phosphoinositide 3′-Kinase γ Facilitates Polyomavirus Infection. *Viruses* **2020**, *12*, 1190. [CrossRef] [PubMed]

49. Vanhaesebroeck, B.; Guillermet-Guibert, J.; Graupera, M.; Bilanges, B. The Emerging Mechanisms of Isoform-Specific PI3K Signalling. *Nat. Rev. Mol. Cell Biol.* **2010**, *11*, 329–341. [CrossRef]

50. Orellana, J.A.; Kwun, H.J.; Artusi, S.; Chang, Y.; Moore, P.S. Sirolimus and Other MTOR Inhibitors Directly Activate Latent Pathogenic Human Polyomavirus Replication. *J. Infect. Dis.* **2020**, *7*, 224. [CrossRef]

51. Fruman, D.A.; Rommel, C. PI3K and Cancer: Lessons, Challenges and Opportunities. *Nat. Rev. Drug Discov.* **2014**, *13*, 140–156. [CrossRef] [PubMed]

52. Yang, J.; Nie, J.; Ma, X.; Wei, Y.; Peng, Y.; Wei, X. Targeting PI3K in Cancer: Mechanisms and Advances in Clinical Trials. *Mol. Cancer* **2019**, *18*, 26. [CrossRef]

53. Oki, Y.; Fanale, M.; Romaguera, J.; Fayad, L.; Fowler, N.; Copeland, A.; Samaniego, F.; Kwak, L.W.; Neelapu, S.; Wang, M.; et al. Phase II Study of an AKT Inhibitor MK2206 in Patients with Relapsed or Refractory Lymphoma. *Brit. J. Haematol.* **2015**, *171*, 463–470. [CrossRef] [PubMed]

54. Blagosklonny, M.V. Rapamycin for Longevity: Opinion Article. *Aging* **2019**, *11*, 8048–8067. [CrossRef]

55. Augustine, J.J.; Bodziak, K.A.; Hricik, D.E. Use of Sirolimus in Solid Organ Transplantation. *Drugs* **2007**, *67*, 369–391. [CrossRef]

56. Kastrati, A.; Mehilli, J.; von Beckerath, N.; Dibra, A.; Hausleiter, J.; Pache, J.; Schühlen, H.; Schmitt, C.; Dirschinger, J.; Schömig, A.; et al. Sirolimus-Eluting Stent or Paclitaxel-Eluting Stent vs Balloon Angioplasty for Prevention of Recurrences in Patients with Coronary In-Stent Restenosis: A Randomized Controlled Trial. *JAMA* **2005**, *293*, 165–171. [CrossRef] [PubMed]

57. Cloughesy, T.F.; Yoshimoto, K.; Nghiemphu, P.; Brown, K.; Dang, J.; Zhu, S.; Hsueh, T.; Chen, Y.; Wang, W.; Youngkin, D.; et al. Antitumor Activity of Rapamycin in a Phase I Trial for Patients with Recurrent PTEN-Deficient Glioblastoma. *PLoS Med.* **2008**, *5*, e8. [CrossRef] [PubMed]

58. Kaeberlein, M.; Galvan, V. Rapamycin and Alzheimer's Disease: Time for a Clinical Trial? *Sci. Transl. Med.* **2019**, *11*, eaar4289. [CrossRef]

59. Peterson, J.N.; Lin, B.; Shin, J.; Phelan, P.J.; Tsichlis, P.; Schwob, J.E.; Bullock, P.A. The Replication of JCV DNA in the G144 Oligodendrocyte Cell Line Is Dependent Upon Akt. *J. Virol.* **2017**, *91*, e00735-17. [CrossRef] [PubMed]

60. Vacante, D.A.; Traub, R.; Major, E.O. Extension of JC Virus Host Range to Monkey Cells by Insertion of a Simian Virus 40 Enhancer into the JC Virus Regulatory Region. *Virology* **1989**, *170*, 353–361. [CrossRef]

61. DuShane, J.K.; Wilczek, M.P.; Mayberry, C.L.; Maginnis, M.S. ERK Is a Critical Regulator of JC Polyomavirus Infection. *J. Virol.* **2018**, *92*, e01529-17. [CrossRef]

62. Mayberry, C.L.; Wilczek, M.P.; Fong, T.M.; Nichols, S.L.; Maginnis, M.S. GRK2 Mediates β-Arrestin Interactions with 5-HT 2 Receptors for JC Polyomavirus Endocytosis. *J. Virol.* **2021**, *95*, e02139-20. [CrossRef] [PubMed]

63. Mayberry, C.L.; Soucy, A.N.; Lajoie, C.R.; DuShane, J.K.; Maginnis, M.S. JC Polyomavirus Entry by Clathrin-Mediated Endocytosis Is Driven by β-Arrestin. *J. Virol.* **2019**, *93*, e01948-18. [CrossRef] [PubMed]

64. DuShane, J.K.; Mayberry, C.L.; Wilczek, M.P.; Nichols, S.L.; Maginnis, M.S. JCPyV-Induced MAPK Signaling Activates Transcription Factors during Infection. *Int. J. Mol. Sci.* **2019**, *20*, 4779. [CrossRef] [PubMed]

65. DuShane, J.K.; Wilczek, M.P.; Crocker, M.A.; Maginnis, M.S. High-Throughput Characterization of Viral and Cellular Protein Expression Patterns During JC Polyomavirus Infection. *Front. Microbiol.* **2019**, *10*, 783. [CrossRef] [PubMed]

66. Wilczek, M.P.; Armstrong, F.J.; Geohegan, R.P.; Mayberry, C.L.; DuShane, J.K.; King, B.L.; Maginnis, M.S. The MAPK/ERK Pathway and the Role of DUSP1 in JCPyV Infection of Primary Astrocytes. *Viruses* **2021**, *13*, 1834. [CrossRef]
67. Robinson, M.D.; McCarthy, D.J.; Smyth, G.K. EdgeR: A Bioconductor Package for Differential Expression Analysis of Digital Gene Expression Data. *Bioinformatics* **2010**, *26*, 139–140. [CrossRef]
68. Kanehisa, M.; Furumichi, M.; Sato, Y.; Ishiguro-Watanabe, M.; Tanabe, M. KEGG: Integrating Viruses and Cellular Organisms. *Nucleic Acids Res.* **2020**, *49*, D545–D551. [CrossRef]
69. Kanehisa, M.; Goto, S. KEGG: Kyoto Encyclopedia of Genes and Genomes. *Nucleic Acids Res.* **2000**, *28*, 27–30. [CrossRef]
70. Kanehisa, M. Toward Understanding the Origin and Evolution of Cellular Organisms. *Protein Sci.* **2019**, *28*, 1947–1951. [CrossRef]
71. Howe, K.L.; Achuthan, P.; Allen, J.; Allen, J.; Alvarez-Jarreta, J.; Amode, M.R.; Armean, I.M.; Azov, A.G.; Bennett, R.; Bhai, J.; et al. Ensembl 2021. *Nucleic Acids Res.* **2020**, *49*, D884–D891. [CrossRef]
72. Larsson, J. Eulerr: Area-Proportional Euler and Venn Diagrams with Ellipses. R Package Version 6.1.1, 2020. Available online: https://CRAN.R-project.org/package=eulerr (accessed on 1 June 2021).
73. Shaul, Y.D.; Seger, R. The MEK/ERK Cascade: From Signaling Specificity to Diverse Functions. *Biochim. Biophys. Acta BBA Mol Cell Res.* **2007**, *1773*, 1213–1226. [CrossRef] [PubMed]
74. Querbes, W.; Benmerah, A.; Tosoni, D.; Fiore, P.P.D.; Atwood, W.J. A JC Virus-Induced Signal Is Required for Infection of Glial Cells by a Clathrin- and Eps15-Dependent Pathway. *J. Virol.* **2003**, *78*, 250–256. [CrossRef] [PubMed]
75. McCubrey, J.A.; Lee, J.T.; Steelman, L.S.; Blalock, W.L.; Moye, P.W.; Chang, F.; Pearce, M.; Shelton, J.G.; White, M.K.; Franklin, R.A.; et al. Interactions between the PI3K and Raf Signaling Pathways Can Result in the Transformation of Hematopoietic Cells. *Cancer Detect Prev.* **2001**, *25*, 375–393.
76. Dangoria, N.S.; Breau, W.C.; Anderson, H.A.; Cishek, D.M.; Norkin, L.C. Extracellular Simian Virus 40 Induces an ERK/MAP Kinase-Independent Signalling Pathway That Activates Primary Response Genes and Promotes Virus Entry. *J. Gen. Virol.* **1996**, *77*, 2173–2182. [CrossRef]
77. Rodriguez-Viciana, P.; Collins, C.; Fried, M. Polyoma and SV40 Proteins Differentially Regulate PP2A to Activate Distinct Cellular Signaling Pathways Involved in Growth Control. *Proc. Natl. Acad. Sci. USA* **2006**, *103*, 19290–19295. [CrossRef]
78. Mendoza, M.C.; Er, E.E.; Blenis, J. The Ras-ERK and PI3K-MTOR Pathways: Cross-Talk and Compensation. *Trends Biochem. Sci.* **2011**, *36*, 320–328. [CrossRef] [PubMed]
79. Hayashi, H.; Tsuchiya, Y.; Nakayama, K.; Satoh, T.; Nishida, E. Down-regulation of the PI3-kinase/Akt Pathway by ERK MAP Kinase in Growth Factor Signaling. *Genes Cells* **2008**, *13*, 941–947. [CrossRef] [PubMed]
80. heon Rhim, J.; Luo, X.; Gao, D.; Xu, X.; Zhou, T.; Li, F.; Wang, P.; Wong, S.T.C.; Xia, X. Cell Type-Dependent Erk-Akt Pathway Crosstalk Regulates the Proliferation of Fetal Neural Progenitor Cells. *Sci. Rep* **2016**, *6*, 26547. [CrossRef] [PubMed]
81. Turke, A.B.; Song, Y.; Costa, C.; Cook, R.; Arteaga, C.L.; Asara, J.M.; Engelman, J.A. MEK Inhibition Leads to PI3K/AKT Activation by Relieving a Negative Feedback on ERBB Receptors. *Cancer Res.* **2012**, *72*, 3228–3237. [CrossRef]
82. Link, A.; Shin, S.K.; Nagasaka, T.; Balaguer, F.; Koi, M.; Jung, B.; Boland, C.R.; Goel, A. JC Virus Mediates Invasion and Migration in Colorectal Metastasis. *PLoS ONE* **2009**, *4*, e8146. [CrossRef] [PubMed]
83. Martini, M.; De Santis, M.C.; Braccini, L.; Gulluni, F.; and Hirsch, E. PI3K/AKT Signaling Pathway and Cancer: An Updated Review. *Ann. Med.* **2014**, *46*, 372–383. [CrossRef] [PubMed]
84. Foote, M.B.; White, J.R.; Jee, J.; Argilés, G.; Wan, J.C.M.; Rousseau, B.; Pessin, M.S.; Diaz, L.A. Association of Antineoplasic Therapy with Decreased SARS-CoV-2 Infection Rates in Patients with Cancer. *JAMA Oncol.* **2021**, *7*. [CrossRef]
85. Karam, B.S.; Morris, R.S.; Bramante, C.T.; Puskarich, M.; Zolfaghari, E.J.; Lotfi-Emran, S.; Ingraham, N.E.; Charles, A.; Odde, D.J.; Tignanelli, C.J. MTOR Inhibition in COVID-19: A Commentary and Review of Efficacy in RNA Viruses. *J. Med. Virol.* **2021**, *93*, 1843–1846. [CrossRef] [PubMed]
86. Kindrachuk, J.; Ork, B.; Hart, B.J.; Mazur, S.; Holbrook, M.R.; Frieman, M.B.; Traynor, D.; Johnson, R.F.; Dyall, J.; Kuhn, J.H.; et al. Antiviral Potential of ERK/MAPK and PI3K/AKT/MTOR Signaling Modulation for Middle East Respiratory Syndrome Coronavirus Infection as Identified by Temporal Kinome Analysis. *Antimicrob. Agents Chemother.* **2015**, *59*, 1088–1099. [CrossRef]
87. Appelberg, S.; Gupta, S.; Akusjärvi, S.S.; Ambikan, A.T.; Mikaeloff, F.; Saccon, E.; Végvári, Á.; Benfeitas, R.; Sperk, M.; Ståhlberg, M.; et al. Dysregulation in Akt/MTOR/HIF-1 Signaling Identified by Proteo-Transcriptomics of SARS-CoV-2 Infected Cells. *Emerg. Microbes Infec.* **2020**, *9*, 1748–1760. [CrossRef] [PubMed]
88. Zhu, N.; Zhang, D.; Wang, W.; Li, X.; Yang, B.; Song, J.; Zhao, X.; Huang, B.; Shi, W.; Lu, R.; et al. A Novel Coronavirus from Patients with Pneumonia in China, 2019. *N. Engl. J Med.* **2020**, *382*, 727–733. [CrossRef]
89. Liacini, A.; Seamone, M.E.; Muruve, D.A.; Tibbles, L.A. Anti-BK Virus Mechanisms of Sirolimus and Leflunomide Alone and in Combination: Toward a New Therapy for BK Virus Infection. *Transplantation* **2010**, *90*, 1450–1457. [CrossRef]
90. Andrabi, S.; Gjoerup, O.V.; Kean, J.A.; Roberts, T.M.; Schaffhausen, B. Protein Phosphatase 2A Regulates Life and Death Decisions via Akt in a Context-Dependent Manner. *Proc. Natl. Acad. Sci. USA* **2007**, *104*, 19011–19016. [CrossRef]
91. Bollag, B.; Hofstetter, C.A.; Reviriego-Mendoza, M.M.; Frisque, R.J. JC Virus Small t Antigen Binds Phosphatase PP2A and Rb Family Proteins and Is Required for Efficient Viral DNA Replication Activity. *PLoS ONE* **2010**, *5*, e10606. [CrossRef]
92. Hirsch, H.H.; Yakhontova, K.; Lu, M.; Manzetti, J. BK Polyomavirus Replication in Renal Tubular Epithelial Cells Is Inhibited by Sirolimus, but Activated by Tacrolimus Through a Pathway Involving FKBP-12. *Am. J. Transplant.* **2016**, *16*, 821–832. [CrossRef]
93. Hirsch, H.H.; Randhawa, P.; the AST Infectious Diseases Community of Practice. BK Polyomavirus in Solid Organ Transplantation. *Am. J. Transplant.* **2013**, *13*, 179–188. [CrossRef]

94. Campistol, J.M.; Eris, J.; Oberbauer, R.; Friend, P.; Hutchison, B.; Morales, J.M.; Claesson, K.; Stallone, G.; Russ, G.; Rostaing, L.; et al. Sirolimus Therapy after Early Cyclosporine Withdrawal Reduces the Risk for Cancer in Adult Renal Transplantation. *J. Am. Soc. Nephrol.* **2006**, *17*, 581–589. [CrossRef] [PubMed]

95. Zoncu, R.; Efeyan, A.; Sabatini, D.M. MTOR: From Growth Signal Integration to Cancer, Diabetes and Ageing. *Nat. Rev. Mol. Cell Biol.* **2011**, *12*, 21–35. [CrossRef]

96. Pollard, S.M.; Yoshikawa, K.; Clarke, I.D.; Danovi, D.; Stricker, S.; Russell, R.; Bayani, J.; Head, R.; Lee, M.; Bernstein, M.; et al. Glioma Stem Cell Lines Expanded in Adherent Culture Have Tumor-Specific Phenotypes and Are Suitable for Chemical and Genetic Screens. *Cell Stem Cell* **2009**, *4*, 568–580. [CrossRef] [PubMed]

97. Louis, N.; Evelegh, C.; Graham, F.L. Cloning and Sequencing of the Cellular–Viral Junctions from the Human Adenovirus Type 5 Transformed 293 Cell Line. *Virology* **1997**, *233*, 423–429. [CrossRef]

98. Berk, A.J. Recent Lessons in Gene Expression, Cell Cycle Control, and Cell Biology from Adenovirus. *Oncogene* **2005**, *24*, 7673–7685. [CrossRef]

99. Moran, E. DNA Tumor Virus Transforming Proteins and the Cell Cycle. *Curr. Opin. Genet. Dev.* **1993**, *3*, 63–70. [CrossRef]

100. Nevins, J.R. E2F: A Link between the Rb Tumor Suppressor Protein and Viral Oncoproteins. *Science* **1992**, *258*, 424–429. [CrossRef]

101. Frisch, S.M.; Mymryk, J.S. Adenovirus-5 E1A: Paradox and Paradigm. *Nat. Rev. Mol. Cell Biol.* **2002**, *3*, 441–452. [CrossRef]

102. Levine, A.J. The P53 Protein and Its Interactions with the Oncogene Products of the Small DNA Tumor Viruses. *Virology* **1990**, *177*, 419–426. [CrossRef]

103. Arroyo, J.D.; Hahn, W.C. Involvement of PP2A in Viral and Cellular Transformation. *Oncogene* **2005**, *24*, 7746–7755. [CrossRef]

104. Sariyer, I.K.; Khalili, K.; Safak, M. Dephosphorylation of JC Virus Agnoprotein by Protein Phosphatase 2A: Inhibition by Small t Antigen. *Virology* **2008**, *375*, 464–479. [CrossRef]

105. Jhanwar-Uniya, M.; Wainwright, J.V.; Mohan, A.L.; Tobias, M.E.; Murali, R.; Gandhi, C.D.; Schmidt, M.H. Diverse Signaling Mechanisms of MTOR Complexes: MTORC1 and MTORC2 in Forming a Formidable Relationship. *Adv. Biol. Regul.* **2019**, *72*, 51–62. [CrossRef]

106. Sarbassov, D.D.; Ali, S.M.; Kim, D.-H.; Guertin, D.A.; Latek, R.R.; Erdjument-Bromage, H.; Tempst, P.; Sabatini, D.M. Rictor, a Novel Binding Partner of MTOR, Defines a Rapamycin-Insensitive and Raptor-Independent Pathway That Regulates the Cytoskeleton. *Curr. Biol.* **2004**, *14*, 1296–1302. [CrossRef]

107. Feldman, M.E.; Apsel, B.; Uotila, A.; Loewith, R.; Knight, Z.A.; Ruggero, D.; Shokat, K.M. Active-Site Inhibitors of MTOR Target Rapamycin-Resistant Outputs of MTORC1 and MTORC2. *PLoS Biol.* **2009**, *7*, e1000038. [CrossRef] [PubMed]

108. Kwun, H.J.; Chang, Y.; Moore, P.S. Protein-Mediated Viral Latency Is a Novel Mechanism for Merkel Cell Polyomavirus Persistence. *Proc. Natl. Acad. Sci. USA* **2017**, *114*, E4040–E4047. [CrossRef] [PubMed]

109. Wu, J.; Su, H.; Yu, Z.; Xi, S.; Guo, C.; Hu, Z.; Qu, Y.; Cai, H.; Zhao, Y.; Zhao, H.; et al. Skp2 Modulates Proliferation, Senescence and Tumorigenesis of Glioma. *Cancer Cell Int.* **2020**, *20*, 71. [CrossRef] [PubMed]

110. Zhang, L.; Wang, C. F-Box Protein Skp2: A Novel Transcriptional Target of E2F. *Oncogene* **2006**, *25*, 2615–2627. [CrossRef]

111. Assetta, B.; Maginnis, M.S.; Ahufinger, I.G.; Haley, S.A.; Gee, G.V.; Nelson, C.D.S.; O'Hara, B.A.; Ramdial, S.A.A.; Atwood, W.J. 5-HT2 Receptors Facilitate JC Polyomavirus Entry. *J. Virol.* **2013**, *87*, 13490–13498. [CrossRef]

112. Wipf, P.; Halter, R.J. Chemistry and Biology of Wortmannin. *Org. Biomol. Chem.* **2005**, *3*, 2053–2061. [CrossRef]

113. Pittini, Á.; Casaravilla, C.; Allen, J.E.; Díaz, Á. Pharmacological Inhibition of PI3K Class III Enhances the Production of Pro- and Anti-Inflammatory Cytokines in Dendritic Cells Stimulated by TLR Agonists. *Int. Immunopharmacol.* **2016**, *36*, 213–217. [CrossRef] [PubMed]

114. Brunn, G.J.; Williams, J.; Sabers, C.; Wiederrecht, G.; Lawrence, J.C.; Abraham, R.T. Direct Inhibition of the Signaling Functions of the Mammalian Target of Rapamycin by the Phosphoinositide 3-kinase Inhibitors, Wortmannin and LY294002. *EMBO J.* **1996**, *15*, 5256–5267. [CrossRef]

115. O'Hara, B.A.; Morris-Love, J.; Gee, G.V.; Haley, S.A.; Atwood, W.J. JC Virus Infected Choroid Plexus Epithelial Cells Produce Extracellular Vesicles That Infect Glial Cells Independently of the Virus Attachment Receptor. *PLoS Pathog.* **2020**, *16*, e1008371. [CrossRef] [PubMed]

116. Lei, Y.; Huang, K.; Gao, C.; Lau, Q.C.; Pan, H.; Xie, K.; Li, J.; Liu, R.; Zhang, T.; Xie, N.; et al. Proteomics Identification of ITGB3 as a Key Regulator in Reactive Oxygen Species-Induced Migration and Invasion of Colorectal Cancer Cells. *Mol. Cell Proteom.* **2011**, *10*, M110.005397. [CrossRef]

117. Fuentes, P.; Sesé, M.; Guijarro, P.J.; Emperador, M.; Sánchez-Redondo, S.; Peinado, H.; Hümmer, S.; y Cajal, S.R. ITGB3-Mediated Uptake of Small Extracellular Vesicles Facilitates Intercellular Communication in Breast Cancer Cells. *Nat. Commun.* **2020**, *11*, 4261. [CrossRef]

118. Van Niel, G.; D'Angelo, G.; Raposo, G. Shedding Light on the Cell Biology of Extracellular Vesicles. *Nat. Rev. Mol. Cell Biol.* **2018**, *19*, 213–228. [CrossRef] [PubMed]

119. Mathieu, M.; Martin-Jaular, L.; Lavieu, G.; Théry, C. Specificities of Secretion and Uptake of Exosomes and Other Extracellular Vesicles for Cell-to-Cell Communication. *Nat. Cell. Biol.* **2019**, *21*, 9–17. [CrossRef]

120. Zhang, N.; Ma, D.; Wang, L.; Zhu, X.; Pan, Q.; Zhao, Y.; Zhu, W.; Zhou, J.; Wang, L.; Chai, Z.; et al. Insufficient Radiofrequency Ablation Treated Hepatocellular Carcinoma Cells Promote Metastasis by Up-Regulation ITGB3. *J. Cancer* **2017**, *8*, 3742–3754. [CrossRef]

121. Anthony, D.C.; Ferguson, B.; Matyzak, M.K.; Miller, K.M.; Esiri, M.M.; Perry, V.H. Differential Matrix Metalloproteinase Expression in Cases of Multiple Sclerosis and Stroke. *Neuropath. Appl. Neurobiol.* **1997**, *23*, 406–415. [CrossRef]
122. Anthony, D.C.; Miller, K.M.; Fearn, S.; Townsend, M.J.; Opdenakker, G.; Wells, G.M.A.; Clements, J.M.; Chandler, S.; Gearing, A.J.H.; Perry, V.H. Matrix Metalloproteinase Expression in an Experimentally-Induced DTH Model of Multiple Sclerosis in the Rat CNS. *J. Neuroimmunol.* **1998**, *87*, 62–72. [CrossRef]
123. Maeda, A.; Sobel, R.A. Matrix Metalloproteinases in the Normal Human Central Nervous System, Microglial Nodules, and Multiple Sclerosis Lesions. *J. Neuropathol. Exp. Neurol.* **1996**, *55*, 300–309. [CrossRef] [PubMed]

Review

SEA and GATOR 10 Years Later

Yahir A. Loissell-Baltazar and Svetlana Dokudovskaya *

CNRS UMR9018, Institut Gustave Roussy, Université Paris-Saclay, 94805 Villejuif, France; y.loissell@gmail.com
* Correspondence: svetlana.dokudovskaya@gustaveroussy.fr

Abstract: The SEA complex was described for the first time in yeast *Saccharomyces cerevisiae* ten years ago, and its human homologue GATOR complex two years later. During the past decade, many advances on the SEA/GATOR biology in different organisms have been made that allowed its role as an essential upstream regulator of the mTORC1 pathway to be defined. In this review, we describe these advances in relation to the identification of multiple functions of the SEA/GATOR complex in nutrient response and beyond and highlight the consequence of GATOR mutations in cancer and neurodegenerative diseases.

Keywords: SEA complex; GATOR complex; mTORC1 pathway; autophagy; amino acid signaling; cancer; epilepsy; neurological disorders

Citation: Loissell-Baltazar, Y.A.; Dokudovskaya, S. SEA and GATOR 10 Years Later. *Cells* **2021**, *10*, 2689. https://doi.org/10.3390/cells10102689

Academic Editors: Jean Christopher Chamcheu, Claudia Bürger and Shile Huang

Received: 31 August 2021
Accepted: 3 October 2021
Published: 8 October 2021

Publisher's Note: MDPI stays neutral with regard to jurisdictional claims in published maps and institutional affiliations.

1. Introduction

The highly conserved mechanistic (or mammalian) target of rapamycin (mTOR) plays a key role in cellular homeostasis. mTOR kinase forms the following two different complexes: mTORC1 and mTORC2, which regulate cellular responses to many stresses [1,2]. In order to maintain optimal growth and metabolism, the mTORC1 pathway integrates signals from a wide variety of intracellular and extracellular cues, which include amino acids, growth factors, energy, oxygen, DNA damaging agents, etc. Depending on the nature of the signal, mTORC1 will drive the cell either to the anabolic pathway, promoting the proliferation and survival, or to the catabolic pathway by controlling autophagy or the ubiquitin-proteasome system. In order to coordinate this vast network, mTORC1 relies on many upstream modulators and downstream effectors. Ten years ago, one of the major upstream regulators of mTORC1 pathway, the SEA/GATOR complex, was identified [3]. Over these years, many advances have been made in our understanding of the SEA/GATOR complex functions and their consequences to the operation of the mTORC1 pathway; however, many questions are still unsolved [4]. Our comprehension of the SEA/GATOR complex regulation and function is particularly important because of the consequences of its dysfunction in diverse pathological settings, especially in cancer and neurodegenerative diseases. This review covers the most important findings about the SEA/GATOR complex that have been made during the last decade.

2. Discovery of the SEA Complex

The SEA complex was initially identified in yeast *Saccharomyces cerevisiae* through an atypical way [3,5–7]. Back in 2007, a multidisciplinary approach was undertaken to solve the structure of one of the largest macromolecular machines in the cell—the nuclear pore complex (NPC) [5,8]. Central to this approach were the collection of many kinds of biophysical and proteomic data, the translation of these data to spatial restrains and the calculation of a final architecture that satisfies all the restrains. This was how the immunopurification of one of the NPC components, nucleoporin Seh1, revealed that this protein did not only co-purify with the Nup84 subcomplex, the major constituent of the NPC scaffold, but also with the following four high-molecular-weight proteins with completely unknown functions at the time: Yjr138p (Iml1), Yol138p (Rtc1), Ydr128p (Mtc5) and Ybl104p [5]. Four years later, in 2011, a paper that described the full SEA complex for the first time was published [3]. The four proteins, which were first observed in Seh1 pullouts in 2007, were given a common name, Sea (for <u>Se</u>h1-<u>a</u>ssociated) and named Sea1 through Sea4, respectively. The following three other protein components completed the full SEA eight-protein complex: Sec13, Npr2 and Npr3. The proteins of the SEA complex appeared to be dynamically associated with the vacuole membrane and have a role in autophagy. The function of Iml1-Npr2-Npr3 in autophagy was also described by the Tu group that same year [9]. Meanwhile, in 2009, Npr2 and Npr3 were shown to form an evolutionary conserved heterodimer, involved in the upstream regulation of TORC1 in response to amino acid starvation in *S. cerevisiae* [10]. This fundamental function of the SEA complex was further confirmed both in yeast and humans by de Virgilio and Sabatini laboratories in 2013 [11,12].

The SEA complex in *S. cerevisiae* consists of two subcomplexes, named SEACIT (SEA subcomplex inhibiting TORC1) and SEACAT (SEA subcomplex activating TORC1) (see below) [11,13,14] (Figure 1). In 2013, these complexes were characterized for the first time in humans by Sabatini's laboratory and were re-named to GATOR1 (GTPase activating protein activity toward RAGA, see below) and GATOR2, respectively [12]. SEACIT is composed of Iml1/Sea1, Npr2 and Npr3 (DEPDC5, NPRL2 and NPRL3 in GATOR1), and SEACAT contains Sea2, Sea3, Sea4, Seh1 and Sec13 (WDR24, WDR59, MIOS, SEH1L, SEC13 in GATOR2) (Figure 1).

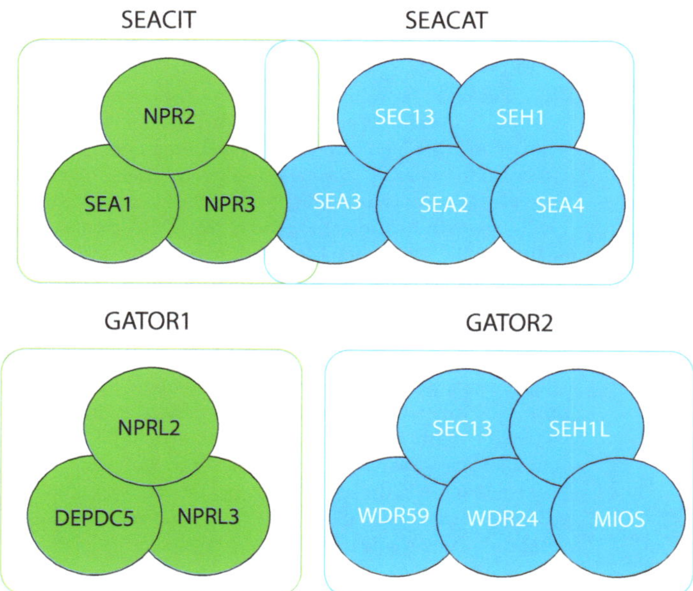

Figure 1. Composition of yeast SEA complex and mammalian GATOR complex. SEACIT subcomplex in yeast can tightly interact with SEACAT, most probably via Npr3/Sea3 connection. GATOR1 and GATOR2 do not form a stable full GATOR complex.

Phylogenetic analyses demonstrated that SEA/GATOR complex subunits are present across various eukaryotic kingdoms, suggesting an origin of these factors before the last common eukaryotic ancestor [3]. Homologs of all eight proteins could be clearly found in the genomes of fungi and metazoans, with some representation in protists, but cannot be identified in plants [3,15]. In 2021, the homologs of the SEA complex and its components were characterized in *Schizosaccharomyces pombe* [16], *Caernorhabditis elegans* [17], *Drosophila* [18], zebrafish [19], mice [20], rats [21] and humans [12]. The majority of the structural and functional studies were usually performed in *S. cerevisiae* and in humans. *Drosophila* was very instrumental for the study of the SEA/GATOR role in development; while the zebrafish, mouse and rat models were used to study different human pathologies.

3. SEA/GATOR Nomenclature

The nomenclature of the SEA complex proteins and subcomplexes in different organisms is somewhat confusing. For example, in *S. pombe*, the complex is called GATOR, but the names of the constituent proteins are the same as in *S. cerevisiae* [22]. One of the *Drosophila* GATOR1 components is called Iml1 (impaired minichromosome loss), as its yeast homologue, but all other proteins are named after their human homologues [18]. Moreover, the yeast protein community has a tendency to drop the name Sea1 and call the protein with its initial name, Iml1. The SEA proteins Npr2 (nitrogen permease regulator 2) and Npr3 (nitrogen permease regulator 3) gave names to their human orthologues NPRL2 (Npr2-like) and NPRL3 (Npr3-like) [10,23]. GATOR2 component MIOS obtained its name from its *Drosophila* orthologue Mio (missing oocyte) [24]. On the other hand, the two GATOR2 components, WDR24 and WDR59, still have their systematic names. In the future, it might be reasonable to revise their names so they reflect their function (currently, these functions are not yet defined). Alternatively, the proteins can be systematically named after their yeast homologues (as in the case of NPRL2, NPRL3), i.e., SEAL2 and SEAL3.

4. Structural Features of the SEA and GATOR Complexes

The overall architecture of SEA/GATOR proteins is evolutionary conserved [3]. DEPDC5 is only 10 amino acid residues longer than Iml1/Sea1, but both have an identical fold arrangement. The human orthologs of Sea2-Sea4, Npr2 and Npr3 are smaller than yeast proteins, mainly because of the deletion of protein regions, predicted to be disordered in yeast. It is quite reasonable to expect that the mammalian GATOR components repertoire would be larger compared to yeast due to the expression of alternative splicing products. For example, bioinformatical predictions revealed that WDR24 has at least two isoforms, one of which is missing about 130 amino acid residues in the N-terminal part [3]. One of the NPRL3 isoforms that lacks the N-terminal part and is highly expressed in red blood cells has just recently been characterized [25]. A splicing variant that led to exon 3 skipping in *NPRL2* was detected in an individual with familial focal epilepsy (see below) [26].

Two subcomplexes of the SEA/GATOR are very different structurally (Figure 2). SEACIT/GATOR1 members have domains, found in proteins that control the functions of small GTPases. SEACAT/GATOR2 components are enriched with domains found in coating assemblies (i.e., COPI and COPII coated vesicles, nuclear pore complex, etc.) (see below). Seh1, Sec13 and the N-termini of Sea4 and Sea2 in *S. cerevisiae* SEACAT appear to form a large cluster of β-propeller domains. Similar arrangements of β-propeller domains have been described at the vertex of the evolutionarily related complexes COPI and COPII [27].

In yeast, SEACAT and SEACIT interact to form the full SEA complex (Figure 1) [13]. A 3D map of the *S. cerevisiae* SEA complex, obtained by a combination of biochemical and computational approaches, suggests that SEACAT and SEACIT are connected by interactions between the N-termini of Sea3 from SEACAT and both Npr3 and Iml1/Sea1 from SEACIT [13]. Similar observations have recently been made in *S. pombe*, where Sea3 anchors other GATOR2 components to GATOR1, although, as expected, Sea3 was not required for the assembly of GATOR1 components [28]. In humans, GATOR1 and GATOR2 do not form a stable GATOR complex [12], yet, similar to yeast, NPRL3 is necessary and sufficient for the interaction with GATOR2 [29].

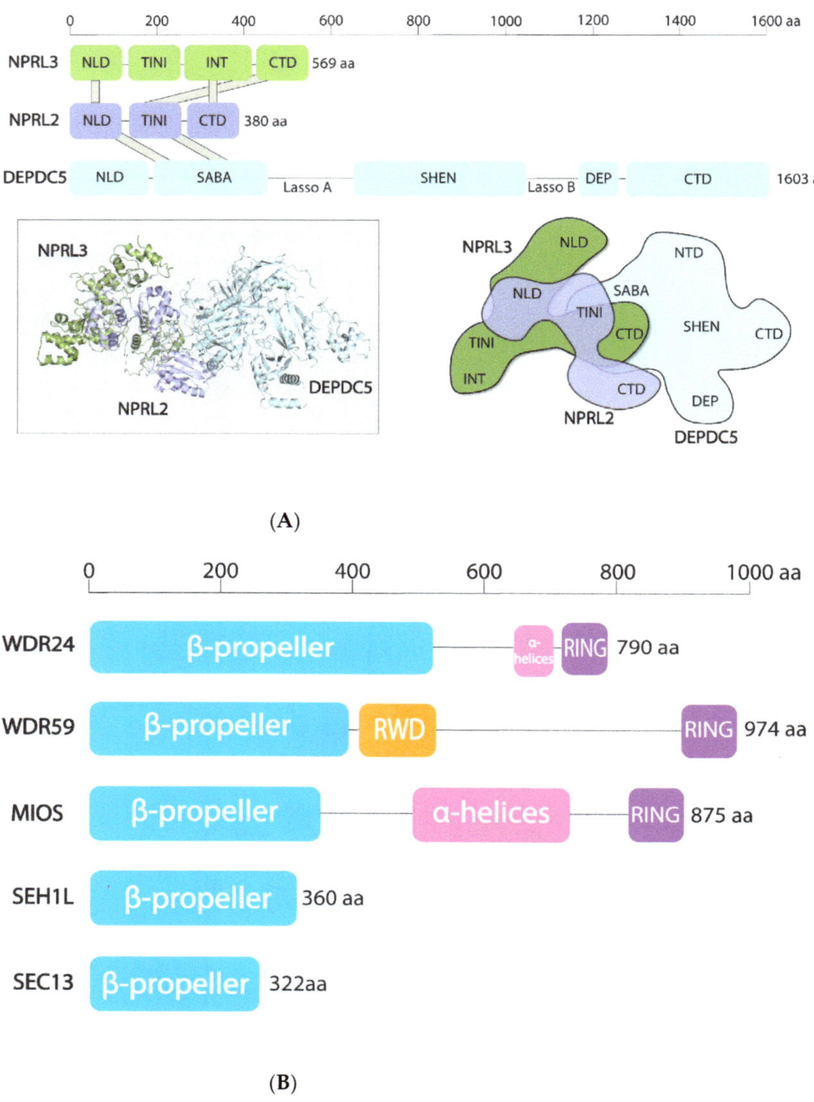

Figure 2. Domain organization of GATOR1 and GATOR2 proteins. (**A**) Domain structure and interaction of GATOR1 proteins (top); atomic model of GATOR1 complex (PDB:6CET), adapted from [29] and modified by PyMOL (bottom left) and cartoon representation of GATOR1 structure with domains indicated (bottom right). (**B**) Schematic representation of GATOR2 components, with domain boundaries according to secondary structure predictions from [3].

Despite the considerable progress in the structural determination of the constituents of the mTORC1 pathway that have been made in the last five years [30], only the structure of human GATOR1 has been solved (Figure 2A). All structural information that is currently available for GATOR2 or for the yeast SEA complex comes from bioinformatic predictions and interactivity assays [3,13]. The lack of high-resolution structures of the GATOR2 and of the entire complex both in yeast and humans are among the major reasons that prevent our full understanding of the SEA/GATOR functions at the present.

4.1. SEACAT/GATOR2

SEACAT and GATOR2 have components that moonlight between functionally unrelated complexes and are structurally connected with vesicle-coating scaffolds. The SEACAT/GATOR2 complex closely resembles the membrane coating assemblies, such as COPII vesicles, nuclear pore complexes and HOPS/CORVET complexes [8,31–33]. It also shares common subunits with both COPII (Sec13/SEC13) and nuclear pore complex (Sec13/SEC13 and Seh1/SEH1L). Sea4/MIOS contains N-terminal WD40 repeats arranged into a β-propeller structure followed by an α-solenoid stretch, which is a structure that is characteristic for proteins that form oligomeric coats (e.g., clathrin and Sec31) in vesicle-coating complexes [3,33] (Figure 2B). Furthermore, every protein in SEACAT contains a β-propeller (and Sea3 probably has two β-propellers), a domain common in coating assemblies [34]. Lastly, there are two dimers, Seh1-Sea4 and Sec13-Sea3 [3,13], that could be analogues to the Sec13-Sec31 dimer, which forms the structural unit of the COPII complex [35]. These dimeric interactions in the SEACAT are most probably conserved, because it was found that the Seh1 in *Drosophila* also directly interacts with Sea4/Mio [36].

Sea4 also contains a C-terminal RING domain, which, together with its β -propeller and α -solenoid motifs, makes it closely resemble several protein subunits of the homotypic fusion and protein sorting (HOPS) and class C core vacuole/endosome tethering (CORVET) complexes, which have been implicated in the tethering of membranes prior to their fusion. HOPS and CORVET are associated with the vacuoles/lysosomes and endosomes, respectively, and play a role in endosomal and vacuolar assembly and trafficking, as well as in nutrient transport and autophagy [32,37]. Sea2/WDR24 and Sea3/WDR59 also have a C-terminal RING domain. Clusters of RING domains are associated with E3 ubiquitin ligase activity, suggesting SEACAT might have such a role. In *S. cerevisiae*, the RING domains appear to be crucial for maintaining the interactions between Sea2, Sea3, Sea4 and the rest of the complex. For example, Sea4 that lacks the RING domain can only interact with Seh1, whereas Sea2 or Sea3 without the RING domain are no longer able to interact with any of the SEACAT complex components [13]. In addition, Sea3 contains an RWD domain that is enriched in β-sheets and common in proteins that also contain a RING motif and a β-propeller [38]. The RWD domain of Sea3 significantly resembles the RWD domain of the GCN2 protein, which is involved in general amino acid sensing and that of ubiquitin-conjugating E2 enzymes [39]. Given that SEACAT contains three proteins with RING domains, as well as numerous β-propeller domains that can mediate the recognition of phospho-substrate within E3 ligase complexes [40], it will be very interesting to investigate whether SEACAT/GATOR2 can act as a E3 ubiquitin ligase, and if this is the case, what are its possible targets.

The presence of the same folds and fold arrangements in both the SEA complex and in coating and tethering assemblies, and the fact that they contain the same "moonlighting" components, are suggestive that these complexes share a common evolutionary origin (see below). The majority of intracellular membranes are likely a result of the evolutionary expansion of an ancestral membrane-curving module—termed the "protocoatomer" complex [31,34]. The SEA complex is a member of the coatomer group, and its existence, thus, provides further evidence that an expansion of the protocoatomer family underpins much of the functional diversity of the endomembrane system.

4.2. SEACIT/GATOR1

The structural profile of the SEACIT/GATOR1 subunits is completely different (Figure 2A). Npr2/NPRL2 is a paralog of Npr3/NPRL3 [10,41] and both proteins possess N-terminal longin domains [42,43]. Iml1/Sea1 and its human homologue DEPDC5 contain a unique composition of domains that are not found in any other proteins. SEACIT components also have PEST motifs that often exist in rapidly degraded proteins [3]. However, PEST motifs are not well preserved in mammalian orthologues and, thus, could be a specific feature of the yeast SEA complex.

The structure of GATOR1, resolved recently by cryo-EM, revealed the architecture of each GATOR1 component [29] (Figure 2A). DEPDC5 has the following five defined domains: N-terminal domain (NTD), followed by SABA (structural axis for binding arrangement), SHEN (steric hinderance for enhancement of nucleotidase activity), DEP (Dishevelled, Egl-10 and Plekstrin) and C-terminal (CTD) domains. Interestingly, NTD, SABA and DEP domains can be found in membrane-associated proteins. For example, a domain similar to NTD exists in the SNARE chaperone Sec18/NSF, the SABA domain—in Sec23 of COPII vesicles (again returning to the theme of coating complexes). The DEP domain, which has diverse functions in signal transduction, is involved in the interactions between the regulator of G protein signaling (RGS) proteins and their membrane-bound receptors, the GPCRs [44]. The DEP domain is also found in a DEPTOR subunit of mTORC1 [45].

NPRL2 and NPRL3 have a similar structure with N-terminal longin domains that heterodimerize (Figure 2A). C-terminal domains of NPRL2 and NPRL3 also form a large contact surface. The SABA domain in DEPDC5 interacts with the NPRL2 TINI domain (tiny intermediary of NPRL2 that interacts (with DEPDC5)). By the way, the domain nomenclature within the GATOR1 complex created a doubtful precedent, where protein domains are named after the first (SHEN) and the last (SABA-TINI) authors of the article that reported the structure [29].

4.3. Posttranslational Modifications of SEA/GATOR

The majority of the information about post-translational modifications of the SEA/GATOR components came from whole proteome studies, essentially in yeast [46–51]. All the SEA and GATOR members are heavily phosphorylated and ubiquitinated (except of Sec13), with many modifications occurring at the disordered regions of proteins. However, there are still very few studies that explore the functional role of these modifications. Several papers, which describe the effect of ubiquitination, are manly focused on the role of this modification on protein stability. Thus, Npr2 in yeast interacts with Grr1, the F-box component of the SCF[Grr1] E3 ubiquitin ligase [52]. Moderately unstable Npr2 is stabilized in *grr1Δ* mutants. In response to amino acids, CUL3-KLH22 E3 ubiquitin ligase induces K48 polyubiquitination on multiple DEPDC5 sites leading to its degradation [53]. Accordingly, DEPDC5 levels are increased during amino acid starvation. In the rich media, NPRL3 is more resistant to proteasome degradation than NPRL2 [54]. The data about the stability of SEA/GATOR proteins during amino acid starvation are contradictory and vary considerably in different species. For example, the level of practically all the SEA members in yeast decreases both during amino acid starvation and rapamycin treatment [13]. In *Drosophila* S2 cell lines, amino acid deprivation increases Nprl3 stability [55], although the reports in human cell lines indicate that the amount of NPRL2 and NPRL3 is not changed at least after 30 min of amino acid starvation [53]. It is reasonable to expect in the following years that we will gain more information about the role of posttranslational modifications not only on the stability of SEA/GATOR members, but also on their function.

5. Function of the SEA and GATOR in Nutrient Sensing and Responding

5.1. Overview of Amino Acid Axis of Signaling to mTORC1

One of the principal roles of SEA and GATOR as upstream regulators of mTORC1 is responding to amino acid availability [11,12] (Figure 3), although the role of both GATOR subcomplexes in glucose sensing has also been reported recently [56]. Effective functioning of the mTORC1 pathway with respect to cellular amino acid levels requires coordinated action of RAG guanosine triphosphatases (RAG-GTPases or RAGs) and their effectors, such as GTPase-activating proteins (GAPs), which stimulate GTP hydrolysis and guanine-nucleotide-exchange factors (GEFs). The major site of mTORC1 activation is the vacuole/lysosomal surface, where mTORC1 is recruited and induced in an RAG-GTPase dependent manner when amino acids are abundant [57,58]. There are the following four RAG GTPases: RAGA and functionally redundant RAGB; RAGC and functionally redundant RAGD (Figure 3A). They exist as obligate heterodimers, e.g., RAGA (or RAGB) with

RAGC (or RAGD). RAGs interact with a pentameric RAGULATOR complex, anchored to the lysosome [57–61]. RAGULATOR also interacts with v-ATPase, a protein pump at the lysosomal membrane. The guanine nucleotide loading is important for RAGs function. In the presence of amino acids, RAGs are active when RAGA/B is loaded with GTP, and RAGC/D is bound to GDP. Reversely, when amino acids are low, RAGs are inactive, and RAGA/B is loaded with GDP and RAGC/D is bound to GTP. Various GAPs and GEFs promote the conversion of RAGs from active to inactive form. This is where the SEACIT and GATOR1 complexes exert their major functions (see below). A RAG-independent induction of mTORC1 by amino acids both at the vacuole/lysosome and Golgi has also been described in yeast and humans [62–65], but will not be thoroughly discussed in this review since neither SEA nor GATOR seem to be involved in this mode of mTORC1 activation. Moreover, a recent study revealed that RAG-independent activation of mTORC1 by amino acids derived from protein degradation in lysosomes required HOPS complex and was negatively regulated by activation of the GATOR-RAGs pathway [37]. Thus, evolutionary related HOPS and GATOR2 [3] have similar but divergent roles in activating mTORC1 in response to different amino acid inputs.

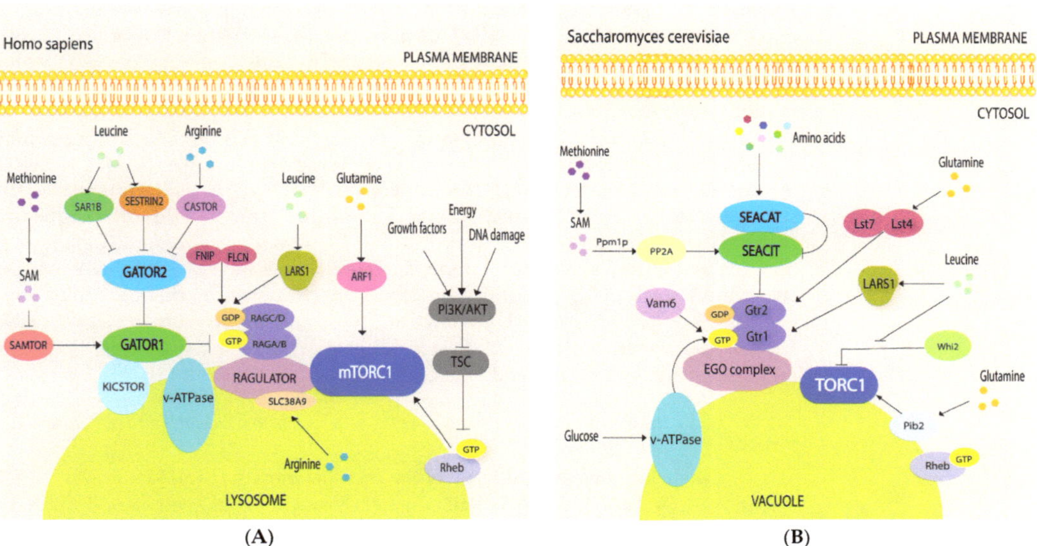

Figure 3. Amino acid signaling. (**A**) mTORC1 signaling in *Homo sapiens*. (**B**) TORC1 signaling in *Saccharomyces cerevisiae*. Yeast and mammalian orthologues are designated with the same color. Arrows and bars represent activation and inhibition, respectively. See text for more details.

When amino acids are scarce, some amino acid sensors (see below) interact with and inhibit the GATOR2 complex, thus preventing inhibition of the GATOR1 by GATOR2 (Figure 3A). A mammalian-specific KICSTOR complex tethers GATOR1 to the lysosomal surface [66,67] where GATOR1 acts as a GAP for RAGA [12], thereby transforming RAGA to its inactive, GDP bound form, which further leads to mTORC1 suppression (Figure 3A).

In the presence of amino acids, RAGULATOR and v-ATPase undergo a conformational change that results in RAGULATOR exerting GEF activity towards RAGA or RAGB [60]. RAGULATOR can also trigger GTP release from RAGC [68]. In parallel, upon arginine binding arginine sensor SLC38A9, which resides at the lysosome, stimulates GDP release from RAGA [68]. A complex between folliculin (FLCN) and folliculin-interacting protein (FNIP) 1 and/or 2 is a GAP for RAGC/D [69]. In addition, leucyl-tRNA synthetase (LeuRS or LARS1 or LRS) also has GAP activity towards RAGD [70]. Active RAGULATOR-RAG stimulates recruitment of mTORC1 to the lysosomal membrane where it is fully activated by small GTPase, RHEB, loaded with GTP [71]. RHEB is under the control of another signaling node—the TSC complex, composed of TSC1, TSC2 and TBC1D7, where TSC2 acts as a GAP to inhibit RHEB. TSC is a nexus of multiple physiological stimuli (e.g., energy status, growth factors, DNA damage) that signal to mTORC1 through PI3K-AKT network [72]. RAG GTPases regulate the recruitment of TSC to the lysosome and its ability to interact with and inhibit RHEB in response to amino acid starvation, growth factors removal and to other stresses that inhibit mTORC1 [73–75]. Both RAGs and RHEB are necessary for mTORC1 activation at the lysosome, as the lone presence of either one is not sufficient. Accordingly, only when both the RAG GTPases and RHEB are inactive mTORC1 fully released from the lysosome [73].

The RAGs and RAGULATOR are conserved both in fission and in budding yeast (Figure 3B) [76,77]. Thus, the orthologue of RAGA/B, a protein called Gtr1 in yeast, forms a heterodimer with Gtr2, which is an orthologue of RAGC/D. Similar to mammals, in order to activate TORC1, GTP-bound Gtr1 and GDP-loaded Gtr2 interact with trimeric *S. cerevisiae* Ego1-Ego3 complex (Lam1-Lam4 in fission yeast) analog of RAGULATOR. Iml1/Sea1 from the SEACIT serves as a GAP for Gtr1 in the absence of amino acids [11]. Interestingly, LARS1 in yeast is the GEF for Gtr1 [78], while in mammalian cells LARS1 was shown to be a GAP for RAGD [70], although a GAP activity was not confirmed in a later study from different laboratory [69]. Lst4-Lst7 complex, an orthologue of mammalian FLCN/FNIP, is GAP for Gtr2 [79]. The GEF for Gtr2 in yeast and for RAGC/D in mammals is still not known.

There are some notable differences between yeast and humans during amino acid signaling to mTORC1 (Figure 3). First, many amino acid sensors (e.g., SAMTOR, SESTRINs) are absent in yeast (see below) [1]. Second, v-ATPase in yeast, which interacts with Gtr1, seems to activate TORC1 in response to glucose [80]. Third, RHEB orthologue in yeast *S. cerevisiae* seems not to be involved in TORC1 signaling, although it is required for arginine and lysine uptake [81]. Fourth, *S. cerevisiae* does not have TSC homologues, thus the entire branch of TSC/RHEB signaling is not conserved in this particular yeast. In contrast, *S. pombe* has both RHEB and TSC, which are involved in mTORC1 activation. How *S. cerevisiae* achieves full TORC1 activation at the vacuole without TSC/RHEB branch is currently not well understood.

5.2. GATOR2 Interactions with Leucine Sensors SESTRINs and SAR1B and Arginine Sensor CASTOR1

Cytosolic leucine can be sensed by the proteins from the SESTRIN family (SESTRINs 1–3) [82–84], by small GTPase SAR1B [85] and by leucyl-tRNA synthetase [70,86,87]. Arginine is sensed by CASTOR1 protein homodimer in the cytoplasm [88,89] and by SLC38A9 together with TM4S5F protein at the lysosomal membrane [90–92].

GATORs can interact directly with several amino acid sensors (Figure 3A). During leucine or arginine starvation, SESTRIN2 [82], SAR1B [85] or CASTOR1, respectively [88,89] interact with and inhibit the GATOR2 complex. WDR24 and SEH1L are essential for interaction with SESTRIN2, but it is not known which component of GATOR2 interacts with SESTRIN2 directly [93,94]. SAR1B directly binds MIOS, but not other GATOR2 subunits [85]. WDR24, SEH1L and MIOS were sufficient for interaction with CASTOR1 [89]; the CASTOR1 N-terminal domain is involved into direct interaction with MIOS [95]. Binding sites for SESTRIN2 and CASTOR1 are located at different parts of GATOR2 [89]. These interactions prevent inhibition of the GATOR1 by GATOR2 [96] and as a consequence, lead to mTORC1 inhibition. Neither SESTRIN2 nor CASTOR1 interact with GATOR1 [89,93,97].

In the presence of amino acids, interaction of leucine to the defined binding pocket in monomeric SESTRIN2 [83] or arginine with its binding pocket at the homodimeric CASTOR1 [88,95,98] results in dissociations of these sensors from GATOR2 and relieves mTORC1 inhibition. It is important to note, however, that SESTRIN2-GATOR2 interactions were initially observed in the cell-lines cultured in leucine-rich conditions [93,97], even if amino acid starvation enhanced this interaction. In vitro addition of leucine reduces the

SESTRIN1-GATOR2 or SESTRIN2-GATOR2 interactions, but it does not affect SESTRIN3-GATOR2 interaction [82,84]. Interestingly, SESTRIN2 and SAR1B detect different parts of leucine; SAR1B recognizes the amino group and side chain of leucine [85], while SESTRIN2 interacts with leucine's amino and carboxyl groups [83].

Interactions of SESTRINs to GATOR2 depends on a cell type and physiological conditions. Thus, in the skeletal muscle of rats, SESTRIN1 is the most abundant isoform, and SESTRIN2 expression is much lower relative to either SESTRIN1 or SESTRIN3. Accordingly, oral administration of leucine to fasted rats induced SESTRIN1-GATOR2 disassembly, but did not affect the interaction of other SESTRIN isoforms with GATOR2 [84]. This suggests that in the rat skeletal muscle, it is probably SESTRIN1 that has a primary role as a leucine sensor and leucine-induced activation of mTORC1 in skeletal muscle happens via SESTRIN1 release from GATOR2. SESTRINs–GATOR2 interactions can also be age dependent. Thus, in the skeletal muscle of young pigs, SESTRIN2 is more abundant than SESTRIN1 but the GATOR2 amounts are the same. Accordingly, during amino acid starvation the abundance of the SESTRIN2–GATOR2 complex reduced more in younger pigs [99].

Recently, GATOR2 was reported to be required for SESTRIN2-induced AKT activation and AKT translocation to plasma membrane [94]. In addition, GATOR2 physically bridges SESTRIN2 with mTORC2 where WDR59's interaction with mTORC2's component RICTOR is essential for the communication between GATOR2 and mTORC2, and WDR24 is crucial for GATOR2-SESTRIN2 interaction. In HeLa cells, GATOR2 promotes AKT activation and facilitates AKT-dependent inhibitory phosphorylation of TSC2 [75]. Thus, although an exact molecular function of GATOR2 has not yet been defined, it is clear that GATOR2 might have a large repertoire of various activities. Solving the structure of GATOR2 alone and in complex with its interactors will provide essential information about how these multiple functions can be exerted.

5.3. GATOR1 Interaction with SAM Sensor, SAMTOR

The SAM sensor, SAMTOR, binds to GATOR1 during SAM or methionine deprivation, and negatively regulates mTORC1 activity [100]. The component of GATOR1 that interacts with SAMTOR is currently unknown. In the presence of SAM, this metabolite occupies its binding pocket in SAMTOR, which disrupts the interaction of an amino acid sensor with GATOR1, promoting mTORC1 activity. SAMTOR and GATOR1 interactions are dependent on KICSTOR (see below). When SAMTOR is bound to SAM, it dissociates from GATOR1–KICSTOR, thus inhibiting GATOR1 and promoting mTORC1 activation [101]. On the other hand, methionine starvation promotes interaction between SAMTOR and the GATOR1–KICKSTOR complex, but weakened the interaction between GATOR1 and GATOR2, thus leading to mTORC1 suppression [100]. SAM levels can be affected by the availability of vitamin B12. Mice NPRL2 KO embryos have significantly reduced methionine levels and demonstrate phenotypes reminiscent of B12 deficiency [20]. It is currently unknown whether methionine can be sensed directly. Interestingly, leucine can also signal to mTORC1 through its metabolite, acetyl-coenzyme A, but in a RAG-independent and cell-specific manner [102].

In a recent study, Jewell laboratory investigated the potency of each amino acid to stimulate mTORC1 in MEF or HEK293 cells [65]. Ten amino acids were able to re-stimulate mTORC1 and promote its lysosomal localization. Glutamine and asparagine signal to mTORC1 through a RAG-independent mechanism via ADP-ribosylation factor ARF1. Eight amino acids (alanine, arginine, histidine, leucine, methionine, serine, threonine and valine) filter through RAGs. While three cytoplasmic sensors for leucine, arginine and methionine (SAM) have been identified, it is not known whether the other five amino acids also have their specific sensors and whether they will interact with GATORs.

5.4. SEACIT and Amino Acid Sensing in Yeast

Amino acid sensing in yeast differs significantly from the mammalian system (Figure 3B). SESTRINs, CASTOR1 and SAMTOR are not conserved in *S. cerevisiae* and

S. pombe, which presumes that the interaction of these amino acid sensors with GATOR complexes arose later in the evolution. Nevertheless, Npr2 does participate in methionine sensing in *S. cerevisiae*, but in a very different way than in mammals. Under normal growth conditions, Ppm1p methyltransferase methylates two subunits of yeast protein phosphatase 2A (PP2A), which promotes Nrp2 dephosphorylation, TORC1 activation and suppression of autophagy [103]. Low methionine level leads to a decreased SAM, which blocks PP2A methylation and its phosphatase activity. As a result, Npr2 accumulates in phosphorylated form, which most probably changes the integrity of the SEACIT complex due to increased interaction between phosphorylated Npr2 and Iml1/Sea1 [9]. Therefore, SEACIT is no longer able to repress TORC1 effectively, resulting in autophagy activation. Interestingly, Npr2-deficient yeast grown in a minimal medium, containing ammonium as a sole nitrogen source and lactate as a nonfermentable carbon source, metabolize glutamine into nitrogen-containing metabolites and maintain high SAM concentrations [104].

As in mammals, yeast also have amino acid sensing pathways parallel to SEA-GTR signaling [105]. For example, Pib2, which resides at the vacuole membrane, interacts with TORC1 complex in a glutamine-sensitive manner, suggesting that Pib2 acts as a part of a putative glutamine sensor [64]. Although both Pib2 and EGO are required for TORC1 tethering to the vacuolar membrane and its activation, they form different complexes with TORC1, ruling out a possibility that the SEA complex can participate in Pib2-dependent amino acid sensing. Even if Pib2 does not have apparent ortholog in mammals, PLEKHF1 protein shares high sequence similarity with Pib2 domains, important for TORC1 activation. However, PLEKHF1 is not involved in the glutamine-dependent regulation of mTORC1 [65]. In addition, Whi2, localized at the cell periphery, specifically senses low amino acid levels in general and leucine levels in particular, and suppresses TORC1 activity independently of the SEA complex [106,107]. The Whi2 homologue in mammals, KCTD11, acts as a negative regulator of mTORC1 during amino acid deprivation [106].

All these recent findings demonstrate that amino acid sensing mechanisms are way more diverse, because not only amino acids themselves, but also their metabolites can be sensed in a RAG-dependent, RAG-independent and cell-specific manner.

Many questions about amino acid sensing ultimately related to SEA and GATOR functions remain unanswered. Does every amino acid have its own sensor? Will all the sensors that work through RAGs interact with GATORs? What are the determinants of the interaction of amino acid sensors with one or another GATOR complex? In other words, why do SESTRIN2 and CASTOR1 interact with GATOR2, and SAMTOR with GATOR1? What are the factors that determine sensing of the same amino acid by different sensors? For example, why does leucine need three sensors (SESTRIN2, SAR1B and LARS1) that function in the same cell types, in the same subcellular location (cytosol), through the same pathway (RAG-dependent)? Leucine can also signal through its catabolite acetyl-CoA and activate mTORC1 via EP-300-mediated acetylation of RAPTOR [102]. Can other amino acids signal both themselves and their metabolites through different sensors? For example, the methionine metabolite SAM is sensed by SAMTOR, does a methionine sensor exist? Amino acid sensing also happens at Golgi, where GATORs, SESTRINS, CASTOR1 and SAMTOR have not been found thus far. How is amino acid sensing is achieved at Golgi? What is the repertoire of cell-type specific sensors? The primary role of aminoacyl tRNA synthetases is binding to cognate amino acids and their attachment onto appropriate tRNAs. Some of them, such as cytosolic LARS1 [70,78] and mitochondrial TARS2 (but not cytosolic TARS) [108], are also implicated in the upstream regulation of mTORC1 pathway. Are other aminoacyl tRNA synthetases also involved in mTORC1 regulation? What are the details of a crosstalk between general amino acid sensing through GCN2 and sensing through the mTORC1 pathway? Finally, what are the main determinants of amino acid sensing in yeast given that many mammalian amino acid sensors discovered thus far do not have yeast homologous, yet the GATOR-RAG-RAGULATOR (SEA-GTR-EGO) system is conserved?

5.5. SEACIT/GATOR1 as GAP for EGO/RAG

Two papers published simultaneously in 2013 reported the results that have dramatically increased the significance of the SEA/GATOR complex in the regulation of mTORC1 pathway. The laboratory of Claudio de Virgilio found that in *S. cerevisiae*, the SEA subcomplex, which was subsequently named SEACIT (SEAC subcomplex inhibiting TORC1 signaling) [14], acts as a GAP for Gtr1 and, thus, inhibits TORC1 [11]. In a parallel study, David Sabatini's laboratory characterized for the first time the human homologue of the SEA complex, and also found the GAP activity of the SEACIT analogue, which received the GATOR1 name (GTPase activating protein activity towards RAGA) [12]. In both yeast and humans, SEACAT/GATOR2 acts upstream of SEACIT/GATOR1, suppressing its GAP activity, thus being "an inhibitor of an inhibitor", although how exactly this suppression is achieved is completely unknown.

A molecular mechanism of how SEACIT/GATOR1 acts as a GAP has been addressed in several functional and structural studies, but a complete consensus of how exactly the GAP function is exerted has not yet been achieved. Indeed, in an initial study by the de Virgilio group, it was demonstrated that in *S. cerevisiae*, Iml1/Sea1 can co-precipitate with Gtr1 in the presence but substantially less in the absence of other SEACIT subunits. Yet, in the in vitro binding and GAP essays, Iml1/Sea1 could directly bind to Gtr1 and promote GTP hydrolysis in the absence of Npr2 and Npr3. GAPs often supply a catalytic amino acid residue (Arg, Asp or Gln) in their active sites, thus forming an "arginine finger" or "Gln/Asn thumb" that can be inserted into nucleotide-binding pocket of a GTPase [109]. In the highly conserved Iml1/Sea1 domain, essential for its GAP activity (aa 929-952), a conserved Arg^{943} was critical for GAP activity both in vitro and in yeast cells. Human DEPDC5 could partially complement TORC1 inhibition defect in *iml1Δ* cells, suggesting a conserved role of Iml1/Sea1 and DEPDC5 across the species. Therefore, when the cryo-EM structure of GATOR1 (Figure 2A) and GATOR1 in the complex with RAG GTPases was solved, it came as a surprise because it revealed a very unexpected mode of interaction between GTPases and GAPs [29].

For the structural studies, GATOR1 was copurified with RAG GTPase heterodimer, containing wild type RAGA and mutant RAGC, which can bind GTP, but not GDP. In addition, this heterodimer was loaded with GDP and non-hydrolysable GTP analogue (GppNHp) to create the most favorable nucleotide-binding configuration for interaction with GATOR1. The structure demonstrated that the overall conformation of the GATOR1 in a complex with RAG GTPases is similar to a free GATOR1 (see above). The SHEN domain of DEPDC5 can contact directly with a site proximal to nucleotide binding pocket of GTP analogue-bound RAGA. However, quite surprisingly, this interaction did not appear to be responsible for the stimulation of GTP hydrolysis. The kinetic analysis of GTP hydrolysis of DEPDC5 alone with RAGA or NPRL2/NPRL3 dimer with RAGA revealed that it is rather NPRL2/NPRL3, which has GAP activity. Moreover, a conserved Arg^{78} localized on the loop of NPRL2 longin domain is the "arginine finger", responsible for GAP activity [110]. However, this Arg^{78} is located far away and is opposite to the RAGs binding interface of DEPDC5. Moreover, an earlier study from Wang laboratory showed that amino acid stimulation enhances the interaction of RAGA with both endogenous DEPDC5 and NPRL3 [111]. To explain these rather contradictory observations, a two-state model of GATOR1 interaction with RAG GTPases was proposed: in the inhibitory mode, DEPDC5 SHEN domain interacts strongly with RAGs and GAP activity of GATOR1 is weak; alternatively, a low affinity interaction dependent on NPRL2/NPRL3 stimulates GAP activity. Such bi-modal activity has not been previously observed between a GAP and a GTPase. Moreover, before this study, longin domains were found to be highly represented in many GEFs, where they would serve as adaptable platforms for GTPases [42]. In addition, in a structure of *Chaetomium thermophilum* Mon1-Ccz1-Ypt7 complex, Mon1-Ccz1 GEF contacts its cognate GTPase Ypt7 through a face of a conserved longin domain heterodimer [112]. NPRL2 and NPRL3 also form a heterodimer using their longin interaction domains; therefore, it is quite intriguing why in case of Mon1-Ccz1 longin heterodimer supports a GEF activity, while NPRL2/NPRL3 longin domains assist to GAP function. One of the plausible explanations might involve a possibility that NPRL2–NPRL3 interaction with RAGs can be sterically compromised by GATOR2, because it is NPRL3, which is necessary and sufficient for interaction with GATOR2. Finally, to add even more complexity, one (and the only) study reported that NPRL2 interacts with RAGD in amino acid scarcity, and with Raptor during amino acid sufficiency to activate mTORC1 [113]. Although the authors explain this behavior by suggesting that NPRL2 may not solely exist as a part of GATOR1, these findings require more clarifications.

It is evident that more structural studies will be necessary to explain this peculiar mode of interaction between GATOR1 with RAG GTPases. For example, a structure of RAGs-NPRL2-NPRL3 would allow to observe the conformation of the active GAP, a task that will not be very easy, given a weak association of NPRL2/NPRL3 heterodimer with RAGs in the absence of DEPDC5. In addition, solving a structure of yeast SEA complex, where the association between SEACAT (GATOR2) and SEACIT (GATOR1) is much stronger and where GAP activity seems to be performed by Iml1/Sea1 (DEPDC5), rather than by other components of the complex, would be absolutely central for the elucidating how SEACIT/GATOR1 exert its GAP function.

5.6. SEA/GATOR Recruitment to the Vacuolar/Lysosomal Membrane

In yeast, both TORC1 and SEA complex localize at the vacuole membrane regardless of the presence or absence of amino acids [3,77,114,115]. Iml1/Sea1 did not require other SEA components to localize to the vacuole membrane in both budding and fission yeast [11,28]. In contrast, Npr2 and Npr3 mutually depend on each other and on Iml1/Sea1 for vacuolar localization [11,28]. Importantly, the deletion of any of the SEACIT components during nitrogen starvation caused the re-localization of Tor1 to the cytoplasm [13].

In mammalian cells, mTORC1 is recruited to the lysosome in the presence of the amino acids, where it is fully activated by RHEB [116]. In addition, the activation of mTORC1 by RHEB can happen at the surface of other organelles, because both RHEB and mTORC1 have been detected at the Golgi apparatus, the peroxisome, the plasma membrane and ER [62,117,118]. Stably expressed GFP-tagged components of GATOR1 (NPRL2 and DEPDC5) and GATOR2 (MIOS and WDR24) localize to the lysosome regardless of the amino acid levels [12,67], although a recent study revealed that during amino acid starvation, WDR24, MIOS and mTOR can be found at a rough ER membrane [119]. Similarly, *Drosophila* GATOR2 components Mio and Seh1 localize to lysosomes in both fed and starved flies. Mammals, however, developed additional mechanisms to maintain GATORs at the lysosomal membrane, which include an interaction with the protein complex KICSTOR, that is not present in non-vertebrates and the regulation of GATOR1-RAGA interaction via ubiquitination.

The mammalian-specific KICSTOR complex identified in 2017 plays a key role in the localization of GATOR1 to its GTPase substrates [66,67]. KICSTOR consist of four proteins, KPTN, ITFG2, C12orf66 and STZ2, whose initial letters gave the complex its name. *C. elegans* only encode a homologue of SZT2, while yeasts and *Drosophila* lack entire KICKSTOR [15,67]. Both GATOR1 and GATOR2 associate with KICKSTOR in an amino-acid insensitive manner. STZ2 is responsible for the interaction of KICKSTOR with GATOR1, since STZ2 knockouts impaired the localization of GATOR1 to the lysosomes, but not GATOR2 or RAG GTPases. SZT2 is also necessary for the coordinated GATOR1 and GATOR2 binding and for GATOR1-dependent inactivation of mTORC1 at the lysosome. SZT2 contains several regions that allow interaction with GATOR1 and GATOR2 [66]. SZT2–DEPDC5 interactions can occur in the absence of other GATOR components [29]. SZT2 does not bind GATOR2 in the absence of NPRL3, once again underlining a crucial role of this protein in GATOR1–GATOR2 interactions. In addition, lysosomal localization of WDR59 is abolished in the absence of SZT2. Thus, KICKSTOR, and especially its largest component 380 kDa SZT2, may facilitate interaction between GATOR1 and GATOR2 and maintain both subcomplexes together. In contrast, in *S. cerevisiae*, both SEA subcomplexes can form a stable complex without other mediating proteins. It is intriguing why, during evolution, mammals acquired a large protein complex to maintain interactions between GATOR1 and GATOR2, which otherwise are quite stable in lower eukaryotes.

GATOR1 is also implicated to the recruitment to the lysosomal surface of another GAP—FLCN/FNIP. GATOR1-dependent control of the RAGA nucleotide state drives FLCN recruitment to lysosomes when amino acids are scarce [120]. Indeed, when amino acids are low, the GAP activity of GATOR1 promotes the GDP-RAGA/B conformation and FLCN/FNIP is recruited to the lysosome to act as a GAP towards RAGC/D. In this study, only knockout of *NPRL3* in HeLa cells were verified, and it is not known whether knockout of other GATOR1 components would have the same effect. Nevertheless, these findings help to resolve the apparent contradiction reported earlier, that FLCN-FNIP heterodimer binds to RAGA/B, but acts as a GAP for RAGC/D [69,121]. Cryo-EM structures of the human FLCN-FNIP-RAG-RAGULATOR complex containing an inactive form of the RAG heterodimer confirmed that the FLCN-FNIP2 heterodimer binds to the GTPase domains of both RAGA and RAGC [122,123].

GATOR1-RAGA interactions are controlled by several kinases and E3 ubiquitin ligases, which are not present in lower eukaryotes. For example, an oncogenic non-receptor tyrosine kinase, SRC, disrupts GATOR1-Rags interactions, promoting mTORC1 recruitment and activation at the lysosomal surface [124]. Currently, it is not known what the mechanisms that activate SRC in response to amino acids are and whether GATOR1 subunits or RAGs can be phosphorylated by SRC. On the other hand, DEPDC5 can be phosphorylated by Pim1 kinase at S1002 and S1530, and by AKT also at S1530 [125]. This phosphorylation seems not to affect the ability of DEPDC5 to interact with neither NPRL2 nor SZT2, but elevated Pim1 expression during amino acid starvation overcame mTORC1 suppression.

Two lysosome localized E3 ligases, RNF152 and SKP2, mediate K63-linked polyubiquitination of RAGA at different sites, which promote GATOR1 recruitment to RAGA and the consequent inactivation of mTORC1 [111,126]. Remarkably, SKP2 ubiquitinates RAGA at K15 during prolonged amino acid stimulation [126], while, quite opposite, RNF152 ubiquitinates RAGA at a different set of lysines (K142, 220, 230, 244) during amino acid starvation [111]. SKP2 provides a negative feedback loop, where RAGA ubiquitination and GATOR1 recruitment restrict mTORC1 activation upon sustained amino acid stimulation. Inversely, during amino acid starvation, it is RNF152-dependent RAGA ubiquitination, which enhances GATOR1–RAGA interaction. Interestingly, RNF152 can also ubiquitinate RHEB, sequestering RHEB in its inactive RHEB-GDP form and promoting its interaction with TSC2, which leads to mTORC1 inactivation [127]. Thus, RNF152 acts a negative mTORC1 regulator in both amino acid and growth factor brunches of mTORC1 signaling.

5.7. SEA/GATOR in Autophagy

One of the major functions of mTORC1 is in the regulation of autophagy, which is induced when mTORC1 is inhibited. Thus, it is not surprising that deletions of SEACIT/GATOR1 components suppress autophagy in yeast [3,9,13,103,104,128,129], *Drosophila* [130], *C. elegans* [131] and mammals [129,132]. Just as the opposite, mutations in GATOR2 may promote autophagy, which can happen even in the absence of nutrient starvation, as it is a case of *wdr24* mutants in *Drosophila* [133]. In contrast, deletions of SEACAT members in yeast seem not to have a drastic effect on autophagy initiation and flux [3]. Interestingly, the nitrogen starvation deletion of *SEA1* or double deletion of *NPR2* and *NPR3* resulted in the inhibition of vacuolar fusion [13]. As the inactivation of TORC1 during nitrogen deprivation promotes vacuole coalescence [134], deletions of any of the SEACIT members increase TORC1 activity during starvation, and, therefore, induce vacuolar fragmentation and defects in autophagy.

Recently a bi-directional feedback loop, which regulates autophagy and involves SEACAT, has been described [50]. In order to control autophagy, TORC1 phosphorylates and inhibits the Atg1 kinase essential for autophagy initiation, but Atg1, in turn, can phosphorylate SEACAT components. Although it is currently not known whether that phosphorylation acts positively or negatively on TORC1 activity, this finding uncovers the important node of convergence between TORC1 and Atg1, with the SEACAT being both the regulator and effector of autophagy.

The SEA complex is also important for specific types of autophagy. Thus, yeast with deletions of SEACIT complex members failed to activate selective degradation of mitochondria via mitophagy (Figure 4) [135,136]. Given the conservation of the SEA/GATOR function, it is reasonable to assume a similar role of GATOR in mammals, although the involvement of GATOR in specific types of autophagy in mammals has not yet been described.

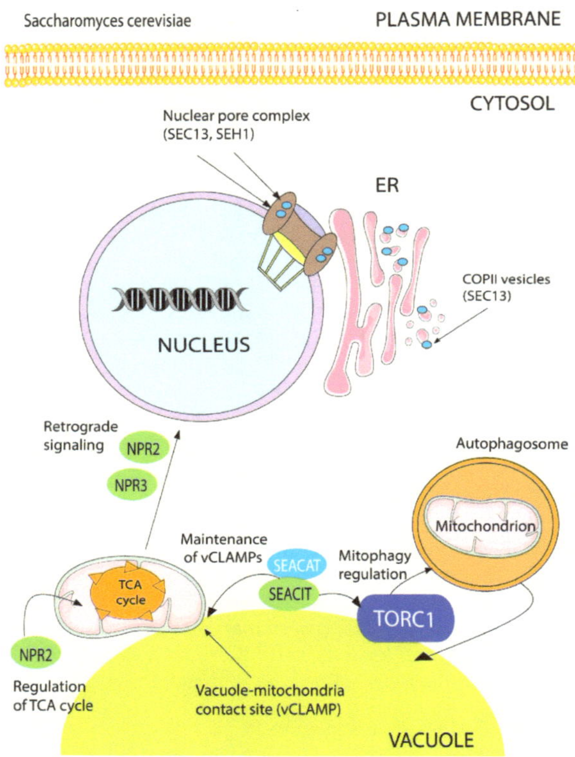

Figure 4. Functions of the *S. cerevisiae* SEA complex and its components beyond nutrient response. Indicated are Seh1 and Sec13 as components of the nuclear pore complex and Sec13 as part of COPII vesicles. Npr2 and Npr3 regulate retrograde signaling. Npr2 is also involved in the regulation of TCA cycle. Finally, SEA complex is involved in the maintenance of the vacuole-mitochondria contact sites (vCLAMPs) and is important for mitophagy.

6. SEA and GATOR Functions beyond Nutrient Responding

6.1. SEA/GATOR Evolution Origin

SEA/GATOR has always been "living double lives" with a number of its components having diverse "moonlighting" functions beyond their role in the regulation of nutrient sensing and responding (Figures 4 and 5). Although the majority of these functions seem to be related to the SEA/GATOR role in the regulation of mTORC1, others are clearly associated with totally different pathways. Accordingly, despite the fact that the main localization site of SEA/GATOR is a vacuole/lysosomal membrane, some of its components can be found in the nucleus, ER, mitochondria, plasma membrane, etc., depending on the functions that they fulfil in different cell types, stages of cell cycle progression and physiological conditions [24,54,94,135,137–140]. The most outstanding examples are Seh1 and Sec13, which together are the members of the Nup84 subcomplex in the nuclear pore complex, with Sec13 also being a component of COPII coated vesicles [3]. This "double life" of Seh1 and "triple life" of Sec13 witnesses the evolution of the endomembrane system. Indeed, the progression from prokaryotic to eukaryotic cells was accompanied by the acquisition of membranous structures, eventually transformed into organelles, which often adopted preexisting molecules and adjusted them for new needs via duplication and neofunctionalization [33]. During this transformation, a central role was played by ancient protocoatomers, which facilitated membrane bending. Not only Seh1 and Sec13, but the entire SEACAT/GATOR2 complex belongs to the large family of protocoatomer-derived complexes that form transport vesicles (COPI, COPII, clathrin), membrane-associated coats (nuclear pore complexes), tethering complexes (HOPS/CORVET) and other membrane associated structures, such as SEACAT/GATOR itself [3,31,34]. These various assemblies have a number of structural similarities, including a hallmark feature—a presence of N-terminal β-propeller, formed by WD40 repeats, and C-terminal α-soleniod composed of α-helices (HEAT repeats). In that view, Sea4/MIOS is the most well preserved protocoatomer descendant, while Sea2/WDR59 and Sea3/WDR24 diverged more profoundly, loosing many α-helices, but still preserving N-terminal β-propellers.

GTPases, with their corresponding GEFs and GAPs, are other important elements of membrane-associated assemblies. SEACIT/GATOR1 carries this functional feature of endomembrane system, being a GAP for RAGA GTPase. In addition, longin domains present in two components of the SEACIT/GATOR1 can also be found in small GTPases and many other proteins involved in assembly, fusion and tethering of membranes [141]. Here, again, paralogs Npr2/NPRL2 and Npr3/NPRL3 evidence that evolution progressed through duplication and divergence, because both proteins seem to have additional functions, apart from mTORC1 regulation.

Remarkably, the entire vacuole/lysosome-associated mTORC1 pathway machinery contains multiple structural elements typical for classical endomembrane systems [30]. For example, the mTORC1 complex has a β-propeller subunit mLST8, structurally very close to Seh1 and Sec13. Similar to other coatomers, another mTORC1 subunit, Kog1/RAPTOR, contains HEAT repeats and β-propeller, but in a "Lego game of evolution" these structural elements switch places with HEAT repeats situated at the N-terminus and β-propeller at the C-terminus. By the way, in the mTORC1 complex, RAPTOR interacts with the HEAT domain of mTOR. Finally, the abundance of small GTPases, GAPs and GEFs that control mTORC1 witness the common evolution origin of the core endomembrane system and its regulators.

6.2. Regulation of Mitochondrial Biogenesis and Quality Control

The mTORC1 pathway plays an essential role in mitochondrial biogenesis, mitochondrial genome repair, the phosphorylation of mitochondrial proteins and the regulation of mitophagy, the selective degradation of mitochondria by autophagy. As a central controller of the mTORC1 pathway, SEA/GATOR is also involved in the regulation of mitochondria function and quality control (Figure 4, Figure 5). The analysis of synthetic genetic interactions in *S. cerevisiae* revealed already in 2011 that SEA genes interact with many mitochondrial genes, with Npr2 located close to the mitochondrial gene cluster [3,142,143]. About 20% of proteins that co-precipitate with SEA components are mitochondrial proteins [13,135] and, inversely, enriched mitochondrial fractions contain SEA proteins [137]. Both C-terminal GFP tagged Iml1/Sea1 and Sea4 can be localized to the mitochondria [135]. Moreover, treatment with rapamycin significantly increases the amount of cells with cytoplasmic and mitochondrial localizations of Iml1/Sea1, although a fraction of Iml1/Sea1 can still be observed at the vacuole [138]. Similarly, in HEK 293T cells NPRL2 can be localized to the mitochondria and many mitochondrial proteins can be found in the proteome of NPRL2 and NPRL3 [54]. Recently, SESTRIN2, which interacts with GATOR2 during leucine starvation (see above), was also found to be localized to mitochondria and silencing of GATOR2 genes considerably reduced the mitochondrial pool of SESTRIN2 [144]. Finally, Sec13 was shown to be interacting with mitochondrial antiviral signal protein (MAVS, also known as VISA) [145,146]. MAVS is localized on the outer membrane of mitochondria, with a small proportion present at mitochondria-associated membranes (MAMs). Sec13 overexpression increases MAVS aggregation and facilitates interferon β production, while low levels of Sec13 result in a weaker host antiviral immune response. Currently, it is not clear whether other proteins from nuclear pore complex or COPII or GATOR2 are also involved in these interactions.

	Yeast	Worm	Fly	Mouse	Human
Amino acid signaling to mTORC1 (SEA/GATOR)	✓	✓	✓	✓	✓
General autophagy (SEACIT/GATOR1)	✓	✓	✓	?	✓
Mitophagy (SEACIT/GATOR1)	✓	?	?	?	?
Retrograde signaling (NPR2, NPR3)	✓	?	?	?	?
Vacuole-mitochondria contact sites (SEA)	✓	?	?	?	?
Nuclear pore complex components (SEC13, SEH1)	✓	✓	✓	✓	✓
COPII vesicles (SEC13)	✓	✓	✓	✓	✓
DNA damage response (NPR2/NPRL2)	?	?	✓	?	✓
Cell division and cell cycle regulation (SEA/GATOR)	?	?	✓	?	✓

Figure 5. Multiple functions of the SEA/GATOR complex.

The deletion of SEA/GATOR components affects mitochondria functions. The total abundance of SEA proteins is increased during respiratory growth and decreased upon nitrogen starvation, *sea2* deletion impairs respiration capacity in *S. cerevisiae* [147]. *npr2Δ* cells have defective mitochondrial-housed metabolic pathways, such as synthesis of amino acids, and an impaired tricarboxylic acid (TCA) cycle activity. *npr2*-deficient cells showed decreased pools of nitrogen-containing intermediates of the TCA cycle and nucleotides. Yet, *npr2Δ* yeast use TCA cycle intermediates for replenishment of biosynthetic pathways to sustain the hypermetabolic state due to mTORC1 constant activation, suggesting a role of SEACIT in the regulation of cataplerotic reactions of the TCA cycle depending on the amino acid and nitrogen status of the cell [148]. This was later supported by another study that demonstrated that skeletal-muscle-specific NPRL2 loss in mice promoted aerobic glycolysis by altering the tuning between the amino acid sensing pathway and TCA cycle function. NPRL2-mKO mice also had less oxidative muscle fibers and more glycolytic muscle fibers, a hallmark of aerobic glycolysis, which highlights the functional role of NPRL2 in vivo in the regulation of glucose entry into the TCA cycle [149].

The function of GATOR1 proteins in mitochondrial health seems not to be limited to NPRL2. A heterozygous mutation in the CTD domain of DEPDC5 gene found in an autistic child was correlated with a significant decrease in mitochondrial complex IV activity and decrease in the overall oxygen consumption rate in peripheral blood mononuclear cells. Therefore, this variant of DEPDC5 can be directly related to an altered mitochondrial function in autistic disease [150]. Mice with skeletal-muscle specific deletion of DEPDC5 showed increased mitochondrial respiratory capacity and TCA cycle activity [151].

SEACIT is also involved in the communication of the mitochondria with other organelles. The mitochondria-to-nucleus communication pathway, known as the retrograde signaling, is triggered by mitochondrial dysfunctions in order to alter the expression of nucleus-encoded mitochondrial genes to effect metabolic reprogramming and to restore cellular fitness [152,153]. *npr2Δ* and *npr3Δ* yeast strains failed to activate the retrograde signaling pathway when grown in media containing ammonia as nitrogen source [10,148]. In order to recruit the substrates for biochemical reactions and export resulting products mitochondria rely on direct transport with organelles through contact sites [154]. The vacuole and mitochondria contact sites, vCLAMPs, are important for lipid exchange [155] and may also serve for the sensing of the integrity and functionality of mitochondria (Figure 4) [135]. Importantly, SEACIT is required for the maintenance of vCLAMPs and the deletion of any SEACIT members drastically reduces the amount of vCLAMPs in yeast cells [135]. Whether GATOR1 has the same functions in mammalian cells remains to be discovered.

6.3. GATOR1 and DNA Damage Response

The notion that Npr2/NPRL2 might have a role in DNA damage response appeared when it was found that mutations in this protein, both in yeast and human, confer resistance to the anticancer drugs cisplatin and doxorubicin (Figures 4 and 5) (see below) [156,157]. These compounds induce high levels of DNA damage, which eventually lead to cell cycle arrest and apoptosis [158,159]. Study of the role of NPRL2 in DNA damage response in non-small-cell-lung cancer cells treated with cisplatin [160] demonstrated that the ectopic expression of NPRL2 activates the DNA damage checkpoint pathway in cisplatin-resistant and NPRL2-negative cells, leading to cell cycle arrest in the G2/M phase and induction of apoptosis. Upon ectopic expression, NPRL2 promotes ROS production via NADPH oxidase (NOX) 2 activation [54]. Overexpressed NPRL2 accumulates in the nucleus, where it interacts with the apoptosis initiation factor, AIF. In addition, NPRL2 expression provokes the phosphorylation of tumor suppressor p53, which, in turn, activates a DNA-damage checkpoint pathway via p21 and CDC2. An excessive amount of NPRL2 results in cell cycle arrest in G1 phase in cells with constitutively p53 and to CHK2-dependent S or G2/M in p53-negative cancer cell lines [54,161]. Currently, it is not known whether these functions are performed by NPRL2 as a part of GATOR1 complex, or separately. *Drosophila* GATOR is also critical to the response to meiotic double strand DNA breaks (DSB) during oogenesis, since depletion of each GATOR1 component fails to repair DSB with nprl3 mutants showing increased sensitivity to genotoxic stress both in germline and somatic cells [162].

6.4. GATOR in Cell Division and Cell Cycle Regulation

GATOR2 is important for both mitotic and meiotic division (Figure 5). Depletion of MIOS in HeLa cells resulted in mitotic defects, such as spindle assembly defects and delay or failure in cytokinesis [163]. MIOS regulates mitotic events through Aurora A kinase and Polo-like kinase 1 (Plk1), which control the localization and function of mitotic spindle. MIOS is important for spindle formation, subsequent chromosome segregation and proper concentration of active Plk1 and Aurora A at centrosomes and spindle poles. SEH1, which forms a complex with MIOS (see above), targets GATOR2 to mitotic chromosomes, required for the localization of chromosomal passenger complex and functions in chromosome alignment and segregation by regulating the centromeric localization of Aurora B [164]. This function of GATOR2 nevertheless seems to be related to its role in mTORC1 activation, because depletion of MIOS causes reduced mTORC1 activity at centromeres in mitotic cells [163].

In *Drosophila*, Mio localizes to oocyte nucleus at the onset of prophase and meiosis I, and is required for the maintenance of the meiotic cycle during oocyte maturation [24]. *Drosophila* Seh1 is also involved in the maintenance of meiotic cycle and regulation of microtubule dynamics in ovarian cysts [36]. Depletion of *iml1* in the female germ line delays mitotic/meiotic transition and ovarian cysts undergo an extra mitotic division [18]. Thus, GATOR1 downregulates TORC1 activity to promote the mitotic/meiotic transition in ovarian cysts, while inhibition of GATOR1 by GATOR2 prevents the constitutive downregulation of TORC1 at the later stages of oogenesis.

6.5. The Role of GATOR in Development

Animal development and growth is closely related to the ability to respond to different nutrient cues. Therefore, it is not surprising that GATOR components are important at different stages of embryonic and somatic development. Various studies in *Drosophila* by Lilly's group demonstrated that mutations of *mio*, resulting in the production of truncated protein, suppresses oocyte growth and differentiation [24]. Seh1 in *Drosophila* is also required in oogenesis, but is dispensable for somatic development [36]. Both Mio and Seh1 promote TORC1 activation in female fly's germ lines, but play a relatively minor role in the activation of TORC1 in many somatic types [18]. Wdr24, which is also required for ovary growth and female fertility, promotes TORC1-dependent cell growth not only in germ line, but also in somatic tissues of *Drosophila* [133]. *nprl2* mutations in *Drosophila* decrease the lifespan in flies, which have an accelerated gastrointestinal tract aging process [165].

In *C. elegans*, NPRL2 and NPRL3 are required for postembryonic development, which is supported by the availability of a specific sphingolipid. When *C. elegans* larvae are placed in the environment lacking this lipid, they suspend growth and cell division, which can be overcome by resupplying the lipid. When this lipid is absent, postembryonic growth and development can be re-initiated by activating TORC1 or inhibiting NPRL2/3 [17]. NPRL3 represses intestinal TORC1 activity at least in part by regulating apical membrane polarity, which is probably the main reason of larval development defects in worms that are not supplied with a sphingolipid [166] In addition, *nprl3*-deficient worms grow slowly due to the lack of the ability to sense vitamin B2 deficiency in their food [131]. NPRL3 deficiency in worms' intestines triggers a gut protease activity, which derives in abnormal behavior and growth impairing [131].

7. Deletion Phenotypes of the SEA/GATOR Components across Different Species

In unicellular yeast *S. cerevisiae*, SEA genes (apart from Sec13) are non-essential [3] and in rich media, SEA deletion mutants grow practically with the same rate as wild type yeast [3]. In fission yeast *S. pombe*, deletion of any GATOR1 as well as GATOR2 component Sea3 results in a severe growth defect [22,28]. Homozygous deletions of *nprl2* and *nprl3* in *Drosophila* are semi-lethal and deletions of *iml* are lethal, with GATOR1 activity required for animals to transit the last stage of pupal development [130]. In addition, *nprl2* null flies have a significantly reduced lifespan [165]. Similarly, *depdc5* knockout in zebrafish resulted in premature death at 2–3 weeks post-fertilization [167]. In mice homozygous knockouts of Seh1 [168], Wdr59 and Wdr24 are embryonically lethal [169]. Constitutive knockout homozygous and heterozygous GATOR1 rodent models differs significantly. Thus, GATOR1 homozygous animals $Nprl2^{-/-}$ mice [20], $Nprl3^{-/-}$ mice [41], $Depdc5^{-/-}$ rats [21] and $Depdc5^{-/-}$ mice [170] are embryonically lethal. Mice embryos deficient for NPRL2 expression show a compromised liver hematopoiesis, which has a negative impact on embryonic viability [20]. Although mutations in GATOR1 genes are associated with epileptic disorders and brain malformations, heterozygous *Depdc5*[+/−] rats and mice did not present spontaneous epileptic seizures, but *Depdc5*[+/−] rats have subtle cortical malformations [21,170]. Several tissue specific knockouts have also been investigated. Neuron-specific conditional homozygous Depdc5 knockout mice lived till adulthood, but had larger brains and exhibited a decreased survival [171]. The hepatic deletion of *Depdc5* in mice resulted in mild liver inflammation and decreased fat level [172]. Skeletal muscle-specific Depdc5 depletion in mice resulted in muscle hypertrophy, but neither the physical nor contractile muscle function of these mice improved [151]. Similarly, mice with Nprl2 deletion in skeletal muscles had larger muscle fibers and exhibited altered running behavior [149]. In conclusion, deletions of SEA/GATOR components in every organism studied thus far provoked severe defects on growth and viability.

8. GATOR in Human Diseases

During the last decade it became increasingly evident that alternations in the expression of GATOR genes can cause various human diseases (Figure 6). Mutations of GATOR2 components can be found in various cancers according to The Cancer Genome Atlas (TCGA) and Cancer Cell Line Encyclopedia (CCLE), COSMIC and cBioPortal databases, although their recurrent mutation frequency is very low [173]. None of the GATOR2 mutations in these cancers were studied on the molecular level and currently there are no data about the involvement of GATOR2 components in other human pathologies [174]. One of the reasons of the low pathogenicity of GATOR2 mutations could be that they would cause an increased, but most probably not complete, suppression of the mTORC1 pathway, which can rather be associated with healthier conditions.

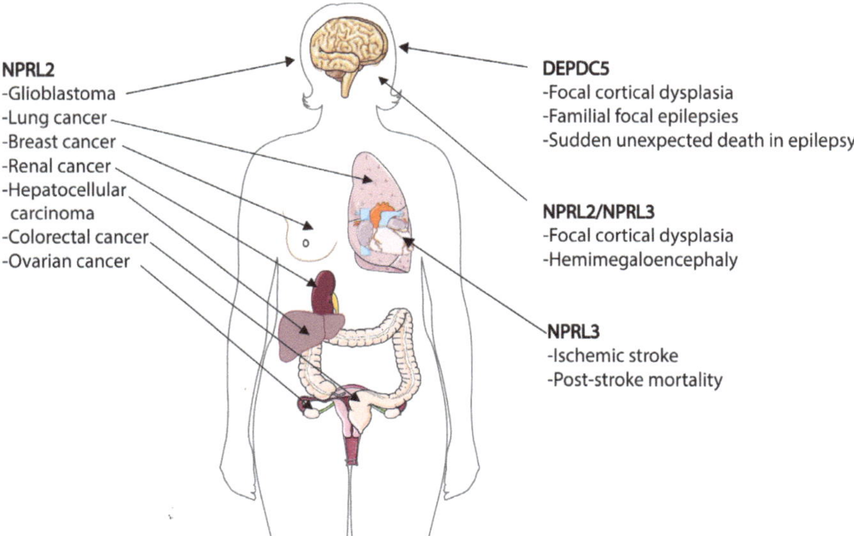

Figure 6. Deregulation of GATOR1 components in different human diseases. Expression of GATOR1 components is downregulated in many cancers and GATOR1-related neurological disorders.

In striking contrast to GATOR2, many pathological mutations in GATOR1 genes have been reported. These mutations are mainly related with two main types of human diseases—cancer and epilepsy. Although the alternations in sequence and gene expression associated with these pathologies have been reported for all three GATOR1 genes, there are striking differences that mark some kind of "preferences" of a gene for a pathology.

Thus, DEPDC5 mutations are more frequent in epilepsies in comparison with mutations in other GATOR1 members. NPRL2 mutations can be found more often in different types of cancers and are associated with resistance to anticancer drugs cisplatin and doxorubicin. Even though NPRL3 is a paralogue of NPRL2, its alternations in cancer are less recurrent. Instead NPRL3 appeared to be required for the normal development of the cardiovascular system. Below we will describe alternations of GATOR1 expression in different diseases.

8.1. Epilepsies and Brain Malformations—DEPDC5 and Others

In 2013, DEPDC5 was reported as the first gene implicated in familial focal epilepsies by Baulac and Scheffer groups [175,176]. In the following years, it became clear that mutations in DEPDC5 are also related with brain malformations, notably with focal cortical dysplasia (FCD), which is a major cause of drug-resistant epilepsy [177] and can be associated with sudden unexpected death in epilepsy (SUDEP) [178]. In 2016, mutations related with focal epilepsies, familial cortical dysplasia and SUDEP were also reported for Nprl2 and Nprl3 [174,179,180]. Since then, more than 140 variants of GATOR1 genes have been found in up to 37% of patients with familial focal and in other forms of epilepsies [181]. These variants include loss-of-function mutations (67%), missense mutations (27%), splice site changes (4%), frameshifts and copy number variants (~1%). Interestingly, the distribution of mutations in an epilepsy cohort differs drastically from the overall distribution of GATOR1 mutations listed in the gnomAD database, where loss-of-function represents only 4% of variants, with the majority (88%) being missense mutations. Importantly, histopathological analysis of brain tissues resected from individuals with GATOR1 gene mutations demonstrate the hyperactivation of mTORC1 pathway, suggesting that mTORC1 signaling plays an important role in brain development [174,179,180].

Nearly 85% of GATOR1 mutations in epilepsies account for changes in DEPDC5 with both somatic and germline mutations detected all through the gene without clustering. Initially, it was not clear how germline *Depdc5* mutations can cause FCD, especially taking into account that these mutations are often dominantly inherited from an asymptomatic carrier parent [181] and that in rodent models *Depdc5*$^{+/-}$ constitutive heterozygous mutations do not exhibit an epileptic phenotype [21,170]. The discovery of second hit somatic mutations in trans, which led to a biallelic inactivation in a subset of brain cells, explained this phenomenon [182,183]. Nprl2 and Nprl3 mutation are less frequent (6% and 9%, respectively), which might be partially related with the fact that their involvement in epilepsies and brain malformations has been tested in a low number of people [26,181]. Cases with simultaneous mutations in different GATOR1 genes have not been described thus far. Several *Nprl2* or *Nprl3* variants found in individuals with FCD or hemimegaloencephaly (HME) have been reported recently [184,185]. Interestingly, *NPRL3* single nucleotide polymorphism has been associated with ischemic stroke susceptibility and post-stroke mortality [186], which can be related with increased mTOR activity, that is known to accelerate brain recovery after stroke. The role of NPRL3 in this disease is most probably related with its function in focal epilepsies that might occur in ischemic cerebrovascular disorders [187]. Finally, genetic alternations of KICSTOR complex, required for GATOR1-mediated repression of mTORC1 signaling (see above), have also been linked to epilepsies and brain malformations [188–190].

Thus, it is evident that GATOR1 plays an essential role in cortical formation and development. Mutations of GATOR1 components became important features of "mTORopathies" —a set of pathological conditions characterized by brain malformations, neurological disorders and mTORC1 hyperactivity due to either gain-of-function mutations in a pathway activators (e.g., *AKT, RHEB, MTOR* itself) or loss-of-function mutations of inhibitors (e.g., *TSC1, TSC2*) [191,192]. However, mutations of GATOR1 genes seem to result in a broader spectrum of neurological disorders than other "mTORopathic" genes. Not only are these mutations highly related with medically intractable epilepsies and, especially SUDEP, but they are also observed in autism spectrum disorders [150] and could be implicated in

Parkinson's disease [193]. Therefore, it was recently proposed to name GATOR1-related neurological disorders as GATORopathies [194].

8.2. Cancer and Anticancer Drug Resistance—NPRL2 and Others

Among GATOR1 components, NPRL2 was the first that was suggested to be a tumor suppressor [195] almost a decade before the GATOR1 complex was described for the first time. NPRL2 has the higher cancer-associated recurrent mutational frequency out of all the GATOR1 genes [173]. For example, missense mutations in metastatic breast cancers are twice more frequent in Nprl2 (1.55%), than in Nprl3 or Depdc5 (0.78%) [196]. Low levels of NPRL2 expression have mostly been detected in solid tumors (Figure 6), including hepatocellular carcinoma [197], glioblastoma [12], as well as in renal [198,199], ovarian [12,199], colorectal [199–202], breast [199,203] and lung cancers [157,160,199,204, 205]. Paradoxically, NPRL2 might also have functions as an oncogene. Recent studies in castration resistant prostate cancer (CRPC) revealed that poor prognosis is associated with high expression of NPRL2 [206].

Alternations of NPRL2 expression is also related to the resistance to a number of anticancer drugs. The most recurrent cases are associated with the resistance to cisplatin and doxorubicin, which has been initially observed in Npr2 deletion mutants in yeast [156] and further confirmed in human lung cancer cell lines [157,160]. The reason of this resistance is still not clear, but it could be related with a role of NPRL2 in DNA damage response (see above) [54,160]. Overexpression of NPRL2 in colon cancer cells increases the sensitivity to a topoisomerase I inhibitor irinotecan (CPT-11) by activation of the DNA damage checkpoints [207]. Genomic alternations of all three GATOR1 components have recently been associated with the resistance to PI3Kα inhibitors in primary and metastatic breast cancer [208]. This resistance is explained by the sustained activation of the mTORC1 pathway due to the loss of function mutations of GATOR1 components. In this case, it is reasonable to expect that concomitant mTOR blockage by rapalogs or mTOR pan-inhibitors might overcome resistance. Inversely, CRPC cells, where NPRL2 expression is elevated, are resistant to everolimus [209].

Surprisingly, during the last decade, not a single article reported a study about the involvement of NPRL3 in cancer and drug resistance, even if in the COSMIC database there are almost three times more somatic cancer mutations listed for NPRL3 than for its paralogue NPRL2.

A low frequency DEPDC5 inactivation mutation has been observed in glioblastoma and ovarian cancer, but was not further investigated [12]. DEPDC5 downregulation was also observed in tumors of breast cancer patients [53], where it is strongly correlated with the upregulation of KLHL22 E3-ubiquitin ligase, responsible for DEPDC5 polyubiquity-lation and degradation (see above). Recently, DEPDC5 inactivation was discovered in gastrointestinal stromal tumors (GIST), one of the most common human sarcomas. Chromosome 22q deletions are observed in ~50% of GIST and recurrent genomic inactivation of DEPDC5 (>16%) makes it the bona-fide tumor suppressor contributing to GIST progression via increased mTORC1 pathway signaling [210]. This is in striking contrast with >250 non-GIST sarcomas where DEPDC5 aberrations are infrequent (~1%). Interestingly, cancer occurrence in epilepsy probands with germline GATOR1 variants is very low and at present it is considered that there is no link between epileptic germline GATOR1 variants and cancer [181].

Currently, >2000 somatic mutations in different tumors are listed for GATOR1 genes in the COSMIC database, none of them have been studied in detail. It is reasonable to expect that in the following years we should gain more information about the molecular mechanisms associated with the tumorogenesis provoked by these mutations.

8.3. Cardiovascular Diseases—NPRL3

In striking contrast to other GATOR1 components, and especially to its paralogue NPRL2, NPRL3 seems to be less important for epilepsy and cancer. Rather it appears as a crucial gene, necessary for the normal development of the cardiovascular system [41]. Mice with the deletion of NPRL3 promoter often have severe embryonal cardiac defects and die in late gestations. A single nucleotide polymorphism of NPRL3 was reported in sickle cell anemia [211], a disease characterized by various hemoglobin abnormalities. These defects are explained by the fact that the introns of NPRL3 contain super-enhancers required for high level expression of the genes encoding the α-globin subunits of hemoglobin in humans and mice [212,213]. These regulation elements appeared to be deeply preserved during evolution. A recent genomic study revealed that the NPRL3 gene carrying a strong regulatory element became linked to at least two different globin genes in ancestral vertebrate, just before the divergence between jawless and jawed vertebrates [214]. Each of these ancestral globin genes evolved in the modern hemoglobin genes, but kept their enhancers in NPRL3, which provide an explanation to a long-standing enigma of how globin genes linked to the same adjacent gene undergo convergent evolution in different species.

Therefore, the pathologies associated with NPRL3 mutations are related with the disturbances of the transcriptional elements in the Nprl3 gene rather than with the function of the protein product in the mTORC1 pathway. Similarly, the higher recurrence of NPRL2 mutations in cancers and DEPDC5 mutations in epilepsies could be related with the specific moonlighting functions of these GATOR1 members beyond the regulation of the mTORC1 pathway.

9. Conclusions

Since its discovery ten years ago, the SEA/GATOR complex has been recognized as an important regulator of the mTORC1 pathway that deals with the cell's response to amino acid and glucose availability, DNA damage, mitochondria impairment, etc. Many studies have also revealed the role of the SEA/GATOR complex in human diseases, especially in cancer and epilepsies. Despite the growing number of discoveries involving the SEA/GATOR complex in many organisms, a lot of questions concerning its function and the mechanisms leading to pathologies are still left unanswered. For example, the role of the GATOR complex in amino acid sensing and response has been already clarified in great detail in several studies; however, it is still unknown whether the SEA complex in yeast can perform sensing functions, given that many amino acid sensors interacting with GATOR are not conserved in yeast. The functions of the SEA complex in autophagy and in the formation of organelle contact sites have been extensively studied in yeast. Whether the GATOR complex has these functions in higher eukaryotes is currently unknown (Figure 5). Finally, the most intriguing problem at the moment concerns the molecular function of the SEACAT/GATOR2 complex, an enigma that has remained unresolved despite these 10 years of research and discoveries. Without any doubt, having a high-resolution structure of this subcomplex with or without its partners (SESTRINs, CASTOR2 and others) will be crucial for understanding its function. It will be also important to figure out the principles of interaction between the two SEA/GATOR subcomplexes in different organisms, which can shed light on how evolution shaped this assembly to adapt for the particular needs of various species. SEA members appeared earlier than GATOR members, similar to crocodiles, which are slightly older than alligators [215]. In the same way with crocodiles and alligators, SEA and GATOR are similar to each other in terms of size, structure (appearance) and function (behavior). On the other hand, both SEA and GATOR have a number of subtle yet significant differences that might be able to explain how they each adjusted to operate optimally in different organisms and environments. For example, as with crocodiles, which are bigger than alligators, SEA components are also bigger than their human homologues. Therefore, it will not be surprising if the structural studies reveal that the shape of the SEA complex will slightly differ from that of the GATOR complex, as the V-shape crocodiles' snout differs from larger U-shape snout of alligators. Despite the slight difference in shape, both reptiles use their snouts to effectively catch and hold the food. Similarly, SEA and GATOR complexes, despite several structural differences, can still respond to the presence of nutrients during regulation of the mTORC1 pathway.

We are, therefore, confident that the next decade of SEA/GATOR research will lead to new exciting discoveries of the structure and function of this complex, that can better characterize its implication in health and diseases.

Funding: This research was funded by La Ligue contre le Cancer 94 (Comité de Val-de-Marne). Y.A.L.B. is a recipient of CONACYT PhD fellowship from Mexican government (Becas al extranjero N708006).

Institutional Review Board Statement: Not applicable.

Informed Consent Statement: Not applicable.

Data Availability Statement: Not applicable.

Acknowledgments: We are grateful to Reynand Canoy for discussions and critical reading of manuscript.

Conflicts of Interest: The authors declare no conflict of interest.

References

1. Liu, G.Y.; Sabatini, D.M. MTOR at the Nexus of Nutrition, Growth, Ageing and Disease. *Nat. Rev. Mol. Cell. Biol.* **2020**, *21*, 183–203. [CrossRef]
2. Szwed, A.; Kim, E.; Jacinto, E. Regulation and Metabolic Functions of MTORC1 and MTORC2. *Physiol. Rev.* **2021**, *101*, 1371–1426. [CrossRef] [PubMed]
3. Dokudovskaya, S.; Waharte, F.; Schlessinger, A.; Pieper, U.; Devos, D.P.; Cristea, I.M.; Williams, R.; Salamero, J.; Chait, B.T.; Sali, A.; et al. A Conserved Coatomer-Related Complex Containing Sec13 and Seh1 Dynamically Associates with the Vacuole in Saccharomyces Cerevisiae. *Mol. Cell. Proteom.* **2011**, *10*, M110.006478. [CrossRef]
4. Dokudovskaya, S.; Rout, M.P. SEA You Later Alli-GATOR–a Dynamic Regulator of the TORC1 Stress Response Pathway. *J. Cell Sci.* **2015**, *128*, 2219–2228. [CrossRef]
5. Alber, F.; Dokudovskaya, S.; Veenhoff, L.M.; Zhang, W.; Kipper, J.; Devos, D.; Suprapto, A.; Karni-Schmidt, O.; Williams, R.; Chait, B.T.; et al. Determining the Architectures of Macromolecular Assemblies. *Nature* **2007**, *450*, 683–694. [CrossRef]
6. Dokudovskaya, S.; Rout, M.P. A Novel Coatomer-Related SEA Complex Dynamically Associates with the Vacuole in Yeast and Is Implicated in the Response to Nitrogen Starvation. *Autophagy* **2011**, *7*, 1392–1393. [CrossRef]
7. Algret, R.; Dokudovskaya, S.S. The SEA Complex—the Beginning. *Biopolym. Cell* **2012**, *28*, 281–284. [CrossRef]
8. Alber, F.; Dokudovskaya, S.; Veenhoff, L.M.; Zhang, W.; Kipper, J.; Devos, D.; Suprapto, A.; Karni-Schmidt, O.; Williams, R.; Chait, B.T.; et al. The Molecular Architecture of the Nuclear Pore Complex. *Nature* **2007**, *450*, 695–701. [CrossRef]
9. Wu, X.; Tu, B.P. Selective Regulation of Autophagy by the Iml1-Npr2-Npr3 Complex in the Absence of Nitrogen Starvation. *Mol. Biol. Cell* **2011**, *22*, 4124–4133. [CrossRef]
10. Neklesa, T.K.; Davis, R.W. A Genome-Wide Screen for Regulators of TORC1 in Response to Amino Acid Starvation Reveals a Conserved Npr2/3 Complex. *PLoS Genet.* **2009**, *5*, e1000515. [CrossRef] [PubMed]
11. Panchaud, N.; Péli-Gulli, M.-P.P.; De Virgilio, C.; Peli-Gulli, M.P.; De Virgilio, C.; Péli-Gulli, M.-P.P.; De Virgilio, C. Amino Acid Deprivation Inhibits TORC1 through a GTPase-Activating Protein Complex for the Rag Family GTPase Gtr1. *Sci. Signal.* **2013**, *6*, ra42. [CrossRef]
12. Bar-Peled, L.; Chantranupong, L.; Cherniack, A.D.; Chen, W.W.; Ottina, K.A.; Grabiner, B.C.; Spear, E.D.; Carter, S.L.; Meyerson, M.; Sabatini, D.M. A Tumor Suppressor Complex with GAP Activity for the Rag GTPases That Signal Amino Acid Sufficiency to MTORC1. *Science* **2013**, *340*, 1100–1106. [CrossRef]
13. Algret, R.; Fernandez-Martinez, J.; Shi, Y.; Kim, S.J.; Pellarin, R.; Cimermancic, P.; Cochet, E.; Sali, A.; Chait, B.T.; Rout, M.P.; et al. Molecular Architecture and Function of the SEA Complex, a Modulator of the TORC1 Pathway. *Mol. Cell. Proteom.* **2014**, *13*, 2855–2870. [CrossRef]
14. Panchaud, N.; Peli-Gulli, M.P.; De Virgilio, C.; Péli-Gulli, M.P.; De Virgilio, C. SEACing the GAP That NEGOCiates TORC1 Activation: Evolutionary Conservation of Rag GTPase Regulation. *Cell Cycle* **2013**, *12*, 1–5. [CrossRef] [PubMed]
15. Wolfson, R.L.; Sabatini, D.M. The Dawn of the Age of Amino Acid Sensors for the MTORC1 Pathway. *Cell Metab.* **2017**, *26*, 301–309. [CrossRef] [PubMed]
16. Ma, N.; Liu, Q.; Zhang, L.; Henske, E.P.; Ma, Y. TORC1 Signaling Is Governed by Two Negative Regulators in Fission Yeast. *Genetics* **2013**, *195*, 457–468. [CrossRef] [PubMed]
17. Zhu, H.; Shen, H.; Sewell, A.K.; Kniazeva, M.; Han, M. A Novel Sphingolipid-TORC1 Pathway Critically Promotes Postembryonic Development in Caenorhabditis Elegans. *eLife* **2013**, *2*, e00429. [CrossRef]
18. Wei, Y.; Reveal, B.; Reich, J.; Laursen, W.J.; Senger, S.; Akbar, T.; Iida-Jones, T.; Cai, W.; Jarnik, M.; Lilly, M.A. TORC1 Regulators Iml1/GATOR1 and GATOR2 Control Meiotic Entry and Oocyte Development in Drosophila. *Proc. Natl. Acad. Sci. USA* **2014**, *111*, E5670–E5677. [CrossRef] [PubMed]

19. de Calbiac, H.; Dabacan, A.; Marsan, E.; Tostivint, H.; Devienne, G.; Ishida, S.; Leguern, E.; Baulac, S.; Muresan, R.C.; Kabashi, E.; et al. Depdc5 Knockdown Causes MTOR-Dependent Motor Hyperactivity in Zebrafish. *Ann. Clin. Transl. Neurol.* **2018**, *5*, 510–523. [CrossRef]

20. Dutchak, P.A.; Laxman, S.; Estill, S.J.; Wang, C.; Wang, Y.Y.; Wang, Y.Y.; Bulut, G.B.; Gao, J.; Huang, L.J.; Tu, B.P. Regulation of Hematopoiesis and Methionine Homeostasis by MTORC1 Inhibitor NPRL2. *Cell Rep.* **2015**, *12*, 371–379. [CrossRef]

21. Marsan, E.; Ishida, S.; Schramm, A.; Weckhuysen, S.; Muraca, G.; Lecas, S.; Liang, N.; Treins, C.; Pende, M.; Roussel, D.; et al. Depdc5 Knockout Rat: A Novel Model of MTORopathy. *Neurobiol. Dis.* **2016**, *89*, 180–189. [CrossRef] [PubMed]

22. Chia, K.H.; Fukuda, T.; Sofyantoro, F.; Matsuda, T.; Amai, T.; Shiozaki, K. Ragulator and GATOR1 Complexes Promote Fission Yeast Growth by Attenuating TOR Complex 1 through Rag GTPases. *eLife* **2017**, *6*, e30880. [CrossRef]

23. Rousselet, G.; Simon, M.; Ripoche, P.; Buhler, J.-M.M. A Second Nitrogen Permease Regulator in Saccharomyces Cerevisiae. *FEBS Lett.* **1995**, *359*, 215–219. [CrossRef]

24. Iida, T.; Lilly, M.A. Missing Oocyte Encodes a Highly Conserved Nuclear Protein Required for the Maintenance of the Meiotic Cycle and Oocyte Identity in Drosophila. *Development* **2004**, *131*, 1029–1039. [CrossRef]

25. Bertuzzi, M.; Tang, D.; Calligaris, R.; Vlachouli, C.; Finaurini, S.; Sanges, R.; Goldwurm, S.; Catalan, M.; Antonutti, L.; Manganotti, P.; et al. A Human Minisatellite Hosts an Alternative Transcription Start Site for *NPRL3* Driving Its Expression in a Repeat Number-dependent Manner. *Hum. Mutat.* **2020**, *41*, 807–824. [CrossRef]

26. Zhang, J.; Shen, Y.; Yang, Z.; Yang, F.; Li, Y.; Yu, B.; Chen, W.; Gan, J. A Splicing Variation in NPRL2 Causing Familial Focal Epilepsy with Variable Foci: Additional Cases and Literature Review. *J. Hum. Genet.* **2021**. Online ahead of print. [CrossRef]

27. Lee, C.; Goldberg, J. Structure of Coatomer Cage Proteins and the Relationship among COPI, COPII, and Clathrin Vesicle Coats. *Cell* **2010**, *142*, 123–132. [CrossRef]

28. Fukuda, T.; Sofyantoro, F.; Tai, Y.T.; Chia, K.H.; Matsuda, T.; Murase, T.; Morozumi, Y.; Tatebe, H.; Kanki, T.; Shiozaki, K. Tripartite Suppression of Fission Yeast TORC1 Signaling by the GATOR1-Sea3 Complex, the TSC Complex, and Gcn2 Kinase. *eLife* **2021**, *10*, e60969. [CrossRef]

29. Shen, K.; Huang, R.K.; Brignole, E.J.; Condon, K.J.; Valenstein, M.L.; Chantranupong, L.; Bomaliyamu, A.; Choe, A.; Hong, C.; Yu, Z.; et al. Architecture of the Human GATOR1 and GATOR1–Rag GTPases Complexes. *Nature* **2018**, *556*, 64–69. [CrossRef] [PubMed]

30. Tafur, L.; Kefauver, J.; Loewith, R. Structural Insights into TOR Signaling. *Genes* **2020**, *11*, 885. [CrossRef] [PubMed]

31. Devos, D.; Dokudovskaya, S.; Alber, F.; Williams, R.; Chait, B.T.; Sali, A.; Rout, M.P. Components of Coated Vesicles and Nuclear Pore Complexes Share a Common Molecular Architecture. *PLoS Biol.* **2004**, *2*, e380. [CrossRef]

32. Balderhaar, H.J.K.; Ungermann, C. CORVET and HOPS Tethering Complexes-Coordinators of Endosome and Lysosome Fusion. *J. Cell Sci.* **2013**, *126*, 1307–1316. [CrossRef]

33. Rout, M.P.; Field, M.C. The Evolution of Organellar Coat Complexes and Organization of the Eukaryotic Cell. *Annu. Rev. Biochem.* **2017**, *86*, 637–657. [CrossRef] [PubMed]

34. Field, M.C.; Sali, A.; Rout, M.P. Evolution: On a Bender—BARs, ESCRTs, COPs, and Finally Getting Your Coat. *J. Cell Biol.* **2011**, *193*, 963–972. [CrossRef]

35. Fath, S.; Mancias, J.D.; Bi, X.; Goldberg, J. Structure and Organization of Coat Proteins in the COPII Cage. *Cell* **2007**, *129*, 1325–1336. [CrossRef]

36. Senger, S.; Csokmay, J.; Akbar, T.; Jones, T.I.; Sengupta, P.; Lilly, M.A.; Tanveer, A.; Jones, T.I.; Sengupta, P.; Lilly, M.A. The Nucleoporin Seh1 Forms a Complex with Mio and Serves an Essential Tissue-Specific Function in Drosophila Oogenesis. *Development* **2011**, *138*, 2133–2142. [CrossRef]

37. Hesketh, G.G.; Papazotos, F.; Pawling, J.; Rajendran, D.; Knight, J.D.R.; Martinez, S.; Taipale, M.; Schramek, D.; Dennis, J.W.; Gingras, A.-C. The GATOR–Rag GTPase Pathway Inhibits MTORC1 Activation by Lysosome-Derived Amino Acids. *Science* **2020**, *370*, 351–356. [CrossRef] [PubMed]

38. Doerks, T.; Copley, R.R.; Schultz, J.; Ponting, C.P.; Bork, P. Systematic Identification of Novel Protein Domain Families Associated with Nuclear Functions. *Genome Res.* **2002**, *12*, 47–56. [CrossRef] [PubMed]

39. Nameki, N.; Yoneyama, M.; Koshiba, S.; Tochio, N.; Inoue, M.; Seki, E.; Matsuda, T.; Tomo, Y.; Harada, T.; Saito, K.; et al. Solution Structure of the RWD Domain of the Mouse GCN2 Protein. *Protein Sci.* **2004**, *13*, 2089–2100. [CrossRef] [PubMed]

40. Patton, E.E.; Willems, A.R.; Sa, D.; Kuras, L.; Thomas, D.; Craig, K.L.; Tyers, M. Cdc53 Is a Scaffold Protein for Multiple Cdc34/Skp1/F Box Protein Complexes That Regulate Cell Division and Methionine Biosynthesis in Yeast. *Genes Dev.* **1998**, *12*, 692–705. [CrossRef]

41. Kowalczyk, M.S.; Hughes, J.R.; Babbs, C.; Sanchez-Pulido, L.; Szumska, D.; Sharpe, J.A.; Sloane-Stanley, J.A.; Morriss-Kay, G.M.; Smoot, L.B.; Roberts, A.E.; et al. Nprl3 Is Required for Normal Development of the Cardiovascular System. *Mamm. Genome* **2012**, *23*, 404–415. [CrossRef]

42. Levine, T.P.; Daniels, R.D.; Wong, L.H.; Gatta, A.T.; Gerondopoulos, A.; Barr, F.A. Discovery of New Longin and Roadblock Domains That Form Platforms for Small GTPases in Ragulator and TRAPP-II. *Small GTPases* **2013**, *4*, 1–8. [CrossRef] [PubMed]

43. Nookala, R.K.; Langemeyer, L.; Pacitto, A.; Donaldson, J.C.; Ochoa-montan, B.; Blaszczyk, B.K.; Chirgadze, D.Y.; Barr, F.A.; Bazan, J.F.; Blundell, T.L.; et al. Crystal Structure of Folliculin Reveals a HidDENN Function in Genetically Inherited Renal Cancer. *Open Biol.* **2012**, *2*, 120071. [CrossRef]

44. Consonni, S.V.; Maurice, M.M.; Bos, J.L. DEP Domains: Structurally Similar but Functionally Different. *Nat. Rev. Mol. Cell Biol.* **2014**, *15*, 357–362. [CrossRef]

45. Caron, A.; Briscoe, D.M.; Richard, D.; Laplante, M. DEPTOR at the Nexus of Cancer, Metabolism, and Immunity. *Physiol. Rev.* **2018**, *98*, 1765–1803. [CrossRef]

46. Albuquerque, C.P.; Smolka, M.B.; Payne, S.H.; Bafna, V.; Eng, J.; Zhou, H. A Multidimensional Chromatography Technology for In-Depth Phosphoproteome Analysis. *Mol. Cell. Proteom.* **2008**, *7*, 1389–1396. [CrossRef]

47. Breitkreutz, A.; Choi, H.; Sharom, J.R.; Boucher, L.; Neduva, V.; Larsen, B.; Lin, Z.Y.; Breitkreutz, B.J.; Stark, C.; Liu, G.; et al. A Global Protein Kinase and Phosphatase Interaction Network in Yeast. *Science* **2010**, *328*, 1043–1046. [CrossRef]

48. Hitchcock, A.L.; Auld, K.; Gygi, S.P.; Silver, P. A Subset of Membrane-Associated Proteins Is Ubiquitinated in Response to Mutations in the Endoplasmic Reticulum Degradation Machinery. *Proc. Natl. Acad. Sci. USA* **2003**, *100*, 12735–12740. [CrossRef] [PubMed]

49. Iesmantavicius, V.; Weinert, B.T.; Choudhary, C. Convergence of Ubiquitylation and Phosphorylation Signaling in Rapamycin-Treated Yeast Cells. *Mol. Cell. Proteom.* **2014**, *13*, 1979–1992. [CrossRef] [PubMed]

50. Hu, Z.; Raucci, S.; Jaquenoud, M.; Hatakeyama, R.; Stumpe, M.; Rohr, R.; Reggiori, F.; De Virgilio, C.; Dengjel, J. Multilayered Control of Protein Turnover by TORC1 and Atg1. *Cell Rep.* **2019**, *28*, 3486–3496.e6. [CrossRef] [PubMed]

51. Martínez-Montañés, F.; Casanovas, A.; Sprenger, R.R.; Topolska, M.; Marshall, D.L.; Moreno-Torres, M.; Poad, B.L.J.; Blanksby, S.J.; Hermansson, M.; Jensen, O.N.; et al. Phosphoproteomic Analysis across the Yeast Life Cycle Reveals Control of Fatty Acyl Chain Length by Phosphorylation of the Fatty Acid Synthase Complex. *Cell Rep.* **2020**, *32*, 108024. [CrossRef]

52. Spielewoy, N.; Guaderrama, M.; Wohlschlegel, J.A.; Ashe, M.; Yates, J.R., 3rd; Wittenberg, C.; Yates, J.R.; Wittenberg, C. Npr2, Yeast Homolog of the Human Tumor Suppressor NPRL2, Is a Target of Grr1 Required for Adaptation to Growth on Diverse Nitrogen Sources. *Eukaryot. Cell* **2010**, *9*, 592–601. [CrossRef]

53. Chen, J.; Ou, Y.; Yang, Y.; Li, W.; Xu, Y.; Xie, Y.; Liu, Y. KLHL22 Activates Amino-Acid-Dependent MTORC1 Signalling to Promote Tumorigenesis and Ageing. *Nature* **2018**, *557*, 585–589. [CrossRef]

54. Ma, Y.; Silveri, L.; LaCava, J.; Dokudovskaya, S. Tumor Suppressor NPRL2 Induces ROS Production and DNA Damage Response. *Sci. Rep.* **2017**, *7*, 15311. [CrossRef] [PubMed]

55. Zhou, Y.; Guo, J.; Wang, X.; Cheng, Y.; Guan, J.; Barman, P.; Sun, M.-A.; Fu, Y.; Wei, W.; Feng, C.; et al. FKBP39 Controls Nutrient Dependent Nprl3 Expression and TORC1 Activity in Drosophila. *Cell Death Dis.* **2021**, *12*, 571. [CrossRef] [PubMed]

56. Orozco, J.M.; Krawczyk, P.A.; Scaria, S.M.; Cangelosi, A.L.; Chan, S.H.; Kunchok, T.; Lewis, C.A.; Sabatini, D.M. Dihydroxyacetone Phosphate Signals Glucose Availability to MTORC1. *Nat. Metab.* **2020**, *2*, 893–901. [CrossRef]

57. Sancak, Y.; Peterson, T.R.; Shaul, Y.D.; Lindquist, R.A.; Thoreen, C.C.; Bar-Peled, L.; Sabatini, D.M. The Rag GTPases Bind Raptor and Mediate Amino Acid Signaling to MTORC1. *Science* **2008**, *320*, 1496–1501. [CrossRef] [PubMed]

58. Kim, E.; Goraksha-Hicks, P.; Li, L.; Neufeld, T.P.; Guan, K.-L. Regulation of TORC1 by Rag GTPases in Nutrient Response. *Nat. Cell Biol.* **2008**, *10*, 935–945. [CrossRef]

59. Zoncu, R.; Bar-Peled, L.; Efeyan, A.; Wang, S.; Sancak, Y.; Sabatini, D.M. MTORC1 Senses Lysosomal Amino Acids through an Inside-out Mechanism That Requires the Vacuolar H(+)-ATPase. *Science* **2011**, *334*, 678–683. [CrossRef]

60. Bar-Peled, L.; Schweitzer, L.D.; Zoncu, R.; Sabatini, D.M. Ragulator Is a GEF for the Rag GTPases That Signal Amino Acid Levels to MTORC1. *Cell* **2012**, *150*, 1196–1208. [CrossRef]

61. Sancak, Y.; Bar-Peled, L.; Zoncu, R.; Markhard, A.L.; Nada, S.; Sabatini, D.M. Ragulator-Rag Complex Targets MTORC1 to the Lysosomal Surface and Is Necessary for Its Activation by Amino Acids. *Cell* **2010**, *141*, 290–303. [CrossRef]

62. Thomas, J.D.; Zhang, Y.J.; Wei, Y.H.; Cho, J.H.; Morris, L.E.; Wang, H.Y.; Zheng, X.F. Rab1A Is an MTORC1 Activator and a Colorectal Oncogene. *Cancer Cell* **2014**, *26*, 754–769. [CrossRef]

63. Jewell, J.L.; Kim, Y.C.; Russell, R.C.; Yu, F.X.; Park, H.W.; Plouffe, S.W.; Tagliabracci, V.S.; Guan, K.L. Metabolism. Differential Regulation of MTORC1 by Leucine and Glutamine. *Science* **2015**, *347*, 194–198. [CrossRef]

64. Ukai, H.; Araki, Y.; Kira, S.; Oikawa, Y.; May, A.I.; Noda, T. Gtr/Ego-Independent TORC1 Activation Is Achieved through a Glutamine-Sensitive Interaction with Pib2 on the Vacuolar Membrane. *PLoS Genet.* **2018**, *14*, e1007334. [CrossRef]

65. Meng, D.; Yang, Q.; Wang, H.; Melick, C.H.; Navlani, R.; Frank, A.R.; Jewell, J.L. Glutamine and Asparagine Activate MTORC1 Independently of Rag GTPases. *J. Biol. Chem.* **2020**, *295*, 2890–2899. [CrossRef] [PubMed]

66. Peng, M.; Yin, N.; Li, M.O. SZT2 Dictates GATOR Control of MTORC1 Signalling. *Nature* **2017**, *543*, 433–437. [CrossRef] [PubMed]

67. Wolfson, R.L.; Chantranupong, L.; Wyant, G.A.; Gu, X.; Orozco, J.M.; Shen, K.; Condon, K.J.; Petri, S.; Kedir, J.; Scaria, S.M.; et al. KICSTOR Recruits GATOR1 to the Lysosome and Is Necessary for Nutrients to Regulate MTORC1. *Nature* **2017**, *543*, 438–442. [CrossRef]

68. Shen, K.; Sabatini, D.M. Ragulator and SLC38A9 Activate the Rag GTPases through Noncanonical GEF Mechanisms. *Proc. Natl. Acad. Sci. USA* **2018**, *115*, 9545–9550. [CrossRef] [PubMed]

69. Tsun, Z.-Y.Y.; Bar-Peled, L.; Chantranupong, L.; Zoncu, R.; Wang, T.; Kim, C.; Spooner, E.; Sabatini, D.M. The Folliculin Tumor Suppressor Is a GAP for the RagC/D GTPases That Signal Amino Acid Levels to MTORC1. *Mol. Cell* **2013**, *52*, 495–505. [CrossRef]

70. Han, J.M.; Jeong, S.J.; Park, M.C.; Kim, G.; Kwon, N.H.; Kim, H.K.; Ha, S.H.; Ryu, S.H.; Kim, S. Leucyl-TRNA Synthetase Is an Intracellular Leucine Sensor for the MTORC1-Signaling Pathway. *Cell* **2012**, *149*, 410–424. [CrossRef]

71. Long, X.; Ortiz-Vega, S.; Lin, Y.; Avruch, J. Rheb Binding to Mammalian Target of Rapamycin (MTOR) Is Regulated by Amino Acid Sufficiency. *J. Biol. Chem.* **2005**, *280*, 23433–23436. [CrossRef] [PubMed]

72. Hoxhaj, G.; Manning, B.D. The PI3K-AKT Network at the Interface of Oncogenic Signalling and Cancer Metabolism. *Nat. Rev. Cancer* **2020**, *20*, 74–88. [CrossRef]

73. Demetriades, C.; Doumpas, N.; Teleman, A.A. Regulation of TORC1 in Response to Amino Acid Starvation via Lysosomal Recruitment of TSC2. *Cell* **2014**, *156*, 786–799. [CrossRef]

74. Demetriades, C.; Plescher, M.; Teleman, A.A. Lysosomal Recruitment of TSC2 Is a Universal Response to Cellular Stress. *Nat. Commun.* **2016**, *7*, 10662. [CrossRef]

75. Yang, S.; Zhang, Y.; Ting, C.-Y.; Bettedi, L.; Kim, K.; Ghaniam, E.; Lilly, M.A. The Rag GTPase Regulates the Dynamic Behavior of TSC Downstream of Both Amino Acid and Growth Factor Restriction. *Dev. Cell* **2020**, *55*, 272–288.e5. [CrossRef]

76. Dubouloz, F.; Deloche, O.; Wanke, V.; Cameroni, E.; De Virgilio, C. The TOR and EGO Protein Complexes Orchestrate Microautophagy in Yeast. *Mol. Cell* **2005**, *19*, 15–26. [CrossRef] [PubMed]

77. Binda, M.; Péli-Gulli, M.-P.; Bonfils, G.; Panchaud, N.; Urban, J.; Sturgill, T.W.; Loewith, R.; De Virgilio, C. The Vam6 GEF Controls TORC1 by Activating the EGO Complex. *Mol. Cell* **2009**, *35*, 563–573. [CrossRef]

78. Bonfils, G.; Jaquenoud, M.; Bontron, S.; Ostrowicz, C.; Ungermann, C.; De Virgilio, C. Leucyl-TRNA Synthetase Controls TORC1 via the EGO Complex. *Mol. Cell* **2012**, *46*, 105–110. [CrossRef]

79. Péli-Gulli, M.P.; Sardu, A.; Panchaud, N.; Raucci, S.; De Virgilio, C. Amino Acids Stimulate TORC1 through Lst4-Lst7, a GTPase-Activating Protein Complex for the Rag Family GTPase Gtr2. *Cell Rep.* **2015**, *13*, 1–7. [CrossRef]

80. Dechant, R.; Saad, S.; Ibáñez, A.J.; Peter, M. Cytosolic PH Regulates Cell Growth through Distinct GTPases, Arf1 and Gtr1, to Promote Ras/PKA and TORC1 Activity. *Mol. Cell* **2014**, *55*, 409–421. [CrossRef] [PubMed]

81. Urano, J.; Tabancay, A.P.; Yang, W.; Tamanoi, F. The Saccharomyces Cerevisiae Rheb G-Protein Is Involved in Regulating Canavanine Resistance and Arginine Uptake. *J. Biol. Chem.* **2000**, *275*, 11198–11206. [CrossRef]

82. Wolfson, R.L.; Chantranupong, L.; Saxton, R.A.; Shen, K.; Scaria, S.M.; Cantor, J.R.; Sabatini, D.M. Sestrin2 Is a Leucine Sensor for the MTORC1 Pathway. *Science* **2016**, *351*, 43–48. [CrossRef]

83. Saxton, R.A.; Knockenhauer, K.E.; Wolfson, R.L.; Chantranupong, L.; Pacold, M.E.; Wang, T.; Schwartz, T.U.; Sabatini, D.M. Structural Basis for Leucine Sensing by the Sestrin2-MTORC1 Pathway. *Science* **2016**, *351*, 53–58. [CrossRef] [PubMed]

84. Xu, D.; Shimkus, K.L.; Lacko, H.A.; Kutzler, L.; Jefferson, L.S.; Kimball, S.R. Evidence for a Role for Sestrin1 in Mediating Leucine-Induced Activation of MTORC1 in Skeletal Muscle. *Am. J. Physiol. Endocrinol. Metab.* **2019**, *316*, E817–E828. [CrossRef] [PubMed]

85. Chen, J.; Ou, Y.; Luo, R.; Wang, J.; Wang, D.; Guan, J.; Li, Y.; Xia, P.; Chen, P.R.; Liu, Y. SAR1B Senses Leucine Levels to Regulate MTORC1 Signalling. *Nature* **2021**, *596*, 281–284. [CrossRef] [PubMed]

86. Lee, M.; Kim, J.H.; Yoon, I.; Lee, C.; Fallahi Sichani, M.; Kang, J.S.; Kang, J.; Guo, M.; Lee, K.Y.; Han, G.; et al. Coordination of the Leucine-Sensing Rag GTPase Cycle by Leucyl-TRNA Synthetase in the MTORC1 Signaling Pathway. *Proc. Natl. Acad. Sci. USA* **2018**, *115*, E5279–E5288. [CrossRef]

87. Kim, S.; Yoon, I.; Son, J.; Park, J.; Kim, K.; Lee, J.-H.; Park, S.-Y.; Kang, B.S.; Han, J.M.; Hwang, K.Y.; et al. Leucine-Sensing Mechanism of Leucyl-TRNA Synthetase 1 for MTORC1 Activation. *Cell Rep.* **2021**, *35*, 109031. [CrossRef]

88. Saxton, R.A.; Chantranupong, L.; Knockenhauer, K.E.; Schwartz, T.U.; Sabatini, D.M. Mechanism of Arginine Sensing by CASTOR1 Upstream of MTORC1. *Nature* **2016**, *536*, 229–233. [CrossRef]

89. Chantranupong, L.; Scaria, S.M.; Saxton, R.A.; Gygi, M.P.; Shen, K.; Wyant, G.A.; Wang, T.; Harper, J.W.; Gygi, S.P.; Sabatini, D.M. The CASTOR Proteins Are Arginine Sensors for the MTORC1 Pathway. *Cell* **2016**, *165*, 153–164. [CrossRef]

90. Wang, S.; Tsun, Z.Y.; Wolfson, R.L.; Shen, K.; Wyant, G.A.; Plovanich, M.E.; Yuan, E.D.; Jones, T.D.; Chantranupong, L.; Comb, W.; et al. Metabolism. Lysosomal Amino Acid Transporter SLC38A9 Signals Arginine Sufficiency to MTORC1. *Science* **2015**, *347*, 188–194. [CrossRef] [PubMed]

91. Rebsamen, M.; Pochini, L.; Stasyk, T.; De Arajo, M.E.G.; Galluccio, M.; Kandasamy, R.K.; Snijder, B.; Fauster, A.; Rudashevskaya, E.L.; Bruckner, M.; et al. SLC38A9 Is a Component of the Lysosomal Amino Acid Sensing Machinery That Controls MTORC1. *Nature* **2015**, *519*, 477–481. [CrossRef]

92. Jung, J.W.; Macalino, S.J.Y.; Cui, M.; Kim, J.E.; Kim, H.-J.; Song, D.-G.; Nam, S.H.; Kim, S.; Choi, S.; Lee, J.W. Transmembrane 4 L Six Family Member 5 Senses Arginine for MTORC1 Signaling. *Cell Metab.* **2019**, *29*, 1306–1319.e7. [CrossRef] [PubMed]

93. Parmigiani, A.; Nourbakhsh, A.; Ding, B.; Wang, W.; Kim, Y.C.; Akopiants, K.; Guan, K.L.; Karin, M.; Budanov, A.V. Sestrins Inhibit MTORC1 Kinase Activation through the GATOR Complex. *Cell Rep.* **2014**, *9*, 1281–1291. [CrossRef]

94. Kowalsky, A.H.; Namkoong, S.; Mettetal, E.; Park, H.-W.; Kazyken, D.; Fingar, D.C.; Lee, J.H. The GATOR2–MTORC2 Axis Mediates Sestrin2-Induced AKT Ser/Thr Kinase Activation. *J. Biol. Chem.* **2020**, *295*, 1769–1780. [CrossRef] [PubMed]

95. Gai, Z.; Wang, Q.; Yang, C.; Wang, L.; Deng, W.; Wu, G. Structural Mechanism for the Arginine Sensing and Regulation of CASTOR1 in the MTORC1 Signaling Pathway. *Cell Discov.* **2016**, *2*, 16051. [CrossRef] [PubMed]

96. Kim, J.S.; Ro, S.H.; Kim, M.; Park, H.W.; Semple, I.A.; Park, H.; Cho, U.S.; Wang, W.; Guan, K.L.; Karin, M.; et al. Sestrin2 Inhibits MTORC1 through Modulation of GATOR Complexes. *Sci. Rep.* **2015**, *5*, 9502. [CrossRef] [PubMed]

97. Chantranupong, L.; Wolfson, R.L.; Orozco, J.M.; Saxton, R.A.; Scaria, S.M.; Bar-Peled, L.; Spooner, E.; Isasa, M.; Gygi, S.P.; Sabatini, D.M. The Sestrins Interact with GATOR2 to Negatively Regulate the Amino-Acid-Sensing Pathway Upstream of MTORC1. *Cell Rep.* **2014**, *9*, 1–8. [CrossRef]

98. Xia, J.; Wang, R.; Zhang, T.; Ding, J. Structural Insight into the Arginine-Binding Specificity of CASTOR1 in Amino Acid-Dependent MTORC1 Signaling. *Cell Discov.* **2016**, *2*, 16035. [CrossRef] [PubMed]

99. Suryawan, A.; Davis, T.A. Amino Acid- and Insulin-Induced Activation of MTORC1 in Neonatal Piglet Skeletal Muscle Involves Sestrin2-GATOR2, Rag A/C-MTOR, and RHEB-MTOR Complex Formation. *J. Nutr.* **2018**, *148*, 825–833. [CrossRef]

100. Gu, X.; Orozco, J.M.; Saxton, R.A.; Condon, K.J.; Liu, G.Y.; Krawczyk, P.A.; Scaria, S.M.; Harper, J.W.; Gygi, S.P.; Sabatini, D.M. SAMTOR Is an S-Adenosylmethionine Sensor for the MTORC1 Pathway. *Science* **2017**, *358*, 813–818. [CrossRef]

101. Rathore, R.; Caldwell, K.E.; Schutt, C.; Brashears, C.B.; Prudner, B.C.; Ehrhardt, W.R.; Leung, C.H.; Lin, H.; Daw, N.C.; Beird, H.C.; et al. Metabolic Compensation Activates Pro-Survival MTORC1 Signaling upon 3-Phosphoglycerate Dehydrogenase Inhibition in Osteosarcoma. *Cell Rep.* **2021**, *34*, 108678. [CrossRef]

102. Son, S.M.; Park, S.J.; Lee, H.; Siddiqi, F.; Lee, J.E.; Menzies, F.M.; Rubinsztein, D.C. Leucine Signals to MTORC1 via Its Metabolite Acetyl-Coenzyme A. *Cell Metab.* **2019**, *29*, 192–201.e7. [CrossRef]

103. Sutter, B.M.; Wu, X.; Laxman, S.; Tu, B.P. Methionine Inhibits Autophagy and Promotes Growth by Inducing the SAM-Responsive Methylation of PP2A. *Cell* **2013**, *154*, 403–415. [CrossRef]

104. Laxman, S.; Sutter, B.M.; Shi, L.; Tu, B.P. Npr2 Inhibits TORC1 to Prevent Inappropriate Utilization of Glutamine for Biosynthesis of Nitrogen-Containing Metabolites. *Sci. Signal.* **2014**, *7*, ra120. [CrossRef] [PubMed]

105. Stracka, D.; Jozefczuk, S.; Rudroff, F.; Sauer, U.; Hall, M.N. Nitrogen Source Activates TOR (Target of Rapamycin) Complex 1 via Glutamine and Independently of Gtr/Rag Proteins. *J. Biol. Chem.* **2014**, *289*, 25010–25020. [CrossRef]

106. Chen, X.; Wang, G.; Zhang, Y.; Dayhoff-Brannigan, M.; Diny, N.L.; Zhao, M.; He, G.; Sing, C.N.; Metz, K.A.; Stolp, Z.D.; et al. Whi2 Is a Conserved Negative Regulator of TORC1 in Response to Low Amino Acids. *PLoS Genet.* **2018**, *14*, e1007592. [CrossRef] [PubMed]

107. Teng, X.; Hardwick, J.M. Whi2: A New Player in Amino Acid Sensing. *Curr. Genet.* **2019**, *65*, 701–709. [CrossRef] [PubMed]

108. Kim, S.-H.; Choi, J.-H.; Wang, P.; Go, C.D.; Hesketh, G.G.; Gingras, A.-C.; Jafarnejad, S.M.; Sonenberg, N. Mitochondrial Threonyl-TRNA Synthetase TARS2 Is Required for Threonine-Sensitive MTORC1 Activation. *Mol. Cell* **2021**, *81*, 398–407.e4. [CrossRef]

109. Wittinghofer, A.; Vetter, I.R. Structure-Function Relationships of the G Domain, a Canonical Switch Motif. *Annu. Rev. Biochem.* **2011**, *80*, 943–971. [CrossRef] [PubMed]

110. Shen, K.; Valenstein, M.L.; Gu, X.; Sabatini, D.M. Arg-78 of Nprl2 Catalyzes GATOR1-Stimulated GTP Hydrolysis by the Rag GTPases. *J. Biol. Chem.* **2019**, *294*, 2970–5944. [CrossRef] [PubMed]

111. Deng, L.; Jiang, C.; Chen, L.; Jin, J.; Wei, J.; Zhao, L.; Chen, M.; Pan, W.; Xu, Y.; Chu, H.; et al. The Ubiquitination of RagA GTPase by RNF152 Negatively Regulates MTORC1 Activation. *Mol. Cell* **2015**, *58*, 804–818. [CrossRef]

112. Kiontke, S.; Langemeyer, L.; Kuhlee, A.; Schuback, S.; Raunser, S.; Ungermann, C.; Kümmel, D. Architecture and Mechanism of the Late Endosomal Rab7-like Ypt7 Guanine Nucleotide Exchange Factor Complex Mon1–Ccz1. *Nat. Commun.* **2017**, *8*, 14034. [CrossRef] [PubMed]

113. Kwak, S.S.; Kang, K.H.; Kim, S.; Lee, S.; Lee, J.H.; Kim, J.W.; Byun, B.; Meadows, G.G.; Joe, C.O. Amino Acid-Dependent NPRL2 Interaction with Raptor Determines MTOR Complex 1 Activation. *Cell Signal.* **2016**, *28*, 32–41. [CrossRef]

114. Urban, J.; Soulard, A.; Huber, A.; Lippman, S.; Mukhopadhyay, D.; Deloche, O.; Wanke, V.; Anrather, D.; Ammerer, G.; Riezman, H.; et al. Sch9 Is a Major Target of TORC1 in Saccharomyces Cerevisiae. *Mol. Cell* **2007**, *26*, 663–674. [CrossRef]

115. Sturgill, T.W.; Cohen, A.; Diefenbacher, M.; Trautwein, M.; Martin, D.E.; Hall, M.N. TOR1 and TOR2 Have Distinct Locations in Live Cells. *Eukaryot. Cell* **2008**, *7*, 1819–1830. [CrossRef]

116. Betz, C.; Hall, M.N. Where Is MTOR and What Is It Doing There? *J. Cell Biol.* **2013**, *203*, 563–574. [CrossRef] [PubMed]

117. Hao, F.; Kondo, K.; Itoh, T.; Ikari, S.; Nada, S.; Okada, M.; Noda, T. Rheb Localized on the Golgi Membrane Activates Lysosome-Localized MTORC1 at the Golgi-Lysosome Contact Site. *J. Cell Sci.* **2017**, *131*, jcs.208017. [CrossRef] [PubMed]

118. Zhang, J.; Kim, J.; Alexander, A.; Cai, S.; Tripathi, D.N.; Dere, R.; Tee, A.R.; Tait-Mulder, J.; Di Nardo, A.; Han, J.M.; et al. A Tuberous Sclerosis Complex Signalling Node at the Peroxisome Regulates MTORC1 and Autophagy in Response to ROS. *Nat. Cell Biol.* **2013**, *15*, 1186–1196. [CrossRef]

119. Zhang, J.; Andersen, J.; Sun, H.; Liu, X.; Sonenberg, N.; Nie, J.; Shi, Y. Aster-C Coordinates with COP I Vesicles to Regulate Lysosomal Trafficking and Activation of MTORC1. *EMBO Rep.* **2020**, *21*, e49898. [CrossRef]

120. Meng, J.; Ferguson, S.M. GATOR1-Dependent Recruitment of FLCN–FNIP to Lysosomes Coordinates Rag GTPase Heterodimer Nucleotide Status in Response to Amino Acids. *J. Cell Biol.* **2018**, *217*, 2765–2776. [CrossRef] [PubMed]

121. Petit, C.S.; Roczniak-Ferguson, A.; Ferguson, S.M. Recruitment of Folliculin to Lysosomes Supports the Amino Acid-Dependent Activation of Rag GTPases. *J. Cell Biol.* **2013**, *202*, 1107–1122. [CrossRef]

122. Shen, K.; Rogala, K.B.; Chou, H.-T.; Huang, R.K.; Yu, Z.; Sabatini, D.M. Cryo-EM Structure of the Human FLCN-FNIP2-Rag-Ragulator Complex. *Cell* **2019**, *179*, 1319–1329.e8. [CrossRef]

123. Lawrence, R.E.; Fromm, S.A.; Fu, Y.; Yokom, A.L.; Kim, D.J.; Thelen, A.M.; Young, L.N.; Lim, C.-Y.; Samelson, A.J.; Hurley, J.H.; et al. Structural Mechanism of a Rag GTPase Activation Checkpoint by the Lysosomal Folliculin Complex. *Science* **2019**, *366*, 971–977. [CrossRef] [PubMed]

124. Pal, R.; Palmieri, M.; Chaudhury, A.; Klisch, T.J.; di Ronza, A.; Neilson, J.R.; Rodney, G.G.; Sardiello, M. Src Regulates Amino Acid-Mediated MTORC1 Activation by Disrupting GATOR1-Rag GTPase Interaction. *Nat. Commun.* **2018**, *9*, 4351. [CrossRef]

125. Padi, S.K.R.; Singh, N.; Bearss, J.J.; Olive, V.; Song, J.H.; Cardó-Vila, M.; Kraft, A.S.; Okumura, K. Phosphorylation of DEPDC5, a Component of the GATOR1 Complex, Releases Inhibition of MTORC1 and Promotes Tumor Growth. *Proc. Natl. Acad. Sci. USA* **2019**, *116*, 20505–20510. [CrossRef] [PubMed]

126. Jin, G.; Lee, S.W.; Zhang, X.; Cai, Z.; Gao, Y.; Chou, P.C.; Rezaeian, A.H.; Han, F.; Wang, C.Y.; Yao, J.C.; et al. Skp2-Mediated RagA Ubiquitination Elicits a Negative Feedback to Prevent Amino-Acid-Dependent MTORC1 Hyperactivation by Recruiting GATOR1. *Mol. Cell* **2015**, *58*, 989–1000. [CrossRef] [PubMed]
127. Deng, L.; Chen, L.; Zhao, L.; Xu, Y.; Peng, X.; Wang, X.; Ding, L.; Jin, J.; Teng, H.; Wang, Y.; et al. Ubiquitination of Rheb Governs Growth Factor-Induced MTORC1 Activation. *Cell Res.* **2019**, *29*, 136–150. [CrossRef] [PubMed]
128. Graef, M.; Nunnari, J. Mitochondria Regulate Autophagy by Conserved Signalling Pathways. *EMBO J.* **2011**, *30*, 2101–2114. [CrossRef] [PubMed]
129. Kira, S.; Tabata, K.; Shirahama-Noda, K.; Nozoe, A.; Yoshimori, T.; Noda, T. Reciprocal Conversion of Gtr1 and Gtr2 Nucleotide-Binding States by Npr2-Npr3 Inactivates TORC1 and Induces Autophagy. *Autophagy* **2014**, *10*, 1565–1578. [CrossRef] [PubMed]
130. Wei, Y.; Reveal, B.; Cai, W.; Lilly, M.A. The GATOR1 Complex Regulates Metabolic Homeostasis and the Response to Nutrient Stress in Drosophila Melanogaster. *G3* **2016**, *6*, 3859–3867. [CrossRef]
131. Qi, B.; Kniazeva, M.; Han, M. A Vitamin-B2-Sensing Mechanism That Regulates Gut Protease Activity to Impact Animal's Food Behavior and Growth. *eLife* **2017**, *6*, e26243. [CrossRef] [PubMed]
132. Luo, S.; Shao, L.; Chen, Z.; Hu, D.; Jiang, L.; Tang, W. NPRL2 Promotes Docetaxel Chemoresistance in Castration Resistant Prostate Cancer Cells by Regulating Autophagy through the MTOR Pathway. *Exp. Cell Res.* **2020**, *390*, 111981. [CrossRef]
133. Cai, W.; Wei, Y.; Jarnik, M.; Reich, J.; Lilly, M.A. The GATOR2 Component Wdr24 Regulates TORC1 Activity and Lysosome Function. *PLoS Genet.* **2016**, *12*, e1006036. [CrossRef] [PubMed]
134. Michaillat, L.; Baars, T.L.; Mayer, A. Cell-Free Reconstitution of Vacuole Membrane Fragmentation Reveals Regulation of Vacuole Size and Number by TORC1. *Mol. Biol. Cell* **2012**, *23*, 881–895. [CrossRef]
135. Ma, Y.; Moors, A.; Camougrand, N.; Dokudovskaya, S. The SEACIT Complex Is Involved in the Maintenance of Vacuole–Mitochondria Contact Sites and Controls Mitophagy. *Cell. Mol. Life Sci.* **2019**, *76*, 1623–1640. [CrossRef]
136. Liu, Y.; Okamoto, K. The TORC1 Signaling Pathway Regulates Respiration-Induced Mitophagy in Yeast. *Biochem. Biophys. Res. Commun.* **2018**, *502*, 76–83. [CrossRef]
137. Elbaz-Alon, Y.; Rosenfeld-Gur, E.; Shinder, V.; Futerman, A.H.; Geiger, T.; Schuldiner, M. A Dynamic Interface between Vacuoles and Mitochondria in Yeast. *Dev. Cell* **2014**, *30*, 95–102. [CrossRef]
138. Chong, Y.T.; Koh, J.L.Y.; Friesen, H.; Kaluarachchi Duffy, S.; Cox, M.J.; Moses, A.; Moffat, J.; Boone, C.; Andrews, B.J. Yeast Proteome Dynamics from Single Cell Imaging and Automated Analysis. *Cell* **2015**, *161*, 1413–1424. [CrossRef] [PubMed]
139. Weill, U.; Yofe, I.; Sass, E.; Stynen, B.; Davidi, D.; Natarajan, J.; Ben-Menachem, R.; Avihou, Z.; Goldman, O.; Harpaz, N.; et al. Genome-Wide SWAp-Tag Yeast Libraries for Proteome Exploration. *Nat. Methods* **2018**, *15*, 617–622. [CrossRef]
140. Orre, L.M.; Vesterlund, M.; Pan, Y.; Arslan, T.; Zhu, Y.; Fernandez Woodbridge, A.; Frings, O.; Fredlund, E.; Lehtiö, J. Sub-CellBarCode: Proteome-Wide Mapping of Protein Localization and Relocalization. *Mol. Cell* **2019**, *73*, 166–182.e7. [CrossRef] [PubMed]
141. De Franceschi, N.; Wild, K.; Schlacht, A.; Dacks, J.B.; Sinning, I.; Filippini, F. Longin and GAF Domains: Structural Evolution and Adaptation to the Subcellular Trafficking Machinery: Structure and Evolution of Longin Domains. *Traffic* **2014**, *15*, 104–121. [CrossRef] [PubMed]
142. Costanzo, M.; Baryshnikova, A.; Bellay, J.; Kim, Y.; Spear, E.D.; Sevier, C.S.; Ding, H.; Koh, J.L.Y.; Toufighi, K.; Mostafavi, S.; et al. The Genetic Landscape of a Cell. *Science* **2010**, *327*, 425–431. [CrossRef] [PubMed]
143. Costanzo, M.; VanderSluis, B.; Koch, E.N.; Baryshnikova, A.; Pons, C.; Tan, G.; Wang, W.; Usaj, M.; Hanchard, J.; Lee, S.D.; et al. A Global Genetic Interaction Network Maps a Wiring Diagram of Cellular Function. *Science* **2016**, *353*, aaf1420. [CrossRef] [PubMed]
144. Kovaleva, I.E.; Tokarchuk, A.V.; Zheltukhin, A.O.; Dalina, A.A.; Safronov, G.G.; Evstafieva, A.G.; Lyamzaev, K.G.; Chumakov, P.M.; Budanov, A.V. Mitochondrial Localization of SESN2. *PLoS ONE* **2020**, *15*, e0226862. [CrossRef] [PubMed]
145. Chen, T.; Wang, D.; Xie, T.; Xu, L.-G. Sec13 Is a Positive Regulator of VISA-Mediated Antiviral Signaling. *Virus Genes* **2018**, *54*, 514–526. [CrossRef]
146. De Falco, F.; Cutarelli, A.; Gentile, I.; Cerino, P.; Uleri, V.; Catoi, A.F.; Roperto, S. Bovine Delta Papillomavirus E5 Oncoprotein Interacts With TRIM25 and Hampers Antiviral Innate Immune Response Mediated by RIG-I-Like Receptors. *Front. Immunol.* **2021**, *12*, 658762. [CrossRef]
147. Perrone, G.G.; Grant, C.M.; Dawes, I.W. Genetic and Environmental Factors Influencing Glutathione Homeostasis in *Saccharomyces Cerevisiae*. *Mol. Biol. Cell* **2005**, *16*, 218–230. [CrossRef] [PubMed]
148. Chen, J.; Sutter, B.M.; Shi, L.; Tu, B.P. GATOR1 Regulates Nitrogenic Cataplerotic Reactions of the Mitochondrial TCA Cycle. *Nat. Chem. Biol.* **2017**, *13*, 1179–1186. [CrossRef]
149. Dutchak, P.A.; Estill-Terpack, S.J.; Plec, A.A.; Zhao, X.; Yang, C.; Chen, J.; Ko, B.; Deberardinis, R.J.; Yu, Y.; Tu, B.P. Loss of a Negative Regulator of MTORC1 Induces Aerobic Glycolysis and Altered Fiber Composition in Skeletal Muscle. *Cell Rep.* **2018**, *23*, 1907–1914. [CrossRef]
150. Burger, B.J.; Rose, S.; Bennuri, S.C.; Gill, P.S.; Tippett, M.L.; Delhey, L.; Melnyk, S.; Frye, R.E. Autistic Siblings with Novel Mutations in Two Different Genes: Insight for Genetic Workups of Autistic Siblings and Connection to Mitochondrial Dysfunction. *Front. Pediatr.* **2017**, *5*, 219. [CrossRef]

151. Graber, T.G.; Fry, C.S.; Brightwell, C.R.; Moro, T.; Maroto, R.; Bhattarai, N.; Porter, C.; Wakamiya, M.; Rasmussen, B.B. Skeletal Muscle–Specific Knockout of DEP Domain Containing 5 Protein Increases MTORC1 Signaling, Muscle Cell Hypertrophy, and Mitochondrial Respiration. *J. Biol. Chem.* **2019**, *294*, 4091–4102. [CrossRef]

152. Guaragnella, N.; Coyne, L.P.; Chen, X.J.; Giannattasio, S. Mitochondria–Cytosol–Nucleus Crosstalk: Learning from Saccharomyces Cerevisiae. *FEMS Yeast Res.* **2018**, *18*, foy088. [CrossRef]

153. Quirós, P.M.; Mottis, A.; Auwerx, J. Mitonuclear Communication in Homeostasis and Stress. *Nat. Rev. Mol. Cell Biol.* **2016**, *17*, 213–226. [CrossRef]

154. Zung, N.; Schuldiner, M. New Horizons in Mitochondrial Contact Site Research. *Biol. Chem.* **2020**, *401*, 793–809. [CrossRef] [PubMed]

155. Hönscher, C.; Mari, M.; Auffarth, K.; Bohnert, M.; Griffith, J.; Geerts, W.; van der Laan, M.; Cabrera, M.; Reggiori, F.; Ungermann, C.; et al. Cellular Metabolism Regulates Contact Sites between Vacuoles and Mitochondria. *Dev. Cell* **2014**, *30*, 86–94. [CrossRef]

156. Schenk, P.W.; Brok, E.; Boersma, A.W.M.; Brandsma, J.A.; Den Dulk, H.; Burger, H.; Stoter, G.; Brouwer, J.; Nooter, K. Anticancer Drug Resistance Induced by Disruption of the Saccharomyces Cerevisiae NPR2 Gene: A Novel Component Involved in Cisplatin- and Doxorubicin-Provoked Cell Kill. *Mol. Pharmacol.* **2003**, *64*, 259–268. [CrossRef]

157. Ueda, K.; Kawashima, H.; Ohtani, S.; Deng, W.-G.G.; Ravoori, M.; Bankson, J.; Gao, B.; Girard, L.; Minna, J.D.; Roth, J.A.; et al. The 3p21.3 Tumor Suppressor NPRL2 Plays an Important Role in Cisplatin-Induced Resistance in Human Non-Small-Cell Lung Cancer Cells. *Cancer Res.* **2006**, *66*, 9682–9690. [CrossRef] [PubMed]

158. Chen, S.-H.; Chang, J.-Y. New Insights into Mechanisms of Cisplatin Resistance: From Tumor Cell to Microenvironment. *Int. J. Mol. Sci.* **2019**, *20*, 4136. [CrossRef]

159. Sritharan, S.; Sivalingam, N. A Comprehensive Review on Time-Tested Anticancer Drug Doxorubicin. *Life Sci.* **2021**, *278*, 119527. [CrossRef] [PubMed]

160. Jayachandran, G.; Ueda, K.; Wang, B.; Roth, J.A.; Ji, L. NPRL2 Sensitizes Human Non-Small Cell Lung Cancer (NSCLC) Cells to Cisplatin Treatment by Regulating Key Components in the DNA Repair Pathway. *PLoS ONE* **2010**, *5*, e11994. [CrossRef]

161. Ma, Y.; Vassetzky, Y.; Dokudovskaya, S. MTORC1 Pathway in DNA Damage Response. *Biochim. Biophys. Acta (BBA)-Mol. Cell Res.* **2018**, *1865*, 1293–1311. [CrossRef] [PubMed]

162. Wei, Y.; Bettedi, L.; Ting, C.-Y.; Kim, K.; Zhang, Y.; Cai, J.; Lilly, M.A. The GATOR Complex Regulates an Essential Response to Meiotic Double-Stranded Breaks in Drosophila. *eLife* **2019**, *8*, e42149. [CrossRef] [PubMed]

163. Platani, M.; Trinkle-Mulcahy, L.; Porter, M.; Arockia Jeyaprakash, A.; Earnshaw, W.C. Mio Depletion Links MTOR Regulation to Aurora A and Plk1 Activation at Mitotic Centrosomes. *J. Cell Biol.* **2015**, *210*, 45–62. [CrossRef] [PubMed]

164. Platani, M.; Samejima, I.; Samejima, K.; Kanemaki, M.T.; Earnshaw, W.C. Seh1 Targets GATOR2 and Nup153 to Mitotic Chromosomes. *J. Cell Sci.* **2018**, *131*, jcs213140. [CrossRef] [PubMed]

165. Xi, J.; Cai, J.; Cheng, Y.; Fu, Y.; Wei, W.; Zhang, Z.; Zhuang, Z.; Hao, Y.; Lilly, M.A.; Wei, Y. The TORC1 Inhibitor Nprl2 Protects Age-Related Digestive Function in Drosophila. *Aging* **2019**, *11*, 9811–9828. [CrossRef] [PubMed]

166. Zhu, H.; Sewell, A.K.; Han, M. Intestinal Apical Polarity Mediates Regulation of TORC1 by Glucosylceramide in *C. Elegans*. *Genes Dev.* **2015**, *29*, 1218–1223. [CrossRef] [PubMed]

167. Swaminathan, A.; Hassan-Abdi, R.; Renault, S.; Siekierska, A.; Riché, R.; Liao, M.; de Witte, P.A.M.; Yanicostas, C.; Soussi-Yanicostas, N.; Drapeau, P.; et al. Non-Canonical MTOR-Independent Role of DEPDC5 in Regulating GABAergic Network Development. *Curr. Biol.* **2018**, *28*, 1924–1937.e5. [CrossRef]

168. Liu, Z.; Yan, M.; Liang, Y.; Liu, M.; Zhang, K.; Shao, D.; Jiang, R.; Li, L.; Wang, C.; Nussenzveig, D.R.; et al. Nucleoporin Seh1 Interacts with Olig2/Brd7 to Promote Oligodendrocyte Differentiation and Myelination. *Neuron* **2019**, *102*, 587–601.e7. [CrossRef]

169. International Mouse Phenotyping Consortium. Available online: https://www.mousephenotype.org (accessed on 31 August 2021).

170. Hughes, J.; Dawson, R.; Tea, M.; McAninch, D.; Piltz, S.; Jackson, D.; Stewart, L.; Ricos, M.G.; Dibbens, L.M.; Harvey, N.L.; et al. Knockout of the Epilepsy Gene Depdc5 in Mice Causes Severe Embryonic Dysmorphology with Hyperactivity of MTORC1 Signalling. *Sci. Rep.* **2017**, *7*, 12618. [CrossRef]

171. Yuskaitis, C.J.; Jones, B.M.; Wolfson, R.L.; Super, C.E.; Dhamne, S.C.; Rotenberg, A.; Sabatini, D.M.; Sahin, M.; Poduri, A. A Mouse Model of DEPDC5-Related Epilepsy: Neuronal Loss of Depdc5 Causes Dysplastic and Ectopic Neurons, Increased MTOR Signaling, and Seizure Susceptibility. *Neurobiol. Dis.* **2018**, *111*, 91–101. [CrossRef] [PubMed]

172. Cho, C.-S.; Kowalsky, A.H.; Namkoong, S.; Park, S.-R.; Wu, S.; Kim, B.; James, A.; Gu, B.; Semple, I.A.; Tohamy, M.A.; et al. Concurrent Activation of Growth Factor and Nutrient Arms of MTORC1 Induces Oxidative Liver Injury. *Cell Discov.* **2019**, *5*, 60. [CrossRef] [PubMed]

173. Grabiner, B.C.; Nardi, V.; Birsoy, K.K.; Possemato, R.; Shen, K.; Sinha, S.; Jordan, A.; Beck, A.H.; Sabatini, D.M. A Diverse Array of Cancer-Associated MTOR Mutations Are Hyperactivating and Can Predict Rapamycin Sensitivity. *Cancer Discov.* **2014**, *4*, 554–563. [CrossRef] [PubMed]

174. Weckhuysen, S.; Marsan, E.; Lambrecq, V.; Marchal, C.; Morin-Brureau, M.; An-Gourfinkel, I.; Baulac, M.; Fohlen, M.; Kallay Zetchi, C.; Seeck, M.; et al. Involvement of GATOR Complex Genes in Familial Focal Epilepsies and Focal Cortical Dysplasia. *Epilepsia* **2016**, *57*, 994–1003. [CrossRef] [PubMed]

175. Ishida, S.; Picard, F.; Rudolf, G.; Noé, E.; Achaz, G.; Thomas, P.; Genton, P.; Mundwiller, E.; Wolff, M.; Marescaux, C.; et al. Mutations of DEPDC5 Cause Autosomal Dominant Focal Epilepsies. *Nat. Genet.* **2013**, *45*, 552–555. [CrossRef]

176. Dibbens, L.M.; de Vries, B.; Donatello, S.; Heron, S.E.; Hodgson, B.L.; Chintawar, S.; Crompton, D.E.; Hughes, J.N.; Bellows, S.T.; Klein, K.M.; et al. Mutations in DEPDC5 Cause Familial Focal Epilepsy with Variable Foci. *Nat. Genet.* **2013**, *45*, 546–551. [CrossRef]

177. Scheffer, I.E.; Heron, S.E.; Regan, B.M.; Mandelstam, S.; Crompton, D.E.; Hodgson, B.L.; Licchetta, L.; Provini, F.; Bisulli, F.; Vadlamudi, L.; et al. Mutations in Mammalian Target of Rapamycin Regulator DEPDC5 Cause Focal Epilepsy with Brain Malformations. *Ann. Neurol.* **2014**, *75*, 782–787. [CrossRef]

178. Nascimento, F.A.; Borlot, F.; Cossette, P.; Minassian, B.A.; Andrade, D.M. Two Definite Cases of Sudden Unexpected Death in Epilepsy in a Family with a *DEPDC5* Mutation. *Neurol. Genet.* **2015**, *1*, e28. [CrossRef]

179. Sim, J.C.; Scerri, T.; Fanjul-Fernández, M.; Riseley, J.R.; Gillies, G.; Pope, K.; Van Roozendaal, H.; Heng, J.I.; Mandelstam, S.A.; McGillivray, G.; et al. Familial Cortical Dysplasia Caused by Mutation in the Mammalian Target of Rapamycin Regulator NPRL3. *Ann. Neurol.* **2016**, *79*, 132–137. [CrossRef]

180. Ricos, M.G.; Hodgson, B.L.; Pippucci, T.; Saidin, A.; Ong, Y.S.; Heron, S.E.; Licchetta, L.; Bisulli, F.; Bayly, M.A.; Hughes, J.; et al. Mutations in the Mammalian Target of Rapamycin Pathway Regulators NPRL2 and NPRL3 Cause Focal Epilepsy. *Ann. Neurol.* **2016**, *79*, 120–131. [CrossRef]

181. Baldassari, S.; Picard, F.; Verbeek, N.E.; van Kempen, M.; Brilstra, E.H.; Lesca, G.; Conti, V.; Guerrini, R.; Bisulli, F.; Licchetta, L.; et al. The Landscape of Epilepsy-Related GATOR1 Variants. *Genet. Med.* **2019**, *21*, 398–408. [CrossRef]

182. Ribierre, T.; Deleuze, C.; Bacq, A.; Baldassari, S.; Marsan, E.; Chipaux, M.; Muraca, G.; Roussel, D.; Navarro, V.; Leguern, E.; et al. Second-Hit Mosaic Mutation in MTORC1 Repressor DEPDC5 Causes Focal Cortical Dysplasia–Associated Epilepsy. *J. Clin. Investig.* **2018**, *128*, 2452–2458. [CrossRef]

183. Lee, W.S.; Stephenson, S.E.M.; Howell, K.B.; Pope, K.; Gillies, G.; Wray, A.; Maixner, W.; Mandelstam, S.A.; Berkovic, S.F.; Scheffer, I.E.; et al. Second-hit *DEPDC5* Mutation Is Limited to Dysmorphic Neurons in Cortical Dysplasia Type IIA. *Ann. Clin. Transl. Neurol.* **2019**, *6*, 1338–1344. [CrossRef] [PubMed]

184. D'Gama, A.M.; Woodworth, M.B.; Hossain, A.A.; Bizzotto, S.; Hatem, N.E.; LaCoursiere, C.M.; Najm, I.; Ying, Z.; Yang, E.; Barkovich, A.J.; et al. Somatic Mutations Activating the MTOR Pathway in Dorsal Telencephalic Progenitors Cause a Continuum of Cortical Dysplasias. *Cell Rep.* **2017**, *21*, 3754–3766. [CrossRef]

185. Chandrasekar, I.; Tourney, A.; Loo, K.; Carmichael, J.; James, K.; Ellsworth, K.A.; Dimmock, D.; Joseph, M. Hemimegalencephaly and Intractable Seizures Associated with the *NPRL3* Gene Variant in a Newborn: A Case Report. *Am. J. Med. Genet.* **2021**, *185*, 2126–2130. [CrossRef]

186. Ryu, C.S.; Bae, J.; Kim, I.J.; Kim, J.; Oh, S.H.; Kim, O.J.; Kim, N.K. MPG and NPRL3 Polymorphisms Are Associated with Ischemic Stroke Susceptibility and Post-Stroke Mortality. *Diagnostics* **2020**, *10*, 947. [CrossRef]

187. Cocito, L.; Loeb, C. Focal Epilepsy as a Possible Sign of Transient Subclinical Ischemia. *Eur. Neurol.* **1989**, *29*, 339–344. [CrossRef] [PubMed]

188. Basel-Vanagaite, L.; Hershkovitz, T.; Heyman, E.; Raspall-Chaure, M.; Kakar, N.; Smirin-Yosef, P.; Vila-Pueyo, M.; Kornreich, L.; Thiele, H.; Bode, H.; et al. Biallelic SZT2 Mutations Cause Infantile Encephalopathy with Epilepsy and Dysmorphic Corpus Callosum. *Am. J. Hum. Genet.* **2013**, *93*, 524–529. [CrossRef]

189. Baple, E.L.; Maroofian, R.; Chioza, B.A.; Izadi, M.; Cross, H.E.; Al-Turki, S.; Barwick, K.; Skrzypiec, A.; Pawlak, R.; Wagner, K.; et al. Mutations in KPTN Cause Macrocephaly, Neurodevelopmental Delay, and Seizures. *Am. J. Hum. Genet.* **2014**, *94*, 87–94. [CrossRef]

190. Trivisano, M.; Rivera, M.; Terracciano, A.; Ciolfi, A.; Napolitano, A.; Pepi, C.; Calabrese, C.; Digilio, M.C.; Tartaglia, M.; Curatolo, P.; et al. Developmental and Epileptic Encephalopathy Due to SZT2 Genomic Variants: Emerging Features of a Syndromic Condition. *Epilepsy Behav.* **2020**, *108*, 107097. [CrossRef] [PubMed]

191. Crino, P.B. MTOR: A Pathogenic Signaling Pathway in Developmental Brain Malformations. *Trends Mol. Med.* **2011**, *17*, 734–742. [CrossRef]

192. Lim, K.C.; Crino, P.B. Focal Malformations of Cortical Development: New Vistas for Molecular Pathogenesis. *Neuroscience* **2013**, *252*, 262–276. [CrossRef] [PubMed]

193. Fang, Y.; Jiang, Q.; Li, S.; Zhu, H.; Xu, R.; Song, N.; Ding, X.; Liu, J.; Chen, M.; Song, M.; et al. Opposing Functions of β-Arrestin 1 and 2 in Parkinson's Disease via Microglia Inflammation and Nprl3. *Cell Death. Differ.* **2021**, *28*, 1822–1836. [CrossRef]

194. Iffland, P.H.; Carson, V.; Bordey, A.; Crino, P.B. GATOR Opathies: The Role of Amino Acid Regulatory Gene Mutations in Epilepsy and Cortical Malformations. *Epilepsia* **2019**, *60*, 2163–2173. [CrossRef] [PubMed]

195. Lerman, M.I.; Minna, J.D. The 630-Kb Lung Cancer Homozygous Deletion Region on Human Chromosome 3p21.3: Identification and Evaluation of the Resident Candidate Tumor Suppressor Genes. The International Lung Cancer Chromosome 3p21.3 Tumor Suppressor Gene Consortium. *Cancer Res.* **2000**, *60*, 6116–6133. [PubMed]

196. Bertucci, F.; Ng, C.K.Y.; Patsouris, A.; Droin, N.; Piscuoglio, S.; Carbuccia, N.; Soria, J.C.; Dien, A.T.; Adnani, Y.; Kamal, M.; et al. Genomic Characterization of Metastatic Breast Cancers. *Nature* **2019**, *569*, 560–564. [CrossRef]

197. Otani, S.; Takeda, S.; Yamada, S.; Sakakima, Y.; Sugimoto, H.; Nomoto, S.; Kasuya, H.; Kanazumi, N.; Nagasaka, T.; Nakao, A. The Tumor Suppressor NPRL2 in Hepatocellular Carcinoma Plays an Important Role in Progression and Can Be Served as an Independent Prognostic Factor. *J. Surg. Oncol.* **2009**, *100*, 358–363. [CrossRef] [PubMed]

198. Tang, Y.Y.; Jiang, L.; Tang, W. Decreased Expression of NPRL2 in Renal Cancer Cells Is Associated with Unfavourable Pathological, Proliferation and Apoptotic Features. *Pathol. Oncol. Res. POR* **2014**, *20*, 829–837. [CrossRef] [PubMed]

199. Li, J.; Wang, F.; Haraldson, K.; Protopopov, A.; Duh, F.M.; Geil, L.; Kuzmin, I.; Minna, J.D.; Stanbridge, E.; Braga, E.; et al. Functional Characterization of the Candidate Tumor Suppressor Gene NPRL2/G21 Located in 3p21.3C. *Cancer Res.* **2004**, *64*, 6438–6443. [CrossRef]

200. Yogurtcu, B.; Hatemi, I.; Aydin, I.; Buyru, N. NPRL2 Gene Expression in the Progression of Colon Tumors. *Genet. Mol. Res.* **2012**, *11*, 4810–4816. [CrossRef]

201. Liu, M.N.; Liu, A.Y.; Pei, F.H.; Ma, X.; Fan, Y.J.; Du, Y.J.; Liu, B.R. Functional Mechanism of the Enhancement of 5-Fluorouracil Sensitivity by TUSC4 in Colon Cancer Cells. *Oncol. Lett.* **2015**, *10*, 3682–3688. [CrossRef]

202. Liu, A.; Qiao, J.; He, L.; Liu, Z.; Chen, J.; Pei, F.; Du, Y. Nitrogen Permease Regulator-Like-2 Exhibited Anti-Tumor Effects and Enhanced the Sensitivity of Colorectal Cancer Cells to Oxaliplatin and 5-Fluorouracil. *OTT* **2019**, *12*, 8637–8644. [CrossRef] [PubMed]

203. Peng, Y.; Dai, H.; Wang, E.; Lin, C.C.J.; Mo, W.; Peng, G.; Lin, S.Y. TUSC4 Functions as a Tumor Suppressor by Regulating BRCA1 Stability. *Cancer Res.* **2015**, *75*, 378–386. [CrossRef] [PubMed]

204. Ji, L.; Nishizaki, M.; Gao, B.N.; Burbee, D.; Kondo, M.; Kamibayashi, C.; Xu, K.; Yen, N.; Atkinson, E.N.; Fang, B.L.; et al. Expression of Several Genes in the Human Chromosome 3p21.3 Homozygous Deletion Region by an Adenovirus Vector Results in Tumor Suppressor Activities in Vitro and in Vivo. *Cancer Res.* **2002**, *62*, 2715–2720. [PubMed]

205. Anedchenko, E.A.; Dmitriev, A.A.; Krasnov, G.S.; Kondrat'eva, O.O.; Kopantsev, E.P.; Vinogradova, T.V.; Zinov'eva, M.V.; Zborovskaya, I.B.; Polotsky, B.E.; Sacharova, O.V.; et al. Downregulation of RBSP3/CTDSPL, NPRL2/G21, RASSF1A, ITGA9, HYAL1, and HYAL2 in Non-Small Cell Lung Cancer. *Mol. Biol.* **2008**, *42*, 859–869. [CrossRef]

206. Chen, Z.; Luo, S.; Chen, Y.; Xie, X.; Du, Z.; Jiang, L. High Expression of NPRL2 Is Linked to Poor Prognosis in Patients with Prostate Cancer. *Hum. Pathol.* **2018**, *76*, 141–148. [CrossRef] [PubMed]

207. Liu, S.; Liu, B. Overexpression of Nitrogen Permease Regulator Like-2 (NPRL2) Enhances Sensitivity to Irinotecan (CPT-11) in Colon Cancer Cells by Activating the DNA Damage Checkpoint Pathway. *Med. Sci. Monit.* **2018**, *24*, 1424–1433. [CrossRef]

208. Cai, Y.; Xu, G.; Wu, F.; Michelini, F.; Chan, C.; Qu, X.; Selenica, P.; Ladewig, E.; Castel, P.; Cheng, Y.; et al. Genomic Alterations in PIK3CA-Mutated Breast Cancer Result in MTORC1 Activation and Limit Sensitivity to PI3Kα Inhibitors. *Cancer Res.* **2021**, *81*, 2470–2480. [CrossRef] [PubMed]

209. Chen, Z.; Jiang, Q.; Zhu, P.; Chen, Y.; Xie, X.; Du, Z.; Jiang, L.; Tang, W. NPRL2 Enhances Autophagy and the Resistance to Everolimus in Castration-Resistant Prostate Cancer. *Prostate* **2019**, *79*, 44–53. [CrossRef]

210. Pang, Y.; Xie, F.; Cao, H.; Wang, C.; Zhu, M.; Liu, X.; Lu, X.; Huang, T.; Shen, Y.; Li, K.; et al. Mutational Inactivation of MTORC1 Repressor Gene *DEPDC5* in Human Gastrointestinal Stromal Tumors. *Proc. Natl. Acad. Sci. USA* **2019**, *116*, 22746–22753. [CrossRef]

211. Milton, J.N.; Rooks, H.; Drasar, E.; McCabe, E.L.; Baldwin, C.T.; Melista, E.; Gordeuk, V.R.; Nouraie, M.; Kato, G.J.R.; Minniti, C.; et al. Genetic Determinants of Haemolysis in Sickle Cell Anaemia. *Br. J. Haematol.* **2013**, *161*, 270–278. [CrossRef]

212. Kowalczyk, M.S.; Hughes, J.R.; Garrick, D.; Lynch, M.D.; Sharpe, J.A.; Sloane-Stanley, J.A.; McGowan, S.J.; De Gobbi, M.; Hosseini, M.; Vernimmen, D.; et al. Intragenic Enhancers Act as Alternative Promoters. *Mol. Cell* **2012**, *45*, 447–458. [CrossRef] [PubMed]

213. Hay, D.; Hughes, J.R.; Babbs, C.; Davies, J.O.J.; Graham, B.J.; Hanssen, L.L.P.; Kassouf, M.T.; Oudelaar, A.M.; Sharpe, J.A.; Suciu, M.C.; et al. Genetic Dissection of the α-Globin Super-Enhancer in Vivo. *Nat. Genet.* **2016**, *48*, 895–903. [CrossRef] [PubMed]

214. Miyata, M.; Gillemans, N.; Hockman, D.; Demmers, J.A.A.; Cheng, J.-F.; Hou, J.; Salminen, M.; Fisher, C.A.; Taylor, S.; Gibbons, R.J.; et al. An Evolutionarily Ancient Mechanism for Regulation of Hemoglobin Expression in Vertebrate Red Cells. *Blood* **2020**, *136*, 269–278. [CrossRef] [PubMed]

215. Everything Reptiles. Available online: https://www.everythingreptiles.com/Alligator-vs-Crocodile (accessed on 31 August 2021).

Article

GSI Treatment Preserves Protein Synthesis in C2C12 Myotubes

Joshua R. Huot [1,2], Brian Thompson [1], Charlotte McMullen [1], Joseph S. Marino [1] and Susan T. Arthur [1,*]

1 Laboratory of Systems Physiology, Department of Kinesiology, University of North Carolina at Charlotte, Charlotte, NC 28223, USA; jrhuot@iu.edu (J.R.H.); brianthompson072@gmail.com (B.T.); charlottemc27@gmail.com (C.M.); joseph.marino@uncc.edu (J.S.M.)
2 Department of Surgery, Indiana University School of Medicine, Indianapolis, IN 46202, USA
* Correspondence: sarthur8@uncc.edu; Tel.: +1-(704)-687-0856

Abstract: It has been demonstrated that inhibiting Notch signaling through γ-secretase inhibitor (GSI) treatment increases myogenesis, AKT/mTOR signaling, and muscle protein synthesis (MPS) in C2C12 myotubes. The purpose of this study was to determine if GSI-mediated effects on myogenesis and MPS are dependent on AKT/mTOR signaling. C2C12 cells were assessed for indices of myotube formation, anabolic signaling, and MPS following GSI treatment in combination with rapamycin and API-1, inhibitors of mTOR and AKT, respectively. GSI treatment increased several indices of myotube fusion and MPS in C2C12 myotubes. GSI-mediated effects on myotube formation and fusion were completely negated by treatment with rapamycin and API-1. Meanwhile, GSI treatment was able to rescue MPS in C2C12 myotubes exposed to rapamycin or rapamycin combined with API-1. Examination of protein expression revealed that GSI treatment was able to rescue pGSK3β Ser9 despite AKT inhibition by API-1. These findings demonstrate that GSI treatment is able to rescue MPS independent of AKT/mTOR signaling, possibly via GSK3β modulation.

Keywords: muscle protein synthesis; GSI; mTOR; AKT

Citation: Huot, J.R.; Thompson, B.; McMullen, C.; Marino, J.S.; Arthur, S.T. GSI Treatment Preserves Protein Synthesis in C2C12 Myotubes. *Cells* **2021**, *10*, 1786. https://doi.org/10.3390/cells10071786

Academic Editors: Jean Christopher Chamcheu, Claudia Bürger and Shile Huang

Received: 27 May 2021
Accepted: 13 July 2021
Published: 15 July 2021

Publisher's Note: MDPI stays neutral with regard to jurisdictional claims in published maps and institutional affiliations.

1. Introduction

Skeletal muscle wasting is a debilitating result of aging and several disease states, which drastically reduces functional capacity and quality of life [1–3]. Loss of skeletal muscle mass can be attributed to increased protein catabolism, impaired muscle regeneration (i.e., myogenesis), and/or reductions in muscle protein synthesis (MPS) [1,4,5]. The protein kinase B (AKT)/mechanistic target of rapamycin (mTOR) cascade is pivotal for several processes within skeletal muscle including survival, autophagy, differentiation, and MPS [6,7]. Interestingly, emerging evidence has identified mTOR as a primary antagonist of lifespan, revealing that administration of rapamycin (a potent inhibitor of mTOR) can increase lifespan, improve aging, and combat age-related disease development [8–13]. However, reduced mTOR signaling in skeletal muscle diminishes myogenic potential, and reduces anabolic potential of exercise and nutrients [14,15]. This poses an interesting dilemma for skeletal muscle researchers, in particular when seeking to maintain muscle mass in diseased and aged populations.

Another pathway strongly implicated in skeletal muscle health and disease is Notch signaling [1]. Notch is activated when one of its four receptors (Notch1–4) binds to one of several Notch ligands (Delta-like protein (DLL)1, DLL3, DLL4, Jagged1, Jagged2), which initiates successive metalloprotease and γ-secretase cleavages [1]. Aside from its developmental regulation, Notch signaling dictates the myogenic response following injury [16,17]. Similar to mTOR, dysfunctional Notch signaling may also occur in atrophic skeletal muscle, blunting skeletal muscle regeneration [16,18–20]. Moreover, aberrant Notch signaling has been implicated in the development of insulin resistance and cachexia [21–23]. Meanwhile, targeting the γ-secretase cleavage via γ-secretase inhibitors (GSIs), a commonly used method to chemically inhibit Notch signaling, has demonstrated that reduced Notch signaling increases myotube formation and muscle growth [22,24,25]. Altogether, these findings make Notch an interesting target to potentially combat muscle atrophy in aging and other skeletal muscle wasting diseases.

Reducing mTOR activity can promote healthy aging, yet at the same time blunts the anabolic potential of skeletal muscle. Thus, identifying and targeting signaling pathways that modulate MPS independent of mTOR may help to sustain skeletal muscle mass in the aging population. One example of this is AKT, which can mediate MPS via glycogen synthase kinase 3 beta (GSK3β) independently of mTOR [26–28]. Interestingly, our lab recently demonstrated that Notch inhibition via GSI treatment elevated MPS in C2C12 myoblasts and myotubes by modulating the phosphatase and tensin homolog (PTEN)/AKT/mTOR signaling cascade [25]. However, it is not known if this mechanism is reliant on mTOR or if GSIs mediate MPS in an AKT-dependent manner.

Thus, in the present study we sought to investigate whether the beneficial effects of GSIs on myotube size and MPS were dependent on AKT/mTOR. Here, we again demonstrate that GSI treatment increases differentiation and MPS in C2C12 myotubes. Inhibition of AKT and mTOR ablated GSI-induced differentiation in C2C12 cells. However, GSI treatment preserved MPS rates in combination with AKT and mTOR inhibition, suggesting that the use of GSIs may be able to augment MPS independent of AKT/mTOR.

2. Materials and Methods

2.1. Cell Culture

For all in vitro experiments, C2C12 skeletal muscle myoblasts (ATCC p3–p8) were cultured in Dulbecco's Modified Eagles Medium (DMEM), supplemented with 10% fetal bovine serum, 10% horse serum (HS), and 1% penicillin/streptomycin (P/S), as performed previously [25]. In particular, we previously demonstrated that γ-secretase inhibitor (GSI-4 μM: L-685,458; Millipore Sigma- dimethyl sulfoxide (DMSO)) treatment on the onset of C2C12 differentiation increases myotube fusion and AKT/mTOR signaling [25]. Thus, to examine if GSI-mediated effects on myotube fusion were dependent on AKT/mTOR, we exposed C2C12s at the onset of differentiation to 4 μM GSI in combination with 100 nM rapamycin (RAP; 13346; Cayman Chemicals in DMSO) or 10 μM 4-Amino-5,8-dihydro-5-oxo-8-b-D-ribofuranosyl-pyrido[2,3-d]pyrimidine-6-carboxamide (API-1; SML1342; Millipore Sigma in DMSO), established inhibitors of mTOR and AKT, respectively [29–31]. Specifically, for the differentiating C2C12 experiments, myoblasts were grown to full confluence, switched to differentiation media (DM: DMEM supplemented with 2% HS and 1% P/S) and treated every 12 h for 96 h in one of the following conditions: (1) GSI, RAP, GSI + RAP, or control (Con: DMSO). (2) GSI, API-1, GSI + API-1, or Con. Myotubes from these experiments were analyzed for indices of fusion and area as outlined below. In addition to demonstrating that GSI treatment enhances C2C12 differentiation, we also previously reported that differentiated myotubes exposed to GSI for 24 h was sufficient to increase MPS [25]. Thus, to determine if the GSI-mediated effects on MPS in differentiated myotubes were reliant on AKT/mTOR signaling, C2C12s were allowed to differentiate for 72 h and were then treated twice (every 12 h) for the next 24 h under one of the following conditions: (1) GSI, RAP, GSI + RAP, or Con; (2) GSI, API-1, GSI + API-1, or Con; (3) GSI, RAP + API-1, GSI + RAP + API-1, or Con. Myotubes from these experiments were analyzed for protein expression, protein synthesis, and myotube diameter, as detailed below.

2.2. Myosin Heavy Chain Staining

Following 96 h of differentiation, myotubes were stained with myosin heavy chain (MHC) and assessed for properties of fusion, area, and diameter, as performed previously [25,32]. Briefly, myotubes were fixed with 70% acetone/30% methanol, serially washed with PBS, blocked for 1 h, and incubated overnight in MHC (MF-20, 1:100; Developmental Studies Hybridoma Bank, Iowa City, IA, USA). Myotubes were subsequently PBS-washed, incubated with an anti-mouse secondary antibody (1:200) and 4′,6-Diamidino-2-Phenylindole, Dihydrochloride (DAPI 1:1000) for 1 h, and mounted with Vectashield.

2.3. Myotube Fusion, Area, and Diameter

Stained myotubes were captured at 20× for indices of fusion and area on an Olympus iX inverted microscope, as performed previously [25]. Following image acquisition, 2 blinded individuals quantified indices of myotube fusion, including myotube number, total nuclei, and fused nuclei using ImageJ. Myotube area was determined from the same images used to calculate fusion index using Adobe Photoshop, as previously described [33]. Briefly, three randomly selected images from each experimental group (Con, GSI, RAP, GSI + RAP or Con, GSI, API-1, GSI + API-1) were used to set accepted tones for MHC (red) and DAPI (blue). The set color range was then subsequently applied to all images in order to obtain measures for total myotube area, area per myotube, and myotube area per fused nuclei. For myotube size, ImageJ software was used to measure the narrowest diameter along the myotube (400 myotubes per condition) [32,34].

2.4. Protein Synthesis

For assessment of protein synthesis, myotubes were treated with 1 μM puromycin (P-1033, A.G. Scientific, San Diego, CA, USA) 30 min prior to cell collection, as previously described [25,35,36]. Puromycin incorporation was subsequently analyzed via western blot, as detailed below.

2.5. Western Blot

To extract protein from C2C12 myotubes, well surfaces were washed two times with cold PBS, mechanically lysed in chilled Radioimmunoprecipitation assay (RIPA) buffer (sc-24948; Santa Cruz Biotechnology, Dallas, TX, USA) containing 1% Triton-x, 2% SDS and protease cocktail inhibitors, as performed previously [25], and centrifuged for 20 min at $20,000 \times g$ (4 °C). Following centrifugation ($20,000 \times g$, 20 min, 4 °C), the supernatant was saved and assessed for protein concentration by the BCA protein assay method (23225; ThermoFisher, Waltham, MA, USA). Samples (20 μg) were loaded and electrophoresed on a 4–12% Bis-Tris gel (3450125; Bio-Rad, Hercules, CA, USA) at 125 V for 2 h, as performed previously [25]. Proteins were then transferred (Towbin Buffer; 10% methanol) onto a 0.22 μM Polyvinylidene difluoride (PVDF) membrane for 1 h at 100 V. Membranes were washed in Tris-buffered saline (TBS), blocked for 1 h in Odyssey blocking buffer (1:1 TBS), and incubated overnight in primary antibodies. Following primary antibody incubation, membranes were serially washed in TBST (TBS: 0.1% Tween 20) and incubated in secondary antibodies (1:10,000 in TBST) for 1 h. Membranes were again serially washed in TBST, and proteins were then visualized and quantified using the Odyssey® Licor CLx System. Antibodies used were as follows: pAKT Thr308 (#13038; 1:500), pAKT Ser473 (#4060; 1:500), AKT (#2920; 1:1000), pmTOR Ser2448 (#5536; 1:500), mTOR (#4517, 1:1000), p4EBP1 Thr37/46 (#2855; 1:500), 4EBP1 (#9644; 1:1000), pp70S6K Thr389 (#9234; 1:500), p70S6K (#2708; 1:1000), peEF2 Thr56 (#2331; 1:500), eEF2 (#2332; 1:1000), pGSK3β Ser9 (#8566; 1:500), GSK3β (#5676; 1:1000), and ABC (#8814; 1:1000) from Cell Signaling; Puromycin (#MABE343; 1:5000) from EMD Millipore; and β-Actin (#A2228; 1:10,000) from Sigma Aldrich, St. Louis, MI, USA.

2.6. Statistical Analysis

One-way analysis of variance (ANOVA) tests were performed to determine differences between experimental groups (1) GSI, RAP, GSI + RAP, or Con. 2) GSI, API-1, GSI + API-1, or Con. 3) GSI, RAP + API-1, GSI + RAP + API-1, or Con). Post-hoc comparisons were accomplished via a Tukey's test, with statistical significance set apriori at $p \leq 0.05$. All statistical analyses and graphs were made using Graphpad Prism 7.03 (GraphPad, San Diego, CA, USA). All data are presented as means ± SD.

3. Results

3.1. Rapamycin Ablates GSI-Mediated Elevations in Myotube Formation

Since our work previously demonstrated that Notch inhibition via GSI treatment was sufficient to increase myotube formation as well as mTOR signaling, we wanted to determine whether GSI-mediated effects on myogenesis were dependent on mTOR [25]. Thus, we treated differentiating C2C12 myotubes with GSI and the commonly used mTOR inhibitor, rapamycin, for 96 h. Similar to our previous reported results, GSI treatment significantly increased several indices of myotube formation compared to all groups, including fused nuclei per field, nuclei per myotube per field, and fusion index per field (Figure 1). GSI treatment also resulted in significantly reduced non-fused nuclei compared to all other groups (Figure 1). Interestingly, RAP significantly reduced fused nuclei, nuclei per myotube, and fusion index compared to Con, but did not differ from GSI + RAP in any of these measures, suggesting that GSI-mediated increases in fusion are dependent on mTOR (Figure 1). Moreover, both RAP and GSI + RAP had significantly elevated non-fused nuclei compared to Con, while also increasing total nuclei per field compared to Con and GSI, yet did not differ from each other (Figure 1). The myotube number was also significantly reduced in RAP and GSI + RAP compared to both Con and GSI (Figure 1). With respect to area measures, GSI treatment increased total myotube area, area per myotube, and myotube area per fused nuclei (Figure 1). Meanwhile, RAP and GSI + RAP both reduced myotube area, area per myotube, and myotube area per fused nuclei compared to Con, further suggesting that GSI-mediated increases in myotube formation are dependent on mTOR (Figure 1).

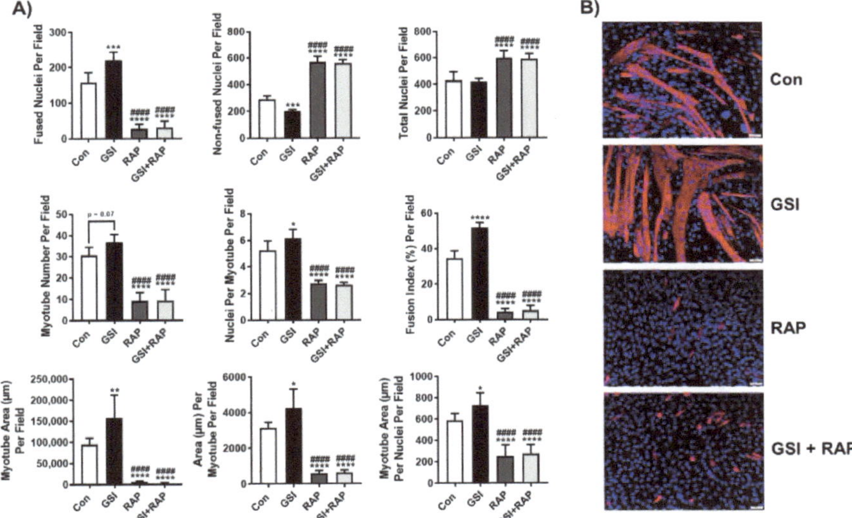

Figure 1. Rapamycin ablates GSI-mediated elevations in myotube formation. (**A**) Indices of myotube fusion and area. Graph order, top left to right: Fused nuclei per field, Non-fused nuclei per field, and Total nuclei per field. Graph order, middle left to right: Myotube number per field, Nuclei per myotube per field, Fusion index per field. Graph order, bottom left to right: Myotube area (μm) per field, Area (μm) per myotube per field, Myotube area (μm) per nuclei per field. (**B**) Representative image of 96-h myotubes co-stained with myosin heavy chain (MHC:red) and DAPI:blue. Images were taken at 20× magnification and the scale bar = 50 μm. At the onset of differentiation C2C12 cells were treated every 12 h with either control (Con), 4 μM γ-secretase inhibitor (GSI), 100 nM rapamycin (RAP), or GSI + RAP co-treatment. All data were analyzed using a one-way ANOVA followed by Tukey's multiple comparison test. * $p < 0.05$, ** $p < 0.01$, *** $p < 0.001$, **** $p < 0.0001$ vs. Con; #### $p < 0.0001$ vs. GSI ($n = 3$ experiments). Data are mean ± SD.

3.2. GSI Treatment Preserves Protein Synthesis in the Presence of Rapamycin

We previously demonstrated that GSI treatment increased MPS in differentiating and differentiated C2C12 myotubes [25]. As the present data demonstrated that the use of rapamycin ablated GSI-augmented fusion, we wanted to determine whether the effects of GSI on MPS were also reliant on mTOR. To do this we differentiated C2C12 myotubes for 72 h and then exposed them to GSI and RAP for 24 h. Confirming our published work, GSI treatment increased protein synthesis compared to all other groups, while RAP exhibited reduced protein synthesis compared to Con (Figure 2A). Interestingly, GSI treatment protected protein synthesis rates and myotube size in the presence of RAP (Figure 2A and Figure S1). In addition, in line with our prior work, GSI treatment increased phosphorylated (p)-mTOR at Ser2448 compared to all other groups, while p-mTOR was reduced compared to Con in both RAP and GSI + RAP treated myotubes, suggesting that GSI preservation of MPS is not dependent on mTOR (Figure 2B). Moreover, despite increasing p-mTOR, GSI treatment did not exert effects directly downstream of mTOR (p-4EBP1, p-p70S6K, p-eEF2), which is comparable to our prior findings (Figure 2C–E) [25]. Meanwhile, RAP and GSI + RAP did not differ in any downstream target of mTOR, showing reduced 4EBP1, reduced p-p70S6K, and increased p-eEF2 (Figure 2B–D). These results suggest that GSI treatment may protect MPS levels in the presence of rapamycin and thus may modulate MPS in mechanisms other than mTOR.

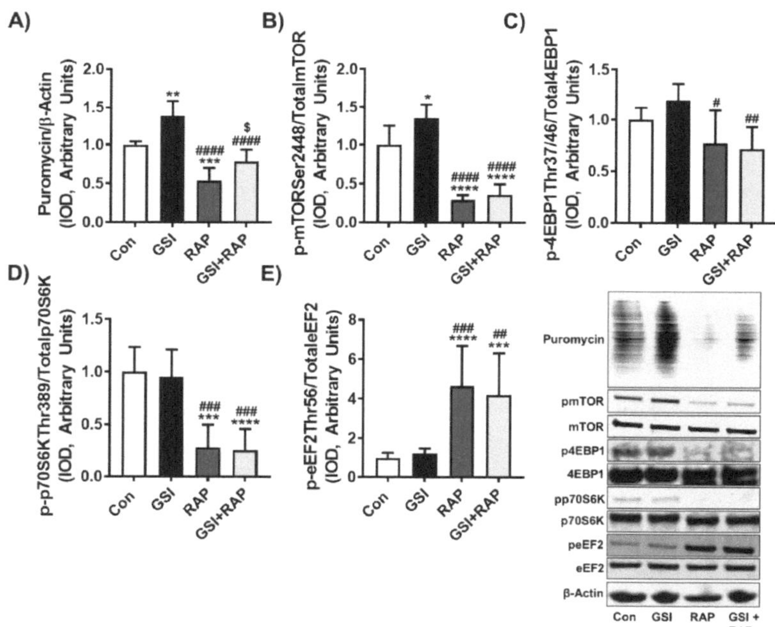

Figure 2. GSI treatment rescues protein synthesis in the presence of rapamycin. Representative western blotting and quantification (expressed as fold change vs. control (Con) for (**A**) Puromycin/β-Actin (**B**) Phospho (p)-mTOR Ser2448/Total mTOR; (**C**) p-4EBP1 Thr37/46/Total 4EBP1; (**D**) p70S6K Thr389/Total p70S6K; and (**E**) p-eEF2 Thr56/Total eEF2. Next, 72 h post-differentiation, C2C12 cells were treated every 12 h with either Con, 4 μM γ-secretase inhibitor (GSI), 100 nM rapamycin (RAP), or GSI + RAP co-treatment until 96 h post-differentiation. Then, 30 min prior to collection, all cells were treated with 1 μM puromycin. All data were analyzed using a one-way ANOVA followed by Tukey's multiple comparison test. $* p < 0.05$, $** p < 0.01$, $*** p < 0.001$, $**** p < 0.0001$ vs. Con; $\# p < 0.05$, $\#\# p < 0.01$, $\#\#\# p < 0.001$, $\#\#\#\# p < 0.0001$ vs. GSI; $\$ p < 0.05$ vs. RAP ($n = 3$ experiments). Data are mean ± SD.

3.3. API-1 Ablates GSI-Mediated Elevations in Myotube Formation

To expand upon our prior findings that GSI treatment promotes fusion as well as increased AKT signaling in C2C12s, we also decided to investigate if AKT was necessary for GSI-mediated effects on myotube formation [25]. To do this we treated differentiating C2C12s with GSI and an AKT inhibitor, API-1. In concert with previous experiments, GSI treatment increased all measured markers of myotube formation and fusion. GSI treatment increased fused nuclei, nuclei per myotube, and fusion index compared to all other groups (Figure 3). API-1 and GSI + API-1 treatment induced lower fused nuclei, nuclei per myotube, and fusion index compared to Con, but did not differ from each other in any parameter (Figure 3). GSI treatment also reduced non-fused nuclei compared to Con; however, there were no differences between the other groups (Con; API-1; GSI + API-1) (Figure 3). The lack of difference in non-fused nuclei between groups is likely due to the significant reduction seen in total nuclei with API-1 and GSI + API-1 treatment (Figure 3). Similar to reductions in total nuclei per field, the myotube number was also reduced in API-1 and GSI + API-1 compared to Con and GSI (Figure 3). Regarding area measures, total myotube area, area per myotube, and myotube area per fused nuclei were increased with GSI treatment compared to all groups (Figure 3). Further suggesting that API-1 treatment ablates myotube formation induced by GSI treatment, API-1 and GSI + API-1 had reduced total myotube area, area per myotube, and myotube area per fused nuclei compared to Con, but did not differ from each other in any parameter (Figure 3).

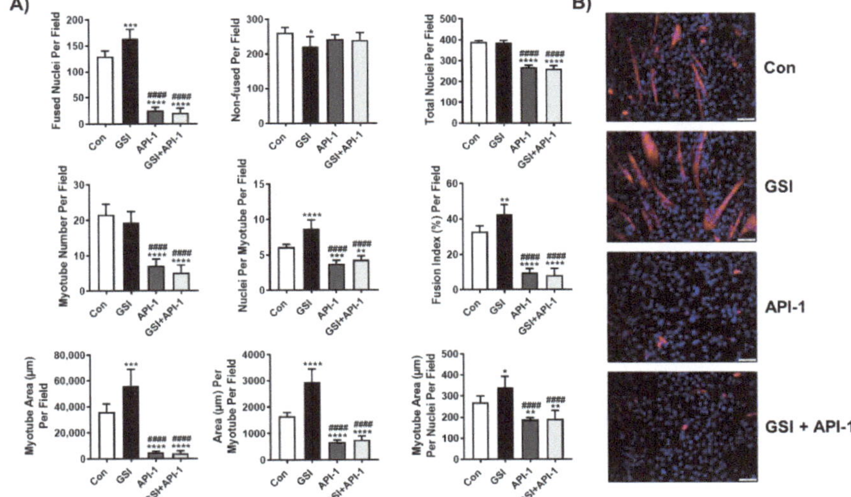

Figure 3. API-1 ablates GSI-mediated elevations in myotube formation. (**A**) Indices of myotube fusion and area. Graph order, top left to right: Fused nuclei per field, Non-fused nuclei per field, and Total nuclei per field. Graph order, middle left to right: Myotube number per field, Nuclei per myotube per field, Fusion index per field. Graph order, bottom left to right: Myotube area (μm) per field, Area (μm) per myotube per field, Myotube area (μm) per nuclei per field. (**B**) Representative image of 96-h myotubes co-stained with myosin heavy chain (MHC:red) and DAPI:blue. Images were taken at 20× magnification and the scale bar = 50 μm. At the onset of differentiation C2C12 cells were treated every 12 h with either control (Con), 4 μM γ-secretase inhibitor (GSI), 10 μM 4-Amino-5,8-dihydro-5-oxo-8-β-D-ribofuranosyl-pyrido[2,3-d]pyrimidine-6-carboxamide (API-1), or GSI + API-1 co-treatment. All data were analyzed using a one-way ANOVA followed by Tukey's multiple comparison test. * $p < 0.05$, ** $p < 0.01$, *** $p < 0.001$, **** $p < 0.0001$ vs. Con; #### $p < 0.0001$ vs. GSI (n = 3 experiments). Data are mean $\pm$ SD.

3.4. GSI Treatment Preserves Protein Synthesis in the Presence of API-1 and Rapamycin

Given our prior findings that GSI-treated myotubes have increased AKT signaling along with increased MPS and our present findings that GSI treatment protected MPS in the presence of rapamycin, we wanted to determine whether GSI-mediated effects on MPS are dependent on AKT (Figure 2) [25]. This was also of interest as AKT can mediate MPS independently of mTOR by way of GSK3β [26–28]. In line with our prior findings, GSI treatment increased MPS and myotube size compared to all groups; however, the use of API-1 was not sufficient to reduce MPS compared to Con (Figure 4A and Figure S2). In addition, MPS with GSI + API-1 was no different than API-1 alone, suggesting that GSI-mediated effects on MPS may be dependent on AKT (Figure 4A). However, despite not reducing MPS, API-1 was sufficient to reduce myotube size, while introduction of GSI in the presence of API-1 preserved the myotube size (Figure S2). Interestingly, though GSI treatment increased phosphorylation of AKT on both Thr308 and Ser473, the use of API-1 was only sufficient to reduce pAKT Ser473 (Figure 4B,C), and GSI + API-1 did not differ from API-1 at either phosphorylation site (Figure 4B,C). Downstream of AKT, GSI treatment significantly elevated pmTORSer2448 compared to all groups (Figure 4D). Intriguingly, and similar to pAKT Thr308, API-1 and GSI + API-1 did not reduce pmTORSer2448 compared to Con, nor were they different from each other, suggesting that GSI treatments elevation of mTOR may be dependent on AKT. In addition, the fact that pmTOR did not reduce with API-1 treatment may also explain the lack of reduction in MPS.

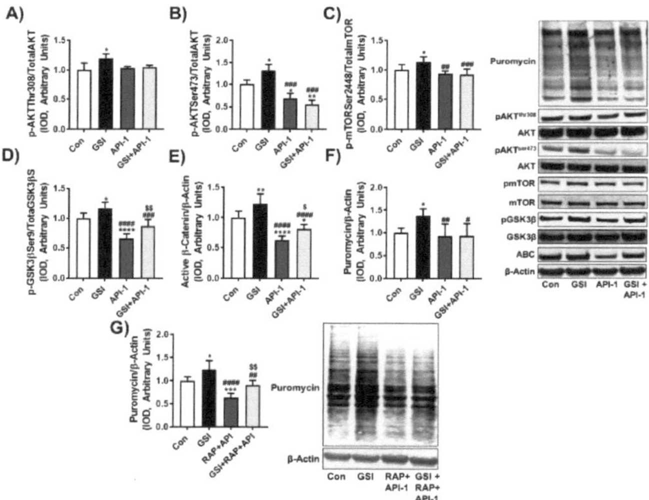

Figure 4. GSI treatment rescues protein synthesis in the presence of API-1 and rapamycin. Representative western blotting and quantification (expressed as fold change vs. control (Con) for (**A**) Puromycin/β-Actin; (**B**) Phospho (p)-AKT Thr308/Total AKT; (**C**) p-AKT Ser473/Total AKT; (**D**) p-mTOR Ser2448/Total mTOR; (**E**) p-GSK3β Ser9/Total GSK3β; (**F**) Active β-Catenin/β-Actin. Then, 72 h post-differentiation, C2C12 cells were treated every 12 h with either Con, 4 µM γ-secretase inhibitor (GSI), 10 µM 4-Amino-5,8-dihydro-5-oxo-8-β-D-ribofuranosyl-pyrido[2,3-d]pyrimidine-6-carboxamide (API-1), or GSI + API-1 co-treatment until 96 h post-differentiation. All cells were treated with 1 µM puromycin 30 min prior to collection. Representative western blotting and quantification for (**G**) Puromycin/β-Actin. Next, 72 h post-differentiation, C2C12 cells were treated every 12 h with either Con, 4 µM GSI, 100 nM rapamycin (RAP) + 10 µM API-1, or GSI + RAP + API-1 co-treatment until 96 h post-differentiation. All cells were treated with 1 µM puromycin 30 min prior to collection. All data were analyzed using a one-way ANOVA followed by Tukey's multiple comparison test.

* $p < 0.05$, ** $p < 0.01$, *** $p < 0.001$, **** $p < 0.0001$ vs. Con; # $p < 0.05$, ## $p < 0.01$, ### $p < 0.001$, #### $p < 0.0001$ vs. GSI; $ $p < 0.05$, $$ $p < 0.01$ vs. API-1 (**D,E**); $$ $p < 0.01$ vs. RAP + API-1 (**G**) ($n = 3$ experiments). Data are mean $\pm$ SD.

Since AKT also mediates protein synthesis independently of mTOR by phosphorylating GSK3β, we wanted to assess if this signaling was changed with GSI and with API-1 treatments. GSI-treated C2C12s demonstrated elevations in pGSK3βSer9 (Figure 4E), and while API-1 did not cause reductions in mTOR, it did reduce pGSK3βSer9 compared to Con (Figure 4E). Intriguingly, pGSK3βSer9 was elevated in GSI + API-1 compared to API-1 and did not differ from Con. Moreover, GSI treatment increased active β-catenin (ABC) compared to all groups and GSI + API-1 was sufficient to preserve ABC compared to API-1 alone (Figure 4F). The preservation of pGSK3βSer9 with GSI treatment in combination with API-1 convinced us to observe protein synthesis conditions in which MPS would surely be reduced. Thus, we conducted an additional experiment set utilizing GSI + RAP + API-1. Again, GSI treatment increased MPS compared to all other groups (Figure 4G). Treatment of C2C12s with RAP + API-1 significantly reduced MPS and myotube size compared to Con, while the introduction of GSI in the presence of RAP + API-1 preserved both MPS and myotube size (Figure 4G and Figure S3).

4. Discussion

Our lab recently demonstrated that Notch inhibition via GSI treatment elevates protein synthesis in C2C12 muscle cells, possibly through modulation of the AKT/mTOR signaling cascade [25]. In the present study, we expand upon our prior findings and demonstrate that GSI treatment may be able to augment MPS independent of AKT/mTOR, as GSI treatment in the presence of mTOR and AKT inhibition was able to provide protection of protein synthesis in C2C12 myotubes.

One goal of the present study was to determine if the GSI-mediated elevations in myotube formation and growth was dependent on mTOR, as mTOR is a pivotal regulator of myogenesis [37–39]. Here, we show that GSI treatment significantly elevates a myriad of fusion and hypertrophy indices (fusion index, area/myotube, and myotube area/fused nuclei) in differentiating C2C12 myotubes and that these elevations are completely ablated by the introduction of rapamycin. This is in full support of the notion that mTOR is required for muscle cell differentiation and demonstrates that GSI-mediated myogenesis enhancement is dependent on mTOR. In contrast, we demonstrated that GSI treatment protects MPS and myotube size in differentiated C2C12s exposed to rapamycin, suggesting that GSI may modulate MPS in an mTOR-independent manner.

Given AKT's ability to regulate myotube differentiation and elevate protein synthesis independent of mTOR through regulation of GSK3β, myotubes were also exposed to the AKT inhibitor, API-1 [26,40,41]. In a similar fashion, API-1 diminished GSI-mediated enhancements on C2C12 differentiation, which is in concert with prior findings that AKT is essential for the initiation of myoblast differentiation [41]. However, API-1 alone was not sufficient to reduce mTOR or MPS in differentiated myotubes. API-1 is a novel small molecule inhibitor that has gained research attention in recent years as a possible anti-cancer therapeutic, and, while evidence shows reduced pAKTSer473, to our knowledge it is unclear if both pAKTThr308 and Ser473 are reduced with API-1 treatment [31,42–44]. Here, we show that API-1 only reduced pAKTSer473 and not pAKTThr308. This may explain why we did not observe detectable reductions in mTOR signaling or MPS with API-1 treatment. Though it is reported that reductions in AKT can blunt protein synthesis, other proteins may influence mTOR independent of AKT, which could explain why the use of API-1 is not sufficient enough to reduce protein synthesis [6,7,45]. In contrast, API-1 treatment did reduce another downstream target of AKT, pGSK3βSer9 and subsequent ABC protein expression, demonstrating that the use of API-1 does negatively influence the AKT function. Furthermore, despite its lack of effect on MPS, API-1 treatment did reduce myotube size. AKT is a known pro-survival signaling pathway and API-1 has shown to induce apoptosis, thus the atrophy observed in the present study is likely due to heighted protein breakdown and not a suppression of MPS [46–48]. However, we did not measure specific indices of protein catabolism in the present study and can only speculate at this time.

An underlying premise of this study was to gain better insight into the mechanisms by with GSI treatment elevates MPS in C2C12 myotubes, and since API-1 alone was not sufficient to reduce protein synthesis, myotubes were exposed to API-1 and RAP together. Intriguingly, GSI treatment was able to rescue protein synthesis rate and myotube size in the presence of both API-1 and RAP, suggesting that GSI may mediate MPS in a mechanism independent of AKT/mTOR. Our data suggests that this rescue of MPS may be via GSK3β. When active, GSK3β phosphorylates and inactivates the translation initiation factor eIF2B [28,40]. However, when GSK3β is phosphorylated on Ser9 by AKT, eIF2B is able to partake in translation initiation [49]. As mentioned above, pGSK3βSer9 was significantly reduced with API-1 treatment. Interestingly, pGSK3βSer9 was rescued with GSI treatment in the presence of API-1, suggesting that GSIs may act on GSK3β independent of AKT. In concert with this finding, ABC was also rescued, indicating that function of GSK3β may be modulated by GSIs independent of AKT. This is not the first time that GSIs or Notch and GSK3β crosstalk have been postulated. In fact, during investigations using HEK293T cells, smooth muscle cells and fibroblasts have identified Notch as a target of GSK3β [50–52]. Interestingly, however, work within skeletal muscle has also suggested a regulatory role of Notch on GSK3β. Brack et al. discussed GSK3β as a mediator between Notch and Wnt during skeletal muscle regeneration, demonstrating GSI-mediated elevations in pGSK3βSer9 similar to our present findings [16]. However, their work did not discuss GSK3β as a mediator of protein synthesis and did not investigate whether GSI treatment impacts MPS. GSK3β is a known regulator of MPS and overexpression of its downstream target eIF2Bε has shown to significantly increase rates of MPS and induce muscle hypertrophy [40,53]. Additionally, Notch signaling has been implicated in regulating muscle hypertrophy in both in vivo and in vitro models [22,54–56]. To our knowledge, we are the first to address the research paradigm of Notch as a regulator of MPS and have shown for the first time that GSI treatment may modulate MPS independently of mTOR/AKT. An interesting yet puzzling finding from our lab's prior work was that 4EBP1 was the only downstream target of mTOR that was altered in a protein synthesis positive fashion with GSI treatment. This may actually support the idea that GSI-mediated elevations or rescue of MPS is through GSK3β. In fact, recent studies within cancer cell lines have implicated GSK3β directly in the phosphorylation of 4EBP1 [57,58]. It is also interesting that GSK3β has been implicated in regulating PTEN stabilization [59,60]. Based on our previous published work demonstrating alterations in the PTEN/AKT/mTOR cascade, it is plausible that GSI treatment's modulation of PTEN/AKT/mTOR is via GSK3β.

Though our present study reveals that GSIs may be able to rescue MPS when combined with mTOR and AKT inhibitors, this study is not without limitations. Though we have validated GSI treatment as a strategy to increase MPS in vitro, whether GSIs augment MPS in vivo requires further investigation. It should be noted that GSI administration did counteract muscle wasting in a setting of cachexia, but in this case MPS was not assessed [22]. Additionally, though GSIs are routinely used to target Notch signaling, they are not specific to Notch, and have several additional target substrates [61,62]. A large majority of γ-secretase substrates are transmembrane receptors, much like the Notch receptor family, and have known roles within skeletal muscle. For example, the insulin receptor, insulin-like growth factor 1 receptor, and growth hormone receptor, all of which partake in anabolic signaling, have been identified as γ-secretase substrates [62]. However, to our knowledge, the regulation of these receptors by GSIs have not been investigated specifically within skeletal muscle. Another γ-secretase substrate, low-density lipoprotein receptor-related protein 6 (LRP6), is a transmembrane receptor crucial to canonical Wnt signaling [62,63]. Crosstalk between Notch and Wnt have been widely discussed for proper skeletal muscle regeneration, while LRP6-mTOR signaling has been investigated in cardiomyocytes and hepatocytes [1,64,65]. Yet, whether LRP6-mTOR signaling occurs in skeletal muscle or whether LRP6 is responsible for changes in MPS is unknown. Recent work has also demonstrated that inhibition of the receptor for advanced glycation end products (RAGE), a γ-secretase substrate, can partially protect against age-associated muscle atrophy [66]. Further, a proteomics approach following GSI treatment in chick myogenic cells identified sonic hedgehog (Shh) as one of the most altered signaling pathways. Interestingly, separate studies have shown that Shh promotes myoblast proliferation, while also increasing myoblast fusion [67,68]. Yet, to our knowledge, the role of Shh on MPS has not been investigated. Thus, given the vast array of γ-secretase substrates, future investigations must delineate whether Notch is the specific GSI target that is responsible for alterations in MPS reported in the present study. Lastly, though we have demonstrated that administration of a GSI is able to promote pGSK3βSer9 in the presence of API-1, and thus are speculating that GSI-mediated rescue of MPS is via GSK3β, future work will be required to confirm this mechanism of action.

5. Conclusions

This study provided additional validation for the use of GSIs to promote or rescue MPS. We demonstrated that GSI treatment can rescue MPS independently of AKT and mTOR, possibly through regulation of GSK3β. These findings warrant further investigation on the role of GSIs in muscle-wasting conditions, in particular instances in which tumor-suppressing drugs (Rapamycin, API-1) are used, or where MPS rates are reduced, as GSIs may elevate MPS and help to sustain skeletal muscle mass.

Supplementary Materials: The following are available online at https://www.mdpi.com/article/10.3390/cells10071786/s1, Figure S1: GSI preserves myotube size in the presence of rapamycin, Figure S2: GSI preserves myotube size in the presence of API-1, Figure S3: GSI preserves myotube size in the presence of rapamycin and API-1.

Author Contributions: Conceptualization, J.R.H.; methodology, J.R.H.; software, J.R.H., B.T. and C.M.; formal analysis, J.R.H. and S.T.A.; investigation, J.R.H.; data curation, J.R.H., B.T. and C.M.; writing—original draft preparation, J.R.H.; writing—review and editing, J.R.H., J.S.M., S.T.A.; visualization, J.R.H.; supervision, S.T.A.; project administration, J.R.H.; funding acquisition, S.T.A. All authors have read and agreed to the published version of the manuscript.

Funding: This research received no external funding but was funded through a UNC Charlotte Faculty Research Grant (S.T.A.).

Institutional Review Board Statement: Not applicable.

Informed Consent Statement: Not applicable.

Data Availability Statement: Data sharing is not applicable to this article.

Conflicts of Interest: The authors declare no conflict of interest.

References

1. Arthur, S.T.; Cooley, I.D. The effect of physiological stimuli on sarcopenia; impact of Notch and Wnt signaling on impaired aged skeletal muscle repair. *Int. J. Biol. Sci.* **2012**, *8*, 731–760. [CrossRef] [PubMed]
2. Lexell, J. Human aging, muscle mass, and fiber type composition. *J. Gerontol. A Biol. Sci. Med. Sci.* **1995**, *50*, 11–16. [PubMed]
3. Morley, J.E.; Baumgartner, R.N.; Roubenoff, R.; Mayer, J.; Nair, K.S. Sarcopenia. *J. Lab. Clin. Med.* **2001**, *137*, 231–243. [CrossRef] [PubMed]
4. Huot, J.R.; Novinger, L.J.; Pin, F.; Narasimhan, A.; Zimmers, T.A.; O'Connell, T.M.; Bonetto, A. Formation of colorectal liver metastases induces musculoskeletal and metabolic abnormalities consistent with exacerbated cachexia. *JCI Insight* **2020**. [CrossRef] [PubMed]
5. Kim, H.G.; Huot, J.R.; Pin, F.; Guo, B.; Bonetto, A.; Nader, G.A. Reduced rDNA transcription diminishes skeletal muscle ribosomal capacity and protein synthesis in cancer cachexia. *FASEB J.* **2021**, *35*, e21335. [CrossRef] [PubMed]
6. Laplante, M.; Sabatini, D.M. mTOR signaling in growth control and disease. *Cell* **2012**, *149*, 274–293. [CrossRef]
7. Shimobayashi, M.; Hall, M.N. Making new contacts: The mTOR network in metabolism and signalling crosstalk. *Nat. Rev. Mol. Cell Biol.* **2014**, *15*, 155–162. [CrossRef]

8.	Antikainen, H.; Driscoll, M.; Haspel, G.; Dobrowolski, R. TOR-mediated regulation of metabolism in aging. *Aging Cell* **2017**, *16*, 1219–1233. [CrossRef]

9.	Harrison, D.E.; Strong, R.; Sharp, Z.D.; Nelson, J.F.; Astle, C.M.; Flurkey, K.; Nadon, N.L.; Wilkinson, J.E.; Frenkel, K.; Carter, C.S.; et al. Rapamycin fed late in life extends lifespan in genetically heterogeneous mice. *Nature* **2009**, *460*, 392–395. [CrossRef]

10.	Lamming, D.W.; Ye, L.; Katajisto, P.; Goncalves, M.D.; Saitoh, M.; Stevens, D.M.; Davis, J.G.; Salmon, A.B.; Richardson, A.; Ahima, R.S.; et al. Rapamycin-induced insulin resistance is mediated by mTORC2 loss and uncoupled from longevity. *Science* **2012**, *335*, 1638–1643. [CrossRef]

11.	Neff, F.; Flores-Dominguez, D.; Ryan, D.P.; Horsch, M.; Schroder, S.; Adler, T.; Afonso, L.C.; Aguilar-Pimentel, J.A.; Becker, L.; Garrett, L.; et al. Rapamycin extends murine lifespan but has limited effects on aging. *J. Clin. Investig.* **2013**, *123*, 3272–3291. [CrossRef] [PubMed]

12.	Wilkinson, J.E.; Burmeister, L.; Brooks, S.V.; Chan, C.C.; Friedline, S.; Harrison, D.E.; Hejtmancik, J.F.; Nadon, N.; Strong, R.; Wood, L.K.; et al. Rapamycin slows aging in mice. *Aging Cell* **2012**, *11*, 675–682. [CrossRef]

13.	Johnson, S.C.; Rabinovitch, P.S.; Kaeberlein, M. mTOR is a key modulator of ageing and age-related disease. *Nature* **2013**, *493*, 338–345. [CrossRef] [PubMed]

14.	Goodman, C.A.; Frey, J.W.; Mabrey, D.M.; Jacobs, B.L.; Lincoln, H.C.; You, J.S.; Hornberger, T.A. The role of skeletal muscle mTOR in the regulation of mechanical load-induced growth. *J. Physiol.* **2011**, *589*, 5485–5501. [CrossRef] [PubMed]

15.	Willett, M.; Cowan, J.L.; Vlasak, M.; Coldwell, M.J.; Morley, S.J. Inhibition of mammalian target of rapamycin (mTOR) signalling in C2C12 myoblasts prevents myogenic differentiation without affecting the hyperphosphorylation of 4E-BP1. *Cell Signal.* **2009**, *21*, 1504–1512. [CrossRef]

16.	Brack, A.S.; Conboy, I.M.; Conboy, M.J.; Shen, J.; Rando, T.A. A temporal switch from notch to Wnt signaling in muscle stem cells is necessary for normal adult myogenesis. *Cell Stem Cell* **2008**, *2*, 50–59. [CrossRef]

17.	Conboy, I.M.; Conboy, M.J.; Smythe, G.M.; Rando, T.A. Notch-mediated restoration of regenerative potential to aged muscle. *Science* **2003**, *302*, 1575–1577. [CrossRef]

18.	Carey, K.A.; Farnfield, M.M.; Tarquinio, S.D.; Cameron-Smith, D. Impaired expression of Notch signaling genes in aged human skeletal muscle. *J. Gerontol. A Biol. Sci. Med. Sci.* **2007**, *62*, 9–17. [CrossRef]

19.	D'Souza, D.M.; Zhou, S.; Rebalka, I.A.; MacDonald, B.; Moradi, J.; Krause, M.P.; Al-Sajee, D.; Punthakee, Z.; Tarnopolsky, M.A.; Hawke, T.J. Decreased Satellite Cell Number and Function in Humans and Mice With Type 1 Diabetes Is the Result of Altered Notch Signaling. *Diabetes* **2016**, *65*, 3053–3061. [CrossRef]

20.	Jiang, C.; Wen, Y.; Kuroda, K.; Hannon, K.; Rudnicki, M.A.; Kuang, S. Notch signaling deficiency underlies age-dependent depletion of satellite cells in muscular dystrophy. *Dis. Models Mech.* **2014**, *7*, 997–1004. [CrossRef]

21.	Bi, P.; Kuang, S. Notch signaling as a novel regulator of metabolism. *Trends Endocrinol. Metab.* **2015**, *26*, 248–255. [CrossRef]

22.	Mu, X.; Agarwal, R.; March, D.; Rothenberg, A.; Voigt, C.; Tebbets, J.; Huard, J.; Weiss, K. Notch Signaling Mediates Skeletal Muscle Atrophy in Cancer Cachexia Caused by Osteosarcoma. *Sarcoma* **2016**, *2016*, 3758162. [CrossRef]

23.	Pajvani, U.B.; Shawber, C.J.; Samuel, V.T.; Birkenfeld, A.L.; Shulman, G.I.; Kitajewski, J.; Accili, D. Inhibition of Notch signaling ameliorates insulin resistance in a FoxO1-dependent manner. *Nat. Med.* **2011**, *17*, 961–967. [CrossRef]

24.	Arya, M.A.; Tai, A.K.; Wooten, E.C.; Parkin, C.D.; Kudryavtseva, E.; Huggins, G.S. Notch pathway activation contributes to inhibition of C2C12 myoblast differentiation by ethanol. *PLoS ONE* **2013**, *8*, e71632. [CrossRef] [PubMed]

25.	Huot, J.R.; Marino, J.S.; Turner, M.J.; Arthur, S.T. Notch Inhibition via GSI Treatment Elevates Protein Synthesis in C2C12 Myotubes. *Biology* **2020**, *9*, 115. [CrossRef]

26.	Cross, D.A.; Alessi, D.R.; Cohen, P.; Andjelkovich, M.; Hemmings, B.A. Inhibition of glycogen synthase kinase-3 by insulin mediated by protein kinase B. *Nature* **1995**, *378*, 785–789. [CrossRef] [PubMed]

27.	Glass, D.J. Signalling pathways that mediate skeletal muscle hypertrophy and atrophy. *Nat. Cell Biol.* **2003**, *5*, 87–90. [CrossRef]

28.	Rommel, C.; Bodine, S.C.; Clarke, B.A.; Rossman, R.; Nunez, L.; Stitt, T.N.; Yancopoulos, G.D.; Glass, D.J. Mediation of IGF-1-induced skeletal myotube hypertrophy by PI(3)K/Akt/mTOR and PI(3)K/Akt/GSK3 pathways. *Nat. Cell Biol.* **2001**, *3*, 1009–1013. [CrossRef] [PubMed]

29.	Li, B.; Ren, H.; Yue, P.; Chen, M.; Khuri, F.R.; Sun, S.Y. The novel Akt inhibitor API-1 induces c-FLIP degradation and synergizes with TRAIL to augment apoptosis independent of Akt inhibition. *Cancer Prev. Res.* **2012**, *5*, 612–620. [CrossRef]

30.	Yoon, M.S.; Chen, J. Distinct amino acid-sensing mTOR pathways regulate skeletal myogenesis. *Mol. Biol. Cell* **2013**, *24*, 3754–3763. [CrossRef]

31.	Henning, R.J.; Sanberg, P.; Jimenez, E. Human cord blood stem cell paracrine factors activate the survival protein kinase Akt and inhibit death protein kinases JNK and p38 in injured cardiomyocytes. *Cytotherapy* **2014**, *16*, 1158–1168. [CrossRef] [PubMed]

32.	Huot, J.R.; Pin, F.; Essex, A.L.; Bonetto, A. MC38 Tumors Induce Musculoskeletal Defects in Colorectal Cancer. *Int. J. Mol. Sci.* **2021**, *22*, 1486. [CrossRef] [PubMed]

33.	Agley, C.C.; Velloso, C.P.; Lazarus, N.R.; Harridge, S.D. An image analysis method for the precise selection and quantitation of fluorescently labeled cellular constituents: Application to the measurement of human muscle cells in culture. *J. Histochem. Cytochem.* **2012**, *60*, 428–438. [CrossRef]

34.	Schneider, C.A.; Rasband, W.S.; Eliceiri, K.W. NIH Image to ImageJ: 25 years of image analysis. *Nat. Methods* **2012**, *9*, 671–675. [CrossRef]

35. Goodman, C.A.; Hornberger, T.A. Measuring protein synthesis with SUnSET: A valid alternative to traditional techniques? *Exerc. Sport Sci. Rev.* **2013**, *41*, 107–115. [CrossRef]

36. Goodman, C.A.; Mabrey, D.M.; Frey, J.W.; Miu, M.H.; Schmidt, E.K.; Pierre, P.; Hornberger, T.A. Novel insights into the regulation of skeletal muscle protein synthesis as revealed by a new nonradioactive in vivo technique. *FASEB J.* **2011**, *25*, 1028–1039. [CrossRef]

37. Ge, Y.; Chen, J. Mammalian target of rapamycin (mTOR) signaling network in skeletal myogenesis. *J. Biol. Chem.* **2012**, *287*, 43928–43935. [CrossRef]

38. Jansen, K.M.; Pavlath, G.K. Molecular control of mammalian myoblast fusion. *Methods Mol. Biol.* **2008**, *475*, 115–133. [CrossRef]

39. Park, I.H.; Chen, J. Mammalian target of rapamycin (mTOR) signaling is required for a late-stage fusion process during skeletal myotube maturation. *J. Biol. Chem.* **2005**, *280*, 32009–32017. [CrossRef] [PubMed]

40. Jefferson, L.S.; Fabian, J.R.; Kimball, S.R. Glycogen synthase kinase-3 is the predominant insulin-regulated eukaryotic initiation factor 2B kinase in skeletal muscle. *Int. J. Biochem. Cell Biol.* **1999**, *31*, 191–200. [CrossRef]

41. Wilson, E.M.; Rotwein, P. Selective control of skeletal muscle differentiation by Akt1. *J. Biol. Chem.* **2007**, *282*, 5106–5110. [CrossRef] [PubMed]

42. Jin, H.; Sanberg, P.R.; Henning, R.J. Human umbilical cord blood mononuclear cell-conditioned media inhibits hypoxic-induced apoptosis in human coronary artery endothelial cells and cardiac myocytes by activation of the survival protein Akt. *Cell Transplant.* **2013**, *22*, 1637–1650. [CrossRef] [PubMed]

43. Karaboga Arslan, A.K.; Yerer, M.B. alpha-Chaconine and alpha-Solanine Inhibit RL95-2 Endometrium Cancer Cell Proliferation by Reducing Expression of Akt (Ser473) and ERalpha (Ser167). *Nutrients* **2018**, *10*, 672. [CrossRef] [PubMed]

44. Saglam, A.S.; Alp, E.; Elmazoglu, Z.; Menevse, E.S. Effect of API-1 and FR180204 on cell proliferation and apoptosis in human DLD-1 and LoVo colorectal cancer cells. *Oncol. Lett.* **2016**, *12*, 2463–2474. [CrossRef]

45. Mirzoev, T.M.; Tyganov, S.A.; Shenkman, B.S. Akt-dependent and Akt-independent pathways are involved in protein synthesis activation during reloading of disused soleus muscle. *Muscle Nerve* **2017**, *55*, 393–399. [CrossRef] [PubMed]

46. Brunet, A.; Bonni, A.; Zigmond, M.J.; Lin, M.Z.; Juo, P.; Hu, L.S.; Anderson, M.J.; Arden, K.C.; Blenis, J.; Greenberg, M.E. Akt promotes cell survival by phosphorylating and inhibiting a Forkhead transcription factor. *Cell* **1999**, *96*, 857–868. [CrossRef]

47. Waldemer-Streyer, R.J.; Chen, J. Myocyte-derived Tnfsf14 is a survival factor necessary for myoblast differentiation and skeletal muscle regeneration. *Cell Death Dis.* **2015**, *6*, e2026. [CrossRef] [PubMed]

48. Ren, H.; Koo, J.; Guan, B.; Yue, P.; Deng, X.; Chen, M.; Khuri, F.R.; Sun, S.Y. The E3 ubiquitin ligases beta-TrCP and FBXW7 cooperatively mediates GSK3-dependent Mcl-1 degradation induced by the Akt inhibitor API-1, resulting in apoptosis. *Mol. Cancer* **2013**, *12*, 146. [CrossRef]

49. Gordon, B.S.; Kelleher, A.R.; Kimball, S.R. Regulation of muscle protein synthesis and the effects of catabolic states. *Int. J. Biochem. Cell Biol.* **2013**, *45*, 2147–2157. [CrossRef]

50. Espinosa, L.; Ingles-Esteve, J.; Aguilera, C.; Bigas, A. Phosphorylation by glycogen synthase kinase-3 beta down-regulates Notch activity, a link for Notch and Wnt pathways. *J. Biol. Chem.* **2003**, *278*, 32227–32235. [CrossRef]

51. Foltz, D.R.; Santiago, M.C.; Berechid, B.E.; Nye, J.S. Glycogen synthase kinase-3beta modulates notch signaling and stability. *Curr. Biol.* **2002**, *12*, 1006–1011. [CrossRef]

52. Guha, S.; Cullen, J.P.; Morrow, D.; Colombo, A.; Lally, C.; Walls, D.; Redmond, E.M.; Cahill, P.A. Glycogen synthase kinase 3 beta positively regulates Notch signaling in vascular smooth muscle cells: Role in cell proliferation and survival. *Basic Res. Cardiol.* **2011**, *106*, 773–785. [CrossRef]

53. Mayhew, D.L.; Hornberger, T.A.; Lincoln, H.C.; Bamman, M.M. Eukaryotic initiation factor 2B epsilon induces cap-dependent translation and skeletal muscle hypertrophy. *J. Physiol.* **2011**, *589*, 3023–3037. [CrossRef]

54. Der Vartanian, A.; Audfray, A.; Al Jaam, B.; Janot, M.; Legardinier, S.; Maftah, A.; Germot, A. Protein O-fucosyltransferase 1 expression impacts myogenic C2C12 cell commitment via the Notch signaling pathway. *Mol. Cell. Biol.* **2015**, *35*, 391–405. [CrossRef]

55. Al Jaam, B.; Heu, K.; Pennarubia, F.; Segelle, A.; Magnol, L.; Germot, A.; Legardinier, S.; Blanquet, V.; Maftah, A. Reduced Notch signalling leads to postnatal skeletal muscle hypertrophy in Pofut1cax/cax mice. *Open Biol.* **2016**, *6*. [CrossRef]

56. Pelisse, M.; Der Vartanian, A.; Germot, A.; Maftah, A. Protein O-Glucosyltransferase 1 Expression Influences Formation of Differentiated Myotubes in C2C12 Cell Line. *DNA Cell Biol.* **2018**, *37*, 359–372. [CrossRef]

57. Ito, H.; Ichiyanagi, O.; Naito, S.; Bilim, V.N.; Tomita, Y.; Kato, T.; Nagaoka, A.; Tsuchiya, N. GSK-3 directly regulates phospho-4EBP1 in renal cell carcinoma cell-line: An intrinsic subcellular mechanism for resistance to mTORC1 inhibition. *BMC Cancer* **2016**, *16*, 393. [CrossRef]

58. Shin, S.; Wolgamott, L.; Tcherkezian, J.; Vallabhapurapu, S.; Yu, Y.; Roux, P.P.; Yoon, S.O. Glycogen synthase kinase-3beta positively regulates protein synthesis and cell proliferation through the regulation of translation initiation factor 4E-binding protein 1. *Oncogene* **2014**, *33*, 1690–1699. [CrossRef]

59. Al-Khouri, A.M.; Ma, Y.; Togo, S.H.; Williams, S.; Mustelin, T. Cooperative phosphorylation of the tumor suppressor phosphatase and tensin homologue (PTEN) by casein kinases and glycogen synthase kinase 3beta. *J. Biol. Chem.* **2005**, *280*, 35195–35202. [CrossRef] [PubMed]

60. Maccario, H.; Perera, N.M.; Davidson, L.; Downes, C.P.; Leslie, N.R. PTEN is destabilized by phosphorylation on Thr366. *Biochem. J.* **2007**, *405*, 439–444. [CrossRef] [PubMed]

61. Rosa de Andrade, I.; Correa, S.; Fontenele, M.; de Oliveira Teixeira, J.D.; Abdelhay, E.; Costa, M.L.; Mermelstein, C. gamma-Secretase Inhibition Induces Muscle Hypertrophy in a Notch-Independent Mechanism. *Proteomics* **2018**, *18*. [CrossRef]
62. Haapasalo, A.; Kovacs, D.M. The many substrates of presenilin/gamma-secretase. *J. Alzheimers Dis.* **2011**, *25*, 3–28. [CrossRef]
63. Ren, Q.; Chen, J.; Liu, Y. LRP5 and LRP6 in Wnt Signaling: Similarity and Divergence. *Front. Cell Dev. Biol.* **2021**, *9*, 670960. [CrossRef]
64. Chen, Z.; Li, Y.; Jiang, G.; Yang, C.; Wang, Y.; Wang, X.; Fang, B.; Zhang, G.; Sun, Y.; Qian, J.; et al. Knockdown of LRP6 activates Drp1 to inhibit survival of cardiomyocytes during glucose deprivation. *Biomed. Pharmacother.* **2018**, *103*, 1408–1414. [CrossRef]
65. Li, L.; Xue, J.; Wan, J.; Zhou, Q.; Wang, S.; Zhou, Y.; Zhao, H.; Wang, X. LRP6 Knockdown Ameliorates Insulin Resistance via Modulation of Autophagy by Regulating GSK3beta Signaling in Human LO2 Hepatocytes. *Front. Endocrinol.* **2019**, *10*, 73. [CrossRef]
66. Davis, H.M.; Essex, A.L.; Valdez, S.; Deosthale, P.J.; Aref, M.W.; Allen, M.R.; Bonetto, A.; Plotkin, L.I. Short-term pharmacologic RAGE inhibition differentially affects bone and skeletal muscle in middle-aged mice. *Bone* **2019**, *124*, 89–102. [CrossRef] [PubMed]
67. Ma, L.; Li, C.; Lian, S.; Xu, B.; Yuan, J.; Lu, J.; Yang, H.; Guo, J.; Ji, H. ActivinA activates Notch1-Shh signaling to regulate proliferation in C2C12 skeletal muscle cells. *Mol. Cell. Endocrinol.* **2021**, *519*, 111055. [CrossRef] [PubMed]
68. Teixeira, J.D.; de Andrade Rosa, I.; Brito, J.; Maia de Souza, Y.R.; Paulo de Abreu Manso, P.; Machado, M.P.; Costa, M.L.; Mermelstein, C. Sonic Hedgehog signaling and Gli-1 during embryonic chick myogenesis. *Biochem. Biophys. Res. Commun.* **2018**, *507*, 496–502. [CrossRef] [PubMed]

Article

Dihydroartemisinin Inhibits mTORC1 Signaling by Activating the AMPK Pathway in Rhabdomyosarcoma Tumor Cells

Jun Luo [1,2,†], Yoshinobu Odaka [1,†,‡], Zhu Huang [1,3,†], Bing Cheng [1], Wang Liu [1], Lin Li [1], Chaowei Shang [1], Chao Zhang [1,4,5], Yang Wu [1,6], Yan Luo [1,6], Shengyong Yang [6], Peter J. Houghton [7], Xiaofeng Guo [2,*] and Shile Huang [1,8,*]

[1]   Department of Biochemistry and Molecular Biology, Louisiana State University Health Sciences Center, Shreveport, LA 71130-3932, USA; junluo@scau.edu.cn (J.L.); odakayu@ucmail.uc.edu (Y.O.); huangzhu@xmu.edu.cn (Z.H.); bing.cheng@lsuhs.edu (B.C.); wliu6@kumc.edu (W.L.); lin.li@lsuhs.edu (L.L.); chaowei.shang@lsuhs.edu (C.S.); zhangchao@ahmu.edu.cn (C.Z.); wuyang@scu.edu.cn (Y.W.); yan.luo@mayo.edu (Y.L.)
[2]   College of Veterinary Medicine, South China Agricultural University, Guangzhou 510642, China
[3]   Research Center of Aquatic Organism Conservation and Water Ecosystem Restoration in Anhui Province, Anqing Normal University, Anqing 246011, China
[4]   Key Laboratory of National Health and Family Planning Commission on Parasitic Disease Control and Prevention, Jiangsu Institute of Parasitic Diseases, Wuxi 214064, China
[5]   Jiangsu Provincial Key Laboratory on Parasite and Vector Control Technology, Jiangsu Institute of Parasitic Diseases, Wuxi 214064, China
[6]   State Key Laboratory of Biotherapy and Cancer Center, West China Hospital, Sichuan University, Chengdu 610041, China; yangsy@scu.edu.cn
[7]   Greehey Children's Cancer Research Institute, University of Texas Health Science Center, San Antonio, TX 78229-3000, USA; houghtonp@uthscsa.edu
[8]   Department of Hematology and Oncology, Louisiana State University Health Sciences Center, Shreveport, LA 71130-3932, USA
*   Correspondence: xfguo@scau.edu.cn (X.G.); shile.huang@lsuhs.edu (S.H.); Tel.: +86-20-38295980 (X.G.); +1-318-675-7759 (S.H.)
†   These authors contributed equally to this work.
‡   Current address: Biology Department, University of Cincinnati Blue Ash College, Blue Ash, OH 45236, USA.

Citation: Luo, J.; Odaka, Y.; Huang, Z.; Cheng, B.; Liu, W.; Li, L.; Shang, C.; Zhang, C.; Wu, Y.; Luo, Y.; et al. Dihydroartemisinin Inhibits mTORC1 Signaling by Activating the AMPK Pathway in Rhabdomyosarcoma Tumor Cells. *Cells* **2021**, *10*, 1363. https://doi.org/10.3390/cells10061363

Academic Editors: Stephen Yarwood and Alexander E. Kalyuzhny

Received: 19 March 2021
Accepted: 29 May 2021
Published: 1 June 2021

Publisher's Note: MDPI stays neutral with regard to jurisdictional claims in published maps and institutional affiliations.

Abstract: Dihydroartemisinin (DHA), an anti-malarial drug, has been shown to possess potent anti-cancer activity, partly by inhibiting the mammalian target of rapamycin (mTOR) complex 1 (mTORC1) signaling. However, how DHA inhibits mTORC1 is still unknown. Here, using rhabdomyosarcoma (RMS) as a model, we found that DHA reduced cell proliferation and viability in RMS cells, but not those in normal cells, which was associated with inhibition of mTORC1. Mechanistically, DHA did not bind to mTOR or FK506 binding protein 12 (FKBP12). In addition, DHA neither inhibited insulin-like growth factor-1 receptor (IGF-1R), phosphoinositide 3-kinase (PI3K), and extracellular signal-regulated kinase $\frac{1}{2}$ (Erk1/2), nor activated phosphatase and tensin homolog (PTEN) in the cells. Rather, DHA activated AMP-activated protein kinase (AMPK). Pharmacological inhibition of AMPK, ectopic expression dominant negative or kinase-dead AMPK, or knockdown of AMPKα attenuated the inhibitory effect of DHA on mTORC1 in the cells. Additionally, DHA was able to induce dissociation of regulatory-associated protein of mTOR (raptor) from mTOR and inhibit mTORC1 activity. Moreover, treatment with artesunate, a prodrug of DHA, dose-dependently inhibited tumor growth and concurrently activated AMPK and suppressed mTORC1 in RMS xenografts. The results indicated that DHA inhibits mTORC1 by activating AMPK in tumor cells. Our finding supports that DHA or artesunate has a great potential to be repositioned for treatment of RMS.

Keywords: dihydroartemisinin; rhabdomyosarcoma; mTOR; AMPK; PTEN; raptor

1. Introduction

Rhabdomyosarcoma (RMS) is the most common soft-tissue sarcoma, which often occurs in the head, neck, bladder, vagina, uterus, arms, legs, and trunk [1]. Approximately

80% of RMS patients are younger than 15 years old [1]. Histologically, RMS manifests in two major types, embryonal (ERMS) and alveolar (ARMS) [2]. Morphologically, the embryonic type resembles to the embryonic muscle cell precursor, whereas the alveolar type has clusters of round cells similar to lung alveoli [1,2]. Approximately 80% of ARMS tumors are characterized with the translocations or expression of the *PAX3/7–FOXO1* fusion transcript, resulting in overexpression of receptor tyrosine kinases such as fibroblast growth factor receptor 4 (FGFR4), hepatocyte growth factor receptor (HGFR, also named MET), and insulin-like growth factor 1 receptor (IGF-1R) [1,2]. In addition, insulin-like growth factor 2 (IGF-2) is upregulated by *PAX3-FOXO1* in RMS, activating the IGF-1R pathway [2]. Hence, the mammalian target of rapamycin (mTOR) pathway is frequently and constitutively activated in ARMS tumors, which have higher propensity for metastasis [1,2].

RMS is generally treated with surgery, radiation therapy, and chemotherapy [1,3]. The 5-year survival rate for children having low-to-intermediate-risk RMS ranges from 50% to 90%, while for high-risk patients (having metastatic or recurrent disease), the 5-year survival rate is less than 30% [1,2]. The standard chemotherapy regimen for RMS is the combination of vincristine, actinomycin D, and cyclophosphamide [1,3]. However, these chemotherapeutic treatments for children have long-term side effects, such as secondary cancers and infertility [1,3]. In recent clinical trials, targeted therapies and immunotherapies have shown improvements in the outcomes in patients with RMS, but the clinical benefit is still limited [4]. Therefore, there is a great need to develop novel systemic treatments, which have better efficacy with long-term safety, for RMS patients.

mTOR is a central controller for protein synthesis, cell growth, proliferation, and survival [5,6]. The dysregulation of the mTOR pathway correlates to tumor development and progression, so mTOR has become a hot target for cancer therapy [5,6]. mTOR functions as two complexes (mTORC1 and mTORC2) in mammalian cells [5,6]. mTORC1 consists of mTOR, mLST8 (mammalian lethal with sec-13 protein 8), raptor (regulatory-associated protein of mTOR), PRAS40 (proline-rich Akt substrate 40), and DEPTOR, whereas mTORC2 is composed of mTOR, mLST8, mSin1, rictor (rapamycin-insensitive companion of mTOR), mSin1 (mammalian stress-activated protein kinase-interacting protein 1), protor (protein observed with rictor), and DEPTOR [5,6]. Of note, raptor is essential for the assembly of mTORC1 and for recruiting mTOR substrates [7]. mTORC1 senses growth factors, nutrients (amino acids), energy, oxygen, and DNA damage, while mTORC2 primarily senses growth factors [5,6]. Both mTORC1 and mTORC2 can be positively regulated by the IGF-IR-phosphatidylinositol-3 kinase (PI3K), which is antagonized by phosphatase and tensin homolog (PTEN) [5,6]. mTORC1 can also be positively regulated by the Ras-Raf-MEK-Erk pathway [5,6]. In addition, mTORC1 is negatively regulated by AMP-activated protein kinase (AMPK) [8]. In response to low energy levels, AMPK is activated, which can phosphorylate tuberous sclerosis complex 2 (TSC2) at multiple sites (including S1387), promoting the formation and activation of TSCs [8], which antagonizes Rheb (Ras homolog enriched in the brain) by hydrolyzing GTP-Rheb to GDP-Rheb, thereby inhibiting Rheb-mediated mTORC1 [5,6]. Besides, activated AMPK can also phosphorylate raptor (S792), resulting in the inhibition of mTORC1 [9]. While p70 S6 kinase 1 (S6K1) and eukaryotic initiation factor 4E (eIF4E) binding protein 1 (4E-BP1) are two well-known substrates of mTORC1, Akt (S473) is the best characterized substrate of mTORC2 [5,6].

Rapamycin and its analogs (e.g., temsirolimus and everolimus) (termed rapalogs) were developed as the first generation of mTOR inhibitors [5,6]. Mechanistically, rapalogs do not impair mTOR's kinase activity per se but first form a complex with the FK506 binding protein 12 (FKBP12) and then bind the FKBP12-rapamycin-binding (FRB) domain of mTOR, disrupting mTORC1 assembly and thus inhibiting mTORC1 [5,6]. However, rapalogs alone lack efficacy in treating most types of cancer, including RMS [5,6,10,11]. This is possibly due to the fact that the phosphorylation of 4E-BP1 (cap-dependent translation) is largely insensitive to rapalogs, and Akt (pro-survival) can be activated by rapalogs via a negative feedback mechanism [5,12–14]. Recently, mTOR kinase inhibitors (e.g., AZD8055 and INK128), called the second generation of mTOR inhibitors, have emerged, which compete with ATP in the catalytic site of mTOR and inhibit both mTORC1 and mTORC2 [5,6]. However, prolonged treatment with these inhibitors can also result in re-activation of Akt [15], highlighting resistance as a key problem that must be tackled by the new generation of mTOR inhibitors [5].

Dihydroartemisinin (DHA) is a derivative of artemisinin originally isolated from the plant *Artemisia annua* [16]. DHA is also the active metabolite of artemisinins, such as artemisinin, artesunate, and artemether [16,17]. Artemisinins have been widely used to treat malaria in children and adults showing high efficacy and safety [16–19]. Increasing evidence has demonstrated that artemisinins also possess potent anticancer effects on diverse tumor cell lines [19]. Artesunate, a water-soluble artemisinin derivative, has been in clinical trials for treatments of lung, cervical, breast, and colon cancers [19]. Multiple anticancer action modes of artemisinins have been described, including the induction of cell cycle arrest, apoptosis, autophagy, as well as the inhibition of cell invasion/motility and angiogenesis [19]. Correspondingly, DHA has been shown to alter the expression/activity of a variety of signaling molecules, such as MYC, cyclin-dependent kinases (CDKs), vascular endothelial growth factor receptor (VEGF), focal adhesion kinase (FAK), and hypoxia-inducible factor 1-alpha (HIF-1α) [19]. Recently, we and others have demonstrated that DHA inhibits mTOR [20–34]. Since many of those molecules (e.g., MYC, CDKs, VEGF, FAK, and HIF-1α) targeted by DHA are also directly or indirectly regulated by mTORC1 [5,6], we proposed that mTORC1 may be a major target of DHA for its anticancer activity, and DHA is a new inhibitor of mTORC1.

To the best of our knowledge, no study has determined how DHA inhibits mTORC1. In this study, we evaluated the anticancer activity of DHA in RMS cells in cell culture and in xenografts in mice. Using RMS as a model, we focused on determining the molecular mechanism by which DHA inhibits mTORC1 in tumor cells.

2. Materials and Methods

2.1. Materials

DHA (purity: >98% by HPLC; TCI America, Portland, OR, USA) was dissolved in dimethyl sulfoxide (DMSO) to prepare a stock solution (10 mM), aliquoted and stored at $-20\,^{\circ}$C. [10$-^{3}$H]-dihydroartemisinin (specificity activity: 2.5 Ci/mmol; concentration: 1.0 mCi/mL in a hexane:ethanol (*v:v*, 7:3) solution; radiochemical purity: 98.5%) was obtained from Moravek Biochemical (Brea, CA, USA). Compound C (EMD Millipore, Burlington, MA, USA) was dissolved in DMSO to prepare a 10 mM stock solution and stored at $-20\,^{\circ}$C. RPMI 1640, Dulbecco's modified Eagle's medium (DMEM) (high glucose), DMEM/F12, 0.05% trypsin-EDTA, and Matrigel membrane matrix were obtained from Corning (Corning, NY, USA), and fetal bovine serum (FBS) was from R&D Systems (Minneapolis, MN, USA). For Western blotting or immunoprecipitation, the following antibodies were used: Erk2, c-Jun, p-c-Jun (Ser63), HIF-1α, REDD1, IGF-1Rβ, p-IGF-1Rβ (Tyr1161), mTOR, S6K1, S6, PI3K, Akt, β-actin, c-Myc, GAPDH (Santa Cruz Biotechnology, Dallas, TX, USA), p-Erk1/2 (Thr202/Tyr204), p38, p-p38 (Thr180/Tyr182), p-S6K1 (Thr389), p-AMPKα (Thr172), AMPKα, p-ACC (Ser79), ACC, p-S6 (Ser235/236), 4E-BP1, p-4E-BP1 (Thr37/46), p-4E-BP1 (Thr70), PDK1, p-PDK1 (Ser241), PTEN, p-PTEN (Ser380/Thr382/383), p-PI3K p85 (Tyr458)/p55 (Tyr199), p-Akt (Ser473), mLST8 (GβL) (Cell Signaling Technology, Danvers, MA, USA), raptor, rictor (Bethyl Laboratories, Montgomery, TX, USA), β-tubulin (Sigma-Aldrich, St. Louis, MO, USA), goat anti-mouse IgG-horseradish peroxidase, and goat anti-rabbit IgG-horseradish peroxidase (Pierce, Rockford, IL, USA). All other chemicals were obtained from Sigma-Aldrich (St. Louis, MO, USA) unless specified elsewhere.

2.2. Cell Lines and Culture

Human RMS (Rh30, RD, Rh18, Rh28, Rh36, and Rh41) and Ewing sarcoma cells (Rh1, also named EW8) which were gifts from Dr. Peter J. Houghton, University of Texas Health Science Center, San Antonio, TX, USA were grown in RMPI 1640 supplemented with 10% FBS. Human primary skeletal muscle cells (#PCS-950-010, American Type Culture

Collection (ATCC), Manassas, VA, USA) were cultured in a Mesenchymal Stem Cell Basal Medium (#PCS-500-030, ATCC) supplemented with Primary Skeletal Muscle Cell Growth Kit (PCS-950-040, ATCC), while human dermal primary fibroblasts (#PCS-201-012, ATCC) were grown in a Fibroblast Basal Medium supplemented with Fibroblast Growth Kit-Low Serum (#PCS-201-041, ATCC); both of them were used within 6 passages. Mouse muscle myoblasts (C2C12 and ATCC), raptor, rictor-inducible knockout (KO) mouse embryonic fibroblasts (MEFs, SV40 large T-antigen-immortalized and expressing the Cre/LoxP system) (gifts from Dr. Michael Hall, University of Basel, Switzerland), and 293A cells (Invitrogen, Calsbad, CA, USA) were cultured in DMEM supplemented with 10% FBS. To induce the KO of raptor or rictor, the MEFs were treated with 1 μM 4-hydroxytamoxifen (Sigma-Aldrich) for 3 days [35]. All cell lines were cultured in a humid incubator (37 °C and 5% CO_2) and trypsinized with a 0.05% trypsin–EDTA solution for subculture or experiments.

2.3. Cell Proliferation and Viability Assays

Cell proliferation and viability were evaluated by cell counting and MTS assay, as described previously [36]. Treatment with DMSO (vehicle) served as a control.

2.4. [³H]-DHA Labeling In Vivo

Rh30 cells were seeded in 100 mm culture dishes (3×10^6 cells/dish) for culture. The next day, the cells were labeled with 10 μCi [³H]-DHA for 21 h. Subsequently, the cells were briefly washed with PBS and lysed in an ice-cold CHAPS lysis buffer (40 mM HEPES, pH 7.4, 120 mM NaCl, 1 mM EDTA, 10 mM pyrophosphate, 10 mM glycerophosphate, 50 mM NaF, 1.5 mM Na_3VO_4, 0.3% (w/v) CHAPS, and a cocktail of protease inhibitors (dilution, 1:1000; Sigma-Aldrich). The cell lysates were sonicated for 20 s and centrifuged at 13,000 rpm and at 4 °C for 3 min. The supernatants were transferred to fresh Eppendorf tubes. The protein concentration in the supernatants was determined using a BCA kit (Pierce). Supernatants with an equal amount (700 μg) of crude protein were incubated with 30 μL of protein A/G agarose beads and 3 μg of antibodies to goat anti-mTOR antibody or normal goat IgG on a rotator overnight at 4 °C. The agarose beads were collected by centrifugation at 3500 rpm and at 4 °C for 3 min and washed once with 1 mL of an ice-cold CHAPS buffer and three additional washes with ice-cold PBS. The relative radioactivity (cpm) of immunoprecipitated products was measured on a Beckman LS6500 scintillation counter (Beckman Coulter, Fullerton, CA, USA).

2.5. Recombinant Adenoviruses, Lentiviral shRNAs, and Infection of Cells

Recombinant adenoviruses expressing green fluorescent protein (GFP) and MYC-tagged dominant negative (DN) AMPKα1 (D157A) (Ad-AMPK-DN) were described previously [37]. Recombinant adenovirus expressing MYC-tagged kinase-dead AMPKα2 (K45R) (Ad-AMPK-KD) [38] was a gift from Dr. Nicholas J. G. Webster (University of California, San Diego, CA). For experiments, the cells were infected with and individual adenovirus at a multiplicity of infection (MOI) of 5 for 24 h. Subsequently, the infected cells were used for experiments. Cells infected with Ad-GFP served as control. The expression of MYC-tagged AMPK-DN or AMPK-KD was determined by Western blotting with antibodies to c-Myc.

Lentiviral shRNAs to human raptor, rictor, AMPKα1, and GFP were described previously [39,40]. For use, monolayer cells, when grown to about 70% confluence, were infected with an individual lentivirus in the presence of 8 μg/mL polybrene for 12 h twice at an interval of 6 h. Uninfected cells were eliminated by exposure to 2 μg/mL puromycin for 48 h before use.

2.6. Western Blotting

Western blotting was performed as described [21].

2.7. Co-Immunoprecipitation of mTOR and In Vitro mTOR Kinase Assay

Rh30 cells were seeded in 100 mm culture dishes (3×10^6 cells/dish) and grown overnight. The cells were then treated with DHA (0–30 μM) for 24 h. After aspirating the used medium, the cells were briefly washed with PBS and lysed in an ice-cold CHAPS lysis buffer, followed by immunoprecipitation with goat anti-mTOR antibody or normal goat IgG (as a control). Finally, to detect the interaction of mTOR with raptor, rictor, and mLST8, the immunoprecipitants were subjected to immunoblotting with antibodies to mTOR, raptor, rictor, and mLST8, as described [41]. To detect the mTORC1 activity, the above immunoprecipitants were utilized for the in vitro mTOR kinase assay, as described [41].

2.8. Molecular Docking

The molecular docking studies were performed using Genetic Optimization of Ligand Docking (GOLD) 5.0 and LibDock [42,43]. The 3-D structures of mTOR and the protein complex of FKBP12 and the FRB domain of mTOR were taken from the PDB database with the PDB entries being 4JT5 and 3FAP, respectively [44,45], while the 3-D structure of the protein complex of AMPK and A-769662 was taken from the PDB database (PDB entry: 4CFF) [46]. Discovery Studio 3.1 (Accelrys, San Diego, CA, USA) software package was used to prepare the protein structures including adding hydrogen atoms to the protein, removing water molecules, and assigning force fields (here the CHARMM force field was adopted). The binding site was defined as a sphere containing residues that remained within 9 Å (for mTOR) or 10 Å (for AMPK) of the ligand, an area large enough to cover the ligand-binding region at the domain of proteins. The binding affinity was estimated using LibDock score and/or GOLD score.

2.9. Study in Rhabdomyosarcoma Xenografts

To study the inhibitory effect of DHA on tumor growth in vivo, artesunate (ART), a pro-drug of DHA, was used. CB17SC scid$^{-/-}$ female mice (Taconic Farms, Germantown, NY) were maintained under barrier conditions, and experiments were conducted using the protocols approved by the institutional animal care and use committee (ethical code number: LSUHSC-S #P20-003; the date of approval: 30 August 2019). Human Rh30 cells (6×10^6 cells resuspended in 100 μL of a 1:1 (*v:v*) solution of serum-free DMEM/matrigel) were injected into the right flank of each mouse. Seven days later, the animals were randomized into 5 groups (10 mice/group). Then, the mice were intraperitoneally (i.p.) injected once daily with artesunate (dissolved in 5% Na_2CO_3 and diluted in 0.9% NaCl) at doses of 25, 50, 100, or 150 mg/kg body weight) or with vehicle control, as described [47]. Tumor volume (calculated as: length $\times$ width^2/2) was determined with a digital caliper every 2–3 days. At the end of the experiment, all animals were sacrificed, and the tumors were collected and analyzed.

To study the in vivo effect of DHA on AMPK and mTORC1, female C.B.17SC scid$^{-/-}$ mice (5–6 weeks old) bearing Rh65 xenografts were treated i.p. with DHA (100 mg/kg body weight). Following treatment for 2, 4, 8, and 24 h, the mice (3 mice per time point) were sacrificed, and the tumor tissues were collected, frozen in liquid nitrogen and stored at –80 °C for further analysis. Non-treatment with DHA served as a control. Tumor lysates were analyzed by Western blotting with indicated antibodies.

2.10. Statistical Analysis

All data were expressed as mean values $\pm$ SD. Data were analyzed using Graph-Pad Prism 6 software (GraphPad Software, La Jolla, CA, USA). Group variability and interaction were compared using Student's *t*-test or one-way ANOVA followed by Bonferroni's post-tests to compare replicate means. A level of $p < 0.05$ was considered to be statistically significant.

3. Results

3.1. DHA Inhibits Cell Proliferation and mTORC1 Signaling in RMS Cells

To reposition DHA for treatment of RMS, six RMS cell lines (Rh30, RD, Rh18, Rh28, Rh36, and Rh41) were employed for the growth inhibition assay. As RMS develops primarily from skeletal muscle cells, normal human primary skeletal muscle cells (HSMCs) and mouse skeletal muscle cells (C2C12) were used as normal controls. As shown in Figure 1A, RMS cell lines tested were sensitive to DHA, with the half maximal inhibitory concentrations (IC_{50}) = 1.89−4.02 µM. In contrast, normal skeletal muscle cells (HSMCs and C2C12) were resistant to DHA ($IC_{50} > 10$ µM). Similar results were observed in normal human primary dermal fibroblasts (HDFs). The results suggested that DHA, at pharmacological concentrations (<10 µM), has little effects on normal cell growth and can selectively target RMS tumor cells.

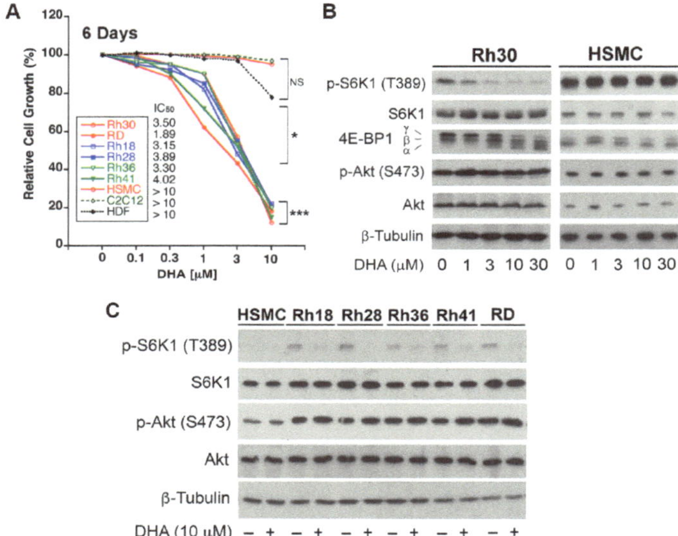

Figure 1. Dihydroartemisinin (DHA) inhibits the proliferation of rhabdomyosarcoma (RMS) cells. (**A**) Indicated cell lines were exposed to DHA (0−10 µM) for 6 days, followed by cell counting using a Beckmann Coulter counter. The results are shown as mean values ($n = 3$). NS, not significant; * $p < 0.05$; *** $p < 0.001$; difference versus vehicle control group. (**B,C**) Indicated cells were treated with DHA at indicated concentrations for 24 h, followed by Western blotting with indicated antibodies.

mTOR is a central controller of cell growth, proliferation, and survival [5,6]. Next, we wondered whether DHA reduction of cell proliferation is related to the inhibition of mTOR. For this, Rh30 cells and HSMCs were treated with DHA (0–30 µM) for 24 h, followed by Western blotting. In line with the above growth inhibitory effect (Figure 1A), the treatment with DHA inhibited mTORC1-mediated phosphorylation of S6K1 and 4E-BP1 in a concentration-dependent manner in RMS (Rh30) cells, but not in normal cells (HMSCs) (Figure 1B). Consistent with our previous findings in Rh1 and C2C12 cells [21], DHA treatment did not impact mTORC2-mediated phosphorylation of Akt (S473) in both Rh30 cells and HSMCs (Figure 1B). Similar results were observed in other RMS cells (Rh18, Rh28, Rh36, Rh41, and RD) (Figure 1C). Of note, the basal phosphorylation levels of S6K1 and Akt were much higher in RMS cells than in normal cells (HSMCs) (Figure 1C), suggesting that the mTOR signaling is hyperactive in RMS cells. The results highlight that DHA is a novel inhibitor of mTORC1.

Raptor and rictor are essential for the activity of mTORC1 and mTORC2, respectively [5,6]. Loss of mTORC1 or mTORC2 function (KO of raptor or rictor) reduces the growth rates of cells [35]. If mTORC1 is the major target for DHA-mediated RMS cell growth suppression, depletion of raptor should confer resistance to DHA. To this end, lentiviral shRNAs to raptor, rictor, and GFP (control) were employed [40]. Consistent with our previous report [40], the infection of Rh30 cells with lentiviral shRNAs to raptor and rictor downregulated the protein levels of raptor and rictor by 90% and 85%, respectively, in the cells compared to in controls. Similar to KO of raptor or rictor in MEFs [35], the knockdown of raptor or rictor in RMS cells also inhibited cell proliferation (Supplementary Figure S1). Of interest, knockdown of raptor, but not rictor, rendered high resistance to DHA-induced cell growth inhibition in Rh30 and RD cells (Figure 2A). To validate the finding, SV40 large T-antigen-immortalized raptor and rictor-inducible KO MEFs [35] were utilized. As expected, the treatment with 1 μM 4-hydroxytamoxifen for 3 days resulted in the deficiency of raptor or rictor in corresponding MEFs (Figure 2B). The KO of raptor inhibited p-S6K1 (T389), while the KO of rictor inhibited p-Akt (S473) in the cells (Figure 2B), indicating the loss of mTORC1 and mTORC2 in these MEFs, respectively. Following 72 h treatment with DHA, raptor-WT (wild-type), rictor-WT, and rictor-KO MEFs were sensitive to DHA (IC_{50} = 3.71–4.02 μM), whereas raptor KO MEFs were highly resistant to DHA (IC_{50} > 20 μM) (Figure 2C). Collectively, these observations support our hypothesis that DHA may execute its anticancer action primarily by targeting mTORC1 signaling.

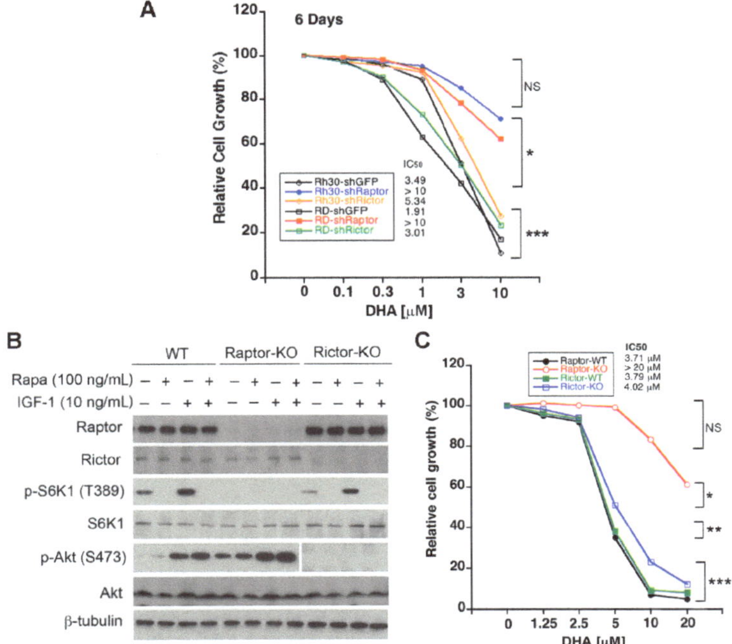

Figure 2. Disruption of mammalian target of rapamycin complex 1 (mTORC1), but not mammalian target of rapamycin complex 2 (mTORC2), confers high resistance to DHA-induced cell growth suppression. (**A**) Rh30 and RD cells, infected with lentiviral shRNAs to raptor, rictor, or GFP (control), were treated with DHA (0−10 μM) for 6 days, followed by cell counting using a Beckmann Coulter counter. (**B**) Raptor or rictor-inducible knockout mouse embryonic fibroblasts (MEFs) were treated with or without 4-hydroxytamoxifen (4-OHT) (1 μM) for 3 days, to generate raptor-WT, raptor-KO, rictor-WT, and rcitor-KO cells. Western blotting was performed with indicated antibodies. (**C**) Indicated cells, seeded in 96-well plates (all at 4×10^3 cells/well), were treated with DHA (0–20 μM) for 72 h, followed by MTS assay. Shown are mean values (n = 3). NS, not significant; * $p < 0.05$; ** $p < 0.01$; *** $p < 0.001$, difference versus vehicle control group (**A,C**).

3.2. DHA Does Not Bind to mTOR or FKBP12

To determine how DHA inhibits mTORC1 signaling, first of all, we investigated whether DHA binds to mTOR. For this, Rh30 cells were labeled with 10 μCi [^{3}H]-DHA for 21 h, followed by immunoprecipitation with mTOR antibodies or normal IgG (control). No significant amount of [^{3}H]-DHA was detected in the immunoprecipitates of mTOR, compared to in normal IgG (Figure 3), suggesting that DHA does not bind to mTOR directly.

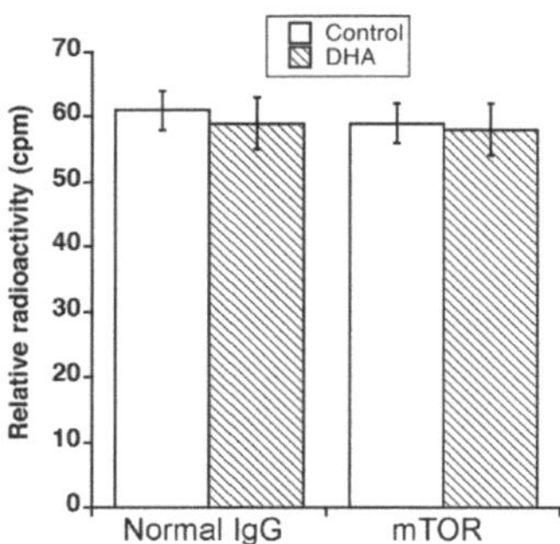

Figure 3. DHA does not directly bind to mTOR or FKBP12. Rh30 cells were pretreated with in vivo labeled with 10 μCi [^{3}H]-DHA for 21 h. The cell lysates were used for immunoprecipitation with anti-mTOR antibodies or normal IgG (control). Shown is the relative radioactivity (cpm) of immunoprecipitation products.

It is known that rapamycin firstly forms a complex with FKBP12 and then binds to the FRB domain of mTOR, inhibiting mTORC1 [5,6]. Next, we wondered whether DHA, such as rapamycin, inhibits mTORC1 by forming a complex with FKBP12. For this, molecular docking was performed. We found that DHA had a possibility to bind the interface cavity of FKBP12 and the FRB domain of mTOR, but the interaction of DHA with the protein complex was much weaker than that of rapamycin (Figure 4A–D), consistent with the calculated scoring function values (e.g., LibDock score: 29.72 for DHA vs. 101.40 for rapamycin) (Table 1). Collectively, our results indicated that DHA does not bind to mTOR or FKBP12 directly, suggesting that DHA inhibits mTORC1 through indirect mechanism(s).

Table 1. Scoring function values of dihydroartemisinin and rapamycin in molecular docking studies, in which the agents were docked into the active pocket of FKBP12 (PDB ID: 3FAP).

Compound	LibDock Score	GOLD Score
Rapamycin	101.40	148.24
Dihydroartemisinin	29.72	92.98

3.3. DHA Does Not Alter the Phosphorylation of IGF-1R/PI3K/PTEN and Erk1/2

Since mTORC1 is positively regulated by the IGF-1R-PI3K-Akt and Ras-Raf-MEK-Erk pathways but negatively regulated by PTEN [5,6], we further tested whether DHA inhibits mTORC1 signaling indirectly by altering these upstream regulators in cells. The treatment with DHA (0–30 μM) for 24 h did not obviously alter the phosphorylation of IGF-1Rβ (Tyr1161), PI3K p85 (Tyr458), PDK1 (Ser241), and PTEN (Ser380/Thr382), as well as total cellular levels of these proteins (Figure 5A) and Akt (Figure 1B). Similarly, DHA did not apparently affect the phosphorylation or total protein level of Erk1/2 in the RMS cells (Figure 5B). Of note, at high concentrations (10–30 μM), DHA slightly inhibited the phosphorylation of p38 MAPK (p-p38) in Rh30 cells but moderately activated p-p38 in RD cells (Figure 5B). In addition, DHA (30 μM) induced the phosphorylation of c-Jun (a substrate of JNK) in both Rh30 and RD cells (Figure 5B). Hence, these data imply that DHA inhibits mTORC1, not by altering the phosphorylation of IGF-1R/PI3K/Akt, PTEN, and Erk1/2.

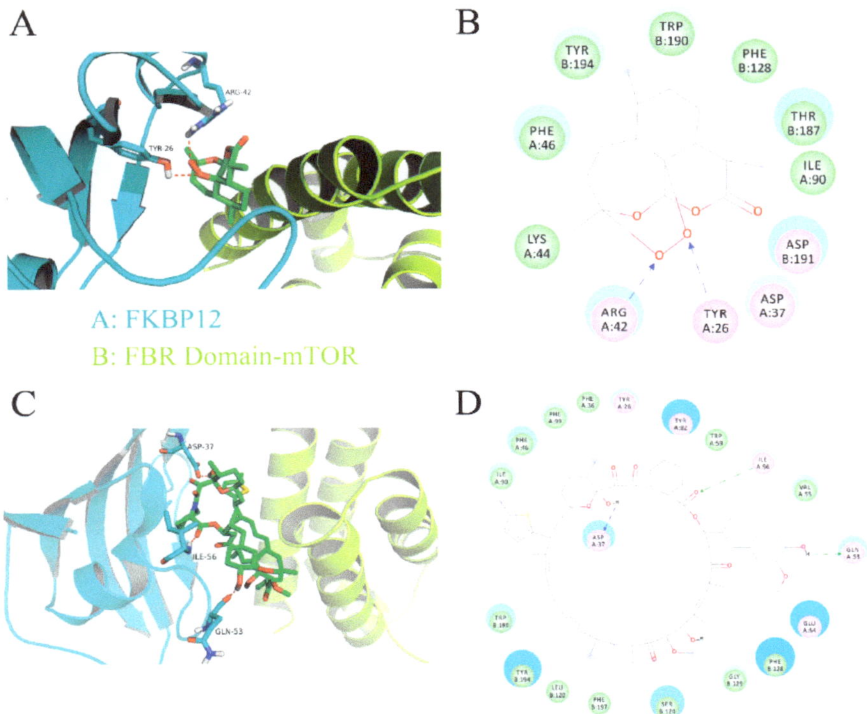

Figure 4. Predicted binding modes of dihydroartemisinin (**A,B**) and rapamycin (**C,D**) in the interface cavity of FKBP12 and the FRB domain of mTOR. The 3-D structure of the protein complex of FKBP12 and the FRB domain was taken from the PDB database (PDB ID: 3FAP). The calculated binding mode of DHA in the interface cavity of FKBP12 and the FRB domain of mTOR is shown in (**A,B**). DHA, namely (3R,12aR)-octahydro-12H-3,12-epoxy[1,2]dioxepino[4,3-i] isochromen-10(3H)-one, is sandwiched between FKBP12 and the FRB domain of mTOR. Two hydrogen bonds are formed between oxygen atoms of DHA and residues ARG42 and TYR26 of FKBP12. The seven-membered ring and the two six-membered rings in DHA form hydrophobic interactions with residues LYS44, PHE46, TYR194, TRP190, PHE128, THR187, and ILE90 in the protein complex of FKBP12 and the FRB domain of mTOR. For comparison, the binding mode of rapamycin in the interface cavity of FKBP12 and FRB domain of mTOR (**C,D**). Obviously, rapamycin forms a much better interaction with FKBP12 than DHA does. Three hydrogen bonds are formed between rapamycin and FKBP12: the first one corresponds to that formed between the carbonyl group of rapamycin and the ILE56 residue in FKBP12, and the other two are between two hydroxyl groups of rapamycin and residues ASP37 and GLN53, respectively. Rapamycin also forms good hydrophobic interactions with residues TRP190, TYR194, LEU120, PHE197, SER124, GLY129, PHE128, TYR82, LEU56, VAL55, TRP59, PHE39, TYR26, PHE46, and Phe99 in mTOR. Collectively, although DHA has the possibility to bind the interface cavity of FKBP12 and the FRB domain of mTOR, the interaction of DHA with the protein complex is much weaker than that of rapamycin, consistent with the calculated scoring function values (see Table 1).

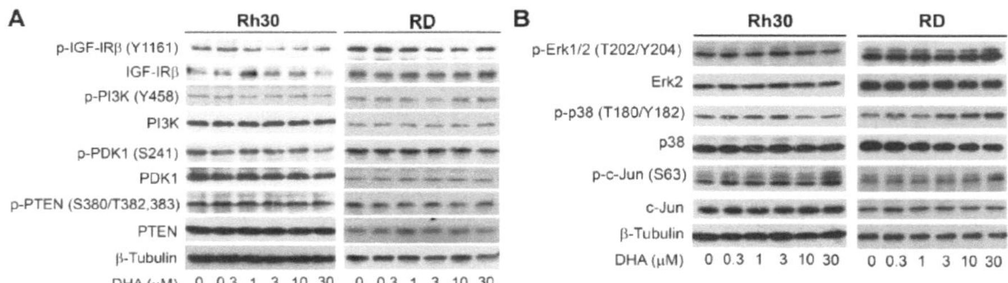

Figure 5. DHA does not alter the phosphorylation of IGF-1R/PI3K/PTEN and Erk1/2 in tumor cells. (**A,B**) Rh30 or RD cells were treated with DHA (0–30 μM) for 24 h, followed by Western blotting with indicated antibodies.

3.4. DHA Does Not Induce HIF-1α/REDD1 Expression, but Triggers AMPK Phosphorylation

As the HIF1-REDD1 and AMPK pathways negatively regulate mTORC1 [8,9,48,49], next, we asked whether DHA inhibits mTORC1 signaling by activating these two pathways. To this end, tumor cells were treated with DHA (0–30 μM) for 24 h, followed by Western blotting. The results showed that the treatment with DHA did not induce the expression of HIF-1α or REDD1 (regulated in development and DNA damage responses 1) in Rh30 cells (Figure 6A). As a positive control [37], the treatment with ciclopirox olamine (CPX) induced the robust expression of HIF-1α and REDD1. Interestingly, DHA treatment induced the phosphorylation of the catalytic subunit of AMPK (p-AMPKα and T172) in Rh30 cells in a dose-dependent manner (Figure 6B). Similar results were also observed in Rh1, Rh18, Rh28, Rh36, Rh41, and RD cells (Figure 6C,E). Since DHA did not inhibit mTORC1 in normal HSMCs (Figure 1B), we also tested whether DHA affects p-AMPKα (T172) in this cell line. As shown in Figure 6D,E, treatment with DHA (0–30 μM) for 24 h had no evident impact on p-AMPKα in HSMCs. The results suggest that DHA inhibits mTORC1 possibly by activating AMPK.

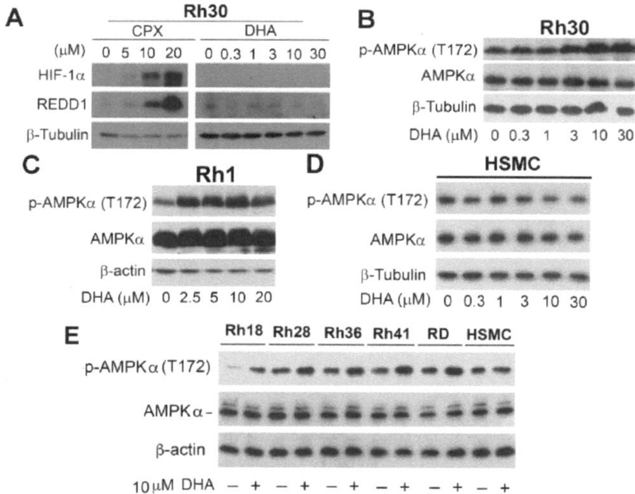

Figure 6. DHA does not induce HIF-1α/REDD1 expression, but triggers AMPK phosphorylation. (**A**) Rh30 cells were treated with DHA (0–30 μM) for 24 h, followed by Western blotting with indicated antibodies. Ciclopirox olamine (CPX) served as a positive control for the induction of HIF1α and REDD1. (**B–E**) Indicated cell lines were treated with DHA at indicated concentrations for 24 h, followed by Western blotting with indicated antibodies.

3.5. DHA-Induced Activation of AMPK Contributes to the Inhibition of mTORC1

To investigate the relationship between the DHA-induced inhibition of mTORC1 and the activation of AMPK, a time-course experiment was performed. When Rh1 cells were treated with DHA (5 μM) for 0–12 h, the phosphorylation level of AMPKα increased in a time-dependent manner. The phosphorylation of AMPKα was modestly induced at 8 h and robustly induced at 12 h, which matched well with the inhibition pattern on mTORC1 (Figure 7A). The results suggest that DHA-induced mTORC1 inhibition may be associated with activation of AMPK.

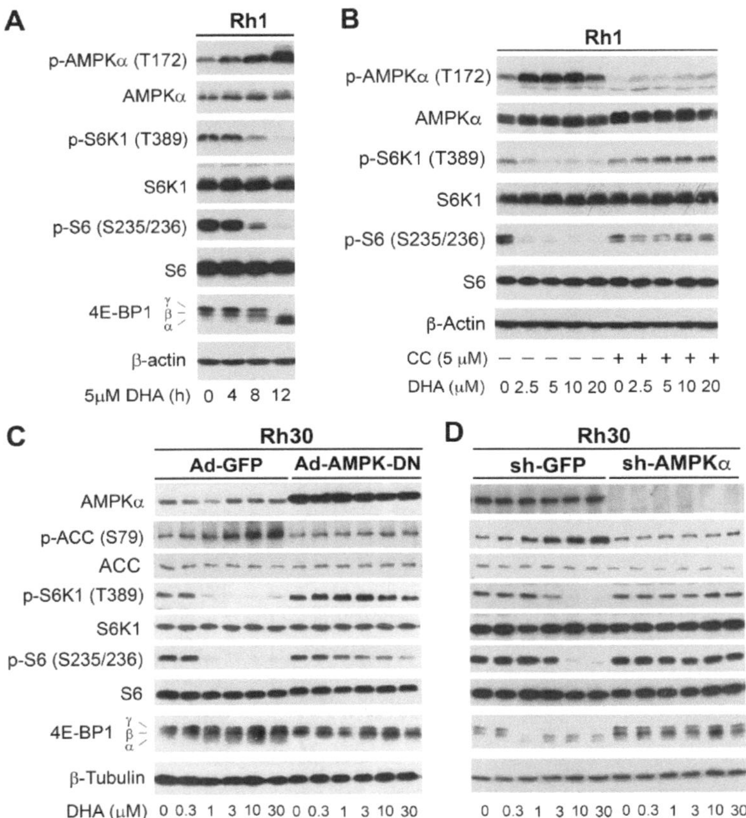

Figure 7. DHA-induced activation of AMPK contributes to inhibition of mTORC1. (**A**) Rh1 cells were treated with DHA (5 μM) for indicated time, followed by Western blotting with indicated antibodies. (**B**) Rh1 cells were pretreated with or without Compound C (CC) (5 μM) for 2 h, and then exposed to with or without DHA (5 μM) for 24 h, followed by Western blotting using indicated antibodies. (**C**) Rh30 cells were infected with recombinant adenovirus expressing myc-tagged dominant negative (DN) AMPK (Ad-AMPK-DN) or GFP (Ad-GFP) for 24 h, and then treated with DHA (0−30 μM) for another 24 h, followed by Western blotting with indicated antibodies. (**D**) Rh30 cells, infected with lentiviral shRNA to human AMPKα1 or GFP, were treated with DHA (0−30 μM) for 24 h, followed by Western blotting with indicated antibodies.

To validate whether DHA-induced activation of AMPK contributes to inhibition of mTORC1, Rh1 cells were pre-treated with or without compound C (a selective inhibitor of AMPK) for 2 h and then exposed to DHA (5 μM) for 24 h. As predicted, pretreatment with compound C remarkably attenuated DHA-induced p-AMPKα (Figure 7B). Of interest, the inhibition of AMPK profoundly prevented DHA from inhibiting the phosphorylation of S6K1 and S6 (Figure 7B). Similar results were observed in Rh30 cells (Supplementary Figure S2). These data support that AMPK activation is involved in DHA-induced mTORC1 inhibition.

To corroborate the above finding, Rh30 cells were infected with a recombinant adenovirus expressing DN AMPKα (Ad-AMPK-DN), kinase-dead AMPKα (Ad-AMPK-KD) [37,38], or GFP (Ad-GFP, control) for 24 h, and then treated with DHA for another 24 h. As anticipated, ectopic expression of AMPK-DN or AMPK-KD, but not GFP, attenuated DHA-induced phosphorylation of ACC (S79), a substrate of AMPK (Figure 7C; Supplementary Figure S3), suggesting that both Ad-AMPK-DN and Ad-AMPK-KD were working well in the cells. Importantly, expression of AMPK-DN or AMPK-KD did render high resistance to the inhibitory effect of DHA on mTORC1 in the cells (Figure 7C; Supplementary Figure S3). Furthermore, similar results were observed in Rh30 cells when AMPKα1 was knocked down with lentiviral shRNA to AMPKα1 (Figure 7D). Together, our results indicate that activation of AMPK plays a critical role in DHA-induced inhibition of mTORC1.

3.6. DHA Dissociates Raptor from mTOR and Inhibits mTORC1 Activity

Rapamycin inhibits mTORC1 through the rapamycin-FKBP12 binding to mTOR, which results in the dissociation of raptor from mTOR, thereby inhibiting the mTORC1 function [7]. Having observed that DHA did not bind to mTOR (Figure 3), we further investigated whether DHA disrupts mTORC1. For this, Rh30 cells were treated with or without DHA (3 μM) for 24 h, or rapamycin (100 ng/mL, positive control) for 2 h, followed by immunoprecipitation with antibodies to mTOR or normal IgG (negative control). By immunoblotting, as expected, rapamycin did not obviously affect the binding of mTOR to mLST8 or rictor but dramatically reduced the interaction of mTOR with raptor (Figure 8A). Interestingly, DHA acted in the same way, although 3 μM of DHA did not cause the dissociation of raptor from mTOR so potently as 100 ng/mL of rapamycin (Figure 8A). The effect of DHA on the mTOR–raptor complex (mTORC1), but not the mTOR–rictor complex (mTORC2), was in line with our finding that DHA inhibits mTORC1-mediated phosphorylation of S6K1 and 4E-BP1 but does not affect mTORC2-mediated phosphorylation of Akt (Figure 1B).

As raptor is essential for the mTORC1 function [5,6], we further assessed the effect of DHA on the mTORC1 activity by in vitro mTOR kinase assay using recombinant 4E-BP1 protein as a substrate. As expected, the treatment with rapamycin (100 ng/mL) for 2 h strongly inhibited the mTORC1 activity in Rh30 cells, as the phosphorylation of 4E-BP1 on T37/46 and T70 was inhibited by approximately 40% and 70%, respectively (Figure 8B,C). Of interest, the treatment with DHA (3 μM) for 24 h inhibited the mTORC1 activity as potently as rapamycin (Figure 8B,C).

3.7. Artesunate Inhibits Tumor Growth, Suppresses mTORC1 and Activates AMPK in RMS Xenografts

Artesunate, a pro-drug of DHA, has been in clinical trials for treatments of lung, colon, breast, and cervical cancer in adults [19]. To assess the potential of DHA for treatments of RMS, we evaluated the anticancer activity of artesunate in Rh30 xenografts in SCID mice. The results showed that treatments with artesunate (i.p. once daily at 25, 50, 100, and 150 mg/kg body weight) for 32 days in a dose-dependent manner inhibited the tumor growth (volume) of Rh30 xenografts in mice, by 23.6, 50.8%, 70.7%, and 80.3%, respectively, compared to the vehicle treatment (Figure 9A; Supplementary Figure S4). Artesunate treatments displayed similar inhibitory effects on the tumor weight (Figure 9B). Of note, no obvious toxicity was observed in all the treated groups except for the 150 mg/kg group, in which the average body weight of mice decreased slightly but not significantly (Figure 9C).

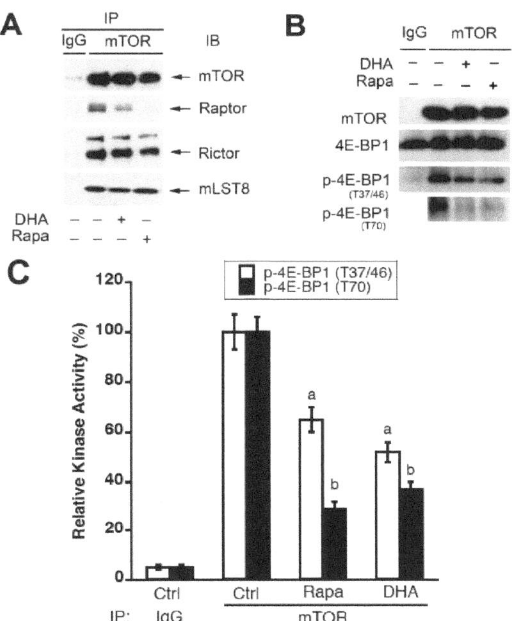

Figure 8. DHA dissociates raptor from mTOR and inhibits mTORC1 activity. (**A,B**) Rh30 cells were treated with or without DHA (3 μM) for 24 h or with rapamycin (Rapa, 100 ng/mL, positive control) for 2 h, followed by immunoprecipitation (IP) using antibodies to mTOR or normal IgG (control). The immunoprecipitates were then subjected to immunoblotting (IB) with indicated antibodies or were used for the in vitro mTOR kinase assay by incubating with recombinant 4E-BP1 protein (as a substrate) at room temperature for 30 min, followed by Western blotting with indicated antibodies. (**C**) Semi-quantitative data (mean ± SD) of three independent experiments for (**B**). [a] indicates the difference with the control group of p-4E-BP1 (T37/46) ($p < 0.05$); [b] indicates the difference with the control group of p-4E-BP1 (T70) ($p < 0.05$).

To study the in vivo effects of DHA on AMPK and mTORC1, SCID mice bearing with Rh65 xenografts were treated i.p. with DHA (100 mg/kg body weight). At 2, 4, 8, and 24 h of post-treatment, the mice were sacrificed, and the tumor tissues were collected. Our Western blotting analysis revealed that treatment with DHA for 8 h remarkably inhibited p-S6 (S235/236) and p-4E-BP1 (T70) but did not apparently affect p-Akt (S473) in the tumors (Figure 9D), in line with our in vitro results (Figure 1B). Interestingly, DHA treatment also time-dependently induced p-AMPK (T172) in vivo (Figure 9D), also consistent with the in vitro data (Figure 7). The results underline that artesunate or DHA has a great potential for RMS therapy.

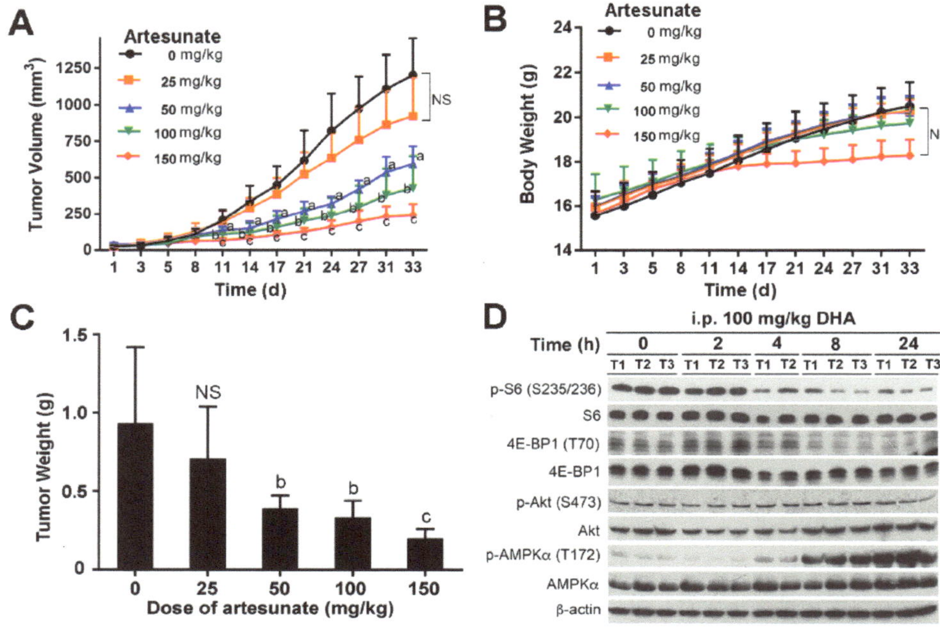

Figure 9. Artesunate inhibits tumor growth, suppresses mTORC1 and activates AMPK in RMS xenografts. (**A–C**) C.B.17SC scid$^{-/-}$ female mice (5–6 weeks old) bearing Rh30 xenografts were treated intraperitoneally (i.p.) with artesunate at the indicated doses or vehicle (0.9% NaCl) once daily. Tumor volume (**A**) and body weight (**C**) were measured at the indicated time. At the end of the experiment, the mice were sacrificed, and the tumor tissues were dissected and weighed (**B**). The data are expressed as the mean ± SD (8–9 mice per group). NS, not significant; [a] $p < 0.05$; [b] $p < 0.05$; [c] $p < 0.001$, difference with the control group (0 mg/kg of artesunate). (**D**) C.B.17SC scid$^{-/-}$ female mice (5–6 weeks old) bearing Rh65 xenografts were treated i.p. with DHA (100 mg/kg). After the treatment for the indicated time, the mice were sacrificed, and the tumor tissues were collected and frozen in liquid N_2. Western blotting was performed with indicated antibodies. Control means non-treatment with DHA; T1, T2, and T3 represent tumors #1, #2, and #3, respectively.

4. Discussion

Increasing evidence suggests that DHA exerts its anticancer activity primarily by inhibiting mTORC1 signaling in tumor cells [20–24]. However, how DHA inhibits mTORC1 is unknown. mTOR can be inhibited due to the binding of compounds (either allosterically by rapalogs or directly by mTOR kinase inhibitors, e.g., INK128 and AZD8055) [5,6]. In addition, mTOR can be inhibited indirectly via multiple mechanisms, including the inhibition of the IGF-1R-PI3K-Akt and Ras-Raf-MEK-Erk pathways and the activation of the HIF1-REDD1 and AMPK pathways [5,6]. Recently, Du et al. has shown that DHA induces autophagy of leukemia cells partly by inhibiting p-mTOR, p-S6K1, and p-S6 and activating p-AMPK in leukemia (HL60 and THP-1) cells [26], but whether DHA-induced mTORC1 inhibition is a consequence of AMPK activation has not been resolved. Here, for the first time, we present evidence that DHA inhibits mTORC1 neither by directly binding to mTOR or FKBP12 nor by indirectly inhibiting the IGF-1R-PI3K-Akt and Erk pathways or activating PTEN, HIF-1α, and REDD1. Instead, DHA inhibits mTORC1 by activating the AMPK pathway in tumor cells. Furthermore, DHA is able to induce the dissociation of raptor from mTOR and inhibit the activity of mTORC1.

Current chemotherapeutic treatments for pediatric RMS may cause long-term side effects, such as secondary cancers and infertility [1,3]. Artemisinins have been widely used for treatment of malaria in children and adults for decades, and their safety is clinically proven [16–18]. Especially, artesunate, a pro-drug of DHA, has been in clinical trials for treatment of lung, colon, breast, and cervical cancer in adults [19]. Therefore, we explored whether DHA can be repositioned for treatment of RMS. Here, our in vitro and in vivo data indicate that DHA is able to potently inhibit the growth of RMS by inhibiting mTORC1 signaling. Our findings support that DHA has a great potential for treatment of RMS.

Here, we found that DHA preferably targets RMS tumor cells, but not normal cells (Figure 1A). The possible reasons are discussed here. One the one hand, it has been documented that the *PAX3/7–FOXO1* fusion in RMS tumor cells results in constitutively active mTOR signaling [1–3]. RMS cells, unlike normal cells (normal HSMCs), mouse skeletal muscle cells (C2C12), and normal HDFs, had hyperactive mTOR signaling (see Figure 1C), so they are apparently addictive to mTOR signaling for growth, proliferation, and survival. This may be one of the major reasons why DHA preferably targets RMS cells but not normal cells. On the other hand, as shown in Figure 6, in response to DHA, AMPK can be activated in RMS cells, but not in normal cells. This may be partly related to the differential effects of DHA on the cellular levels of ATP in RMS cells and normal cells. It is well known that AMPK can be activated in response to energy stress [5,6]. In our study, we observed that treatment with DHA for 18 h reduced the cellular ATP level in Rh30 cells (Supplementary Figure S5C). Of note, the 24 h treatment with 30 µM of DHA reduced the intracellular ATP level by 80% (compared to the control) in tumor cells (Rh30), whereas the same treatment only reduced the ATP level by ~30% in normal cells (C2C12) (Supplementary Figure S5D), in line with tumor-selective effects of DHA on cell proliferation/viability and mTORC1. At this stage, we could not rule out other possibilities, such as differential expression levels of drug efflux transporters (e.g., P-glycoprotein and multidrug resistance proteins).

In this study, we noticed that the treatment with artesunate dose-dependently inhibited the tumor growth of Rh30 xenografts in mice. The treatment with artesunate (50 and 100 mg/kg) potently inhibited the tumor growth (by 58% and 65%, respectively; $p < 0.01$), but had no marked effect on the body weight of the animals, reinforcing the good safety of artesunate. However, we have to acknowledge that single treatment with artesunate, even at 150 mg/kg, failed to result in tumor regression (Figure 9), implying that combination treatments are necessary. Given that the standard chemotherapy for RMS is the combination of vincristine, actinomycin D, and cyclophosphamide [1,3], it is worthy to further study whether artesunate synergizes with these chemotherapeutic agents in RMS.

Here, we found that DHA (0−30 µM) did not influence the phosphorylation of Akt (S473) in RMS (Rh18, Rh28, Rh36, Rh30, Rh41, and RD) cells (Figure 1), which is consistent with our previous observation in Ewing sarcoma (Rh1) cells [21]. Rapamycin has been shown to inhibit mTORC1 but induces p-Akt (S473) in RMS cells [40]. These findings suggest that the effect of DHA on p-Akt is different from that of rapamycin. It is known that rapalogs inhibit mTORC1 but activate Akt through the S6K1-IRS1 negative feedback mechanism [12–14]. Rapalogs-activated Akt is regarded as a major drawback contributing to their mild anticancer activity in most clinical settings, as activated Akt can promote cancer cell survival [5,6]. A number of clinical trials of artesunate are ongoing for treatments of various cancers [19]. Since rapalogs alone lack efficacy in treatments of most types of cancer, including RMS [5,6,10,11], it would be interesting to determine whether DHA or artesunate, as an anti-cancer agent, is clinically superior to rapalogs.

Of note, a recent report has shown that a treatment with 40 µM of DHA reduces the protein levels of HIF-1α and p-Akt (S473) in prostate LNCaP cells [30]. This is in contrast to our results that a treatment with 0.3–30 µM of DHA did not alter the levels of HIF-1α (Figure 5A) and p-Akt in both Rh30 (Figure 1B) and Rh1 cells [21]. Whether the discrepancy is due to different concentrations of DHA used remains to be determined. We have noticed that curcumin at high concentrations (>20 µM), which is clinically irrelevant, is able to inhibit both mTORC1 and mTORC2 [41].

It has been described that AMPK activation induces phosphorylation of p53 on serine 15 [50] and p53 activation can inhibit mTORC1 [51]. In the present study, we found that

DHA was able to inhibit mTORC1 in multiple cell lines, of which Rh30, RD and Rh1 cells expressed mutant p53 alleles (Rh30 Arg273→Cys; RD Arg248→Trp; Rh1: Tyr220→Cys), losing the function of p53 [21]. Thus, our results suggest that the AMPK-p53 pathway is dispensable for DHA-induced mTORC1 inhibition. Since activated AMPK can also inhibit mTORC1 by activating the formation of TSC1/2 complex and/or phosphorylating raptor (S792) [8,9], to better understand how DHA inhibits mTORC1, further research is required to define whether the AMPK-TSC and AMPK-raptor pathways are involved in DHA-induced mTORC1 inhibition.

A new question is that how DHA activates AMPK. Our molecular docking indicates that although DHA has the possibility to bind the interface cavity of the carbohydrate-binding module (CBM, also known as the glycogen-binding domain) and the kinase domain of AMPK, the interactions of DHA with the protein complex are much weaker than those of A-769662 (a known AMPK activator) [46], consistent with the calculated scoring function (GOLD scores: 44.73 for DHA vs. 75.77 for A-769662). Further research is needed to confirm whether DHA is able to bind to α, β, or γ subunit of AMPK or not.

AMPKα (T172) can be activated by liver kinase B1 (LKB1) in response to low levels of energy (ATP), by transforming growth factor β-activated kinase 1 (TAK1) due to increased cytokines and/or by calmodulin-dependent kinase kinase β (CaMKKβ) upon elevated intracellular Ca^{2+} levels [52–54]. Besides, in response to various stimuli (e.g., oxidative stress, glucose, tumor necrosis factor-α, and palmitate), AMPKα (T172) can be dephosphorylated and inactivated by protein phosphatase 1 (PP1) [55], protein phosphatase 2A (PP2A) [56–58], protein phosphatase 2B (PP2B, also calcineurin) [59], protein phosphatase 2C (PP2C) [58], and protein phosphatase 5 (PP5) [60]. In our study, we noticed that the treatment with DHA (0–30 μM) for 24 h did not alter intracellular Ca^{2+} levels in Rh30 and RD cells. Of interest, 8 h or 24 h treatments with DHA (0−30 μM) induced ROS in Rh30 cells in a dose-dependent manner (Supplementary Figure S5A,B). In addition, the treatment with DHA (10 or 30 μM) for 18–24 h significantly reduced the cellular ATP levels in Rh30 and RD cells (Supplementary Figure S5C,D). Therefore, it would be interesting to figure out whether DHA-induced activation of AMPK is mediated by any of these kinases and/or phosphatases.

In the present study, we also observed that DHA induced the dissociation of raptor from mTOR (Figure 8A). AMPK-mediated phosphorylation of both TSCs and raptor does not cause disassembly of mTORC1 [5,6]. It has been shown that GRp58/ERp57 is involved in the assembly of mTORC1 and positively regulates mTORC1 signaling at the cytosol and the cytosolic side of the ER [61]. Further research is needed to address whether DHA disrupts mTORC1 by targeting GRp58/ERp57.

5. Conclusions

Here, we showed that DHA inhibited mTORC1 in tumor cells not through direct binding to mTOR or FKBP12, but via indirect mechanisms. Apparently, DHA altered neither the phosphorylation of IGF-IR/PI3K/Akt/PTEN and Erk1/2, nor the expression of HIF1α/REDD1 in tumor cells. Instead, DHA inhibited mTORC1 by activating the AMPK pathway (Figure 10). Additionally, DHA was able to induce dissociation of raptor from mTOR and inhibit the activity of mTORC1. To our knowledge, this is the first study to unveil how DHA inhibits mTORC1 in tumor cells. In addition, our in vitro and in vivo data demonstrate that DHA has a great potential for RMS treatment. To facilitate repurposing this anti-malaria agent for RMS treatment, further research is warranted to determine whether artesunate (alone or in combination with other anticancer agents) is more effective than rapalogs in mouse tumor models.

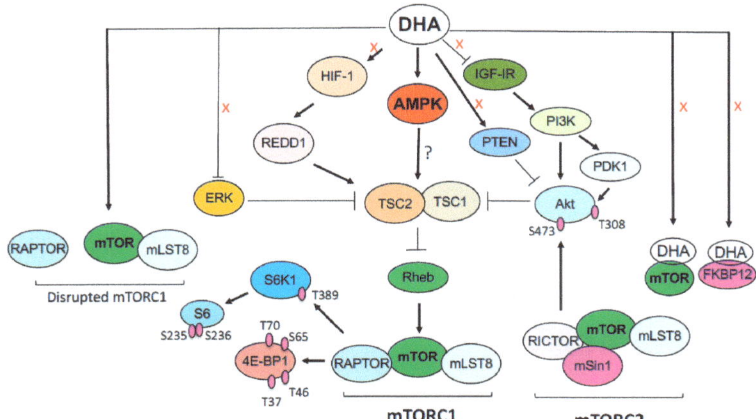

Figure 10. A proposed model of mTORC1 inhibition by DHA. DHA does not directly bind to mTOR or FKBP12. Besides, DHA alters neither the phosphorylation of IGF-IR/PI3K/Akt/PTEN and Erk1/2, nor the expression of HIF1α/REDD1 in tumor cells. Instead, DHA inhibits mTORC1 by activating the AMPK pathway. In addition, DHA induces the dissociation of raptor from mTOR and inhibit the activity of mTORC1. Abbreviations: 4E-BP1, eukaryotic initiation factor 4E (eIF4E) binding protein 1; Akt, protein kinase B; AMPK, AMP-activated protein kinase; DHA, dihydroartemisinin; ERK, extracellular signal-regulated kinases; FKBP12, FK506 binding protein 12; HIF1, hypoxia-inducible factor 1; IGF-1R, insulin-like growth factor-1 receptor; mLST8, mammalian lethal with sec-13 protein 8; mSin1, mammalian stress-activated protein kinase-interacting protein 1; mTOR, mammalian target of rapamycin; mTORC1, mTOR complex 1; mTORC2, mTOR complex 2; PDK1, PI3K, phosphoinositide 3-kinase; PTEN, phosphatase and tensin homolog; RAPTOR, regulatory-associated protein of mTOR; REDD1, Rheb, Ras homolog enriched in brain; RICTOR, rapamycin-insensitive companion of mTOR; S, serine; S6, ribosomal protein S6; S6K1, p70 S6 kinase 1; T, threonine; TSC1, tuberous sclerosis complex 1; TSC2, tuberous sclerosis complex 2.

Supplementary Materials: The following are available online at https://www.mdpi.com/article/10.3390/cells10061363/s1, Figure S1: Knockdown of raptor or rictor inhibits RMS cell growth. Figure S2: Inhibition of AMPK attenuates DHA-induced inhibition of mTORC1 in Rh30 cells. Figure S3: Ectopic expression of kinase-dead AMPK attenuates DHA-induced inhibition of mTORC1 in Rh30 cells. Figure S4: Artesunate inhibits the tumor growth of Rh30 xenografts in SCID mice. Figure S5: DHA increases cellular ROS and reduces ATP levels in tumor cells.

Author Contributions: Conception and design, J.L., Y.O., Z.H., X.G., and S.H.; development of methodology, J.L., Y.O., Z.H., S.Y., and S.H.; acquisition of data, J.L., Y.O., Z.H., B.C., W.L., L.L., C.S., C.Z., Y.W., Y.L., S.Y., and P.J.H.; analysis and interpretation of data, J.L., Y.O., Z.H., W.L., S.Y., and S.H.; writing, review, and/or revision of the manuscript, J.L., Y.O., Z.H., P.J.H., X.G., and S.H.; study supervision, S.H. All authors have read and agreed to the published version of the manuscript.

Funding: This work was supported by the National Institutes of Health (CA115414, S.H.), American Cancer Society (RSG-08-135-01-CNE, S.H.), the Feist-Weiller Cancer Center of LSU Health Sciences Center in Shreveport (Y.O., S.H.), Graduate Student Overseas Study Program of South China Agricultural University (2017LHPY028, J.L.), Major Special Project of Anhui Science and Technology Department (KJ2020A0516, Z.H.), and 2019 University Excellent Talent Training Project (GXGWFX2019038, Z.H.).

Institutional Review Board Statement: The animal experiments in this study were conducted in compliance with the guidelines set forth by the Guide for the Care and Use of Laboratory Animals, and approved by the institutional animal care and use committee of LSU Health Sciences Center-Shreveport (Protocol Code Number: LSUHSC-S #P20-003; the date of approval: 30 August 2019).

Informed Consent Statement: Not applicable.

Data Availability Statement: All the data presented in this study are included in this article.

Acknowledgments: We thank Michael N. Hall for providing raptor and rictor-inducible knockout mouse embryonic fibroblasts, Nicholas J. G. Webster for providing a recombinant adenovirus expressing myc-tagged kinase-dead AMPKα2, and Xiuping Yu for technical assistance in xenograft studies.

Conflicts of Interest: The authors declare no competing financial interests.

References

1. Skapek, S.X.; Ferrari, A.; Gupta, A.A.; Lupo, P.J.; Butler, E.; Shipley, J.; Barr, F.G.; Hawkins, D.S. Rhabdomyosarcoma. *Nat. Rev. Dis. Primers.* **2019**, *5*, 1. [CrossRef]
2. Pal, A.; Chiu, H.Y.; Taneja, R. Genetics, epigenetics and redox homeostasis in rhabdomyosarcoma: Emerging targets and therapeutics. *Redox Biol.* **2019**, *25*, 101124. [CrossRef] [PubMed]
3. Miwa, S.; Yamamoto, N.; Hayashi, K.; Takeuchi, A.; Igarashi, K.; Tsuchiya, H. Recent Advances and Challenges in the Treatment of Rhabdomyosarcoma. *Cancers* **2020**, *12*, 1758. [CrossRef]
4. Ingley, K.M.; Cohen-Gogo, S.; Gupta, A.A. Systemic therapy in pediatric-type soft-tissue sarcoma. *Curr. Oncol.* **2020**, *27*, 6–16. [CrossRef]
5. Liu, G.Y.; Sabatini, D.M. mTOR at the nexus of nutrition, growth, ageing and disease. *Nat. Rev. Mol. Cell Biol.* **2020**, *21*, 183–203. [CrossRef] [PubMed]
6. Mossmann, D.; Park, S.; Hall, M.N. mTOR signalling and cellular metabolism are mutual determinants in cancer. *Nat. Rev. Cancer* **2018**, *18*, 744–757. [CrossRef]
7. Kim, D.H.; Sarbassov, D.D.; Ali, S.M.; King, J.E.; Latek, R.R.; Erdjument-Bromage, H.; Tempst, P.; Sabatini, D.M. mTOR interacts with raptor to form a nutrient-sensitive complex that signals to the cell growth machinery. *Cell* **2002**, *110*, 163–175. [CrossRef]
8. Inoki, K.; Zhu, T.; Guan, K.-L. TSC2 mediates cellular energy response to control cell growth and survival. *Cell* **2003**, *115*, 577–590. [CrossRef]
9. Gwinn, D.M.; Shackelford, D.B.; Egan, D.F.; Mihaylova, M.M.; Mery, A.; Vasquez, D.S.; Turk, B.E.; Shaw, R.J. AMPK Phosphorylation of Raptor Mediates a Metabolic Checkpoint. *Mol. Cell* **2008**, *30*, 214–226. [CrossRef]
10. Spunt, S.L.; Grupp, S.A.; Vik, T.A.; Santana, V.M.; Greenblatt, D.J.; Clancy, J.; Berkenblit, A.; Krygowski, M.; Ananthakrishnan, R.; Boni, J.P.; et al. Phase I study of temsirolimus in pediatric patients with recurrent/refractory solid tumors. *J. Clin. Oncol.* **2011**, *29*, 2933–2940. [CrossRef]
11. Geoerger, B.; Kieran, M.W.; Grupp, S.; Perek, D.; Clancy, J.; Krygowski, M.; Ananthakrishnan, R.; Boni, J.P.; Berkenblit, A.; Spunt, S.L. Phase II trial of temsirolimus in children with high-grade glioma, neuroblastoma and rhabdomyosarcoma. *Eur. J. Cancer* **2012**, *48*, 253–262. [CrossRef] [PubMed]
12. Shah, O.J.; Wang, Z.; Hunter, T. Inappropriate activation of the TSC/Rheb/mTOR/S6K cassette induces IRS1/2 depletion, insulin resistance, and cell survival deficiencies. *Curr. Biol.* **2004**, *14*, 1650–1656. [CrossRef]
13. Sun, S.Y.; Rosenberg, L.M.; Wang, X.; Zhou, Z.; Yue, P.; Fu, H.; Khuri, F.R. Activation of Akt and eIF4E survival pathways by rapamycin-mediated mammalian target of rapamycin inhibition. *Cancer Res.* **2005**, *65*, 7052–7058. [CrossRef] [PubMed]
14. O'Reilly, K.E.; Rojo, F.; She, Q.B.; Solit, D.; Mills, G.B.; Smith, D.; Lane, H.; Hofmann, F.; Hicklin, D.J.; Ludwig, D.L.; et al. mTOR inhibition induces upstream receptor tyrosine kinase signaling and activates Akt. *Cancer Res.* **2006**, *66*, 1500–1508. [CrossRef] [PubMed]
15. Rodrik-Outmezguine, V.S.; Chandarlapaty, S.; Pagano, N.C.; Poulikakos, P.I.; Scaltriti, M.; Moskatel, E.; Baselga, J.; Guichard, S.; Rosen, N. mTOR kinase inhibition causes feedback-dependent biphasic regulation of AKT signaling. *Cancer Discov.* **2011**, *1*, 248–259. [CrossRef]
16. Li, Y. Qinghaosu (artemisinin): Chemistry and pharmacology. *Acta Pharmacol. Sin.* **2012**, *33*, 1141–1146. [CrossRef] [PubMed]
17. O'Neill, P.M.; Barton, V.E.; Ward, S.A. The Molecular Mechanism of Action of Artemisinin—The Debate Continues. *Molecules.* **2010**, *15*, 1705–1721. [CrossRef]
18. Wang., J.; Xu, C.; Lun, Z.R.; Meshnick, S.R. Unpacking 'Artemisinin Resistance'. *Trends Pharmacol. Sci.* **2017**, *38*, 506–511. [CrossRef]
19. Efferth, T. From ancient herb to modern drug: Artemisia annua and artemisinin for cancer therapy. *Semin. Cancer Biol.* **2017**, *46*, 65–83. [CrossRef]
20. Zhao, Y.G.; Wang, Y.; Guo, Z.; Gu, A.D.; Dan, H.C.; Baldwin, A.S.; Hao, W.; Wan, Y.Y. Dihydroartemisinin ameliorates inflammatory disease by its reciprocal effects on Th and regulatory T cell function via modulating the mammalian target of rapamycin pathway. *J. Immunol.* **2012**, *189*, 4417–4425. [CrossRef]
21. Odaka, Y.; Xu, B.; Luo, Y.; Shen, T.; Shang, C.; Wu, Y.; Zhou, H.; Huang, S. Dihydroartemisinin inhibits the mammalian target of rapamycin-mediated signaling pathways in tumor cells. *Carcinogenesis* **2014**, *35*, 192–200. [CrossRef]
22. Feng, X.; Li, L.; Jiang, H.; Jiang, K.; Jin, Y.; Zheng, J. Dihydroartemisinin potentiates the anticancer effect of cisplatin via mTOR inhibition in cisplatin-resistant ovarian cancer cells: Involvement of apoptosis and autophagy. *Biochem. Biophys. Res. Commun.* **2014**, *444*, 376–381. [CrossRef]
23. Mi, Y.J.; Geng, G.J.; Zou, Z.Z.; Gao, J.; Luo, X.Y.; Liu, Y.; Li, N.; Li, C.L.; Chen, Y.Q.; Yu, X.Y.; et al. Dihydroartemisinin inhibits glucose uptake and cooperates with glycolysis inhibitor to induce apoptosis in non-small cell lung carcinoma cells. *PLoS ONE* **2015**, *10*, e0120426. [CrossRef]

24. Jin, H.; Jiang, A.Y.; Wang., H.; Cao, Y.; Wu, Y.; Jiang, X.F. Dihydroartemisinin and gefitinib synergistically inhibit NSCLC cell growth and promote apoptosis via the Akt/mTOR/STAT3 pathway. *Mol. Med. Rep.* **2017**, *16*, 3475–3481. [CrossRef] [PubMed]
25. Liu, X.; Wu, J.; Fan, M.; Shen, C.; Dai, W.; Bao, Y.; Liu, J.H.; Yu, B.Y. Novel dihydroartemisinin derivative DHA-37 induces autophagic cell death through upregulation of HMGB1 in A549 cells. *Cell Death Dis.* **2018**, *9*, 1048. [CrossRef] [PubMed]
26. Du, J.; Wang, T.; Li, Y.; Zhou, Y.; Wang, X.; Yu, X.; Ren, X.; An, Y.; Wu, Y.; Sun, W.; et al. DHA inhibits proliferation and induces ferroptosis of leukemia cells through autophagy dependent degradation of ferritin. *Free Radic. Biol. Med.* **2019**, *131*, 356–369. [CrossRef]
27. Wang, L.; Li, J.; Shi, X.; Li, S.; Tang, P.M.; Li, Z.; Li, H.; Wei, C. Antimalarial Dihydroartemisinin triggers autophagy within HeLa cells of human cervical cancer through Bcl-2 phosphorylation at Ser70. *Phytomedicine* **2019**, *52*, 147–156. [CrossRef]
28. Shi, X.; Wang, L.; Ren, L.; Li, J.; Li, S.; Cui, Q.; Li, S. Dihydroartemisinin, an antimalarial drug, induces absent in melanoma 2 inflammasome activation and autophagy in human hepatocellular carcinoma HepG2215 cells. *Phytother. Res.* **2019**, *33*, 1413–1425. [CrossRef]
29. Zou, J.; Ma, Q.; Sun, R.; Cai, J.; Liao, H.; Xu, L.; Xia, J.; Huang, G.; Yao, L.; Cai, Y.; et al. Dihydroartemisinin inhibits HepG2.2.15 proliferation by inducing cellular senescence and autophagy. *BMB Rep.* **2019**, *52*, 520–524. [CrossRef] [PubMed]
30. Zhu, W.; Li, Y.; Zhao, D.; Li, H.; Zhang, W.; Xu, J.; Hou, J.; Feng, X.; Wang, H. Dihydroartemisinin suppresses glycolysis of LNCaP cells by inhibiting PI3K/AKT pathway and downregulating HIF-1alpha expression. *Life Sci.* **2019**, *233*, 116730. [CrossRef] [PubMed]
31. Liu, J.; Ren, Y.; Hou, Y.; Zhang, C.; Wang, B.; Li, X.; Sun, R.; Liu, J. Dihydroartemisinin Induces Endothelial Cell Autophagy through Suppression of the Akt/mTOR Pathway. *J. Cancer* **2019**, *10*, 6057–6064. [CrossRef] [PubMed]
32. Xia, M.; Liu, D.; Tang, X.; Liu, Y.; Liu, H.; Liu, Y.; Chen, G.; Liu, H. Dihydroartemisinin inhibits the proliferation of IgAN mesangial cells through the mTOR signaling pathway. *Int. Immunopharmacol.* **2020**, *80*, 106125. [CrossRef]
33. Chen, X.; He, L.Y.; Lai, S.; He, Y. Dihydroartemisinin inhibits the migration of esophageal cancer cells by inducing autophagy. *Oncol. Lett.* **2020**, *20*, 94. [CrossRef] [PubMed]
34. Zhu, L.; Chen, X.; Zhu, Y.; Qin, J.; Niu, T.; Ding, Y.; Xiao, Y.; Jiang, Y.; Liu, K.; Lu, J.; et al. Dihydroartemisinin Inhibits the Proliferation of Esophageal Squamous Cell Carcinoma Partially by Targeting AKT1 and p70S6K. *Front. Pharmacol.* **2020**, *11*, 587470. [CrossRef]
35. Cybulski, N.; Zinzalla, V.; Hall, M.N. Inducible raptor and rictor knockout mouse embryonic fibroblasts. *Methods Mol. Biol.* **2012**, *821*, 267–278.
36. Sohretoglu, D.; Zhang, C.; Luo, J.; Huang, S. ReishiMax inhibits mTORC1/2 by activating AMPK and inhibiting IGFR/PI3K/Rheb in tumor cells. *Signal Transduct. Target. Ther.* **2019**, *4*, 21. [CrossRef] [PubMed]
37. Shang, C.; Zhou, H.; Liu, W.; Shen, T.; Luo, Y.; Huang, S. Iron chelation inhibits mTORC1 signaling involving activation of AMPK and REDD1/Bnip3 pathways. *Oncogene* **2020**, *39*, 5201–5213. [CrossRef]
38. Lu, M.; Tang, Q.; Olefsky, J.M.; Mellon, P.L.; Webster, N.J. Adiponectin activates adenosine monophosphate-activated protein kinase and decreases luteinizing hormone secretion in LβT2 gonadotropes. *Mol. Endocrinol.* **2008**, *22*, 760–771. [CrossRef] [PubMed]
39. Chen, L.; Xu, B.; Liu, L.; Luo, Y.; Yin, J.; Zhou, H.; Chen, W.; Shen, T.; Han, X.; Huang, S. Hydrogen peroxide inhibits mTOR signaling by activation of AMPKα leading to apoptosis of neuronal cells. *Lab. Invest.* **2010**, *90*, 762–773. [CrossRef]
40. Liu, L.; Chen, L.; Chung, J.; Huang, S. Rapamycin inhibits F-actin reorganization and phosphorylation of focal adhesion proteins. *Oncogene* **2008**, *27*, 4998–5010. [CrossRef]
41. Beevers, C.S.; Chen, L.; Liu, L.; Luo, Y.; Webster, N.J.; Huang., S. Curcumin disrupts the Mammalian target of rapamycin-raptor complex. *Cancer Res.* **2009**, *69*, 1000–1008. [CrossRef] [PubMed]
42. McConkey, B.J.; Sobolev, V.; Edelman, M. The performance of current methods in ligand–protein docking. *Curr. Sci.* **2002**, *83*, 845–856.
43. Rao, S.N.; Head, M.S.; Kulkarni, A.; LaLonde, J.M. Validation studies of the site-directed docking program LibDock. *J. Chem. Inf. Model.* **2007**, *47*, 2159–2171. [CrossRef]
44. Yang, H.; Rudge, D.G.; Koos, J.D.; Vaidialingam, B.; Yang, H.J.; Pavletich, N.P. mTOR kinase structure, mechanism and regulation. *Nature* **2013**, *497*, 217–223. [CrossRef] [PubMed]
45. Liang, J.; Choi, J.; Clardy, J. Refined structure of the FKBP12–rapamycin–FRB ternary complex at 2.2 Å resolution. *Acta Crystallogr. D Biol. Crystallogr.* **1999**, *55*, 736–744. [CrossRef]
46. Xiao, B.; Sanders, M.J.; Carmena, D.; Bright, N.J.; Haire, L.F.; Underwood, E.; Patel, B.R.; Heath, R.B.; Walker, P.A.; Hallen, S.; et al. Structural basis of AMPK regulation by small molecule activators. *Nat. Commun.* **2013**, *4*, 3017. [CrossRef] [PubMed]
47. Du, J.H.; Zhang, H.D.; Ma, Z.J.; Ji, K.M. Artesunate induces oncosis-like cell death in vitro and has antitumor activity against pancreatic cancer xenografts in vivo. *Cancer Chemother. Pharmacol.* **2010**, *65*, 895–902. [CrossRef] [PubMed]
48. Brugarolas, J.; Lei, K.; Hurley, R.L.; Manning, B.D.; Reiling, J.H.; Hafen, E.; Witters, L.A.; Ellisen, L.W.; Kaelin, W.G., Jr. Regulation of mTOR function in response to hypoxia by REDD1 and the TSC1/TSC2 tumor suppressor complex. *Genes Dev.* **2004**, *18*, 2893–2904. [CrossRef] [PubMed]
49. DeYoung, M.P.; Horak, P.; Sofer, A.; Sgroi, D.; Ellisen, L.W. Hypoxia regulates TSC1/2-mTOR signaling and tumor suppression through REDD1-mediated 14-3-3 shuttling. *Genes Dev.* **2008**, *22*, 239–251. [CrossRef]

50. Jones, R.G.; Plas, D.R.; Kubek, S.; Buzzai, M.; Mu, J.; Xu, Y.; Birnbaum, M.J.; Thompson, C.B. AMP-activated protein kinase induces a p53-dependent metabolic checkpoint. *Mol. Cell* **2005**, *18*, 283–293. [CrossRef]

51. Feng, Z.; Zhang, H.; Levine, A.J.; Jin, S. The coordinate regulation of the p53 and mTOR pathways in cells. *Proc. Natl. Acad. Sci. USA* **2005**, *102*, 8204–8209. [CrossRef] [PubMed]

52. Steinberg, G.R.; Carling, D. AMP-activated protein kinase: The current landscape for drug development. *Nat. Rev. Drug Discov.* **2019**, *18*, 527–551. [CrossRef]

53. Carretero, J.; Medina, P.P.; Pio, R.; Montuenga, L.M.; Sanchez-Cespedes, M. Novel and natural knockout lung cancer cell lines for the LKB1/STK11 tumor suppressor gene. *Oncogene* **2004**, *23*, 4037–4040. [CrossRef] [PubMed]

54. Herrero-Martín, G.; Høyer-Hansen, M.; García-García, C.; Fumarola, C.; Farkas, T.; López-Rivas, A.; Jäättelä, M. TAK1 activates AMPK-dependent cytoprotective autophagy in TRAIL-treated epithelial cells. *EMBO J.* **2009**, *28*, 677–685. [CrossRef]

55. Garcia-Haro, L.; Garcia-Gimeno, M.A.; Neumann, D.; Beullens, M.; Bollen, M.; Sanz, P. The PP1-R6 protein phosphatase holoenzyme is involved in the glucose-induced dephosphorylation and inactivation of AMP-activated protein kinase, a key regulator of insulin secretion, in MIN6 beta cells. *FASEB J.* **2010**, *24*, 5080–5091. [PubMed]

56. Wu, Y.; Song, P.; Xu, J.; Zhang, M.; Zou, M.H. Activation of protein phosphatase 2A by palmitate inhibits AMP-activated protein kinase. *J. Biol. Chem.* **2007**, *282*, 9777–9788. [CrossRef] [PubMed]

57. Joseph, B.K.; Liu, H.Y.; Francisco, J.; Pandya, D.; Donigan, M.; Gallo-Ebert, C.; Giordano, C.; Bata, A.; Nickels, J.T., Jr. Inhibition of AMP Kinase by the Protein Phosphatase 2A Heterotrimer, PP2APpp2r2d. *J. Biol. Chem.* **2015**, *290*, 10588–10598. [CrossRef] [PubMed]

58. Ching, Y.P.; Kobayashi, T.; Tamura, S.; Hardie, D.G. Specificity of different isoforms of protein phosphatase-2A and protein phosphatase-2C studied using site-directed mutagenesis of HMG-CoA reductase. *FEBS Lett.* **1997**, *411*, 265–268. [CrossRef]

59. Park, H.G.; Yi, H.; Kim, S.H.; Yu, H.S.; Ahn, Y.M.; Lee, Y.H.; Roh, M.S.; Kim, Y.S. The effect of cyclosporine A on the phosphorylation of the AMPK pathway in the rat hippocampus. *Prog. Neuropsychopharmacol. Biol. Psychiatry* **2011**, *35*, 1933–1937. [CrossRef]

60. Chen, Y.L.; Hung, M.H.; Chu, P.Y.; Chao, T.I.; Tsai, M.H.; Chen, L.J.; Hsiao, Y.J.; Shih, C.T.; Hsieh, F.S.; Chen, K.F. Protein phosphatase 5 promotes hepatocarcinogenesis through interaction with AMP-activated protein kinase. *Biochem. Pharmacol.* **2017**, *138*, 49–60. [CrossRef]

61. Ramírez-Rangel, I.; Bracho-Valdés, I.; Vázquez-Macías, A.; Carretero-Ortega, J.; Reyes-Cruz, G.; Vázquez-Prado, J. Regulation of mTORC1 complex assembly and signaling by GRp58/ERp57. *Mol. Cell Biol.* **2011**, *31*, 1657–1671. [CrossRef] [PubMed]

Review

TFEB Biology and Agonists at a Glance

Mingyue Chen [†]**, Yashuang Dai** [†]**, Siyu Liu, Yuxin Fan, Zongxian Ding and Dan Li** *

Collaborative Innovation Center of Yangtze River Delta Region Green Pharmaceuticals, Zhejiang University of Technology, Hangzhou 310000, China; mingyuechen0725@gmail.com (M.C.); daiyashuang09@gmail.com (Y.D.); liusiyu1102@gmail.com (S.L.); yuxinfandaisy@gmail.com (Y.F.); dingzongxian@gmail.com (Z.D.)
* Correspondence: lidan@zjut.edu.cn
† These authors contributed equally to this work.

Abstract: Autophagy is a critical regulator of cellular survival, differentiation, development, and homeostasis, dysregulation of which is associated with diverse diseases including cancer and neurodegenerative diseases. Transcription factor EB (TFEB), a master transcriptional regulator of autophagy and lysosome, can enhance autophagic and lysosomal biogenesis and function. TFEB has attracted a lot of attention owing to its ability to induce the intracellular clearance of pathogenic factors in a variety of disease models, suggesting that novel therapeutic strategies could be based on the modulation of TFEB activity. Therefore, TFEB agonists are a promising strategy to ameliorate diseases implicated with autophagy dysfunction. Recently, several TFEB agonists have been identified and preclinical or clinical trials are applied. In this review, we present an overview of the latest research on TFEB biology and TFEB agonists.

Keywords: TFEB agonists; autophagy; lysosome; rapamycin; resveratrol

Citation: Chen, M.; Dai, Y.; Liu, S.; Fan, Y.; Ding, Z.; Li, D. TFEB Biology and Agonists at a Glance. *Cells* **2021**, *10*, 333. https://doi.org/10.3390/cells10020333

Academic Editors: Stephen Yarwood, Jean Christopher Chamcheu, Claudia Büerger and Shile Huang
Received: 13 December 2020
Accepted: 25 January 2021
Published: 5 February 2021

Publisher's Note: MDPI stays neutral with regard to jurisdictional claims in published maps and institutional affiliations.

1. Introduction

Autophagy occurs in all types of eukaryotic cells, which is a major cellular pathway for degradation of long-lived proteins and cytoplasmic organelles [1]. The degradation of proteins and organelles via autophagy plays an important role in maintaining cell homeostasis and responding to certain environmental stresses [2]. Many diseases, such as cancer and neurodegenerative diseases, are closely related to the failure of autophagy regulation mechanisms.

The transcriptional factor EB (TFEB) is a member of the microphthalmia-transcription factor (MiTF)/TFE family of leucine zipper transcription factors, a signal regulator to promote autophagy [3]. TFEB, generally isolated from the cytoplasm, will be translocated to the nucleus to coordinate the expression and regulation of lysosomes if activated. It is considered the main activator of autophagy-lysosomal gene expression [4]. Hence, TFEB agonists are emerging as novel therapeutic strategies to the treatment of diseases with autophagy-lysosomal dysfunction. The regulation mechanism of TFEB is a very complex process, and the research on TFEB agonists is increasing. So far, TFEB agonist, such as rapamycin, has been approved for the treatment of cancer [5]. Many TFEB agonists are still in preclinical trials. This review mainly elaborates on TFEB agonists and TFEB biology to contribute to the latest TFEB research trends.

2. Autophagy and the Lysosome

Autophagy is a lysosomal-dependent degradation pathway characterized by cytoplasmic vacuolization. Autophagy degrades the damaged structures in the cytoplasm to produce amino acids and free fatty acids for protein and energy synthesis, so that cells can adapt to different environments such as hypoxia and starvation [6]. In mammalian cells, autophagy is divided into macroautophagy, microautophagy, and chaperone-mediated autophagy [1]. Under physiological conditions, autophagy activity is required to maintain the homeostasis of the system. For example, autophagy related 9 (ATG 9), the sole transmembrane protein, mediates lipid scrambling and plays a crucial role in the lipid transport system that enables phagophore expansion [7,8]. Additionally, ATG 16L1 contributes to cell survival in the nutrient depletion state [9]. Under pathophysiological conditions, autophagy will be upregulated as a protection mechanism. For example, autophagy can be activated by serum starvation treatment in endothelial cells. During early starvation, autophagy could protect endothelial cell barrier from breakdown by inhibiting the reactive oxygen species (ROS) production [10].

The survival or death of cells is determined by the degree of autophagy. Physiological levels of autophagy, generally at a relatively low level, promote survival, whereas insufficient or excessive levels of autophagy promote death [11]. Hence, regulating the level of autophagy has been a therapeutic method for diseases related to autophagy. Autophagy can be activated as autophagosomes in HepG2 cells treated with 1.0mg/mL matrine. Matrine can inhibit the proliferation of HepG2 cells in a dose-and time-dependent manner [12]. Apoptosis and autophagy in T-cell acute lymphoblastic leukemia cells can be induced by resveratrol [13]. Puerarin can inhibit hyperglycemia-induced NLRP3 inflammasome activation in endothelial cells via regulating autophagy processes [14]. Apoptosis induced by cisplatin can be enhanced by insulin in human esophageal squamous cell carcinoma EC9706 cells, which is related to the inhibition of autophagy [15].

Autophagy is a defense mechanism of cells against adverse environment stressors, but the failure of regulation mechanisms is closely related to many diseases. Autophagy in cancer cells shows a dual effect, which simultaneously inhibits effect and promotes cancer development. For instance, an ATG 9A dependent autophagic pathway also plays an important role in Beclin 2 negatively regulating tumor development [16]. But starvation-induced autophagy can promote the invasion and migration of bladder cancer cells [17]. Autophagy also sustains pancreatic cancer growth and promotes the survival of dormant breast cancer cells [18,19]. Therefore, reagents that may induce autophagy are immediate areas of cancer treatment research.

Autophagy participates in the degradation of abnormal proteins to prevent the accumulation in neurons. Hence, the occurrence of some neurodegenerative diseases, characterized by the change of intracellular protein degradation system, is related to autophagy. For example, the accumulation of synuclein proteins in Parkinson's disease is related to the decrease of autophagy activity. Autophagy plays an important role in nutritional metabolism, blocking the process will lead to a series of metabolic diseases. For example, autophagy is necessary to maintain the structure, number, and function of islet β cells and plays a protective role under stress conditions [20,21]. Autophagy dysfunction will lead to decreased islet function [22]. For lipid metabolism, autophagy disorders can lead to the accumulation of cellular lipids, resulting in obesity, dyslipidemia, fatty liver, and other diseases [23].

Lysosomes, the cell's recycling center, are acidic compartments filled with more than 60 different types of hydrolases in mammalian cells. Lysosomes are mainly responsible for breaking down endocytosis and autophagy substrates, such as membranes, proteins, and lipids, into their basic components. The digested products are transported out of lysosomes via specific catabolic derivatives or vesicles [24]. Lysosome-mediated signal pathways and transcriptional procedures can sense the state of cell metabolism and control the switch between anabolism and catabolism by regulating lysosomal biogenesis and autophagy [25]. Lysosomes also play a role in cellular metabolism, immunity, regulation of hormone secretion, and so on. For example, melanin secretion of melanocytes, bone tissue reabsorption by osteoclasts, proteolytic enzyme secretion of natural killer cells, regulation of cell membrane repair by calcium ions, antigen presentation by macrophages, and B lymphocytes are all related to the activity of lysosomes [26,27].

Pieces of evidence show that lysosomal dysfunction is related to a variety of human diseases. Lysosomal storage disease (LSD) is a group of metabolic diseases caused by lysosomal hydrolase gene mutations, lysosomal membrane protein mutations, or non-lysosomal protein mutations. These mutations affect catabolism, catabolic output or membrane transport, and lysosome function indirectly, respectively. A prominent feature of LSD is the primary and secondary excessive accumulation of undigested lipids in lysosomes, leading to lysosomal dysfunction and cell death, ultimately resulting in pathological symptoms of various tissues and organs [28]. Fabry disease, an inherited lysosomal storage disorder, is caused by a mutation in the *GLA* gene, which results in a deficiency of the α-galactosidase that is a lysosomal acid hydrolase. This deficiency causes multi-organ pathology with high morbidity and a reduced life expectancy [29]. Neuronal ceroid lipofuscinoses, characterized by a heterogeneous origin of the storage material, are caused by mutations in the *CLN3* gene, which encodes a multi-pass lysosomal membrane protein involved in lysosome homeostasis [30]. Other studies have shown that lysosomes are also associated with diseases such as gout, Gaucher's disease, Parkinson's disease, and cancer [31–33].

3. TFEB as a Drug Target

3.1. The Structure of TFEB

TFEB was first cloned from a human B-cell cDNA due to its capacity of binding the adenoviral major late promoter [34]. TFEB is a protein composed of 476 amino acid residues, mainly including glutamine-rich, helix-loop-helix, leucine-zipper, and proline-rich domains. TFEB pertains to the MiTF/TFE family of basic helix-loop-helix-leucine zipper transcription factors [35].

3.2. TFEB Function

Lysosomes undergo nutrient-sensitive adaptive changes in function and biogenesis [36]. Nearly all receptors serving lysosome biogenesis are under the transcriptional control of TFEB, a master regulator of the lysosomal system. It is discovered that there is a consensus DNA sequence in the promoters of 96 lysosomal genes, which is termed the coordinated lysosomal expression and regulation motif. It has been proved that TFEB specifically targets the Coordinated Lysosomal Expression and Regulation motif to up-regulate genes for lysosomal biogenesis and function [37]. TFEB coordinates the expression of lysosomal hydrolases, lysosomal membrane proteins, and autophagy proteins in response to pathways sensing lysosomal stress and the nutritional conditions of the cell among other stimuli [38]. TEFB also plays an important role in regulating lysosomal exocytosis [39]. Lysosomal exocytosis is a process by which lysosomal membrane fuses with the plasma membrane and the content of the lysosome is excluded from the cell. And the exocytosis of conventional lysosomes is regulated by Ca^{2+}. Mucolipin 1 (MCOLN1)/TRPML1 is the principal Ca^{2+} release channel on the lysosomal membrane [36]. TFEB activates the calcium channel protein MCOLN1, which promotes calcium influx and the fusion of lysosome and plasma membrane [40].

TFEB is of great importance in various cellular physiological processes. The role of TFEB in regulating autophagy is dependent on the cellular localization of TFEB, which is controlled by its phosphorylation status. In nutrient-rich conditions, TFEB is phosphorylated and retained in cytoplasm. Upon starvation or under conditions of lysosomal dysfunction, TFEB is dephosphorylated and translocated from cytoplasm to the nucleus, where it is active to regulate the expression of target genes for cargo recognition, autophagosome formation, vesicle fusion, and substrate degradation. For example, TFEB overexpression activates the biogenesis of autophagosomes [41]. Also, it has been discovered that post-modification of TFEB contributes to its transcriptional activity. TFEB transcription activity is enhanced with histone deacetylase inhibitor suberoylanilide hydroxamic acid, which results in the activation of lysosomal function in human cancer cells [42]. MiTF has been proved to be subject to small ubiquitin-like-modifier modification, as were the related family members TFEB. Sumoylation affects TFEB transcriptional activity in a manner dependent on the promoter elements present in the target genes [43].

A series of in vivo roles of TFEB has been identified. TFEB regulates lipid breakdown in the liver via peroxisome proliferator-activated receptor γ coactivator 1 α (PGC-1α) and peroxisome proliferator activated receptor α (PPAR-α). TFEB and proper lysosomal function are necessary to endoderm differentiation [44]. There is a RANKL-dependent signaling pathway taking place in differentiated osteoclasts. It culminates in the activation of TFEB to enhance lysosomal biogenesis which is necessary for proper bone resorption [45]. Endothelial TFEB promotes glucose metabolism via upregulation of Insulin Receptor Substrate 1 and Insulin Receptor Substrate 2 [46]. TFEB plays an important role in resisting intestinal epithelial cell injury in vivo, potentially mediated by APOA1 [47].

4. Mechanisms of TFEB Activation

4.1. Ca^{2+}-Dependent Mechanism

There is a calcium signaling mechanism that begins with lysosomes and controls autophagy through calcineurin-mediated TFEB induction. Figure 1 shows a model that TFEB is regulated by mTOR- or calcineurin-related pathways under starvation or physical exercise. The lysosomal calcium channels on the lysosome membrane are involved in basic cellular processes. Recent discovery suggests that calcium microdomains probably locate on the surface of lysosomes. Calcium microdomains mediate local calcium signaling in several intracellular compartments, such as mitochondria [48].

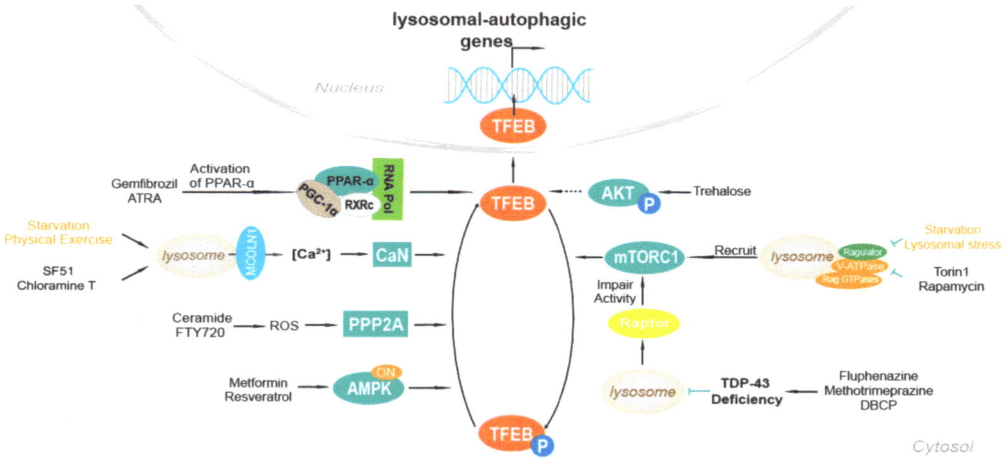

Figure 1. Schematic of mechanisms of TFEB agonists. The schematic shows the model of Ca^{2+} mediated regulation of TFEB, the model of lysosomal sensing and lysosome-to-nucleus signaling by TFEB and mTOR as well as the model of PP2A stimulating regulation of TFEB. AKT, loss of FLCN modulates the autophagy-lysosomal pathway via TFEB. PPAR-α-RXRα-PGC-1α complex and knocking down TDP-43 can promote the transcriptional activation of TFEB. Abbreviation: CaN, Calcineurin; AKT, protein kinase B.

The Ca^{2+} signaling mechanism of lysosome controls the activity of calcineurin phosphatase and TFEB. MCOLN1 is a direct transcriptional target of TFEB. There is a positive feedback loop in TFEB that promotes TFEB activity through Ca^{2+}-mediated calcineurin activation by regulating MCOLN1 expression. After the release of lysosome Ca^{2+} through MCOLN1, the establishment of Ca^{2+} microdomains leads to a decrease in TFEB phosphorylation rate as mTORC1 is inhibited. TFEB dephosphorylation is induced by calcineurin, resulting in nuclear translocation of TFEB. Dephosphorylated TFEB is no longer able to bind 14-3-3 proteins and can freely move to the nucleus where it transcriptionally activates the lysosomal and autophagy pathway [49].

Furthermore, various oxidants including chloramine T, H_2O_2, and m-chlorophenylhydrazone can activate the MCOLN1/TRPML1, the major Ca^{2+} release channel on the lysosome membrane [36]. Additionally, the use of MCOLN1-specific agonist, SF51 can promote TFEB nuclear translocation in a calcineurin-dependent manner [49].

4.2. AKT

The serine and threonine kinase AKT, also known as protein kinase B integrates inputs from growth factors and metabolic effectors to control key multifunctional signaling hubs through direct phosphorylation of substrates that control cell growth, proliferation, and survival signal hub.

In the AKT-mediated autophagy-lysosomal pathway, AKT phosphorylates TFEB on S467, which is an evolutionarily conserved serine among vertebrate species and a close homolog of MiTF family. The TFEB phosphate mutant (TFEB S467A), in which the AKT phosphate receptor site is replaced by alanine to prevent AKT-mediated phosphorylation, increases nuclear localization and stability, and enhances the ability to activate TFEB downstream target genes. The use of pharmacological inhibitors or trehalose that inhibits AKT activity can promote nuclear translocation of TFEB and activation of the autophagy-lysosomal pathway, and enhance autophagy and lysosomal substrate clearance [50].

4.3. mTOR

The pathway regulating autophagy is mediated by calcium ion dependent phosphatase calcineurin. Figure 1 shows a model of TFEB regulated by mTOR or calcineurin-related pathways in the context of starvation and physical exercise. MTORC1 is a major kinase complex that plays an active role on the lysosome surface in mediating TFEB phosphorylation, positively regulating cell growth, and negatively regulating autophagy. MTOR mediated phosphorylation of TFEB-serine residues S142 and S211 promotes the interaction between TFEB and 14-3-3 protein, leading to cytoplasmic localization [51]. In contrast, conditions that lead to mTOR inhibition, such as starvation and lysosomal stress, promote TFEB nuclear translocation and transcriptional activation of lysosomal and autophagy genes.

The phosphorylation status of TFEB and its subcellular localization are entirely determined by the activation state of the Rag GTPases, which regulate mTORC1 activity downstream of amino acids. At any given time, some of TFEB rapidly and transiently binds to the surface of lysosomes, where phosphorylated and kept in the cytoplasm by mTORC1. When there are full nutrients without stress of lysosomal, the complex formed by V-ATPase, Regulator, and Rag GTPases is in the active state and recruits mTORC1 to the lysosomal surface, where mTORC1 is activated. At the lysosome, mTORC1 binds to and phosphorylates TFEB. TFEB then cycles between the cytoplasm and the surface of the lysosome. Under the phosphorylation conducted by mTORC1, TFEB remains in the cytoplasm and is prevented from being translocated to the nucleus. Starvation, v-ATPase inhibition, or lysosomal stress switches the Rags off, making mTORC1 depart from the lysosome and leading it to inactivation [52]. Hence, TFEB can be activated by Torin1, Rapamycin, and other inhibitors of mTOR.

4.4. PPAR-α Activation

Nuclear receptor peroxisome proliferator-activated receptor alpha (PPAR-α) is a PPAR subtype, a transcription factor that interacts with cis-acting DNA elements. PPAR-α agonists can induce the recruitment of the PPAR-α-RXRα-PGC-1α complex on the TFEB promoter and regulate the TFEB promoter region to enhance the transcriptional activation of TFEB, upregulating lysosomal biogenesis. After treating mouse primary astrocytes and neurons with gemfibrozil and ATRA in serum-free medium for 24 h, the overall level of TFEB is increased by 4 times, while the localization of TFEB in the nucleus is increased 5–6 times [53].

4.5. TDP-43 Loss of Function

TAR DNA-binding protein 43 (TDP-43) is considered a major component of the pathogenesis of ALS, FTLD, and other neurodegenerative diseases. TDP-43 regulates the localization of TFEB by targeting raptor, but not Rag GTPases. In TDP-43 knock-down cells, TFEB is translocated from the cytosol to the nucleus and formed lysosomal puncta in cytoplasm, whereas mTOR had a diffusively cytoplasmic distribution, suggesting involvement of mTORC1 in TDP-43-mediated TFEB nuclear translocation. Moreover, TDP-43-induced redistribution of TFEB and mTOR depends on raptor [54]. This regulation in turn enhances overall gene expression in the autophagy-lysosomal pathway (ALP) and increases autophagosome and lysosome biogenesis. However, the loss of TDP-43 also impaired the fusion of autophagosomes and lysosomes through the down-regulation of dynein 1, resulting in immature autophagic vesicle accumulation and overwhelming ALP function. Fluphenazine, methotrimeprazine, and 10-(4′-(N-diethylamino)butyl)-2-chlorophenoxazine (DBCP) have been proved to decrease TDP-43 protein level and activate autophagy in a live cell autophagic flux assay [55].

4.6. AMPK, FLCN, ERK

AMP-activated protein kinase (AMPK), a heterotrimeric enzyme, is an evolutionarily conserved energy sensor that functions to maintain energy homeostasis through coordinating effective metabolic responses to reduced energy availability [56,57]. AMPK not only elicits acute metabolic responses but also promotes metabolic reprogramming and adaptations through regulation of specific transcription factors and coactivators [56]. Folliculin (FLCN) is a binding partner negatively regulating AMPK. The interaction of FLCN with AMPK was regulated by two homologous FLCN-binding proteins FNIP1 and FNIP2. The ablation of FLCN expression or loss of the interaction of FLCN to AMPK causes constitutive AMPK activation, which is associated with metabolic transformation [56]. AMPK plays a pivotal role in regulating TFEB. Pharmacological activation of AMPK promotes TFEB localization, while phosphorylation status of mTOR shows no significant changes. The regulation of TFEB/TFE3 by FLCN is evolutionarily conserved and independent of mTOR pathway [57].

It has been shown that drugs including metformin and resveratrol activate AMPK indirectly by increasing cellular AMP and ADP through inhibiting mitochondrial ATP synthesis [58]. AMPK can also regulate extracellular signal-regulated kinase (ERK) levels, and cells with reduced phospho-ERK levels have elevated phospho-AMPK. Therefore, it can be concluded that inhibition of ERK can up-regulate TFEB [59].

4.7. PP2A Stimulation

Protein phosphatase 2A (PP2A) is a heterotrimeric enzyme with a scaffolding subunit "A", regulatory subunit "B", and catalytic subunit "C". It is a major serine/threonine phosphatase that functions in regulation of many cellular processes including cell cycle, growth, metabolism, and apoptosis. Thus, PP2A is involved in many diseases including neurodegenerative disorders, cardiovascular pathologies, and cancer [60]. PP2A can dephosphorylate TFEB at several residues, including S109, S114, S122, and S211 in response to oxidative stress. This is a mTOR-independent pathway. Depletion of either PP2A catalytic subunits or the regulatory subunit, alone or in combination with the catalytic subunits, reduces TFEB-S211 dephosphorylation significantly. Ceramide or FTY720 can cause TFEB nuclear translocation through PP2A activation [61].

5. TFEB and Diseases

The current enzyme replacement therapy (ERT) for LSD has been proved successful in reversing cardiac abnormalities but has a limitation in skeletal muscle abnormalities [62]. Besides, the replacement enzymes are not easily accessible to target tissues. Compared to ERT, modulation of TFEB exploits lysosomal exocytosis to expel the storage material into the extracellular space, which solves the problem of enzyme delivery efficiency. TFEB is also proved to promote autophagosome-lysosomal fusion [40].

Pompe disease (PD) is a metabolic myopathy, characterized by the deficiency of acid alpha-glucosidase, which results in the excessive lysosomal glycogen storage. PD is also characterized by the secondary accumulation of autophagic debris. Overexpression of TFEB can stimulate fusion between lysosomes and autophagosomes, resulting in the formation of autolysosomes and increased exocytosis in muscle cells, in isolated live muscle fibers in PD mice [63,64].

Huntington's disease (HD) is characterized by an expanded polyglutamine (PolyQ) chain in the HTT proteins, which is caused by the repeat expansion of CAG trinucleotide in the first exon of the *HTT* gene. PolyQ-expanded proteins misfold forming aggregates. Enhancing the autophagy-lysosomal pathway through TFEB overexpression can reduce HTT protein aggregation in cells. In a mouse model of HD, TFEB can improve neurological function when overexpressed [65,66].

Parkinson's disease (PD) without fully defined pathogenesis has major pathological changes including the progressive death of nigral dopamine neurons and the accumulations of the pathogenic protein SNCA/α-synuclein [67]. The accumulation of SNCA invalidates the autophagy pathway, further leading to pathogenic protein aggregation [68]. In general, enhancing autophagy-mediated degradation of SNCA through TFEB regulation is a promising strategy for PD prevention and treatment.

Alzheimer's disease (AD) is a neurodegenerative disease with main pathological features of β-amyloid peptides (Aβ) deposition in the brain and intracellular neurofibrillary tangle formation. In AD, the maturation of autophagosome-lysosome and retrograde transport are blocked, resulting in large swellings along dystrophic and degenerating neurites. The activated autophagy-lysosomal pathway is a key degradative pathway that can partially protect neurons from pathogenic protein aggregation [69].

6. Methods to Screen TFEB Agonists

There are plenty of methods to identify TFEB agonists. But there are no absolute criteria for determining autophagic status for each biological or experimental context. It is because some assays are inappropriate, problematic or may not work at all in particular cells, tissues, or organisms [70]. Table 1 shows the different methods to screen TFEB agonists. TFEB activation can be reflected by its subcellular localization. TFEB localization [71] can be directly revealed by experiments such as immunofluorescence. Real-time fluorescent quantitative PCR [72] shows that the expression of TFEB and TFEB downstream targets are up regulated after TFEB agonist treatment with high dynamic and accuracy. TFEB agonists can also be determined by the TFEB- promoter binding activity [53]. TFEB agonists are supposed to activate autophagy, and the change of autophagic flow can be detected by electron microscopy [73,74]. Therefore, the activity of TFEB agonists can be determined. When TFEB is activated, the protein expression of TFEB targets is upregulated, which can be assessed by Western Blot [75]. The electrophysiology technique [76] can measure the transmembrane current on phospholipid membranes to verify the effect of TFEB agonists as well. A nanotechnology-enabled high-throughput screen (HTS) [77] to identify small-molecule agonists of TFEB has been reported. It is an experimental method based on both cellular and molecular levels, so it has the characteristics of trace, rapid, sensitive, and accurate.

Table 1. Methods to screen TFEB agonists.

Method	Function	Detection Indicator	Reference
Nuclear translocation	When TFEB is activated, it is displaced into the nucleus, so it is possible to determine whether TFEB is activated by observing the nucleus translocation.	DAPI, GFP, FITC, EB	Najibi, et al., 2016 [71]
Gene expression (qPCR)	Fluorescence chemicals were used to measure the total amount of products after each PCR cycle	CT value	Bao, et al., 2016 [72]
Promoter activation	The initiation of transcription is the key stage of gene expression, and promoter activation increases the level of gene expression.	Target protein	Ghosh, et al., 2015 [53]
Autophagy flux	Observing the accumulation of autophagosomes can reflect the induction of autophagosomes and the reduction of autophagosome consumption	mRFP-GFP-LC3, LC3-I, LC3-II	Castillo, et al., 2013 [73] Wang, et al., 2009 [74]
Protein expression (WB)	TFEB is activated and the amount of target protein produced increases.	Target protein	Dai, et al., 2020 [75]
Electrophysiology	Membrane potential changes as substances cross cell membranes, and TFEB changes are observed by recording the electrical activity of cells in the body.	Transmembrane current, Action potential	Kosacka, et al., 2013 [76]
A UPS-enabled high-throughput screen	Trace, fast, sensitive, and accurate screening the TFEB agonists.	PH, LAMP1	Wang, et al., 2017 [77]

7. TFEB Agonists

TFEB agonists that have been discovered so far are shown in Table 2.

Table 2. TFEB agonists.

TFEB Agonists	Structure	Mechanism	Clinical Trials and Preclinical Trials	Reference
Resveratrol (D)	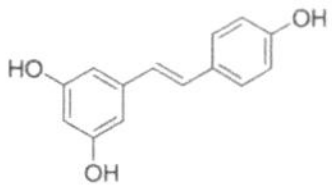	How RSV regulates autophagy has not been discovered.	Clinical trials	Zhou, et al., 2019 [78] Elnaz, et al., 2020 [79]
Curcumin analog C1 (D)	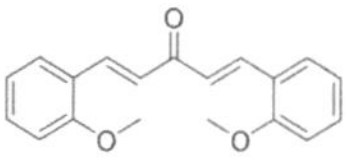	C1 binds specifically to TFEB, facilitating TFEB entry into the nucleus.	Preclinical trials	Song, et al., 2016 [80] Song, et al., 2020 [81]
Progestin R5020 (D)	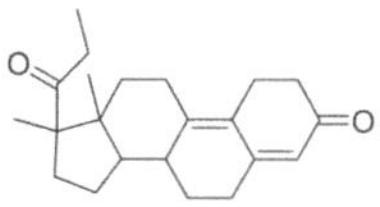	The interaction between PRB and TFEB increased autophagy in McF-7 breast cancer cells.	Preclinical trials	Tan, et al., 2019 [82]
Potential TFEB agonists (D)	A class of compounds related to Ca^{2+} that are structurally diverse	Ca^{2+} dependent mechanisms	Approved drug digoxin, ikarugamycin, alexidine dihydrochloride	Wang, et al., 2017 [77] Settembre, et al., 2013 [83]
Torin1 (In)	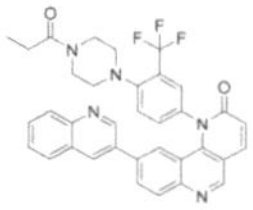	Inhibits mTOR	Preclinical trials	Thoreen, et al., 2009 [84] Cheng, et al., 2016 [85]

Table 2. *Cont.*

TFEB Agonists	Structure	Mechanism	Clinical Trials and Preclinical Trials	Reference
3,4-Dimet-hoxychalc-one (In)		Inhibits mTOR	Preclinical trials	Chen, et al., 2019 [86]
Fisetin (In)		Inhibits mTOR and is related to ALP	Preclinical trials	Kim, et al., 2016 [87]
Rapamycin (In)		Inhibits mTOR	Approved drug: Sirolimus	Abe, et al., 2019 [88] Cui, et al., 2020 [89]

D = Direct TFEB agonists; In = Indirect TFEB agonists.

7.1. Direct TFEB Agonists

7.1.1. Resveratrol

Resveratrol (3,4′,5-trihydroxystilbene, RSV) is a natural polyphenolic compound commonly extracted from grapes. In the study, RSV pretreatment in PA-treated human umbilical vein endothelial cells efficiently up-regulates LC3-II expression and down-regulates p62 expression, which indicates that RSV activates autophagy. It is proved that TFEB is partly a target of RSV. RSV promotes TFEB translocation into nucleus, evaluates TFEB and TFEB-downstream gene expression and thus promotes autophagy. It is clear that the expression of TFEB, lysosomal associated membrane protein 1 (LAMP1), and downregulating of p62 were significantly inhibited by siTFEB transfection in PA-treated human umbilical vein endothelial cells [78].

7.1.2. Curcumin Analog C1

A synthesized curcumin derivative termed C1 has been proved as a new direct TFEB agonist. C1 specifically binds to TFEB and promotes the TFEB's entry to the nucleus. Interestingly, TFEB promotes phosphorylation of RPS6KB1 and mTOR, which confirms that C1 activates TFEB without inhibiting mTOR pathway. At the same time, it is found that C1 has almost no effects on the activity of mTOR related kinases, which play fundamental roles in regulating autophagy. It has been proved that TFEB is especially required when C1 promotes autophagy through the experiments in which key autophagy genes *Atg5* (autophagy related 5), *Becn1* (beclin 1) and *TFEB* in N2a cells were knocked down. C1 is of structural stability and good blood-brain barrier permeability, so it is a potential drug for neurodegenerative diseases [80].

The effects of C1 on three AD animal models, which represent beta-amyloid precursor protein (APP) pathology (5xFAD mice), tauopathy (P301S mice) and the APP/Tau combined pathology (3xTg-AD mice), were investigated. And C1 efficiently promotes autophagy and lysosomal activity through TFEB activation. Thus APP, APP C-terminal fragments (CTF-β/α), β-amyloid peptides, and Tau aggregates are reduced [81]. The function of C1 in AD models indicates that TFEB agonists can be a new therapeutic target for such diseases.

7.1.3. Progestin R5020

Prolonged treatment with progestin R5020 upregulates autophagy in MCF-7 human breast cancer cells via a novel interplay between progesterone receptor B and TFEB [82]. In addition, R5020 enhances the co-recruitment of progesterone receptor B and TFEB to mutually promote the verification of TFEB. Once in the nucleus, TFEB induces autophagy expression and lysosomal genes, which enhances autophagy. Since R5020-induced autophagy is independent of Akt-mTOR signaling, it is speculated that R5020 may affect TFEB activation, thereby enhancing autophagy in MCF-7 cells [82]. So, progestin R5020 is one of the TFEB agonists.

7.1.4. Potential TFEB Agonists

Starvation or calorie restriction can activate TFEB. There are three novel compounds that promote autophagolysosomal activity, including the clinically approved drug, digoxin (DG); the marine-derived natural product, ikarugamycin (IKA); and the synthetic compound, alexidine dihydrochloride, which is known to act on a mitochondrial target. They participate in the TFEB activation mechanism through three different Ca^{2+} sources and Ca^{2+}. And all three compounds promote autophagy flux and activate TFEB in a dose-dependent manner.

These Ca^{2+} sources selectively participate in the global Ca^{2+}-sensing CaMKKβ-AMPK pathway or the local Ca^{2+}-sensing MCOLN1-calcineurin pathway to release "brake" (inhibition of mTORC1 through AMPK) and/or promote TFEB activation "promoter" (activation of TFEB phosphatase). DG-induced lysosomal calcium released through MCOLN1 depends on the activity of an uncharacterized Ca^{2+}-reactive TFEB phosphatase. Alexidine dihydrochloride-induced TFEB activation is realized by the direct perturbation of the mitochondrial protein PTPMT1 and the ROS-dependent MCOLN1-calcineurin pathway. Unlike DG and alexidine dihydrochloride, IKA triggers conventional ER-mediated Ca^{2+} release to activate the CaMKKβ-AMPK pathway [77].

In animals, TFEB plays a key role in promoting lipid metabolism during starvation, at least in part through global transcriptional activation of PGC-1α and PPAR-α [83]. Consistent with physiologically relevant TFEB activation, DG, alexidine dihydrochloride, and IKA significantly improves oleic acid-induced lipid accumulation in human hepatocytes. Newly discovered small-molecule TFEB agonists reduce metabolic syndrome and prolong lifespan of the body, addressing aging and age-related diseases.

7.2. Indirect TFEB Agonists

7.2.1. Torin1

Previous studies have shown that TFEB is regulated by mTORC1 [90]. However, TFEB regulation by mTORC1 is complex and cell context-dependent [91]. Nutritional deprivation leads to the inhibition of mTORC1, the reduction of TFEB phosphorylation, and the promotion of TFEB nuclear translocation and lysosomal gene expression. It is reported that mTORC1 phosphorylates TFEB on S211 and on S142 [52]. The phosphorylation of S211 by mTORC1 creates a high affinity binding site for YWHA protein, which leads to the cytoplasmic retention of TFEB. When mTORC1 was inhibited, S211 is dephosphorylated and TFEB enters the nucleus.

Torin1 is an mTOR inhibitor [84]. When there are abundant nutrients and growth factors, TFEB is mainly cytoplasmic. However, after treatment with Torin1, TFEB was dephosphorylated and mainly localized in the nucleus. Hence, Torin1 treatment can transform TFEB into a rapidly migrating form of low phosphorylation, changing the distribution of TFEB from the whole cell diffusion mode to almost all nuclear cells. This phenomenon has been observed in many cell types [92]. TFEB regulation via Torin1 has been attributed to the change of S211 phosphorylation. The mutation from S211 to A increases the TFEB nuclear translocation [93]. Intra-articular injection of Torin 1 can reduce degeneration of articular cartilage in collagenase-induced osteoarthritis (OA), at least partially by autophagy activation, and it actives TFEB by inhibiting mTOR [85].

7.2.2. Rapamycin

Rapamycin is an inhibitor that interferes with the inhibition of the mTOR signaling pathway, so it is a TFEB agonist [94]. In TNF/ zVAD-treated cells, rapamycin increases nuclear localization of TFEB and promotes autolysosome formation in a TFEB-dependent manner [88]. Rapamycin also suppresses TNF/zVAD-induced RIP1-S166 phosphorylation and increases phosphorylation of RIP1-S320, which is an inhibitory phosphorylation site [88]. Since TFEB is phosphorylated before activation, when the inhibit phosphorylation sites are suppressed, TFEB is more likely to be activated.

7.2.3. 3,4-Dimethoxychalcone

Caloric restriction mimetics (CRMs) are natural or synthetic compounds that mimic the health-promoting and longevity-extending effects of caloric restriction. CRMs provoke the deacetylation of cellular proteins coupled to an increase in autophagic flux in the absence of toxicity [86]. 3,4-Dimethoxychalcone (3,4-DC), a member of CRM, is a TFEB agonist. When added to several different human cell lines, 3,4-DC induces the deacetylation of cytoplasmic proteins and stimulates autophagic flux [86]. Unlike other well-characterized CRMs, 3,4-DC, requires TFEB- and TFE3-dependent gene transcription and mRNA translation to trigger autophagy [86]. 3,4-DC stimulates the translocation of TFEB and TFE3 into nuclei both in vitro and in vivo, in hepatocytes and cardiomyocytes. GFP-TFEB transfers from cytoplasm to nucleus when treated with 3,4-DC, which can be observed by immunofluorescence or western blotting [86]. Tuberous sclerosis complex (*TSC2*) knockout, an operation that leads to structural activation of mTOR, inhibits 3,4-DC-induced TFEB translocation. *TSC2* gene knockout eliminates 3,4-DC induction of p62 and LC3-II protein levels. Moreover, like Torin1, 3,4-DC induces dephosphorylation of TFEB at S211, which is consistent with its ability to inhibit phosphorylation of another mTOR substrate, P70S6K. In vitro and in vivo research identifies 3,4-DC as a novel TFEB/TFE3 agonist through mTOR inhibition [95].

7.2.4. Fisetin

Fisetin is an organic flavonoid found in a variety of fruits and vegetables, including strawberries, mangoes and cucumbers. Initially, it was identified in a screening of flavonoids. Fisetin activates autophagy, as well as TFEB and Nrf2 [87]. The activation of autophagy including TFEB is likely due to fisetin-mediated mammalian target of mTORC1 inhibition since the phosphorylation levels of p70S6K and 4E-BP1 are decreased in the presence of fisetin. Indeed, fisetin-induced phosphorylated tau degradation is attenuated by chemical inhibitors of the autophagy-lysosomal pathway. To examine whether TFEB is involved in fisetin-induced decrease of phosphorylated tau, cytosolic and nuclear fractions were prepared from cortical cells treated with vehicle only 5 or 10 μM of fisetin because upon activation TFEB enters nucleus [87]. Increases in the level of TFEB were observed in nuclear fractions of cells treated with 5 or 10 μM of fisetin compared to control cells treated with vehicle only. And mRNA levels of TFEB-downstream genes such as *ATG9B* and *LAMP1* were significantly increased in both cortical cells and primary neurons treated with fisetin.

8. Clinical Trials and Preclinical Trials of TFEB Agonists

The use of animal models, both lower organisms and mammals, has been very helpful to further elucidate TFEB function. In a mouse model of diet-induced fatty liver disease, TFEB agonists including digoxin, ikarugamycin, and alexidine dihydrochloride have been shown to improve lipid metabolism and overcome insulin resistance. These molecules represent clues to the development of treatment strategies for metabolic syndrome, aging and age-related diseases [77]. Among Akt modulators, MK2206, which is currently undergoing preclinical and phase I and clinical studies, is an effective oral inhibitor of Akt [89,96]. Intraperitoneal injection of MK2206 leads to inhibition of Akt activity and to TFEB nuclear translocation in a mouse brain, which in turn promotes the up-regulation of lysosomal and autophagic genes. It provides the evidence of enhanced pharmacological activation of TFEB for autophagy-lysosomal pathway in vitro and in vivo [97].

For the clinical trials of resveratrol, taking a dose of up to 5 g/day within a month is safe and well tolerated. However, dose-related mild to moderate side effects occur, coupled with resveratrol's ability to alter the activity of drug metabolizing enzymes, which leads to the dose limitation used in future studies to <1.0 g/day. In human, oral resveratrol can be effectively absorbed. These effects should be further studied to help determine the optimal dose for the next phase of clinical trials for clinical efficacy evaluation [98]. A new research containing sixteen clinical trials shows that resveratrol supplementation significantly increases Glutathione Peroxidase serum levels. Hence, further large prospective clinical trials are needed to confirm the effect of resveratrol supplement on oxidative stress markers [79].

Curcumin analog C1 activates TFEB by directly binding to TFEB and promotes its entry into the nucleus, without affecting TFEB phosphorylation or inhibiting mTORC1 and MAPK1/ERK2 activity. Curcumin analog C1 works on the homozygous human P301S tau transgenic mice and homozygous 3xTg mice, and is still in the preclinical trials [81].

Rapamycin has already been an approved drug named "Sirolimus", which is used to treat cancer [5]. Besides, progestin R5020, 3,4-Dimethoxychalcone, fisetin and Torin 1 are still in the process of preclinical trials [82,84,86,87].

9. Conclusions

The identification of TFEB as a global regulation of genes involved in the lysosomal–autophagic pathway, has provided new insights into the therapeutic role of TFEB. Owing to the broad number of diseases that potentially benefit from promoting lysosome and autophagy function, modulating the activity of TFEB represents an appealing therapeutic target. Genetic modification of TFEB has shown protection effects in several animal disease models. However, long-term effects of such treatments have not been evaluated. Studying the role and TFEB agonists in diseases will help provide a new perspective for the treatment and the development of new drugs. In this review, we summarize currently identified TFEB agonists, however, the effects of these agonists in preclinical or clinical trials still requires further investigations.

Author Contributions: Conceptualization, D.L.; resources, M.C., Y.D., S.L., Y.F.; Z.D. and D.L.; data curation, D.L., M.C., Y.D., S.L., Y.F. and Z.D.; writing—original draft preparation, M.C., Y.D., Y.F. and D.L.; writing—review and editing, D.L., M.C., Y.D., S.L., Y.F. and Z.D.; supervision, D.L.; project, D.L. All authors have read and agreed to the published version of the manuscript.

Funding: This work was supported by a grant from National Natural Science Foundation, China (31600823 to D.L.).

Institutional Review Board Statement: Not applicable.

Informed Consent Statement: Not applicable.

Data Availability Statement: Not applicable.

Conflicts of Interest: The authors declare no conflict of interest.

References

1. Lamark, T.; Svenning, S.; Johansen, T. Regulation of selective autophagy: The p62/SQSTM1 paradigm. *Essays Biochem.* **2017**, *61*, 609–624.
2. Yoshimori, T. Autophagy: A regulated bulk degradation process inside cells. *Biochem. Biophys. Res. Commun.* **2004**, *313*, 453–458. [CrossRef]
3. Chao, X.; Wang, S.; Zhao, K.; Li, Y.; Williams, J.A.; Li, T.; Chavan, H.; Krishnamurthy, P.; He, X.C.; Li, L.; et al. Impaired TFEB-Mediated Lysosome Biogenesis and Autophagy Promote Chronic Ethanol-Induced Liver Injury and Steatosis in Mice. *Gastroenterology* **2018**, *155*, 865–879.e12. [CrossRef]
4. Sha, Y.; Rao, L.; Settembre, C.; Ballabio, A.; Eissa, N.T. STUB1 regulates TFEB-induced autophagy-lysosome pathway. *EMBO J.* **2017**, *36*, 2544–2552.
5. Wu, C.; Wang, Q.; Xu, D.; Li, M.; Zeng, X. Sirolimus for patients with connective tissue disease-related refractory thrombocytopenia: A single-arm, open-label clinical trial. *Rheumatology* **2020**, keaa645. [CrossRef]
6. Yang, B.; Ding, L.; Chen, Y.; Shi, J. Augmenting Tumor-Starvation Therapy by Cancer Cell Autophagy Inhibition. *Adv. Sci.* **2020**, *7*, 1902847. [CrossRef]

7.   Matoba, K.; Kotani, T.; Tsutsumi, A.; Tsuji, T.; Mori, T.; Noshiro, D.; Sugita, Y.; Nomura, N.; Iwata, S.; Ohsumi, Y.; et al. Atg9 is a lipid scramblase that mediates autophagosomal membrane expansion. *Nat. Struct. Mol. Biol.* **2020**, *27*, 1185–1193. [CrossRef]

8.   Maeda, S.; Yamamoto, H.; Kinch, L.N.; Garza, C.M.; Takahashi, S.; Otomo, C.; Grishin, N.V.; Forli, S.; Mizushima, N.; Otomo, T. Structure, lipid scrambling activity and role in autophagosome formation of ATG9A. *Nat. Struct. Mol. Biol.* **2020**, *27*, 1194–1201. [CrossRef]

9.   Chen, F.; Amgalan, D.; Kitsis, R.N.; Pessin, J.E.; Feng, D. ATG16L1 autophagy pathway regulates BAX protein levels and programmed cell death. *J. Biol. Chem.* **2020**, *295*, 15045–15053. [CrossRef]

10. Yang, Z.; Huang, C.; Wu, Y.; Chen, B.; Zhang, W.; Zhang, J. Autophagy Protects the Blood-Brain Barrier Through Regulating the Dynamic of Claudin-5 in Short-Term Starvation. *Front. Physiol.* **2019**, *10*, 2. [CrossRef]

11. Liu, K.; Sun, X.; Chen, W.; Sun, Y. Autophagy: A double-edged sword for neuronal survival after cerebral ischemia. *Neural Regen. Res.* **2014**, *9*, 1210–1216. [CrossRef]

12. Zhang, J.; Li, Y.; Liu, T.; He, W.; Chen, Y.; Chen, X.; Li, X.; Zhou, W.; Yi, J.; Ren, Z. Antitumor effect of matrine in human hepatoma G2 cells by inducing apoptosis and autophagy. *World J. Gastroenterol.* **2010**, *16*, 4281–4290. [CrossRef]

13. Ge, J.; Liu, Y.; Guo, X.; Gu, L.; Ma, Z.G.; Zhu, Y.P. Resveratrol Induces Apoptosis and Autophagy in T-cell Acute Lymphoblastic Leukemia Cells by Inhibiting Akt/mTOR and Activating p38-MAPK. *Biomed. Environ. Sci.* **2013**, *26*, 902–911.

14. Cai, D.; Han, R.; Liu, J.; Zhang, X.; Lian, D.; Lai, X.; Chen, Y. Puerarin inhibits hyperglycemia-induced NLRP3 inflammasome activation in endothelial cells via regulating autophagy process. *Chin. J. Pharm. Toxicol.* **2019**, *33*, 795.

15. Yang, Y.; Wen, F.; Dang, L.; Fan, Y.; Liu, N.; Wu, K.; Zhao, S. Insulin enhances apoptosis induced by cisplatin in human esophageal squamous cell carcinoma EC9706 cells related to inhibition of autophagy. *Chin. Med. J.* **2014**, *127*, 353–358.

16. Zhu, M.; Deng, G.; Tan, P.; Xing, C.; Guan, C.; Jiang, C.; Zhang, Y.; Ning, B.; Li, C.; Yin, B.; et al. Beclin 2 negatively regulates innate immune signaling and tumor development. *J. Clin. Investig.* **2020**, *130*, 5349–5369. [CrossRef]

17. Tong, H.; Yin, H.; Hossain, M.A.; Wang, Y.; Wu, F.; Dong, X.; Gao, S.; Zhan, K.; He, W. Starvation-induced autophagy promotes the invasion and migration of human bladder cancer cells via TGF-beta1/Smad3-mediated epithelial-mesenchymal transition activation. *J. Cell. Biochem.* **2019**, *120*, 5118–5127.

18. Yang, A.; Herter-Sprie, G.; Zhang, H.; Lin, E.Y.; Biancur, D.; Wang, X.; Deng, J.; Hai, J.; Yang, S.; Wong, K.-K.; et al. Autophagy Sustains Pancreatic Cancer Growth through Both Cell-Autonomous and Nonautonomous Mechanisms. *Cancer Discov.* **2018**, *8*, 276–287. [CrossRef]

19. Vera-Ramirez, L.; Vodnala, S.K.; Nini, R.; Hunter, K.W.; Green, J.E. Autophagy promotes the survival of dormant breast cancer cells and metastatic tumour recurrence. *Nat. Commun.* **2018**, *9*, 1–12. [CrossRef]

20. Jung, H.S.; Chung, K.W.; Kim, J.W.; Kim, J.; Komatsu, M.; Tanaka, K.; Nguyen, Y.H.; Kang, T.M.; Yoon, K.-H.; Kim, J.-W.; et al. Loss of autophagy diminishes pancreatic beta cell mass and function with resultant hyperglycemia. *Cell Metab.* **2008**, *8*, 418–424.

21. Han, D.; Yang, B.; Olson, L.K.; Greenstein, A.; Baek, S.-H.; Claycombe, K.J.; Goudreau, J.L.; Yu, S.-W.; Kim, E.-K. Activation of autophagy through modulation of 5′-AMP-activated protein kinase protects pancreatic beta-cells from high glucose. *Biochem. J.* **2010**, *425*, 541–551.

22. Kuma, A.; Mizushima, N. Physiological role of autophagy as an intracellular recycling system: With an emphasis on nutrient metabolism. *Semin. Cell Dev. Biol.* **2010**, *21*, 683–690. [CrossRef]

23. Reiners, J.J.; Agostinis, P.; Berg, K.; Oleinick, N.L.; Kessel, D. Assessing autophagy in the context of photodynamic therapy. *Autophagy* **2010**, *6*, 7–18. [CrossRef]

24. Haoxing, X.; Dejian, R. Lysosomal physiology. *Annu. Rev. Physiol.* **2015**, *77*, 57–80.

25. Ballabio, A.; Bonifacino, J.S. Lysosomes as dynamic regulators of cell and organismal homeostasis. *Nat. Rev. Mol. Cell Biol.* **2020**, *21*, 101–118. [CrossRef]

26. Gerasimenko, J.V.; Gerasimenko, O.V.; Petersen, O.H. Membrane repair Ca^{2+}-elicited lysosomal exocytosis. *Curr. Biol.* **2001**, *11*, R971–R974.

27. Blott, E.J.; Griffiths, G.M. Secretory lysosomes. *Nat. Rev. Mol. Cell Biol.* **2002**, *3*, 122–131.

28. Samie, M.A.; Xu, H. Lysosomal exocytosis and lipid storage disorders. *J. Lipid Res.* **2014**, *55*, 995–1009. [CrossRef]

29. Wanner, C.; Arad, M.; Baron, R.; Burlina, A.P.; Elliott, P.M.; Feldt-Rasmussen, U.; Fomin, V.V.; Germain, D.; Hughes, A.D.; Jovanovic, A.; et al. European expert consensus statement on therapeutic goals in Fabry disease. *Mol. Genet. Metab.* **2018**, *124*, 189–203. [CrossRef]

30. Carcel-Trullols, J.; Kovacs, A.D.; Pearce, D.A. Cell biology of the NCL proteins: What they do and don't do. *Biochim. Biophys. Acta* **2015**, *1852*, 2242–2255.

31. Pitcairn, C.; Wani, W.Y.; Mazzulli, J.R. Dysregulation of the autophagic-lysosomal pathway in Gaucher and Parkinson's disease. *Neurobiol. Dis.* **2018**, *122*, 72–82. [CrossRef]

32. Kazuyoshi, S.; Yoshiya, T. Recent topics on new biological agents for treatment of rheumatoid arthritis. *Nihon Naika Gakkai Zasshi J. Jpn. Soc. Intern. Med.* **2008**, *97*, 2418–2423.

33. Foghsgaard, L.; Wissing, D.; Mauch, D.; Lademann, U.; Bastholm, L.; Boes, M.; Elling, F.; Leist, M.; Jäättelä, M. Cathepsin B acts as a dominant execution protease in tumor cell apoptosis induced by tumor necrosis factor. *J. Cell Biol.* **2001**, *153*, 999–1010.

34. Carr, C.S.; Sharp, P.A. A helix-loop-helix protein related to the immunoglobulin E box-binding proteins. *Mol. Cell. Biol.* **1990**, *10*, 4384–4388. [CrossRef]

35. Steingrimsson, E.; Copeland, N.G.; Jenkins, N.A. Melanocytes and the *Microphthalmia* transcription factor network. *Annu. Rev. Genet.* **2004**, *38*, 365–411.
36. Zhang, X.; Yu, L.; Xu, H. Lysosome calcium in ROS regulation of autophagy. *Autophagy* **2016**, *12*, 1954–1955. [CrossRef]
37. Sardiello, M.; Palmieri, M.; Di Ronza, A.; Medina, D.L.; Valenza, M.; Gennarino, V.A.; Di Malta, C.; Donaudy, F.; Embrione, V.; Polishchuk, R.S.; et al. A Gene Network Regulating Lysosomal Biogenesis and Function. *Science* **2009**, *325*, 473–477. [CrossRef]
38. Bajaj, L.; Lotfi, P.; Pal, R.; di Ronza, A.; Sharma, J.; Sardiello, M. Lysosome biogenesis in health and disease. *J. Neurochem.* **2018**, *148*, 573–589. [CrossRef]
39. Rodriguez, A.; Webster, P.; Ortego, J.; Andrews, N.W. Lysosomes behave as Ca^{2+}-regulated exocytic vesicles in fibroblasts and epithelial cells. *J. Cell Biol.* **1997**, *137*, 93–104.
40. Palmieri, M.; Impey, S.; Kang, H.; Di Ronza, A.; Pelz, C.; Sardiello, M.; Ballabio, A. Characterization of the CLEAR network reveals an integrated control of cellular clearance pathways. *Hum. Mol. Genet.* **2011**, *20*, 3852–3866. [CrossRef]
41. Settembre, C.; Ballabio, A. TFEB regulates autophagy: An integrated coordination of cellular degradation and recycling processes. *Autophagy* **2011**, *7*, 1379–1381. [CrossRef]
42. Zhang, J.; Wang, J.; Zhou, Z.; Park, J.E.; Wang, L.; Wu, S.; Sun, X.; Lu, L.; Wang, T.; Lin, Q.; et al. Importance of TFEB acetylation in control of its transcriptional activity and lysosomal function in response to histone deacetylase inhibitors. *Autophagy* **2018**, *14*, 1–17. [CrossRef]
43. Miller, A.J.; Levy, C.; Davis, I.J.; Razin, E.; Fisher, D.E. Sumoylation of MITF and Its Related Family Members TFE3 and TFEB. *J. Biol. Chem.* **2005**, *280*, 146–155. [CrossRef]
44. Young, N.P.; Kamireddy, A.; Van Nostrand, J.L.; Eichner, L.J.; Shokhirev, M.N.; Dayn, Y.; Shaw, R.J. AMPK governs lineage specification through Tfeb-dependent regulation of lysosomes. *Genes Dev.* **2016**, *30*, 535–552. [CrossRef]
45. Ferron, M.; Settembre, C.; Shimazu, J.; Lacombe, J.; Kato, S.; Rawlings, D.J.; Ballabio, A.; Karsenty, G. A RANKL-PKC-TFEB signaling cascade is necessary for lysosomal biogenesis in osteoclasts. *Genes Dev.* **2013**, *27*, 955–969. [CrossRef]
46. Sun, J.; Lu, H.; Liang, W.; Zhao, G.; Ren, L.; Hu, D.; Chang, Z.; Liu, Y.; Garcia-Barrio, M.T.; Zhang, J.; et al. Endothelial TFEB (Transcription Factor EB) Improves Glucose Tolerance via Upregulation of IRS (Insulin Receptor Substrate) 1 and IRS2. *Arter. Thromb. Vasc. Biol.* **2020**, 120315310. [CrossRef]
47. Murano, T.; Najibi, M.; Paulus, G.L.C.; Adiliaghdam, F.; Valencia-Guerrero, A.; Selig, M.; Wang, X.; Jeffrey, K.; Xavier, R.J.; Xavier, R.J.; et al. Transcription factor TFEB cell-autonomously modulates susceptibility to intestinal epithelial cell injury in vivo. *Sci. Rep.* **2017**, *7*, 13938. [CrossRef]
48. Rizzuto, R.; Pozzan, T. Microdomains of Intracellular Ca^{2+}: Molecular Determinants and Functional Consequences. *Physiol. Rev.* **2006**, *86*, 369–408. [CrossRef]
49. Tong, Y.; Song, F. Intracellular calcium signaling regulates autophagy via calcineurin-mediated TFEB dephosphorylation. *Autophagy* **2015**, *11*, 1192–1195.
50. Palmieri, M.; Pal, R.; Sardiello, M. AKT modulates the autophagy-lysosome pathway via TFEB. *Cell Cycle* **2017**, *16*, 1237–1238. [CrossRef]
51. Medina, D.L.; Di Paola, S.; Peluso, I.; Armani, A.; De Stefani, D.; Venditti, R.; Montefusco, S.; Rosato, A.S.; Prezioso, C.; Forrester, A.; et al. Lysosomal calcium signalling regulates autophagy through calcineurin and TFEB. *Nat. Cell Biol.* **2015**, *17*, 288–299. [CrossRef]
52. Settembre, C.; Zoncu, R.; Medina, D.L.; Vetrini, F.; Erdin, S.; Erdin, S.; Huynh, T.; Ferron, M.; Karsenty, G.; Vellard, M.C.; et al. A lysosome-to-nucleus signalling mechanism senses and regulates the lysosome via mTOR and TFEB. *EMBO J.* **2012**, *31*, 1095–1108. [CrossRef]
53. Ghosh, A.; Jana, M.; Modi, K.; Gonzales, F.J.; Sims, K.B.; Berry-Kravis, E.; Pahan, K. Activation of peroxisome proliferator-activated receptor alpha induces lysosomal biogenesis in brain cells: Implications for lysosomal storage disorders. *J. Biol. Chem.* **2015**, *290*, 10309–10324.
54. Xia, Q.; Wang, H.; Hao, Z.; Fu, C.; Hu, Q.; Gao, F.; Ren, H.; Chen, D.; Han, J.; Ying, Z.; et al. TDP-43 loss of function increases TFEB activity and blocks autophagosome–lysosome fusion. *EMBO J.* **2015**, *35*, 121–142. [CrossRef]
55. Brown, D.G.; Shorter, J.; Wobst, H.J. Emerging small-molecule therapeutic approaches for amyotrophic lateral sclerosis and frontotemporal dementia. *Bioorgan. Med. Chem. Lett.* **2020**, *30*, 126942. [CrossRef]
56. Collodet, C.; Foretz, M.; Deak, M.; Bultot, L.; Metairon, S.; Viollet, B.; Lefebvre, G.; Raymond, F.; Parisi, A.; Civiletto, G.; et al. AMPK promotes induction of the tumor suppressor FLCN through activation of TFEB independently of mTOR. *FASEB J.* **2019**, *33*, 12374–12391. [CrossRef]
57. El-Houjeiri, L.; Possik, E.; Vijayaraghavan, T.; Paquette, M.; Martina, J.A.; Kazan, J.M.; Ma, E.H.; Jones, R.; Blanchette, P.; Puertollano, R.; et al. The Transcription Factors TFEB and TFE3 Link the FLCN-AMPK Signaling Axis to Innate Immune Response and Pathogen Resistance. *Cell Rep.* **2019**, *26*, 3613–3628. [CrossRef]
58. Hardie, D.G.; Ross, F.A.; Hawley, S.A. AMPK: A nutrient and energy sensor that maintains energy homeostasis. *Nat. Rev. Mol. Cell Biol.* **2012**, *13*, 251–262. [CrossRef]
59. Wang, J.; Whiteman, M.W.; Lian, H.; Wang, G.; Singh, A.; Huang, D.; Denmark, T. A non-canonical MEK/ERK signaling pathway regulates autophagy via regulating Beclin 1. *J. Biol. Chem.* **2009**, *284*, 21412–21424. [CrossRef]
60. O'Connor, C.M.; Perl, A.; Leonard, D.; Sangodkar, J.; Narla, G. Therapeutic targeting of PP2A. *Int. J. Biochem. Cell Biol.* **2018**, *96*, 182–193. [CrossRef]

61. Martina, J.A.; Puertollano, R. Protein phosphatase 2A stimulates activation of TFEB and TFE3 transcription factors in response to oxidative stress. *J. Biol. Chem.* **2018**, *293*, 12525–12534. [CrossRef]
62. Angelini, C.; Semplicini, C. Enzyme Replacement Therapy for Pompe Disease. *Curr. Neurol. Neurosci. Rep.* **2011**, *12*, 70–75. [CrossRef]
63. Spampanato, C.; Feeney, E.; Li, L.; Cardone, M.; Lim, J.-A.; Annunziata, F.; Zare, H.; Polishchuk, R.; Puertollano, R.; Parenti, G.; et al. Transcription factor EB (TFEB) is a new therapeutic target for Pompe disease. *EMBO Mol. Med.* **2013**, *5*, 691–706. [CrossRef]
64. Feeney, E.J.; Spampanato, C.; Puertollano, R.; Ballabio, A.; Parenti, G.; Raben, N. What else is in store for autophagy? Exocytosis of autolysosomes as a mechanism of TFEB-mediated cellular clearance in Pompe disease. *Autophagy* **2013**, *9*, 1117–1118. [CrossRef]
65. Tsunemi, T.; Ashe, T.D.; Morrison, B.E.; Soriano, K.R.; Au, J.; Vazquez Roque, R.A.; Lazarowski, E.R.; Damian, V.A.; Masliah, E.; La Spada, A.R. PGC-1alpha rescues Huntington's disease proteotoxicity by preventing oxidative stress and promoting TFEB function. *Sci. Transl. Med.* **2012**, *4*, 142ra97.
66. La Spada, A.R. PPARGC1A/PGC-1alpha, TFEB and enhanced proteostasis in Huntington disease: Defining regulatory linkages between energy production and protein-organelle quality control. *Autophagy* **2012**, *8*, 1845–1847.
67. Hartmann, A.; Hunot, S.; Hirsch, E.C. Inflammation and dopaminergic neuronal loss in Parkinson's disease: A complex matter. *Exp. Neurol.* **2003**, *184*, 561–564.
68. Singh, P.K.; Kotia, V.; Ghosh, D.; Mohite, G.M.; Kumar, A.; Maji, S.K. Curcumin Modulates α-Synuclein Aggregation and Toxicity. *ACS Chem. Neurosci.* **2013**, *4*, 393–407. [CrossRef]
69. Zhang, Y.-D.; Zhao, J.-J. TFEB Participates in the Aβ-Induced Pathogenesis of Alzheimer's Disease by Regulating the Autophagy-Lysosome Pathway. *DNA Cell Biol.* **2015**, *34*, 661–668.
70. Klionsky, D.J. The autophagosome is overrated! *Autophagy* **2011**, *7*, 353–354. [CrossRef]
71. Najibi, M.; Labed, S.A.; Visvikis, O.; Irazoqui, E.J. An Evolutionarily Conserved PLC-PKD-TFEB Pathway for Host Defense. *Cell Rep.* **2016**, *15*, 1728–1742. [CrossRef]
72. Bao, J.; Zheng, L.; Zhang, Q.; Li, X.; Zhang, X.; Li, Z.; Bai, X.; Zhang, Z.; Hou, W.; Zhao, X.; et al. Deacetylation of TFEB promotes fibrillar Abeta degradation by upregulating lysosomal biogenesis in microglia. *Protein Cell* **2016**, *7*, 417–433.
73. Castillo, K.; Valenzuela, V.; Matus, S.; Nassif, M.; Onate, M.G.; Fuentealba, Y.; Encina, G.; Irrazabal, T.; Parsons, G.; Court, A.F.; et al. Measurement of autophagy flux in the nervous system in vivo. *Cell Death Dis.* **2013**, *4*, e917. [CrossRef]
74. Wang, A.L.; Boulton, M.E.; Dunn, J.W.A.; Rao, H.V.; Cai, J.; Lukas, T.J.; Neufeld, A.H.; A Dunn, W. Using LC3 to Monitor Autophagy Flux in the Retinal Pigment Epithelium. *Autophagy* **2009**, *5*, 1190–1193. [CrossRef]
75. Dai, Y.; Li, K.; Wu, W.; Wu, K.; Yi, H.; Li, W.; Xiao, Y.; Zhong, Y.; Cao, Y.; Tian, L. Steroid hormone 20-hydroxyecdysone induces the transcription and complex assembly of V-ATPases to facilitate autophagy in Bombyx mori. *Insect Biochem. Mol. Biol.* **2020**, *116*, 103255. [CrossRef]
76. Kosacka, J.; Nowicki, M.; Blüher, M.; Baum, P.; Stockinger, M.; Toyka, K.; Klöting, I.; Stumvoll, M.; Serke, H.; Bechmann, I.; et al. Increased autophagy in peripheral nerves may protect Wistar Ottawa Karlsburg W rats against neuropathy. *Exp. Neurol.* **2013**, *250*, 125–135. [CrossRef]
77. Wang, C.; Niederstrasser, H.; Douglas, P.M.; Lin, R.; Jaramillo, J.; Li, Y.; Oswald, N.W.; Zhou, A.; McMillan, E.A.; Mendiratta, S.; et al. Small-molecule TFEB pathway agonists that ameliorate metabolic syndrome in mice and extend *C. elegans* lifespan. *Nat. Commun.* **2017**, *8*, 2270. [CrossRef]
78. Zhou, X.; Yang, J.; Zhou, M.; Zhang, Y.; Liu, Y.; Hou, P.; Zeng, X.; Yi, L.; Mi, M. Resveratrol attenuates endothelial oxidative injury by inducing autophagy via the activation of transcription factor EB. *Nutr. Metab.* **2019**, *16*, 1–12. [CrossRef]
79. Omidian, M.; Abdolahi, M.; Daneshzad, E.; Sedighiyan, M.; Aghasi, M.; Abdollahi, H.; Omidian, P.; Dabiri, S.; Mahmoudi, M.; Hadavi, S.; et al. The Effects of Resveratrol on Oxidative Stress Markers: A Systematic Review and Meta-Analysis of Randomized Clinical Trials. *Endocr. Metab. Immune Disord. Drug Targets (Former. Curr. Drug Targets Immune Endocr. Metab. Disord.)* **2020**, *20*, 718–727. [CrossRef]
80. Song, J.-X.; Sun, Y.-R.; Peluso, I.; Zeng, Y.; Yu, X.; Lu, J.-H.; Xu, Z.; Wang, M.-Z.; Liu, L.-F.; Huang, Y.-Y.; et al. A novel curcumin analog binds to and activates TFEB in vitro and in vivo independent of MTOR inhibition. *Autophagy* **2016**, *12*, 1372–1389. [CrossRef]
81. Song, J.X.; Malampati, S.; Zeng, Y.; Durairajan, S.S.K.; Yang, C.-B.; Tong, B.C.-K.; Iyaswamy, A.; Shang, W.-B.; Sreenivasmurthy, S.G.; Zhu, Z.; et al. A small molecule transcription factor EB activator ameliorates beta-amyloid precursor protein and Tau pathology in Alzheimer's disease models. *Aging Cell* **2020**, *19*, e13069.
82. Tan, S.; Bajalovic, N.; Wong, E.S.P.; Lin, V.C.L. Ligand-activated progesterone receptor B activates transcription factor EB to promote autophagy in human breast cancer cells. *Exp. Cell Res.* **2019**, *382*, 111433. [CrossRef]
83. Settembre, C.; De Cegli, R.; Mansueto, G.; Saha, P.K.; Vetrini, F.; Visvikis, O.; Huynh, T.; Carissimo, A.; Palmer, D.; Klisch, T.J.; et al. TFEB controls cellular lipid metabolism through a starvation-induced autoregulatory loop. *Nat. Cell Biol.* **2013**, *15*, 647–658.
84. Thoreen, C.C.; Kang, S.A.; Chang, J.W.; Liu, Q.; Zhang, J.; Gao, Y.; Reichling, L.J.; Sim, T.; Sabatini, D.M.; Gray, N.S. An ATP-competitive Mammalian Target of Rapamycin Inhibitor Reveals Rapamycin-resistant Functions of mTORC1. *J. Biol. Chem.* **2009**, *284*, 8023–8032. [CrossRef]
85. Cheng, N.-T.; Guo, A.; Cui, Y.-P. Intra-articular injection of Torin 1 reduces degeneration of articular cartilage in a rabbit osteoarthritis model. *Bone Jt. Res.* **2016**, *5*, 218–224. [CrossRef]

86. Chen, G.; Xie, W.; Nah, J.; Sauvat, A.; Liu, P.; Pietrocola, F.; Sica, V.; Carmona-Gutierrez, D.; Zimmermann, A.; Pendl, T.; et al. 3,4-Dimethoxychalcone induces autophagy through activation of the transcription factors TFE 3 and TFEB. *EMBO Mol. Med.* **2019**, *11*, e10469. [CrossRef]

87. Kim, S.; Choi, K.J.; Cho, S.-J.; Yun, S.-M.; Jeon, J.-P.; Koh, Y.H.; Song, J.; Johnson, G.V.W.; Jo, C. Fisetin stimulates autophagic degradation of phosphorylated tau via the activation of TFEB and Nrf2 transcription factors. *Sci. Rep.* **2016**, *6*, 24933. [CrossRef]

88. Abe, K.; Yano, T.; Tanno, M.; Miki, T.; Kuno, A.; Sato, T.; Kouzu, H.; Nakata, K.; Ohwada, W.; Kimura, Y.; et al. mTORC1 inhibition attenuates necroptosis through RIP1 inhibition-mediated TFEB activation. *Biochim. Biophys. Acta (BBA) Mol. Basis Dis.* **2019**, *1865*, 165552. [CrossRef]

89. Cui, H.; Cheng, Y.; He, Y.; Cheng, W.; Zhao, W.; Zhao, H.; Zhou, F.H.; Wang, L.; Dong, J.; Cai, S. The AKT inhibitor MK2206 suppresses airway inflammation and the pro-remodeling pathway in a TDI-induced asthma mouse model. *Mol. Med. Rep.* **2020**, *22*, 3723–3734. [CrossRef]

90. Peña-Llopis, S.; Vega-Rubin-De-Celis, S.; Schwartz, J.C.; Wolff, N.C.; Tran, T.A.T.; Zou, L.; Xie, X.-J.; Corey, D.R.; Brugarolas, J. Regulation of TFEB and V-ATPases by mTORC1. *EMBO J.* **2011**, *30*, 3242–3258. [CrossRef]

91. Peña-Llopis, S.; Brugarolas, J. TFEB, a novel mTORC1 effector implicated in lysosome biogenesis, endocytosis and autophagy. *Cell Cycle* **2011**, *10*, 3987–3988. [CrossRef]

92. Martina, J.A.; Puertollano, R. Rag GTPases mediate amino acid–dependent recruitment of TFEB and MITF to lysosomes. *J. Cell Biol.* **2013**, *200*, 475–491. [CrossRef]

93. Roczniak-Ferguson, A.; Petit, C.S.; Froehlich, F.; Qian, S.; Ky, J.; Angarola, B.; Walther, T.C.; Ferguson, S.M. The Transcription Factor TFEB Links mTORC1 Signaling to Transcriptional Control of Lysosome Homeostasis. *Sci. Signal.* **2012**, *5*, ra42. [CrossRef]

94. Morran, D.C.; Wu, J.; Jamieson, N.B.; Mrowinska, A.; Kalna, G.; Karim, S.A.; Au, A.Y.M.; Scarlett, C.J.; Chang, D.K.; Pajak, M.Z.; et al. Targeting mTOR dependency in pancreatic cancer. *Gut* **2014**, *63*, 1481–1489. [CrossRef]

95. Kepp, O.; Chen, G.; Carmona-Gutierrez, D.; Madeo, F.; Kroemer, G. A discovery platform for the identification of caloric restriction mimetics with broad health-improving effects. *Autophagy* **2019**, *16*, 188–189. [CrossRef]

96. Lara, P.N.; Longmate, J.; Mack, P.C.; Kelly, K.; Socinski, M.A.; Salgia, R.; Gitlitz, B.J.; Li, T.; Koczywas, M.; Reckamp, K.L.; et al. Phase II Study of the AKT Inhibitor MK-2206 plus Erlotinib in Patients with Advanced Non–Small Cell Lung Cancer Who Previously Progressed on Erlotinib. *Clin. Cancer Res.* **2015**, *21*, 4321–4326. [CrossRef]

97. Palmieri, M.; Pal, R.; Nelvagal, H.R.; Lotfi, P.; Stinnett, G.R.; Seymour, M.L.; Chaudhury, A.; Bajaj, L.; Bondar, V.V.; Bremner, L.; et al. Corrigendum: mTORC1-independent TFEB activation via Akt inhibition promotes cellular clearance in neurodegenerative storage diseases. *Nat. Commun.* **2017**, *8*, 15793. [CrossRef]

98. Patel, K.R.; Scott, E.; Brown, V.A.; Gescher, A.J.; Steward, W.P.; Brown, A.K. Clinical trials of resveratrol. *Ann. N. Y. Acad. Sci.* **2011**, *1215*, 161–169. [CrossRef]

Article

mTORC2 Is Involved in the Induction of RSK Phosphorylation by Serum or Nutrient Starvation

Po-Chien Chou [1], Swati Rajput [1], Xiaoyun Zhao [2], Chadni Patel [1], Danielle Albaciete [1], Won Jun Oh [1], Heineken Queen Daguplo [1], Nikhil Patel [1], Bing Su [2], Guy Werlen [1] and Estela Jacinto [1,*]

1 Department of Biochemistry and Molecular Biology, Rutgers-Robert Wood Johnson Medical School, Piscataway, NJ 08854, USA; pcchou303@gmail.com (P.-C.C.); sr1208@scarletmail.rutgers.edu (S.R.); chadni.patel@gsbs.rutgers.edu (C.P.); dalbaciebio@gmail.com (D.A.); wjoh97@gmail.com (W.J.O.); hbd15@scarletmail.rutgers.edu (H.Q.D.); ndp85@scarletmail.rutgers.edu (N.P.); guy.werlen@rutgers.edu (G.W.)
2 Department of Immunology and Microbiology, Shanghai Jiao Tong University School of Medicine, Shanghai 200240, China; zhaoxiaoyunrf@163.com (X.Z.); bingsu@sjtu.edu.cn (B.S.)
* Correspondence: jacintes@rutgers.edu; Tel.: +1-(732)-235-4476

Received: 3 May 2020; Accepted: 23 June 2020; Published: 27 June 2020

Abstract: Cells adjust to nutrient fluctuations to restore metabolic homeostasis. The mechanistic target of rapamycin (mTOR) complex 2 responds to nutrient levels and growth signals to phosphorylate protein kinases belonging to the AGC (Protein Kinases A,G,C) family such as Akt and PKC. Phosphorylation of these AGC kinases at their conserved hydrophobic motif (HM) site by mTORC2 enhances their activation and mediates the functions of mTORC2 in cell growth and metabolism. Another AGC kinase family member that is known to undergo increased phosphorylation at the homologous HM site (Ser380) is the p90 ribosomal S6 kinase (RSK). Phosphorylation at Ser380 is facilitated by the activation of the mitogen-activated protein kinase/extracellular signal regulated kinase (MAPK/ERK) in response to growth factor stimulation. Here, we demonstrate that optimal phosphorylation of RSK at this site requires an intact mTORC2. We also found that RSK is robustly phosphorylated at Ser380 upon nutrient withdrawal or inhibition of glycolysis, conditions that increase mTORC2 activation. However, pharmacological inhibition of mTOR did not abolish RSK phosphorylation at Ser380, indicating that mTOR catalytic activity is not required for this phosphorylation. Since RSK and SIN1β colocalize at the membrane during serum restimulation and acute glutamine withdrawal, mTORC2 could act as a scaffold to enhance RSK HM site phosphorylation. Among the known RSK substrates, the CCTβ subunit of the chaperonin containing TCP-1 (CCT) complex had defective phosphorylation in the absence of mTORC2. Our findings indicate that the mTORC2-mediated phosphorylation of the RSK HM site could confer RSK substrate specificity and reveal that RSK responds to nutrient fluctuations.

Keywords: RSK; mTORC2; p90 ribosomal s6 kinase; nutrients; AGC kinases; MAPK/ERK; CCTβ; CCT/TRiC; chaperonin; starvation; metabolism

---

1. Introduction

mTOR orchestrates metabolic processes in response to levels of nutrients in order to promote cell growth or survival [1–3]. It forms two distinct signaling complexes; mTOR complex 1 (mTORC1) and complex 2 (mTORC2). mTORC1 is composed of the evolutionarily conserved components mTOR, raptor, and mLST8 while mTORC2 contains mTOR, rictor, SIN1, and mLST8. In higher eukaryotes mTOR also associates with other proteins distinct from mTORC1 and mTORC2 [4,5].

mTOR is a serine/threonine protein kinase and its activity is modulated by its protein partners. The best-characterized substrate of mTORC2 is AKT which is a member of the AGC family of protein kinases [6]. Members of this family including AKT are phosphorylated at the kinase activation loop by PDK1 (phosphoinositide-dependent kinase 1) [7]. They are also phosphorylated at one or more sites at the two conserved motifs turn motif (TM) and hydrophobic motif (HM), which are adjacent to the kinase domain. There is accumulating evidence supporting that mTOR either as part of mTORC1 or mTORC2 phosphorylates directly or indirectly the TM and HM of AGC kinases [8–15]. mTORC2 phosphorylates the HM site (Ser473) of AKT in response to growth factors [10]. Recently we and others have also shown that this phosphorylation is enhanced upon nutrient withdrawal [16–18]. On the other hand, mTORC2 mediates phosphorylation of the TM of AKT as well as the HM/TM of PKCs constitutively [9,11,13,19,20]. These observations suggest that specificity of mTORC2 activity towards these targets is likely to be modulated compartmentally in response to levels of growth signals or intracellular metabolites. Indeed we found that the TM phosphorylation of AKT occurs during translation when nascent AKT is associated with translating ribosomes [19]. Identification of other downstream targets or effectors of mTORC2 should help unravel the precise mechanisms involved in mTORC2 signaling

The p90 ribosomal S6 kinase (RSK), another member of the AGC kinase family functions in translation, metabolism, cell adhesion/migration and becomes deregulated in diseases such as cancer [21–26]. RSK has different isoforms, RSK1–4, with distinct as well as overlapping functions. RSK1–4 consists of two kinase domains, the N-Terminal kinase domain (NTKD), which is homologous to the catalytic domain of AGC kinase family and another at the carboxyl terminus (CTKD), which is homologous to the calcium/calmodulin-dependent protein kinase (CaMK) family (Figure 1A). The CTKD and NTKD promote autophosphorylation and substrate phosphorylation, respectively [27,28]. The MAPK family member, ERK1/2, facilitates the activation of RSK. It docks at the C-terminal end and phosphorylates Thr573 of the CTKD activation loop [29]. ERK1/2 is also linked to phosphorylation of Ser363 at the TM, which is located at the linker region between the two kinase domains. This linker region harbors the conserved TM and HM of AGC kinases. Phosphorylation of Ser380 at the HM serves as a docking site for PDK1 that then phosphorylates Ser221 of the NTKD, resulting in full activation of RSK [30]. While HM site phosphorylation is strongly linked to ERK activation and could occur via autophosphorylation or ERK, the role of other kinases has not been excluded [26,31,32]. The combined removal of the ERK docking site and membrane targeting of RSK enhances HM phosphorylation and RSK activation, suggesting that the HM site is phosphorylated at the membrane [33]. In response to growth signals and mitogens, activated RSK phosphorylates a plethora of substrates [22,26]. Despite overlapping functions of RSK and other mTOR-regulated AGC kinases in a variety of cellular processes, the role of mTOR in RSK regulation remains unclear. In the present studies, we determined if mTORC2 could be involved in the regulation of RSK since the RSK NTKD harbors the homologous HM site that is targeted by mTORC2 in Akt, PKC, and SGK1 [9,11–14,34]. We unraveled that mTORC2 is required for optimal RSK HM site phosphorylation in the presence of growth signals but this function of mTORC2 does not require its catalytic activity. Importantly, we also found that RSK responds to nutrient starvation and this response is also mediated by mTORC2.

2. Materials and Methods

2.1. Plasmids and Antibodies

pKFLAG-CCTβ was obtained from Dr. John Blenis (Weill Cornell) [35] and HA-tagged avian RSK1 and Ser381 Ala mutant were obtained from Dr. Philippe Roux (IRIC, Univ. of Montreal) [29,33]. All other antibodies are listed in Table 1.

Table 1. List of Antibodies used.

Target	Catalog No.	Source
pS380 RSK	AF889	R&D Systems (Minneapolis, MN, USA)
pS380 RSK (pHM)	12032	Cell Signaling Tech. (Danvers, MA, USA)
RSK	9347, 2765	Cell Signaling Tech.
pT359/S363 RSK (pTM)	9344	Cell Signaling Tech.
pT573 RSK	9346	Cell Signaling Tech.
pS221 RSK	AF892	R&D Systems
pThr202/Tyr204 ERK	4370, 2331	Cell Signaling Tech.
ERK	Sc-13003	Sta. Cruz Biotech. (Dallas, TX, USA)
pS235/236 S6	4856	Cell Signaling Tech.
S6	2317	Cell Signaling Tech.
pS473 Akt (pHM)	4060	Cell signaling Tech
Akt	9272	Cell Signaling Tech.
pT638/641 PKCα/βII (pHM)	9375	Cell Signaling Tech.
rictor	9476	Cell Signaling Tech.
β-actin	sc-47778	Sta. Cruz Biotech.
α-tubulin	sc-53029	Sta. Cruz Biotech.
pT172 AMPK	2535	Cell Signaling Tech.
AMPK	5831	Cell Signaling Tech.
pS112 BAD	9291	Cell Signaling Tech.
pS21/9 GSK3α/β	9331	Cell Signaling Tech.
pS366 eEF2K	2331	Cell Signaling Tech.
pT56 eEF2	2331	Cell Signaling Tech.
pS240/244 S6	2215	Cell Signaling Tech.
p-Akt substrate (pAS)	9614	Cell Signaling Tech.
CCTb	sc-13874	Sta. Cruz Biotech.
Flag	F7425	Sigma (St. Louis, MO, USA)
HA	3724	Cell Signaling Tech.
DiI	D282	Invitrogen (Carlsbad, CA, USA)
CD3ε	100302	Biolegend (San Diego, CA, USA)

2.2. Cell Culture, Stimulation, Transfection and Harvest

HeLa, WT and SIN1$^{-/-}$ MEFs were cultured in complete DMEM (Sigma D-6546)(St. Louis, MO, USA) [containing 10% FBS, 2 mM glutamine (Gibco 25030-164) (Gaithersburg, MD, USA), penicillin/streptomycin (Gibco 15140-122)]. After culturing for 24 h reaching approximately 70–80% confluency, cells were resuspended either in fresh complete media or starvation media (glucose starvation media-Corning 17-207-CV (Corning, NY, USA); glutamine starvation media-Corning 15-017-CV) as described previously [16]. 10% dialyzed FBS (Hyclone SH30079.03) (Marlborough, MA, USA), 25 mM glucose or 2 mM glutamine were supplemented in resuspension media as indicated. Cells were harvested with CHAPS lysis buffer (40 mM HEPES pH 7.5, 120 mM NaCl, 1 mM EDTA, 0.3% CHAPS) or RIPA lysis buffer (50 mM Tris-HCl pH 8.0, 100 mM NaCl, 5 mM EDTA, 0.2% SDS, 0.5% sodium deoxycholate, 1.0% Triton X-100) containing protease and phosphatase inhibitors. For transient transfections, plasmid constructs were transfected into MEFs (at about 60% confluency) using Lipofectamine Reagent (Invitrogen, Carlsbad, CA, USA) following the manufacturer's protocol. After 24 h, cells were resuspended in DMEM lacking serum and incubated overnight. Cells were then resuspended in PBS for 30 min, then either harvested or restimulated with complete media containing 10% FBS followed by cell lysis. For siRNA transfections, cells were resuspended in Opti-MEM and incubated for 6–7 h. siRNA (mTOR siRNA Dharmacon L-003008-00; scramble Dharmacon D-001810-01-05) (Lafayette, CO, USA) was transfected using Oligofectamine (Invitrogen) following the manufacturer's protocol. Twenty four hours after transfection, cells were harvested using RIPA lysis buffer.

2.3. Mice and Thymocyte Stimulation

The generation of mice with T cell-specific rictor deletion and thymocyte harvest were described previously [36]. For stimulation, thymocytes were incubated with 10 μg/mL CD3ε antibody and/or 10 ng/mL PMA. Handling and experimentation protocols have been reviewed and used in accordance with the Institutional Animal Care and Use Committee regulations of Rutgers University.

2.4. Immunoblotting and Immunoprecipitations

Protein concentrations of cell lysates were determined by Bradford assay and samples (10–30 μg) were subjected to SDS-PAGE. Proteins were transferred onto Immobilon-PVDF (Millipore)(Burlington, MA, USA). Blots were incubated with primary antibodies overnight followed by washing in PBS-Tween. After incubation with secondary antibody for 1 h, blots were washed again. Images were visualized with a SuperSignal ECL detection kit (ThermoFisher, Waltham, MA, USA) and captured using a Amersham Imager 600 (GE)(Marlborough, MA, USA). For immunoprecipitations, lysates were Pre-Cleared by adding Protein G Sepharose beads (GE Healthcare, Marlborough, MA, USA), then allowed to tumble for 1 h at 4 °C. Supernatants were recovered then incubated with antibody overnight followed by an additional 1 h incubation with Protein G Sepharose beads at 4 °C. Beads containing immunoprecipitates were washed 3× with TBS (50 mM Tris-HCl pH 7.4, 150 mM NaCl).

2.5. Immunofluorescence

WT MEFs were Co-Transfected with GFP-SIN1β and HA-RSK1-WT or HA-RSK1-SA mutant by PEI (Polyscience, 2,3966-1) (Niles, IL, USA). After 24 h, cells were changed into DMEM with 10%FBS. 24 h later, cells were resuspended in media with or without serum and incubated for 15 min. Cells were fixed using 1% PFA, blocked and stained for HA and DiI in PBS containing 1% BSA. Images were acquired on a Leica SP8 confocal Laser-Scanning microscope and processed using Bitplane Imaris 9.1.2 (Zurich, Switzerland).

3. Results

3.1. mTORC2 Is Required for Optimal Phosphorylation of RSK at the Hydrophobic Motif Site, Ser380

The phosphorylation of Ser380 at the HM of RSK is induced by growth factors and mediated by ERK [29]. Since this site is homologous to the HM site of Akt, we examined how the HM phosphorylation of RSK is affected in the absence of mTORC2. Using the mTORC2-disrupted cell line, SIN1$^{-/-}$ murine embryonic fibroblasts (MEFs) [34], we examined RSK phosphorylation at Ser380 (based on human RSK1 numbering) (Figure 1A) (herein referred to as phospho-Hydrophobic Motif; pHM). Under basal conditions, pHM RSK was slightly lower in SIN1$^{-/-}$ cells (Figure 1B). Upon resuspension in fresh media with serum, pHM RSK increased robustly from 5–30 min in WT while a blunted response occurred in SIN1$^{-/-}$ MEFs. As expected, pHM of Akt and PKCα/βII was abolished in SIN1$^{-/-}$ cells [9,34]. Since RSK phosphorylation at the HM site is dependent on activated ERK, which docks near the CTKD of RSK [37,38], we next examined ERK1/2 activation using the phosphorylation of ERK at Thr202/Tyr204 (pERK1/2) as the readout. ERK1/2 was activated upon serum restimulation in both WT and SIN1$^{-/-}$ MEFs although this activation was slightly weaker in SIN1$^{-/-}$ cells. Knockdown of mTOR in HeLa cells also diminished pHM RSK (Figure 1C). We also examined pHM RSK phosphorylation during the disruption of rictor, the other mTORC2 component. Using mice with specific deletion of rictor in thymocytes, we found that whereas pHM RSK was present in rictor$^{+/+}$ and rictor$^{+/-}$, it was abolished in the rictor$^{-/-}$ thymocytes, similar to pHM AKT under basal conditions (Figure 1D). We then cultured thymocytes ex vivo and induced signaling downstream of the T cell receptor (TCR) by ligation of the CD3 subunit of the TCR using anti-CD3 antibody. pHM RSK was enhanced by 5 min of anti-CD3 stimulation in WT whereas it remained low in rictor$^{-/-}$ thymocytes (Figure 1E). ERK1/2 was robustly stimulated in both WT and rictor$^{-/-}$ thymocytes. Together, these findings indicate that mTORC2 modulates phosphorylation of the RSK HM phosphosite.

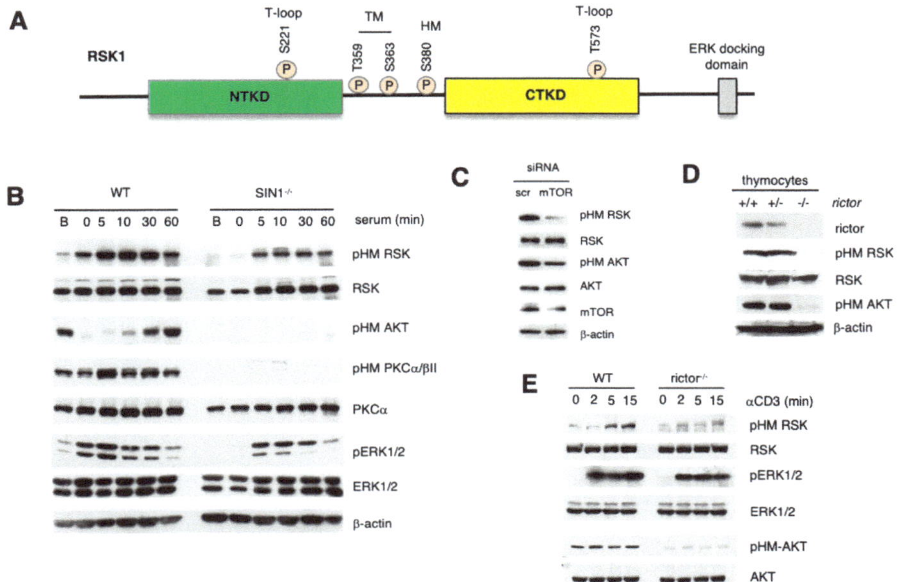

Figure 1. Phosphorylation of RSK at the hydrophobic motif site is diminished in mTORC2-disrupted cells. (**A**) RSK has two conserved catalytic domains, N-Terminal (NTKD) and C-Terminal (CTKD) kinase domains that are phosphorylated at each of their activation loops (T-Loop). These two domains flank a linker region that is conserved among AGC kinases containing the conserved turn (TM) and hydrophobic motifs (HM) that become phosphorylated at the indicated sites. (**B**) Wild type (WT) or SIN1$^{-/-}$ MEFs were grown in complete DMEM (Basal; B) or grown then serum-starved overnight. Serum was re-added and cells were incubated for the indicated times (min) before harvest. Total lysates were prepared with CHAPS containing buffer and subjected to SDS-PAGE and immunoblotting using indicated antibodies. (**C**) HeLa cells were transfected with scramble control (scr) or siRNA targeting mTOR. Cells were lysed using RIPA and processed as in **B**. (**D**) Thymocytes with wild type (+/+), heterozygous (+/−) or homozygous (−/−) deletion of rictor were lysed and subjected to SDS-PAGE and immunoblotting. (**E**) Wild type (WT) or rictor-deficient thymocytes were non-stimulated (0) or stimulated with CD3ε antibody (10 μg/mL) for the indicated times (min).

3.2. ERK Activation Is Essential but Not Sufficient for HM Site Phosphorylation of RSK

Since HM RSK phosphorylation was reduced but not abolished during stimulation of mTORC2-disrupted cells, we analyzed the contribution of ERK1/2 to this phosphorylation. We used the specific MEK inhibitor, U0126, which blocks ERK activation. This inhibitor diminished the serum-induced phosphorylation of the RSK HM site in HeLa, consistent with abrogation of ERK phosphorylation (Figure 2A). It also reduced pHM RSK in WT MEFs and abolished it in SIN1$^{-/-}$ MEFs, correlating with abrogated ERK1/2 phosphorylation in the latter cells (Figure 2B). Thus, ERK1/2 activation, in addition to mTORC2-mediated phosphorylation is required for full RSK phosphorylation at the HM site. We next asked if enhancing activation of the ERK pathway by stimulation with the potent mitogen, phorbol myristate acetate (PMA), could rescue pHM RSK in SIN1-deficient cells. However, ERK activation and pHM RSK remained lower upon PMA stimulation in SIN1$^{-/-}$ cells (Figure 2C). The phosphorylation of the RSK substrate, S6 at Ser235/236 in SIN1$^{-/-}$ MEFs did not change significantly.

We next examined whether pHM RSK in rictor$^{-/-}$ thymocytes would be potentiated when we stimulate T cells with anti-CD3 and PMA. However, pHM RSK remained lower compared to WT

(Figure 2D). On the other hand, ERK phosphorylation in the rictor$^{-/-}$ thymocytes was comparable to WT. These findings indicate that although ERK activation is required, it is not sufficient and that mTORC2 is needed for optimal RSK HM site phosphorylation.

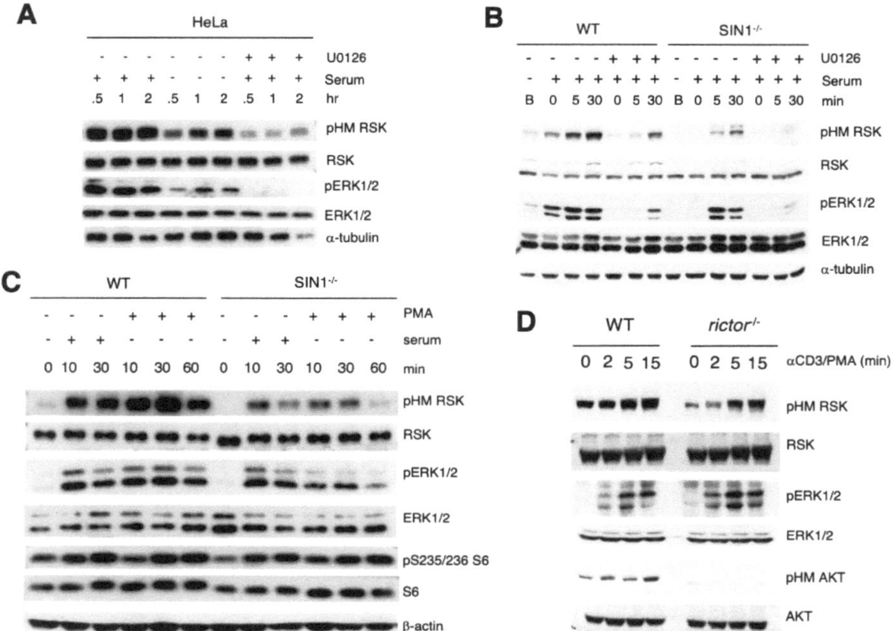

Figure 2. ERK activation is essential but not sufficient for full RSK HM site phosphorylation. (**A**) HeLa cells were resuspended in complete media without serum in the presence or absence of 15 μM U2016 for the indicated times. 1× FBS (serum) was added as indicated, at the last 0.5 h before harvest. Cells were harvested in CHAPS lysis buffer and protein extracts were subjected to SDS-PAGE and immunoblotting using indicated antibodies. (**B**) Wild type (WT) or SIN1$^{-/-}$ MEFs were grown in complete DMEM (Basal; B) or grown then serum-starved overnight. Serum was re-added and cells were incubated for the indicated times (min). Cells were harvested and processed as in (**A**). (**C**) WT or SIN1$^{-/-}$ MEFs were grown overnight in the absence of serum, followed by addition of serum or PMA for the indicated times. (**D**) WT or rictor$^{-/-}$ thymocytes were non-stimulated (0) or stimulated with Anti-CD3 and 10 ng/mL PMA for the indicated times (min).

3.3. RSK HM Site Phosphorylation Is Increased during Nutrient Withdrawal via mTORC2

Since we have previously shown that mTORC2 is activated upon nutrient withdrawal [16], we then investigated how RSK-HM phosphorylation could be modulated by nutrients. We resuspended HeLa cells in media containing or lacking glucose, incubated them for 0.5–3 h with the addition of dialyzed serum for the last half hour before harvest. We found that pHM was more robust in the absence of glucose than when glucose was present, occurring transiently at 0.5 h (Figure 3A). The addition of serum did not further increase pHM when glucose was absent. Next, we incubated cells in the absence or presence of glutamine. Although RSK phosphorylation was low in the absence of both glutamine and serum, it was more robustly increased in the absence of glutamine upon re-addition of serum (Figure 3B). We next combined withdrawal of both glucose and glutamine in the presence or absence of serum. pHM was robustly increased when both nutrients were withdrawn in HeLa (Figure 3C) and WT MEFs (Figure 3D). Furthermore, withdrawal of all amino acids as well as glucose also led to a more robust HM phosphorylation in the absence or presence of serum (Figure S1). The increase in pHM also coincided with increased phosphorylation at the TM site as well as the NTKD and CTKD activation loop phosphosites (Figure S1). In all conditions tested (Figure 3A–D,Figure S1), the increase in RSK phosphorylation was accompanied by augmented ERK1/2 phosphorylation as well.

Since glucose withdrawal robustly enhanced RSK phosphorylation, we then examined whether inhibition of glycolysis using 2-Deoxyglucose (2-DG) could also have the same effect. 2-DG did not affect RSK pHM in the presence of glucose but further increased RSK phosphorylation in the absence of glucose (Figure 3E). The strong effect of glucose starvation and glycolysis inhibition was also evident from the robust AMPK phosphorylation at 0.5–1 h. Together, these findings reveal that RSK HM site phosphorylation is robustly increased during nutrient withdrawal.

We next examined whether the increase in HM RSK phosphorylation during nutrient starvation is also mediated by mTORC2. Compared to WT MEFs, pHM RSK was reduced in SIN1$^{-/-}$ cells cultured for 1 h in the presence of glucose, glutamine and serum (Figure 3F). Upon withdrawal of both glucose and glutamine, pHM RSK was also lower in SIN1$^{-/-}$ compared to the WT. However, when both glucose and glutamine were withdrawn in the absence of serum, pHM RSK was abrogated in SIN1$^{-/-}$. Surprisingly ERK1/2 phosphorylation remained robust upon nutrient or serum withdrawal in SIN1$^{-/-}$. Hence, pHM RSK is responsive to nutrient withdrawal and this response occurs via mTORC2.

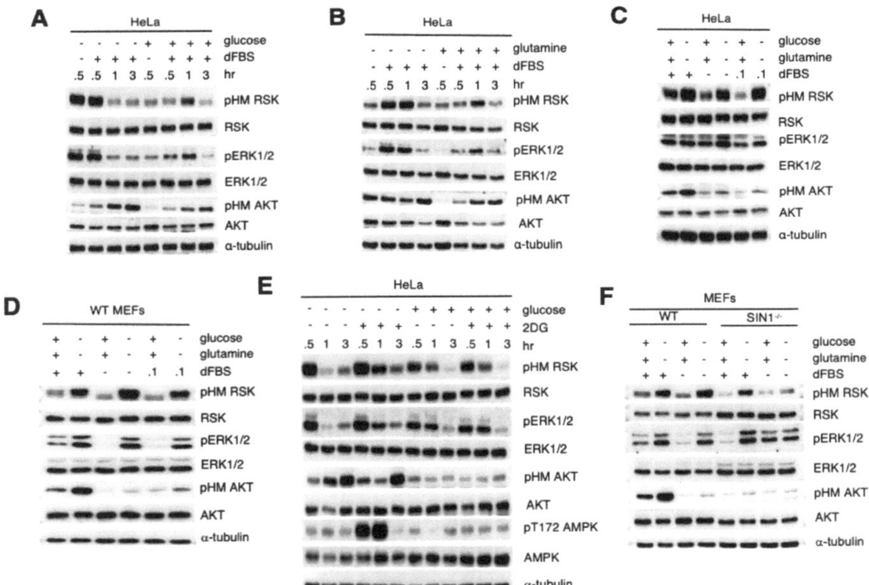

Figure 3. RSK HM site phosphorylation is increased during nutrient withdrawal. (**A–D**). Growing HeLa (**A–C**) or WT MEFs (**D**) were resuspended in media with or without 1X dialyzed FBS (dFBS) (or 1× dFBS as indicated in (**C,D**) and lacking (−) or containing (+) either glucose (**A**), glutamine (**B**) or both glucose and glutamine (**C,D**) at the indicated times. In (**C,D**), cells were harvested after 1 h. Cells were lysed with RIPA buffer and processed for SDS-PAGE and immunoblotting. (**E**) Growing cells were resuspended in media with dFBS in the absence or presence of glucose and/or 500 μM 2-Deoxyglucose (2DG) and incubated for the indicated times. (**F**) Growing WT or SIN1$^{-/-}$ MEFs were resuspended in media lacking or containing glucose, glutamine and/or dFBS and incubated for 1 h before harvest. Cells were harvested and processed for SDS-PAGE and immunoblotting.

3.4. Increased RSK HM Phosphorylation during Nutrient Withdrawal Is Uncoupled from S6 Phosphorylation

RSK phosphorylates the ribosomal protein S6 at Ser235/236 to modulate translation [39]. Indeed, upon serum restimulation of MEFs, we observed a robust phosphorylation of S6 that coincided with increased RSK HM phosphorylation (Figure 4A). When either glucose or glutamine were withdrawn for up to 6 h in WT MEFs, S6 phosphorylation remained robust (Figure 4B). However, when both glucose and glutamine were withdrawn from the culture media, S6 phosphorylation was greatly reduced whereas RSK HM phosphorylation was sustained for up to 3 h. We also examined the effect of combined glucose and glutamine withdrawal in HeLa and found that S6 phosphorylation was abolished by 3 h starvation whereas pHM RSK phosphorylation remained high at this point (Figure 4C). Furthermore, whereas S6 phosphorylation was similarly induced upon serum restimulation of serum-starved vs. serum/nutrient-starved cells, pHM RSK was more robust in the latter condition (Figure S1). Thus, the increased RSK HM phosphorylation that occurs during nutrient starvation is not linked to the phosphorylation of the RSK substrate, S6.

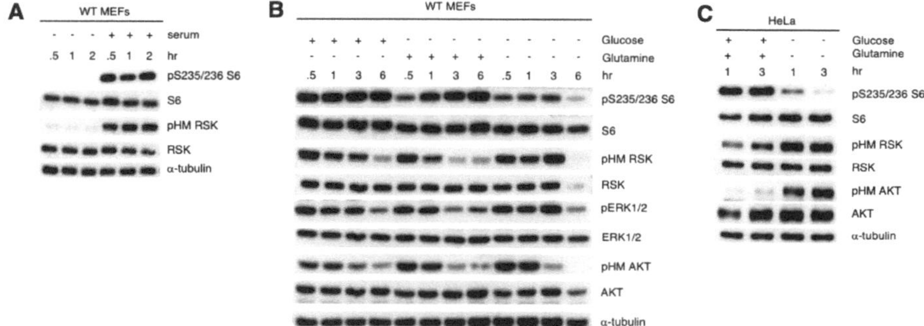

Figure 4. Phosphorylation of S6 is diminished despite increased RSK HM phosphorylation during nutrient withdrawal. (**A**) Growing WT MEFs were resuspended in complete media in the presence (+) or absence (−) of serum for the indicated times. Cells were harvested and processed for SDS-PAGE and immunoblotting. (**B,C**) Growing WT MEFs (**B**) or HeLa (**C**) were resuspended in media with dFBS and lacking either glucose, glutamine or both glucose and glutamine and incubated for the indicated times.

3.5. The Catalytic Activity of mTOR Is Not Required for RSK HM Site Phosphorylation

To further define how RSK HM phosphorylation is modulated by mTORC2, we used the mTOR ATP-Competitive inhibitor, Torin1, which blocks all mTOR complex activity. Treatment of WT MEFs with Torin1 in the presence of serum did not diminish RSK HM phosphorylation (Figure 5A). There was also no effect of Torin1 on serum-induced pHM RSK nor ERK phosphorylation in HeLa cells (Figure 5B). On the other hand, Torin1 inhibited the phosphorylation of S6 and Akt (Figure 5A,B). As expected, rapamycin, which allosterically inhibits mTORC1, also failed to block the serum-induced HM RSK phosphorylation (Figure S2) [40]. Next, we examined if mTOR inhibition would affect nutrient starvation-induced pHM RSK. In WT MEFs, Torin1 did not abolish the phosphorylation induced by glutamine withdrawal (Figure 5C). Similarly, pHM RSK was also higher and even sustained up to 24 h under glucose starvation in Torin1-Treated WT MEFs (Figure 5D). Hence, the catalytic activity of mTOR is not essential in promoting RSK HM phosphorylation and that inhibition of mTOR kinase activity could instead sustain phosphorylation at this site during prolonged starvation.

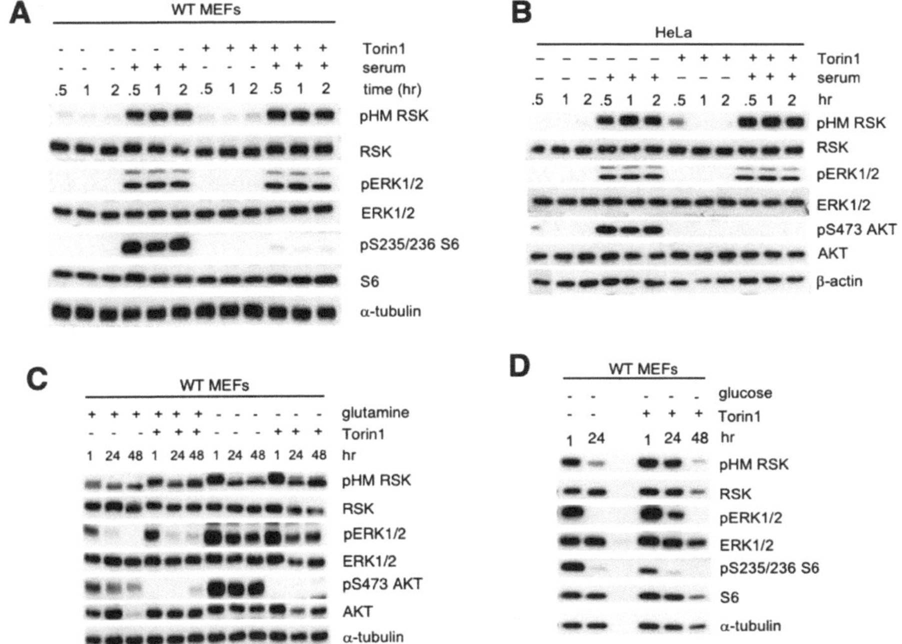

Figure 5. The catalytic activity of mTOR is not required for RSK HM site phosphorylation. (**A,B**) Growing WT MEFs (**A**) or HeLa cells (**B**) were supplemented with either 1 μM Torin1, 1× FBS (serum) or both and incubated for the indicated times. (**C,D**) Growing WT MEFs were resuspended in media with dFBS, lacking (−) or containing) glutamine (**C**) or glucose (**D**). Torin1 (1 μM) or vehicle (−) was added during resuspension and incubated at the indicated times.

3.6. RSK and SIN1 Colocalize at the Plasma Membrane

To further examine the role of the mTORC2-mediated RSK phosphorylation, we co-expressed GFP-SIN1β with either the wild type avian HA-RSK1 (HA-RSK1-WT) or the phospho-deficient mutant HA-RSK1-Ser381Ala (HA-RSK1-SA) that was devoid of HM phosphorylation even after serum restimulation (Figure S3). In the absence of serum, cells were more rounded in morphology (Figure 6). GFP-SIN1β localized to the membrane and the nucleus whereas HA-RSK1-WT localized to the plasma membrane and perinuclear area. Upon serum stimulation, there was more cell-spreading in HA-RSK1-WT-expressing cells. Whereas GFP-SIN1β predominantly localized on the plasma membrane in the presence of serum, HA-RSK1-WT was more diffused and present throughout the cell. On the other hand, when the mutant HA-RSK1-SA was expressed, cells remained round with less cell-spreading even in the presence of serum. Although GFP-SIN1β still localized to the membrane, HA-RSK1-SA localization was less diffused and predominated on the membrane. Next, we withdrew glutamine from the media. HA-RSK and GFP-SIN1 colocalized at the plasma membrane at 30 min glutamine withdrawal (Figure S4). HA-RSK-WT localization was more diffused in the presence of glutamine or by 6 h starvation. On the other hand, HA-RSK-SA mutant had diffused localization at all time points whereas SIN1 remained present at the membrane. These findings indicate that SIN1 and RSK colocalize at the membrane during serum restimulation and acute glutamine withdrawal. We then examined if RSK and SIN1 could interact. Immunoprecipitated Myc-SIN1 had increased association with HA-RSK1 upon withdrawal of both glucose and glutamine (Figure S5). Together, these findings suggest that mTORC2 could act as a scaffold to allow RSK phosphorylation in the membrane in response to serum or nutrient starvation.

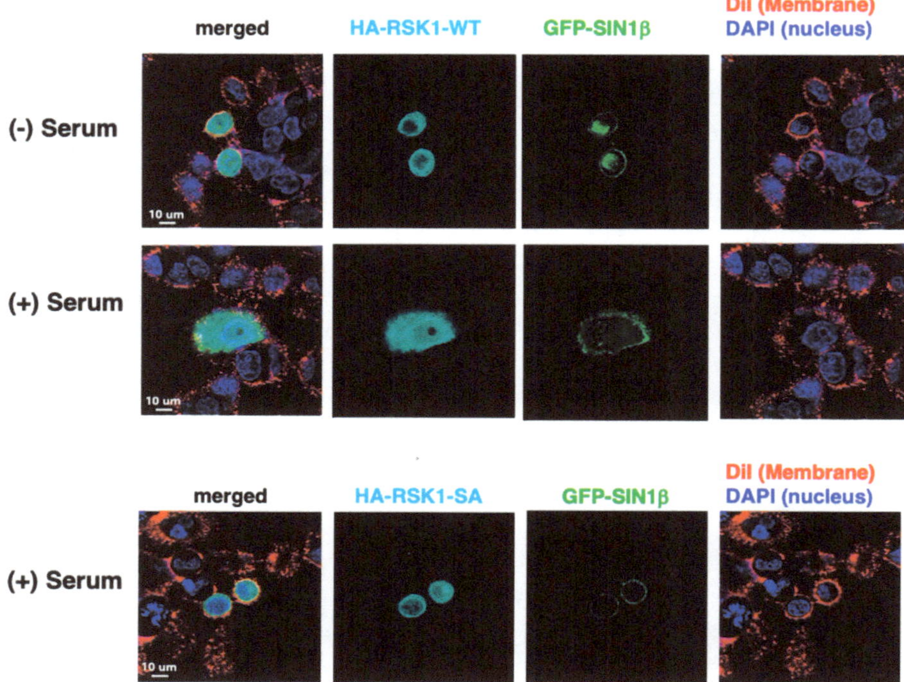

Figure 6. RSK and SIN1 colocalize at the plasma membrane. WT MEFs were co-transfected with GFP-SIN1β and avian HA-RSK-WT or HA-RSK-SA mutant (Ser381 Ala). After 48 h, cells were restimulated with serum for 15 min. Staining and microscopy were performed to detect GFP, HA as well as DiI (membrane) and DAPI (nucleus).

3.7. The RSK Substrate, CCTβ, Has Defective Phosphorylation in the Absence of mTORC2

RSK phosphorylates various substrates in response to growth stimuli. To address how mTORC2 could affect phosphorylation of known RSK targets, we used the mTORC2-disrupted cells and compared phosphorylation of some of these targets. In the Rictor-deficient murine thymocytes, we found that only the phosphorylation of the apoptosis regulator, BAD, was diminished as compared to wild type cells (Figure 7A). Other known RSK targets including pS21/9 GSK3α/β, pS366 eEF2K and pT56 eEF2 [22] did not have discernible changes in phosphorylation. Interestingly, the phosphorylation of S6 at the RSK-targeted site, Ser235/236, was upregulated in the rictor$^{-/-}$ thymocytes whereas the S6K1/mTORC1-mediated phosphorylation, Ser240/244 S6, was not altered. Another target of RSK that undergoes phosphorylation at the consensus Akt phosphorylation motif (RXRXXpS/pT) is CCTβ, a subunit of the chaperonin T-Complex protein-1 ring complex (TRiC/CCT) [35]. We first analyzed the expression of CCTβ in SIN1$^{-/-}$ MEFs and found that its total protein levels were diminished in these cells (Figure 7B). To facilitate analysis and comparison of CCTβ phosphorylation in WT vs. SIN1$^{-/-}$ MEFs, we overexpressed Flag-CCTβ, then analyzed the phosphorylation of immunoprecipitated CCTβ. Using the Phospho-Akt substrate antibody (P-AS), which detects phosphorylated epitope corresponding to the consensus RXRXXpS/T, we found that whereas CCTβ from the WT increased its phosphorylation upon serum addition, immunoprecipitated CCTβ from the SIN1$^{-/-}$ MEFs had no discernible phosphorylation even up to 30 min of serum stimulation (Figure 7C). Hence, the phosphorylation of CCTβ at the RSK-targeted site is dependent on mTORC2.

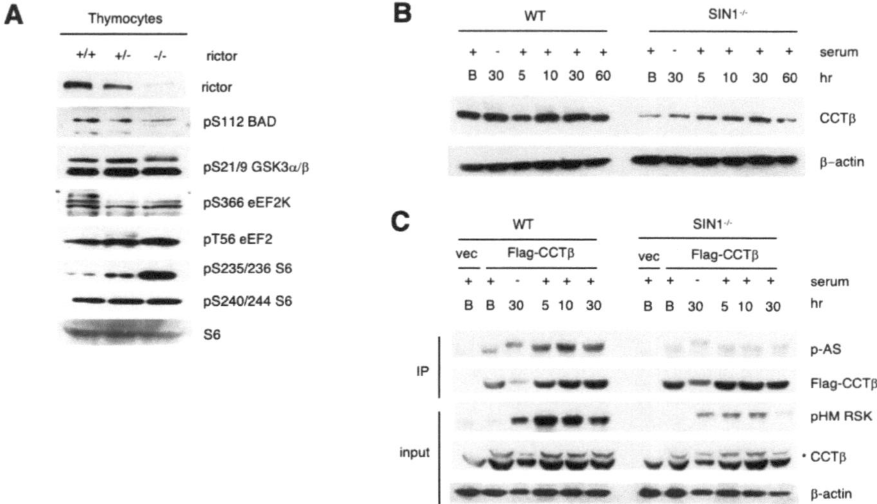

Figure 7. The RSK substrate CCTβ has defective phosphorylation and expression in the absence of mTORC2. (**A**) Thymocyte lysates from wild type (rictor$^{+/+}$), rictor$^{+/-}$, or rictor$^{-/-}$ littermates were resolved by SDS-PAGE and analyzed for phosphorylation of known RSK substrates by immunoblotting. (**B**) Wild type (WT) or SIN1$^{-/-}$ MEFs were grown in complete DMEM (Basal; B) or grown then serum-starved overnight. Cells were then left starved (–) or serum was re-added and cells were incubated for the indicated times. (**C**) WT or SIN1$^{-/-}$ MEFs were transfected with Flag-CCTβ plasmid or vector control (vec). Cells were starved and re-stimulated with serum as in **B**. Lysates were subjected to immunoprecipitation using Flag antibody. Immunoprecipitates were fractionated and blotted for CCTβ phosphorylation (using the Phospho-Akt consensus motif substrate antibody (P-AS)) or Flag. Total extracts (input) were also fractionated by SDS-PAGE and immunoblotted for pHM RSK or CCTβ. (*) indicates exogenous Flag-CCTβ, lower band is endogenous CCTβ.

4. Discussion

mTORC2 regulates cell growth and metabolism in response to levels of growth factors and nutrients. Among its known targets are protein kinases that are members of the AGC kinase family including AKT, PKC, and SGK1 [8,41]. Here, we found that mTORC2 could also mediate the phosphorylation of another AGC kinase, RSK, at the conserved HM site Ser380 (Figure 8). However, unlike its role in modulating other AGC kinases, the catalytic activity of mTORC2 is not required to enhance the RSK HM site phosphorylation and may instead act as a scaffold to allow RSK phosphorylation at the membrane. Among some of the known substrates of RSK that we examined, only the Pro-Apoptotic protein, BAD, and the chaperonin complex subunit, CCTβ, had diminished phosphorylation in the absence of mTORC2. Similar to the regulation of AKT by mTORC2, RSK phosphorylation is also increased during nutrient withdrawal, suggesting a role for RSK in metabolic reprogramming during nutrient-limiting conditions.

First, we found that mTORC2 is required for optimal phosphorylation of RSK at Ser380, which is located at the HM. This site is flanked by two catalytic domains, the NTKD and CTKD, which are phosphorylated at the activation loop (T-Loop) by PDK1 and ERK, respectively [22,26]. These two domains are linked by a region that harbors the TM and HM that are common to many AGC kinases. Phosphorylation of the RSK HM site is believed to be via autophosphorylation, mediated by the CTKD [33]. Stimulation of cells with growth factors or mitogens activates ERK and triggers its phosphorylation of the RSK CTKD activation loop. Previous studies have revealed the importance of ERK activation to pHM but whether another kinase is involved in enhancing this phosphorylation has not been excluded. The phosphorylation of the CTKD by ERK may facilitate membrane translocation of RSK, where it can be further activated in this compartment [33]. Loss of the ERK docking site located at the C-terminus abolishes RSK activity but a C-terminally truncated form that is targeted to the membrane results in HM phosphorylation and RSK activation [33], suggesting a critical role of membrane targeting for RSK regulation. Our studies here reveal that mTORC2 could mediate the regulation of RSK at the membrane. Although the catalytic activity of mTORC2 is not required for the HM site phosphorylation (Figure 5), mTORC2 could instead recruit or anchor RSK at the membrane to allow phosphorylation at this site. Phosphorylated HM serves as a docking site for PDK1, which then phosphorylates the T-Loop of the NTKD, leading to full RSK activation [42,43]. Upon activation at the membrane, RSK redistributes to other cellular compartments including the nucleus where it targets its many substrates [33]. mTORC2 has been found to associate with membrane compartments including the plasma membrane, lysosomes, nucleus, ER, and at the mitochondria-associated ER membrane [44–47]. In particular, SIN1 contains a pleckstrin homology (PH) domain that binds phosphatidylinositol 3,4,5-trisphosphate (PIP3) and as we show here localizes predominantly at the plasma membrane (Figure 6) [48]. The mTOR complexes, either directly or indirectly, mediate the phosphorylation of a number of the AGC kinases in a membrane compartment. In this regard, the subcellular localization of mTORC2 has been shown to impart specificity towards AGC kinase phosphorylation [49]. Here, we have shown that SIN1 and HM-phosphorylated RSK colocalize at the membrane during serum restimulation and acute glutamine withdrawal (Figures 6 and S4). We also found increased association of RSK1 with SIN1 during nutrient starvation (Figure S5). Whether RSK associates with SIN1 and other mTORC2 components at other membrane compartments remains to be examined further. It is also possible that mTORC2 could mediate RSK phosphorylation via modulation of ERK signaling [50,51]. SIN1 physically associates with Ras and there is accumulating evidence on the interaction of mTORC2 with Ras/MAPK signaling [52–55]. Further studies are needed to elucidate how these two pathways cooperate to modulate RSK and other downstream targets to promote cell growth and survival. Our findings here expand the mTORC2 effectors among the AGC kinase family that become phosphorylated in a membrane compartment.

Second, we report here that RSK HM phosphorylation is robustly increased by nutrient withdrawal, a condition that also activates mTORC2 (Figure 3). We have previously demonstrated that the increase in mTORC2 activation during prolonged starvation enhances or maintains flux through critical metabolic pathways [16,56]. mTOR is a key signaling protein that responds to nutrient fluctuations. Whereas mTORC1 activation is promoted by the presence of nutrients, the mode of mTORC2 activation is context dependent. mTORC2 has basal activity that allows constitutive phosphorylation of the TM site of AKT and PKC [9,11,19]. On the other hand, its activation is enhanced by either restimulation of starved cells with growth factors or by withdrawal of nutrients. As we have shown here, RSK HM site phosphorylation also follows this mode of regulation. It is noteworthy that a previous study reported that RSK and ERK phosphorylation was enhanced upon amino acid restimulation of cells that have been starved of both serum and amino acids for prolonged periods [57]. We have also observed a more robust increase in phosphorylation of RSK at the HM, TM and activation loop sites upon restimulation of cells starved of both serum and amino acids as compared to serum-starved alone (Figure S1). However, what we found here was that the withdrawal of nutrients (either glucose, glutamine or both) could transiently increase RSK HM phosphorylation (Figures 3 and 4B). Declining glucose levels were also shown to enhance RSK phosphorylation as well as ERK activation [18]. The findings by Casas-Terradellas et al. are not necessarily contradictory to our results and those from Shin et al. [18,57]. It is possible that similar to AKT which responds to either the increase or decrease of growth signals [16–18,34,58], RSK is also modulated by nutrient fluctuations and suggest that RSK is involved in remodeling metabolic processes. RSK has been previously shown to regulate

PFKFB2 (6-phosphofructo-2-kinase/fructose-2,6-bisphosphatase 2) to maintain flux through glycolysis in melanoma cells [21]. Highly proliferating cells such as cancer cells upregulate signaling pathways that control nutrient availability and flux through metabolic pathways in order to meet the increased demand for macromolecules [59]. Hence, the activation of RSK during nutrient withdrawal is likely relevant during metabolic reprogramming of cancer cells. Future studies should address additional targets of RSK in nutrient acquisition and metabolism.

Third, we uncover that the RSK substrate CCTβ is also modulated by mTORC2. CCTβ, which is part of the multi-protein chaperonin complex (TRiC/CCT) is involved in folding of nascent polypeptides such as those involved in the cytoskeleton [60,61]. Although RSK phosphorylates various substrates, including transcription factors, translational regulators, enzymes, and structural proteins [26], we found that among the RSK substrates that we examined, only CCTβ displayed a profound defect in mTORC2-disrupted cells. Both the phosphorylation of CCTβ and its expression levels were diminished in the absence of mTORC2. CCTβ Ser260 was previously identified as the RSK-targeted site [35]. However, other AGC kinases including S6K1 and Akt could also phosphorylate this site depending on stimulatory conditions. Since Akt activation is also defective in SIN1$^{-/-}$ MEFs, whereas S6K1 phosphorylation is not [34], it is possible that the suboptimal Akt activation contributes to the aberrant CCTβ phosphorylation. How phosphorylation of CCTβ by these mTORC-regulated kinases affects CCT function distinctly remains to be further investigated. Recently, CCT was also shown to mediate the assembly of the mTOR complexes [62]. Hence, it is also possible that there is feedback regulation between the mTORCs and CCT.

RSK and mTOR are important drug targets for diseases including cancer, cardiovascular and neurological disorders [22,63]. Our studies unravel that mTORC2 modulates RSK. Future studies should unravel the precise mechanisms that promote RSK activation by mTORC2 since they could have therapeutic value for targeting diseases with deregulated RSK signaling.

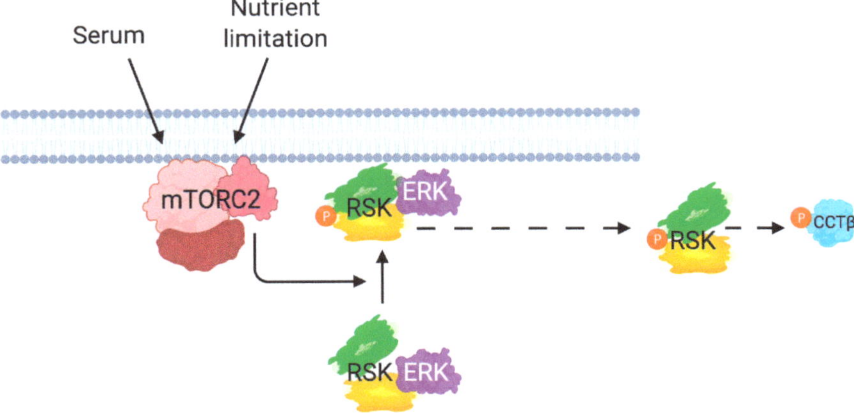

Figure 8. Model of mTORC2 regulation of RSK. In response to serum/growth factor stimulation or upon nutrient withdrawal, RSK phosphorylation (shown in orange) at the hydrophobic motif site (Ser380), located at the linker region between the NTKD (green) and CTKD (yellow), requires an intact mTORC2 but not mTOR catalytic activity. SIN1 associates with and colocalizes with RSK at the plasma membrane. mTORC2 could serve as a scaffold to potentiate the CTKD-mediated RSK HM phosphorylation. The CTKD is activated by ERK. mTORC2 is also required for the phosphorylation of the RSK substrate CCTβ.

Supplementary Materials: The following are available online at http://www.mdpi.com/2073-4409/9/7/1567/s1, Figure S1: There is increased RSK HM site phosphorylation upon removal of amino acids, Figure S2: RSK HM site phosphorylation is not inhibited by rapamycin, Figure S3: RSK-Ser381Ala mutant is not phosphorylated during serum restimulation, Figure S4: RSK-WT colocalizes with SIN1 at the membrane during early nutrient withdrawal, Figure S5: There is increased association of RSK and SIN1 during glucose/glutamine withdrawal.

Author Contributions: Conceptualization, G.W. and E.J.; Data curation, P.-C.C., S.R., X.Z., C.P., D.A., W.J.O. and H.Q.D., N.P.; Formal analysis, P.-C.C., S.R., X.Z., C.P., D.A., W.J.O., H.Q.D., B.S., G.W. and E.J.; Funding acquisition, E.J.; Investigation, P.-C.C., S.R., C.P., D.A., W.J.O., H.Q.D., N.P., G.W. and E.J.; Methodology, P.-C.C., X.Z., C.P., D.A., W.J.O., H.Q.D., B.S. and G.W.; Project administration, E.J.; Supervision, G.W. and E.J.; Validation, P.-C.C., S.R., C.P. and H.Q.D.; Writing—original draft, E.J.; Writing—review and editing, B.S., G.W. and E.J. All authors have read and agreed to the published version of the manuscript.

Funding: This work was supported by NIH grants GM079176, CA154674, GM137493 and New Jersey Commission on Cancer Research Bridge Grant (DFHS18CRF008) to E.J.

Acknowledgments: We thank J.B. and P.R. for generously sharing plasmids.

Conflicts of Interest: The authors declare no competing interests.

References

1. Magaway, C.; Kim, E.; Jacinto, E. Targeting mTOR and Metabolism in Cancer: Lessons and Innovations. *Cells* **2019**, *8*, 1584. [CrossRef]

2. Mossmann, D.; Park, S.; Hall, M.N. mTOR signalling and cellular metabolism are mutual determinants in cancer. *Nat. Rev. Cancer* **2018**, *18*, 744–757. [CrossRef]

3. Liu, G.Y.; Sabatini, D.M. mTOR at the nexus of nutrition, growth, ageing and disease. *Nat. Rev. Mol. Cell Biol.* **2020**, *21*, 183–203. [CrossRef]

4. Nguyen, J.T.; Ray, C.; Fox, A.L.; Mendonca, D.B.; Kim, J.K.; Krebsbach, P.H. Mammalian EAK-7 activates alternative mTOR signaling to regulate cell proliferation and migration. *Sci. Adv.* **2018**, *4*, eaao5838. [CrossRef]

5. Smithson, L.J.; Gutmann, D.H. Proteomic analysis reveals GIT1 as a novel mTOR complex component critical for mediating astrocyte survival. *Genes. Dev.* **2016**, *30*, 1383–1388. [CrossRef]

6. Jacinto, E.; Lorberg, A. TOR regulation of AGC kinases in yeast and mammals. *Biochem. J.* **2008**, *410*, 19–37. [CrossRef]

7. Pearce, L.R.; Komander, D.; Alessi, D.R. The nuts and bolts of AGC protein kinases. *Nat. Rev. Mol. Cell Biol.* **2010**, *11*, 9–22. [CrossRef] [PubMed]

8. Su, B.; Jacinto, E. Mammalian TOR signaling to the AGC kinases. *Crit. Rev. Biochem. Mol. Biol.* **2011**, *46*, 527–547. [CrossRef] [PubMed]

9. Facchinetti, V.; Ouyang, W.; Wei, H.; Soto, N.; Lazorchak, A.; Gould, C.; Lowry, C.; Newton, A.C.; Mao, Y.; Miao, R.Q.; et al. The mammalian target of rapamycin complex 2 controls folding and stability of Akt and protein kinase C. *EMBO J.* **2008**, *27*, 1932–1943. [CrossRef] [PubMed]

10. Sarbassov, D.D.; Guertin, D.A.; Ali, S.M.; Sabatini, D.M. Phosphorylation and regulation of Akt/PKB by the rictor-mTOR complex. *Science* **2005**, *307*, 1098–1101. [CrossRef]

11. Ikenoue, T.; Inoki, K.; Yang, Q.; Zhou, X.; Guan, K.L. Essential function of TORC2 in PKC and Akt turn motif phosphorylation, maturation and signalling. *EMBO J.* **2008**, *27*, 1919–1931. [CrossRef]

12. Garcia-Martinez, J.M.; Alessi, D.R. mTOR complex 2 (mTORC2) controls hydrophobic motif phosphorylation and activation of serum- and glucocorticoid-induced protein kinase 1 (SGK1). *Biochem. J.* **2008**, *416*, 375–385. [CrossRef] [PubMed]

13. Tobias, I.S.; Kaulich, M.; Kim, P.K.; Simon, N.; Jacinto, E.; Dowdy, S.F.; King, C.C.; Newton, A.C. Protein kinase C zeta exhibits constitutive phosphorylation and phosphatidylinositol-3,4,5-triphosphate-independent regulation. *Biochem. J.* **2016**, *473*, 509–523. [CrossRef]

14. Gan, X.; Wang, J.; Wang, C.; Sommer, E.; Kozasa, T.; Srinivasula, S.; Alessi, D.; Offermanns, S.; Simon, M.I.; Wu, D. PRR5L degradation promotes mTORC2-mediated PKC-delta phosphorylation and cell migration downstream of Galpha12. *Nat. Cell Biol.* **2012**, *14*, 686–696. [CrossRef]

15. Yang, C.S.; Melhuish, T.A.; Spencer, A.; Ni, L.; Hao, Y.; Jividen, K.; Harris, T.E.; Snow, C.; Frierson, H.F., Jr.; Wotton, D.; et al. The protein kinase C super-family member PKN is regulated by mTOR and influences differentiation during prostate cancer progression. *Prostate* **2017**, *77*, 1452–1467. [CrossRef] [PubMed]

16. Moloughney, J.G.; Kim, P.K.; Vega-Cotto, N.M.; Wu, C.C.; Zhang, S.; Adlam, M.; Lynch, T.; Chou, P.C.; Rabinowitz, J.D.; Werlen, G.; et al. mTORC2 Responds to Glutamine Catabolite Levels to Modulate the Hexosamine Biosynthesis Enzyme GFAT1. *Mol. Cell.* **2016**, *63*, 811–826. [CrossRef] [PubMed]

17. Kazyken, D.; Magnuson, B.; Bodur, C.; Acosta-Jaquez, H.A.; Zhang, D.; Tong, X.; Barnes, T.M.; Steinl, G.K.; Patterson, N.E.; Altheim, C.H.; et al. AMPK directly activates mTORC2 to promote cell survival during acute energetic stress. *Sci. Signal.* **2019**, *12*, eaav3249. [CrossRef]

18. Shin, S.; Buel, G.R.; Wolgamott, L.; Plas, D.R.; Asara, J.M.; Blenis, J.; Yoon, S.O. ERK2 Mediates Metabolic Stress Response to Regulate Cell Fate. *Mol. Cell.* **2015**, *59*, 382–398. [CrossRef]

19. Oh, W.J.; Wu, C.C.; Kim, S.J.; Facchinetti, V.; Julien, L.A.; Finlan, M.; Roux, P.P.; Su, B.; Jacinto, E. mTORC2 can associate with ribosomes to promote cotranslational phosphorylation and stability of nascent Akt polypeptide. *EMBO J.* **2010**, *29*, 3939–3951. [CrossRef]

20. Li, X.; Gao, T. mTORC2 phosphorylates protein kinase Czeta to regulate its stability and activity. *EMBO Rep.* **2014**, *15*, 191–198. [CrossRef]

21. Houles, T.; Gravel, S.P.; Lavoie, G.; Shin, S.; Savall, M.; Meant, A.; Grondin, B.; Gaboury, L.; Yoon, S.O.; St-Pierre, J.; et al. RSK Regulates PFK-2 Activity to Promote Metabolic Rewiring in Melanoma. *Cancer Res.* **2018**, *78*, 2191–2204. [CrossRef] [PubMed]

22. Houles, T.; Roux, P.P. Defining the role of the RSK isoforms in cancer. *Semin. Cancer Biol.* **2017**, *48*, 53–61. [CrossRef] [PubMed]

23. Meant, A.; Gao, B.; Lavoie, G.; Nourreddine, S.; Jung, F.; Aubert, L.; Tcherkezian, J.; Gingras, A.C.; Roux, P.P. Proteomic Analysis Reveals a Role for RSK in p120-catenin Phosphorylation and Melanoma Cell-Cell Adhesion. *Mol. Cell Proteomics.* **2020**, *19*, 50–64. [CrossRef] [PubMed]

24. Samson, S.C.; Elliott, A.; Mueller, B.D.; Kim, Y.; Carney, K.R.; Bergman, J.P.; Blenis, J.; Mendoza, M.C. p90 ribosomal S6 kinase (RSK) phosphorylates myosin phosphatase and thereby controls edge dynamics during cell migration. *J. Biol. Chem.* **2019**, *294*, 10846–10862. [CrossRef]

25. GN, M.H.; da Silva, F.F.; de Bellis, B.; Lupinacci, F.C.S.; Bellato, H.M.; Cruz, J.R.; Segundo, C.N.C.; Faquini, I.V.; Torres, L.C.; Sanematsu, P.I.; et al. Aberrant expression of RSK1 characterizes high-grade gliomas with immune infiltration. *Mol. Oncol.* **2020**, *14*, 159–179.

26. Anjum, R.; Blenis, J. The RSK family of kinases: Emerging roles in cellular signalling. *Nat. Rev. Mol. Cell Biol.* **2008**, *9*, 747–758. [CrossRef]

27. Fisher, T.L.; Blenis, J. Evidence for two catalytically active kinase domains in pp90rsk. *Mol. Cell Biol.* **1996**, *16*, 1212–1219. [CrossRef]

28. Bjorbaek, C.; Zhao, Y.; Moller, D.E. Divergent functional roles for p90rsk kinase domains. *J. Biol. Chem.* **1995**, *270*, 18848–18852. [CrossRef]

29. Roux, P.P.; Richards, S.A.; Blenis, J. Phosphorylation of p90 ribosomal S6 kinase (RSK) regulates extracellular signal-regulated kinase docking and RSK activity. *Mol. Cell Biol.* **2003**, *23*, 4796–4804. [CrossRef]

30. Frodin, M.; Antal, T.L.; Dummler, B.A.; Jensen, C.J.; Deak, M.; Gammeltoft, S.; Biondi, R.M. A phosphoserine/threonine-binding pocket in AGC kinases and PDK1 mediates activation by hydrophobic motif phosphorylation. *EMBO J.* **2002**, *21*, 5396–5407. [CrossRef]

31. Dalby, K.N.; Morrice, N.; Caudwell, F.B.; Avruch, J.; Cohen, P. Identification of regulatory phosphorylation sites in mitogen-activated protein kinase (MAPK)-activated protein kinase-1a/p90rsk that are inducible by MAPK. *J. Biol. Chem.* **1998**, *273*, 1496–1505. [CrossRef] [PubMed]

32. Frodin, M.; Jensen, C.J.; Merienne, K.; Gammeltoft, S. A phosphoserine-regulated docking site in the protein kinase RSK2 that recruits and activates PDK1. *EMBO J.* **2000**, *19*, 2924–2934. [CrossRef] [PubMed]

33. Richards, S.A.; Dreisbach, V.C.; Murphy, L.O.; Blenis, J. Characterization of regulatory events associated with membrane targeting of p90 ribosomal S6 kinase 1. *Mol. Cell Biol.* **2001**, *21*, 7470–7480. [CrossRef]

34. Jacinto, E.; Facchinetti, V.; Liu, D.; Soto, N.; Wei, S.; Jung, S.Y.; Huang, Q.; Qin, J.; Su, B. SIN1/MIP1 Maintains rictor-mTOR Complex Integrity and Regulates Akt Phosphorylation and Substrate Specificity. *Cell* **2006**, *127*, 125–137. [CrossRef] [PubMed]

35. Abe, Y.; Yoon, S.O.; Kubota, K.; Mendoza, M.C.; Gygi, S.P.; Blenis, J. p90 ribosomal S6 kinase and p70 ribosomal S6 kinase link phosphorylation of the eukaryotic chaperonin containing TCP-1 to growth factor, insulin, and nutrient signaling. *J. Biol. Chem.* **2009**, *284*, 14939–14948. [CrossRef] [PubMed]

36. Chou, P.C.; Oh, W.J.; Wu, C.C.; Moloughney, J.; Ruegg, M.A.; Hall, M.N.; Jacinto, E.; Werlen, G. Mammalian Target of Rapamycin Complex 2 Modulates alphabetaTCR Processing and Surface Expression during Thymocyte Development. *J. Immunol.* **2014**, *193*, 1162–1170. [CrossRef]

37. Gavin, A.C.; Nebreda, A.R. A MAP kinase docking site is required for phosphorylation and activation of p90(rsk)/MAPKAP kinase-1. *Curr. Biol.* **1999**, *9*, 281–284. [CrossRef]

38. Smith, J.A.; Poteet-Smith, C.E.; Malarkey, K.; Sturgill, T.W. Identification of an extracellular signal-regulated kinase (ERK) docking site in ribosomal S6 kinase, a sequence critical for activation by ERK in vivo. *J. Biol. Chem.* **1999**, *274*, 2893–2898. [CrossRef]

39. Roux, P.P.; Shahbazian, D.; Vu, H.; Holz, M.K.; Cohen, M.S.; Taunton, J.; Sonenberg, N.; Blenis, J. RAS/ERK signaling promotes site-specific ribosomal protein S6 phosphorylation via RSK and stimulates cap-dependent translation. *J. Biol. Chem.* **2007**, *282*, 14056–14064. [CrossRef]

40. Chung, J.; Kuo, C.J.; Crabtree, G.R.; Blenis, J. Rapamycin-FKBP specifically blocks growth-dependent activation of and signaling by the 70 kd S6 protein kinases. *Cell* **1992**, *69*, 1227–1236. [CrossRef]

41. Leroux, A.E.; Schulze, J.O.; Biondi, R.M. AGC kinases, mechanisms of regulation and innovative drug development. *Semin. Cancer Biol.* **2018**, *48*, 1–17. [CrossRef] [PubMed]

42. Jensen, C.J.; Buch, M.B.; Krag, T.O.; Hemmings, B.A.; Gammeltoft, S.; Frodin, M. 90-kDa ribosomal S6 kinase is phosphorylated and activated by 3-phosphoinositide-dependent protein kinase-1. *J. Biol. Chem.* **1999**, *274*, 27168–27176. [CrossRef] [PubMed]

43. Richards, S.A.; Fu, J.; Romanelli, A.; Shimamura, A.; Blenis, J. Ribosomal S6 kinase 1 (RSK1) activation requires signals dependent on and independent of the MAP kinase ERK. *Curr. Biol.* **1999**, *9*, 810–820. [CrossRef]

44. Ebner, M.; Sinkovics, B.; Szczygiel, M.; Ribeiro, D.W.; Yudushkin, I. Localization of mTORC2 activity inside cells. *J. Cell Biol.* **2017**, *216*, 343–353. [CrossRef]

45. Boulbes, D.R.; Shaiken, T.; Sarbassov dos, D. Endoplasmic reticulum is a main localization site of mTORC2. *Biochem. Biophys Res. Commun.* **2011**, *413*, 46–52. [CrossRef] [PubMed]

46. Betz, C.; Stracka, D.; Prescianotto-Baschong, C.; Frieden, M.; Demaurex, N.; Hall, M.N. Feature Article: mTOR complex 2-Akt signaling at mitochondria-associated endoplasmic reticulum membranes (MAM) regulates mitochondrial physiology. *Proc. Natl Acad Sci. USA* **2013**, *110*, 12526–12534. [CrossRef]

47. Rosner, M.; Hengstschlager, M. Cytoplasmic and nuclear distribution of the protein complexes mTORC1 and mTORC2: Rapamycin triggers dephosphorylation and delocalization of the mTORC2 components rictor and sin1. *Hum. Mol. Genet.* **2008**, *17*, 2934–2948. [CrossRef]

48. Gan, X.; Wang, J.; Su, B.; Wu, D. Evidence for Direct Activation of mTORC2 Kinase Activity by Phosphatidylinositol 3,4,5-Trisphosphate. *J. Biol. Chem.* **2011**, *286*, 10998–11002. [CrossRef]

49. Gleason, C.E.; Oses-Prieto, J.A.; Li, K.H.; Saha, B.; Situ, G.; Burlingame, A.L.; Pearce, D. Phosphorylation at distinct subcellular locations underlies specificity in mTORC2-mediated activation of SGK1 and Akt. *J. Cell Sci.* **2019**, *132*, jcs224931. [CrossRef]

50. Jindra, P.T.; Jin, Y.P.; Jacamo, R.; Rozengurt, E.; Reed, E.F. MHC class I and integrin ligation induce ERK activation via an mTORC2-dependent pathway. *Biochem. Biophys Res. Commun.* **2008**, *369*, 781–787. [CrossRef]

51. Fourneaux, B.; Chaire, V.; Lucchesi, C.; Karanian, M.; Pineau, R.; Laroche-Clary, A.; Italiano, A. Dual inhibition of the PI3K/AKT/mTOR pathway suppresses the growth of leiomyosarcomas but leads to ERK activation through mTORC2: Biological and clinical implications. *Oncotarget* **2017**, *8*, 7878–7890. [CrossRef] [PubMed]

52. Kovalski, J.R.; Bhaduri, A.; Zehnder, A.M.; Neela, P.H.; Che, Y.; Wozniak, G.G.; Khavari, P.A. The Functional Proximal Proteome of Oncogenic Ras Includes mTORC2. *Mol. Cell.* **2019**, *73*, 830–844. [CrossRef]

53. Lone, M.U.; Miyan, J.; Asif, M.; Malik, S.A.; Dubey, P.; Singh, V.; Singh, K.; Mitra, K.; Pandey, D.; Haq, W.; et al. Direct physical interaction of active Ras with mSIN1 regulates mTORC2 signaling. *BMC Cancer* **2019**, *19*, 1236. [CrossRef] [PubMed]

54. Yao, C.A.; Ortiz-Vega, S.; Sun, Y.Y.; Chien, C.T.; Chuang, J.H.; Lin, Y. Association of mSin1 with mTORC2 Ras and Akt reveals a crucial domain on mSin1 involved in Akt phosphorylation. *Oncotarget* **2017**, *8*, 63392–63404. [CrossRef]

55. Khanna, A.; Lotfi, P.; Chavan, A.J.; Montano, N.M.; Bolourani, P.; Weeks, G.; Shen, Z.; Briggs, S.P.; Pots, H.; Van Haastert, P.J.; et al. The small GTPases Ras and Rap1 bind to and control TORC2 activity. *Sci. Rep.* **2016**, *6*, 25823. [CrossRef] [PubMed]

56. Moloughney, J.G.; Vega-Cotto, N.M.; Liu, S.; Patel, C.; Kim, P.K.; Wu, C.C.; Albaciete, D.; Magaway, C.; Chang, A.; Rajput, S.; et al. mTORC2 modulates the amplitude and duration of GFAT1 Ser-243 phosphorylation to maintain flux through the hexosamine pathway during starvation. *J. Biol. Chem.* **2018**, *293*, 16464–16478. [CrossRef]

57. Casas-Terradellas, E.; Tato, I.; Bartrons, R.; Ventura, F.; Rosa, J.L. ERK and p38 pathways regulate amino acid signalling. *Biochim. Biophys Acta.* **2008**, *1783*, 2241–2254. [CrossRef]

58. Tato, I.; Bartrons, R.; Ventura, F.; Rosa, J.L. Amino acids activate mammalian target of rapamycin complex 2 (mTORC2) via PI3K/Akt signaling. *J. Biol. Chem.* **2011**, *286*, 6128–6142. [CrossRef] [PubMed]

59. Zhu, J.; Thompson, C.B. Metabolic regulation of cell growth and proliferation. *Nat. Rev. Mol. Cell Biol.* **2019**, *20*, 436–450. [CrossRef]

60. Gestaut, D.; Limatola, A.; Joachimiak, L.; Frydman, J. The ATP-powered gymnastics of TRiC/CCT: An asymmetric protein folding machine with a symmetric origin story. *Curr. Opin. Struct. Biol.* **2019**, *55*, 50–58. [CrossRef]

61. Roh, S.H.; Kasembeli, M.; Bakthavatsalam, D.; Chiu, W.; Tweardy, D.J. Contribution of the Type II Chaperonin, TRiC/CCT, to Oncogenesis. *Int. J. Mol. Sci.* **2015**, *16*, 26706–26720. [CrossRef] [PubMed]

62. Cuellar, J.; Ludlam, W.G.; Tensmeyer, N.C.; Aoba, T.; Dhavale, M.; Santiago, C.; Bueno-Carrasco, M.T.; Mann, M.J.; Plimpton, R.L.; Makaju, A.; et al. Structural and functional analysis of the role of the chaperonin CCT in mTOR complex assembly. *Nat. Commun.* **2019**, *10*, 2865. [CrossRef] [PubMed]

63. Ludwik, K.A.; Lannigan, D.A. Ribosomal S6 kinase (RSK) modulators: A patent review. *Expert Opin. Ther. Pat.* **2016**, *26*, 1061–1078. [CrossRef] [PubMed]

Article

Mutation-Associated Phenotypic Heterogeneity in Novel and Canonical PIK3CA Helical and Kinase Domain Mutants

Arman Ali Ghodsinia [†], J-Ann Marie T. Lego and Reynaldo L. Garcia * 

Disease Molecular Biology and Epigenetics Laboratory, National Institute of Molecular Biology and Biotechnology, University of the Philippines Diliman, Quezon City 1101, Philippines; arman.ghodsinia@oncology.ox.ac.uk (A.A.G.); jtlego@up.edu.ph (J-A.M.T.L.)
* Correspondence: reygarcia@mbb.upd.edu.ph
† Present address: MRC Oxford Institute for Radiation Oncology, Department of Oncology, Old Road Campus Research Building, Roosevelt Drive, University of Oxford, Oxford OX3 7DQ, UK.

Received: 4 March 2020; Accepted: 29 April 2020; Published: 30 April 2020

Abstract: Phosphatidylinositol 3-kinase, catalytic subunit alpha (PIK3CA) is an oncogene often mutated in colorectal cancer (CRC). The contribution of PIK3CA mutations in acquired resistance to anti-epidermal growth factor receptor (EGFR) therapy is well documented, but their prognostic and predictive value remain unclear. Domain- and exon-specific mutations are implicated in either favorable or poor prognoses, but there is paucity in the number of mutations characterized outside of the mutational hotspots. Here, two novel non-hotspot mutants—Q661K in exon 13 and C901R in exon 19—were characterized alongside the canonical exon 9 E545K and exon 20 H1047R mutants in NIH3T3 and HCT116 cells. Q661K and E545K both map to the helical domain, whereas C901R and H1047R map to the kinase domain. Results showed variable effects of Q661K and C901R on morphology, cellular proliferation, apoptosis resistance, and cytoskeletal reorganization, with both not having any effect on cellular migration. In comparison, E545K markedly promoted proliferation, survival, cytoskeletal reorganization, migration, and spheroid formation, whereas H1047R only enhanced the first three. In silico docking suggested these mutations negatively affect binding of the p85 alpha regulatory subunit to PIK3CA, thereby relieving PIK3CA inhibition. Altogether, these findings support intra-domain and mutation-specific variability in oncogenic readouts, with implications in degree of aggressiveness.

Keywords: PIK3CA; colorectal cancer; EGFR pathway; tumor heterogeneity

---

1. Introduction

Phosphatidylinositol 3-kinase, catalytic subunit alpha (PIK3CA) is an oncogene mutated in 10% to 25% of colorectal cancers (CRCs) [1–4]. Gain-of-function mutations in PIK3CA allow it to activate several signaling pathways, most notably the PIK3CA– Protein Kinase B alpha (AKT) pathway downstream of the epidermal growth factor receptor (EGFR), to promote tumorigenesis. Hotspot mutations in PIK3CA lie in exon 9 (E542K and E545K) in the helical domain and exon 20 (H1047R) in the catalytic kinase domain. These mutations may co-exist with Kirsten Rat Sarcoma Viral Oncogene Homolog (KRAS) and B-Raf Proto-Oncogene, Serine/Threonine Kinase (BRAF) mutations to exert different oncogenic effects and responses to anti-EGFR therapy [5]. Due to change in protein structure induced by the mutations, exon 9 mutations rely on RAS- Guanosine Triphosphate (RAS-GTP) binding to induce transformation, whereas exon 20 mutations are independent of RAS-GTP binding and rely on p85 binding instead [6]. Studies have also correlated exon 20 mutations but not exon 9 mutations to poor response to anti-EGFR therapy in metastatic colorectal cancer [2,7,8]. In uterine endometrial adenocarcinomas, PIK3CA mutations in the kinase domain (exon 20) are associated with adverse prognostic parameters [9].

Various studies suggest that the phenotypic outcome of helical and kinase domain mutations are cell type- and mutation-specific [10–12]. Pang et al. [10] reported that the E545K helical domain mutation made cells more significantly chemotactic in the MDA-MB-231 Human Caucasian breast adenocarcinoma cell line, as well as increasing rate of intravasation and extravasation in vivo compared to the H1047R kinase domain mutation. In normal human urothelial cells, the E545K mutation enhanced cellular proliferation and survival more than H1047R [11]. Meyer et al. [12] observed that H1047R was more efficient in forming mice mammary tumors compared to E545K. Indeed, there is a wealth of information regarding exon 9 and 20 mutations and most sequencing studies point to these mutations as occurring most frequently in the population. However, the majority of these studies have been performed in developed nations, and hence the mutational spectrum identified has most likely been biased towards the populations under study [13–16]. Consequently, non-hotspot mutations have not been characterized as much [17]. Studies that have functionally characterized non-hotspot mutations have mostly been limited to the assessment of colony forming ability, lipid kinase activity, and AKT phosphorylation [16,18–21]. Given the myriad of signaling pathways affected by PIK3CA, which include both AKT-dependent and -independent mechanisms [22–26], there is a need to phenotypically characterize the oncogenic readouts of non-hotspot mutations in more cancer hallmarks. These mutations may engage different signaling pathways, confer different cancer phenotypes, and may or may not be associated with resistance to anti-EGFR therapy.

This study reports the phenotypic characterization of two novel non-hotspot PIK3CA mutations, namely, Q661K (NM_006218.1: c.1981C > A) located in exon 13 and C901R (Pfam entry PF00454) located in exon 19 [26,27], alongside the canonical mutants E545K (exon 9) and H1047R (exon 20). Q661K has been reported in the Catalogue of Somatic Mutations in Cancer (COSMIC) database [28]. C901R is found in the Human Cancer Proteome Variation [29] and the Epithelial-to-Mesenchymal Gene [30] databases, and is also a distinct and separate entry from the mutants C901F and C901Y [29]. Both Q661K and C901R remain uncharacterized. Notably, Q661K was also identified from an ongoing targeted next generation sequencing study of young-onset sporadic colorectal cancer tissue samples obtained from the Philippine General Hospital, University of the Philippines, Manila [31]. Similar to E545K, Q661K maps to the helical domain of PIK3CA. C901R and H1047R both map to the kinase domain. The mutations were characterized and compared for their effects on cellular proliferation, survival, migration, spheroid formation, general morphology, and cytoskeletal organization.

2. Materials and Methods

2.1. Cloning and Site-Directed Mutagenesis of Wild Type and PIK3CA Mutants

Wild type (WT) PIK3CA (NM_006218.1) was amplified from cDNA template generated from human kidney 2 cells (HK2; cat. no. CRL-2190, American Type Culture Collection (ATCC), Manassas, Virginia, USA) RNA. The forward (F) and reverse (R) primers used to amplify the PIK3CA constructs are listed in Table 1.

Table 1. Primers used for generation of phosphatidylinositol 3-kinase, catalytic subunit alpha (PIK3CA) wild type and mutant constructs.

Primer Name	Sequence
WT-PIK3CA-F	5′-ATGCCTCCACGACCATCATCAGGTGAACTG-3′
WT-PIK3CA-R	5′-TCAGTTCAATGCATGCTGTTTAATTGTGTGGAAG-3′
E545K-F	5′CTCTCTCTGAAATCACT**A**AGCAGGAGAAAGATTTTCTATG-3′
E545K-R	5′-CATAGAAAATCTTTCTCCTGCT**T**AGTGATTTCAGAGAGAG-3′
H1047R-R	5′-GTGACTACTAGTTCAGTTCAATGCATGCTGTTTAATTGTGTGGA AGATCCAATCCATTTTTGTTGTCCAGCCACCATGA**C**GTGCATC-3′
Q661K-F	5′-GAAAGCATTGACTAAT**A**AAAGGATTGGGCACTTTTTCTTTTG-3′
Q661K-R	5′-GAAAAAGTGCCCAATCCTTT**T**ATTAGTCAATGCTTTCTTC-3′
C901R-F	5′-CCTGTTTACACGTTCA**C**GTGCTGGATACTGTGTAGCTACC-3′
C901R-R	5′-GGTAGCTACACAGTATCCAGCAC**G**TGAACGTGTAAACAGG-3′

Note: Bold letters correspond to mutated nucleotide.

Each PCR mixture contained a final concentration of 1X PCR buffer (Titanium Taq PCR buffer, Clontech Laboratories, Inc., Mountain View, CA, USA), 0.125 mM of each deoxynucleoside triphosphate (iNtRON Biotechnology, Sangdaewon-Dong, Jungwon-Seongnam, Korea), 2 mM each of the appropriate forward and reverse primers, 1X Taq polymerase (Titanium Taq polymerase, Clontech Laboratories), and 50 ng of the appropriate template (cDNA for WT PIK3CA, and cloned pTargeT-PIK3CA WT for the mutations). All PCR reactions were carried out using the C1000 Touch Thermal Cycler (Bio-Rad Laboratories, Inc., Hercules, CA, USA). The PCR mixtures were initially denatured at 95 °C for 5 min. Afterwards, they were subjected to 25 cycles of denaturation at 95 °C for 30 s, annealing at 55 °C for 30 s, and extension at 72 °C for 1.5 min.

For each mutant construct, fragments from the first round PCR were fused via splicing-by-overlap extension PCR. Briefly, 25 ng of each fragment was used as template with WT-PIK3CA-F and WT-PIK3CA-R as primers, using similar PCR conditions as described above, except for two changes: the extension time was adjusted to 3 min, and an additional final extension step at 72 °C for 10 min was added. The final extension step was added to ensure the addition of A-overhangs for TA-ligation into the pTargeT mammalian expression vector (Promega Corporation, Madison, Wisconsin, USA). All constructs were verified as error-free via Sanger sequencing.

2.2. Cell Culture and Transfection

NIH3T3 mouse fibroblast cells (cat. no. CRL-1658; ATCC) were cultured in Dulbecco's modified Eagle's medium (DMEM; Gibco; Thermo Fisher Scientific, Inc., Waltham, MA, USA) supplemented with 10% newborn calf serum (NBCS; Gibco; Thermo Fisher Scientific, Inc.). HCT116 human colon cancer cells (cat. no. CCL-247; ATCC) were cultured in Roswell Park Memorial Institute 1640 medium (RPMI-1640; Gibco; Thermo Fisher Scientific, Inc.) supplemented with 10% fetal bovine serum (FBS), 50 U/mL penicillin/streptomycin, and 2.0 g/L sodium bicarbonate. Both cell lines were incubated in a 37 °C humidified chamber with 5% CO_2.

NIH3T3 cells were transfected using the 10 μL Neon Transfection System (Invitrogen, Thermo Fisher Scientific Inc.) according to the manufacturer's instructions. Briefly, 900,000 cells were mixed with 2 μg of plasmid and electroporated at a pulse voltage of 1400 volts, pulse width of 20 ms, and pulse number of 2. The cells were then seeded in a 12-well plate containing DMEM supplemented with 10% NBCS and left to adhere overnight in a 37 °C humidified chamber with 5% CO_2. Transfection parameters were optimized to achieve 80–90% efficiency for all experiments.

HCT116 cells were transfected using FugeneHD (Promega) according to the manufacturer's instructions. Transfection efficiencies of 70–80% for all experiments were consistently obtained using the following plasmid to transfection reagent ratio: 2000 ng DNA/5 μL FugeneHD in 24-well plates, and 200 ng DNA/0.5 μL FugeneHD in 96-well and 48-well plates.

Proof of expression of the constructs is included in Figure S1.

2.3. Morphological Characterization

Forty-eight hours post-transfection, transfected NIH3T3 cells were photographed at 100× magnification using an Olympus IX71 inverted fluorescence microscope (Olympus Corporation, Tokyo, Japan). For each transfected well, at least four random fields containing a minimum of 100 cells were imaged. Cells with prominent filopodial extensions and cells that were reduced in size, round, and refringent were counted using ImageJ software [32]. The total number of cells showing prominent cellular protrusions and cells exhibiting roundedness and birefringence were divided by the total cell count determined via the cell counter plugin of ImageJ software [32].

To measure the length of cellular protrusions, the Simple Neurite Tracer plugin of Fiji software [33] was used. Briefly, each protrusion was traced from the center of the nucleus to the tip of the extension. The total length of all cellular protrusions in each field was divided by the corresponding field's total cell count determined via the cell counter plugin of ImageJ software [32], and the resulting mean lengths were compared to wild type and vector-only control.

2.4. Actin Cytoskeletal Staining and Analysis

Twenty-four hours post-transfection, transfected NIH3T3 cells were trypsinized from the 12-well plate and seeded in an 8-well glass chamber slide (EMD Millipore, Burlington, MA, USA) at a density of 8000 cells per well. Forty-eight hours post-transfection, the cells were fixed with 4% paraformaldehyde for 20 min. They were then permeabilized with 0.1% Triton X-100 in 1× Phosphate Buffered Saline (PBS) for 20 min, and blocked with 1% bovine serum albumin (Sigma-Aldrich; Merck KGaA, Darmstadt, Germany) in 1× PBS for 20 min. The cells were then stained with 100 μL of 0.165 μM of Alexa Flour 488 Phalloidin (Thermo Fisher Scientific, Inc.) in 1× PBS for 30 min, and counterstained with 100 μL of 1 μg/mL Hoechst 33342 for 15 min. Stained cells were mounted in 70% glycerol in 1× PBS. All steps were performed at room temperature, and the cells were washed thrice with 1× PBS in between steps.

Slides were observed under a fluorescence microscope (Olympus IX83; Olympus Corporation, Tokyo, Japan) at 400× magnification, using the green fluorescent filter (λex/λem: 490/525 nm) to visualize stained filamentous actin structures, and the blue fluorescent filter (λex/λem: 355/465 nm) to visualize the nuclei.

2.5. Scratch Wound Healing Assay

NIH3T3 cells were transfected in a 12-well plate. Twenty-four hours post-transfection, cells were reseeded in a 96-well half area plate at a density of 15,000 cells per well in triplicate. Upon reaching full confluence, a scratch was made in each well using a sterile pipette tip. The cells were then maintained in DMEM supplemented with 2.5% NBCS. Migration of cells into the wound area was photographed every hour for 21 h via a motorized time-lapse microscope (Olympus IX83; Olympus Corporation, Tokyo, Japan).

HCT116 cells were seeded at a density of 10,000 cells per well in a 96-well plate and then transfected 36 h later. Upon reaching full confluence, a scratch was made in each well using a sterile pipette tip. The cells were then maintained in RPMI-1640 supplemented with 2% FBS. Migration of cells into the wound area was photographed at 0 h and 24 h post-scratch using Olympus IX71 inverted fluorescence microscope (Olympus Corporation, Tokyo, Japan).

The gap area for each time point was determined using ImageJ software [32]. The migration rate (μm^2/h) for each setup was determined by obtaining the slope of the trend line generated after the gap area (μm^2) was plotted against time (h).

2.6. Cell Proliferation Assay

NIH3T3 cells were transfected in a 12-well plate. Twenty-four hours post-transfection, cells were reseeded in 96-well plates at a density of 2500 cells per well in triplicate. The cells were then maintained in DMEM supplemented with 2.5% NBCS.

HCT116 cells were seeded at a density of 2500 cells per well in 48-well plates and transfected 36 h later. Twenty-four hours post-transfection, cells were maintained in RPMI-1640 supplemented with 0.5% FBS.

The number of metabolically active cells per setup was determined at 48 and 72 h post-transfection for NIH3T3 and at 48, 72, and 96 h post-transfection for HCT116, by incubating each well with 10 μL of CellTiter 96 Aqueous One Solution Cell Proliferation Assay reagent (Promega Corporation) until color development. Absorbance values at 490 nm of each setup were measured with a colorimetric plate reader (FLUOstar Omega microplate reader; BMG Labtech, Cary, NC, USA). Cell counts were calculated from a standard curve (number of cells vs. A490) generated using serial dilutions of an untransfected cell suspension. Mean cell counts were calculated per setup for each time-point.

2.7. Apoptosis (Caspase 3/7) Assay

NIH3T3 cells were transfected in a 12-well plate. Twenty-four hours post-transfection, cells were reseeded in 96-well plates at a density of 2500 cells per well in triplicates. Cells were then maintained in DMEM supplemented with 2.5% NBCS.

HCT116 cells were seeded at a density of 2500 cells per well in a 96-well plate and transfected 36 h later. Twenty-four hours post-transfection, cells were then maintained in RPMI-1640 supplemented with 0.5% FBS.

The level of caspase 3/7 activity for each setup was determined at 48 h post-transfection for NIH3T3 and at 96 h post-transfection for HCT116 by incubating the cells with 10 μL of Caspase-Glo 3/7 assay reagent (Promega Corporation). The plates were incubated at room temperature for 3 h and then the luminescence per well was measured using a plate reader (FLUOstar Omega microplate reader; BMG Labtech). Luminescence readings were normalized to the number of viable cells determined via a cell proliferation assay performed concurrently.

2.8. Western Blot Analyses

NIH3T3 and HCT116 cells were transfected in 12-well plates. Transfected NIH3T3 cells were cultured in DMEM supplemented with 2.5% NBCS at 24 h post-transfection. At 60 h post-transfection, transfected NIH3T3 cells were further serum starved by culturing them in DMEM supplemented with 1% NBCS. Transfected HCT116 cells were cultured in RPMI-1640 supplemented with 10% FBS throughout.

At 72 h post-transfection, proteins were harvested for SDS-PAGE and Western blotting. Cells were lysed with radioimmunoprecipitation lysis buffer (150 mM NaCl, 0.5% sodium deoxycholate, 0.1% sodium dodecyl sulphate, 50 mM Tris (pH 8.0)) supplemented with Halt protease and phosphatase inhibitor cocktail (Thermo Fisher Scientific Inc.). For gel electrophoresis, 20 μg of protein were run on a 7.5% SDS-PAGE gel at 100 V for 1 h. Proteins were blotted for 24 h at a constant current of 40 mA. Membranes were blocked with 5% w/v non-fat dry milk in Tris-buffered saline with 0.1% v/v Tween-20 (TBST) for 1 h at room temperature; probed overnight with primary antibodies obtained from Cell Signaling Technology (Danvers, MA, USA): anti-PIK3CA (1:1000, cat. no. 4249S), anti-N-cadherin (1:1000, cat. no. 14215S), anti-E-cadherin (1:1000, cat. no. 3195S), and anti- Glyceraldehyde 3-phosphate dehydrogenase (GAPDH; 1:1000, cat. no. 2118S); washed with TBST; and incubated with the appropriate secondary antibodies obtained from Thermo Fisher Scientific Inc.: goat anti-mouse Immunoglobulin G (IgG, H + L; 1:3000, cat. no. 31430) and goat anti-rabbit IgG (1:3000, cat. no. 31460) for 1 h at room temperature. Signals were developed with enhanced chemiluminescence substrate and imaged using the ChemiDoc Touch Imaging System (Bio-Rad Laboratories, Inc., Hercules, CA, USA). Densitometric analysis of digitized band intensities was performed using GelQuant.NET software (v1.8.2) provided by biochemlabsolutions.com. Gene expression levels were normalized against GAPDH expression.

2.9. Spheroid Formation Assay

Twenty-four hours post-transfection, transfected NIH3T3 cells were trypsinized from the 12-well plate and seeded in a 96-well round bottom ultra-low-attachment spheroid plate (Corning Incorporated, New York City, NY, USA) at a density of 2500 cells per well in triplicate. The cells were maintained in DMEM supplemented with 2.5% NBCS. The spheroids were photographed at 40× magnification using an Olympus IX71 inverted fluorescence microscope (Olympus Corp., Tokyo, Japan) at 96 h post-transfection. The diameter of the spheroids was then measured using ImageJ software [32].

2.10. Bioinformatics-Based Prediction of the Functional Impact of PIK3CA Mutations

The functional impact of the PIK3CA Q661K and C901R mutations was predicted through four sequence-based algorithms: Polymorphism Phenotyping 2 (Polyphen-2; version 2) (accessed on 20 August 2019; http://genetics.bwh.harvard.edu/pph2/) [34], Mutation Assessor (release 3) (accessed on 20 August 2019; http://mutationassessor.org/r3/) [35], Functional Analysis Through Hidden Markov Models (FATHMM; version 2.3) (accessed on 20 August 2019; http://fathmm.biocompute.org.uk/) [36], and Sorting Tolerant From Intolerant (SIFT; version 5.2.2) (accessed on 20 August 2019; https://sift.bii.a-star.edu.sg/www/SIFT_seq_submit2.html) [37].

To generate homology models of the H1047R, E545K, Q661K, and C901R mutations, the Swiss-model [38] webserver (accessed on 20 August 2019; https://swissmodel.expasy.org/interactive) was utilized. The template used for homology modelling was the crystal structure of native PIK3CA in complex with niSH2 (N-terminal Src Homology 2 / inter Src Homology 2) domain of p85 alpha (Protein Data Bank or PDB ID: 4L1B) [39]. Using BIOVIA Discovery Studio Visualizer (Dassault Systèmes BIOVIA, San Diego, CA, USA), the models of the mutations were superimposed to the wild type and the global main chain root mean square distance (RMSD), and overlay similarity (OS) were calculated.

The impact of the mutations on the ability of the niSH2 domain of p85 alpha to form a dimer with PIK3CA was assessed using the pyDOCK webserver (accessed 22 August 2019; https://life.bsc.es/pid/pydockweb) [40,41]. The pyDock scores of the best poses were compared across all mutations. The interactions between the interface residues of the PIK3CA–p85 alpha dimer were analyzed with PDBePISA (version 1.52) (accessed on 25 August 2019; https://www.ebi.ac.uk/pdbe/pisa/) [42].

2.11. Statistical Analyses

All data were analyzed using unpaired one-tailed *t*-test to compare differences between the vector only control/wild type and each of the mutants. Data from quantitative experiments were represented as mean ± standard deviation. Significant values were represented as * $p < 0.05$, ** $p < 0.01$, and *** $p < 0.001$.

3. Results

3.1. The PIK3CA Mutations Had Variable Effects on Proliferative Rates of NIH3T3 and HCT116 Cells

To determine if expression of the PIK3CA mutants can promote cellular proliferation, the number of viable cells per setup was determined at 24, 48, and 72 h post-transfection for NIH3T3 cells and at 48, 72, and 96 h for HCT116 cells. The results in HCT116 were generally consistent with those obtained in NIH3T3 cells (Figure 1A,B). The canonical mutants E545K and H1047R as well as the novel mutant Q661K enhanced proliferative capacity. C901R enhanced proliferation only in HCT116. The effect of the wild type construct in the two cellular backgrounds, however, showed a marked difference. In NIH3T3 cells, WT had no apparent effect on proliferation and was indistinguishable from that of the vector-only control. In HCT116 cells, WT overexpression was able to enhance proliferative capacity. There are at least two plausible explanations for this. HCT116 harbors an endogenous KRAS G13D mutation and it is highly likely that it is able to hyperactivate wild type PIK3CA, which is downstream of KRAS in the signaling pathway; hence, the observed enhanced proliferation. Alternatively, the presence of the endogenous PIK3CA H1047R (in addition to KRAS G13D) and the overexpression of wild type PIK3CA may have a synergistic effect that could have led to enhanced proliferation.

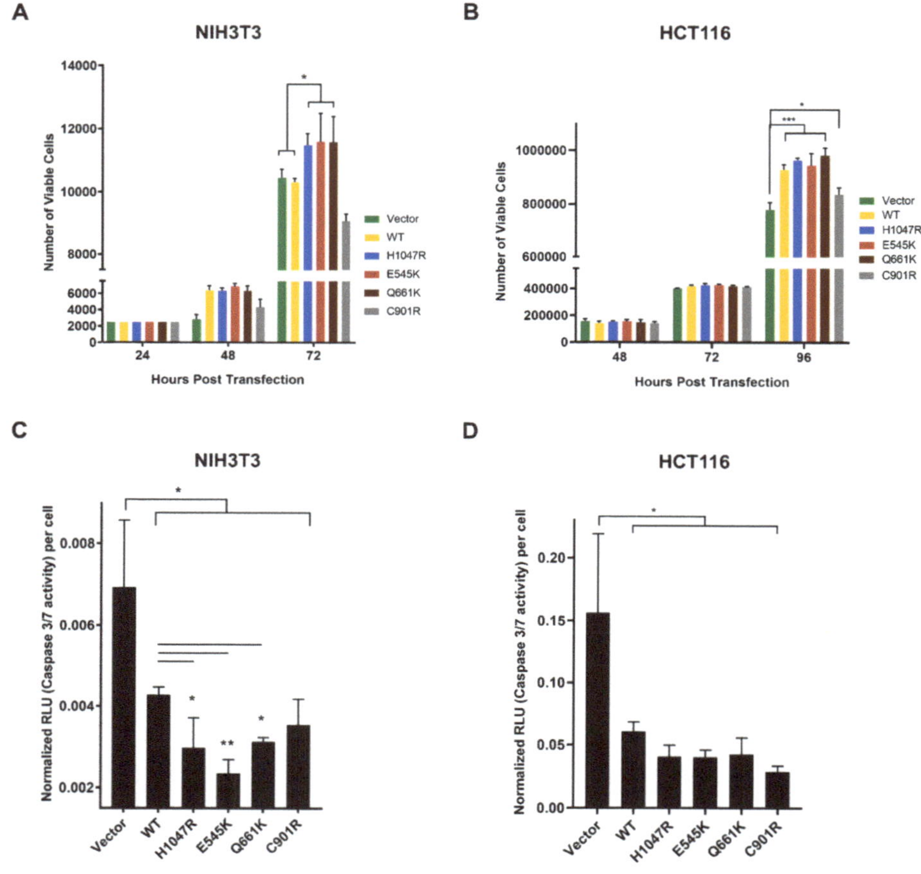

Figure 1. Variable effects of wild type (WT), canonical, and novel PIK3CA mutants on proliferative capacity and apoptosis resistance in NIH3T3 and HCT116 cells. Proliferation rates of (**A**) NIH3T3 and (**B**) HCT116 cells, and caspase 3/7 activity in (**C**) NIH3T3 and (**D**) HCT116 cells transfected with empty vector, wild type PIK3CA, or PIK3CA mutants. Data presented are representative of three independent trials in triplicates and expressed as mean ± standard deviation. * $p < 0.05$, ** $p < 0.01$ and *** $p < 0.001$. WT: wild type.

3.2. Variable Effects of the Canonical Mutants E545K and H1047R, and the Novel Mutants Q661K and C901R on Apoptosis Resistance in NIH3T3 and HCT116 Cells

PIK3CA is known to promote cell survival [43,44]. To test the capacity of the PIK3CA mutants to inhibit apoptosis, the activity of caspase 3/7 was assessed in transfected cells using the caspase-Glo 3/7 assay. In NIH3T3 cells, overexpression of the Q661K novel mutant and the H1047R and E545K canonical mutants led to a significant reduction in caspase 3/7 activity, indicating resistance to apoptosis (Figure 1C). Among all mutants, E545K had the lowest level of caspase 3/7 activity. Cells overexpressing wild type PIK3CA and the novel C901R mutant showed the highest level of caspase 3/7 activity but still demonstrated resistance to apoptosis compared to vector-only control. In HCT116 cells, the wild type and all mutant constructs also induced resistance to apoptosis, although the degree of inhibition did not vary widely among the different setups (Figure 1D). The NIH3T3 cell line is usually preferred in characterizing oncogenes and their mutant variants because they do not require cooperative complementary mutations to express a transformed phenotype [45]. In addition to the non-cancerous background, this may explain the more resolved differences in degree of resistance to apoptosis among the wild type and mutant setups in NIH3T3 compared with HCT116.

3.3. Novel and Canonical PIK3CA Mutants Induced Gross Morphological Alterations and Enhanced Formation of Pseudopodial Extensions

Gross morphological alterations can be indicative of oncogenic transformation. Transformed NIH3T3 cells typically show decreased size, refringency, pronounced pseudopods, and increased cellular protrusions [46,47]. To determine if the canonical and novel PIK3CA mutations can induce morphological alterations, transfected cells were observed under a brightfield microscope. The percentage of cells showing pronounced extensions, the average length of these cellular extensions, and the percentage of cells exhibiting roundedness and birefringence were also quantified.

Results indicate that overexpression of the PIK3CA mutants caused decreased size and rounding of cells, refringency, and formation of cellular protrusions (Figure 2). In cells overexpressing the E545K and H1047R hotspot mutants, there was a dramatic increase in the percentage of cells showing pronounced cellular protrusions compared to the rest of the setups. These cells also had the longest average length of cellular protrusions. Overexpression of C901R enhanced formation of cellular protrusions to levels comparable with that of H1047R and showed an average protrusion length similar to Q661K and wild type setups. Q661K had marginal increase in the number of protrusions and average length compared to the wild type and vector only control. Consistent with these findings, overexpression of all four mutants also increased the percentage of round and refringent cells compared to the vector-only control.

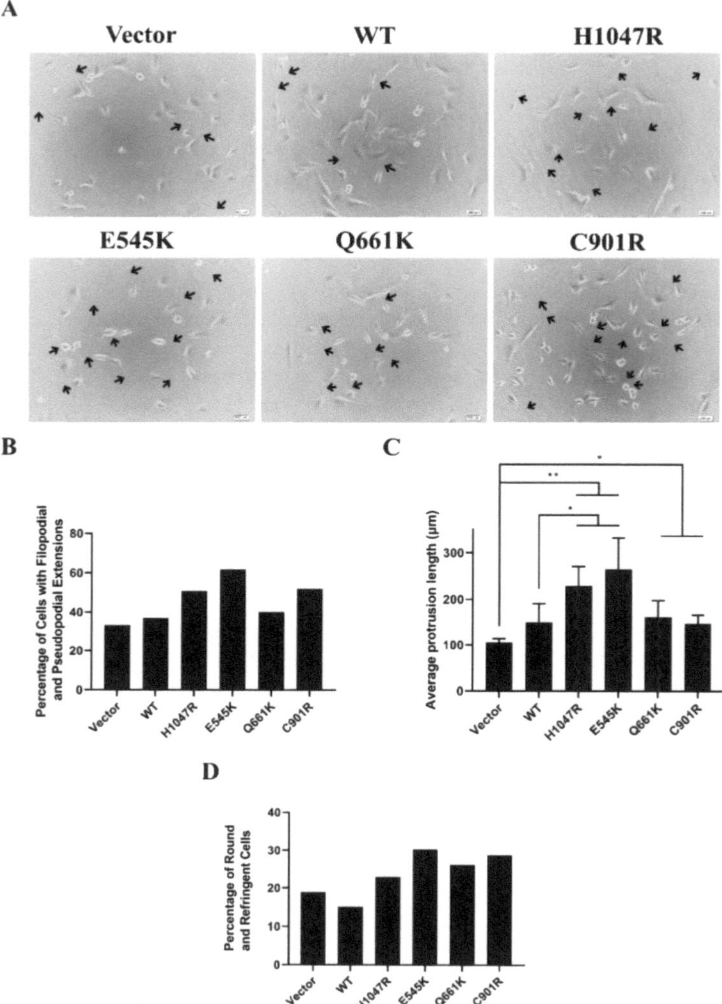

Figure 2. Altered gross morphology of NIH3T3 cells overexpressing PIK3CA mutants. (**A**) Prominent cellular protrusions (black arrows) were observed in cells overexpressing PIK3CA mutants. Scale bar: 200 µm. (**B**) NIH3T3 cells with prominent cellular protrusions were counted and divided by total number of fibroblasts. (**C**) The lengths of cellular protrusions were measured from the center of the nucleus to the tip of the extension. The total length was then divided by the total number of fibroblasts. (**D**) Cells showing decreased size, roundedness, and birefringence were quantified and divided by the total cell count. The percentage of birefringent cells and cells with cellular protrusions, as well as the average length of these protrusions, increased in cells overexpressing the PIK3CA mutations. Data presented are representative of three independent trials in triplicate, and expressed as mean ± standard deviation. * $p < 0.05$ and ** $p < 0.01$. WT: wild type. PIK3CA: phosphatidylinositol 3-kinase, catalytic subunit alpha.

3.4. Cells Overexpressing Wild Type PIK3CA as well as Novel and Canonical PIK3CA Mutants Were Highly Depolarized with Long Cellular Protrusions

To determine if wild type PIK3CA and the novel and canonical PIK3CA mutants can induce cytoskeletal reorganization, F-actin in transfected cells were visualized via phalloidin staining and fluorescence microscopy. In both wild type and mutant setups, the cells were dramatically depolarized, with prominent leading edge, lamellipodia, and long cellular protrusions (Figure 3). All these cytoplasmic changes suggest a highly motile phenotype propelled by dynamic actin networks.

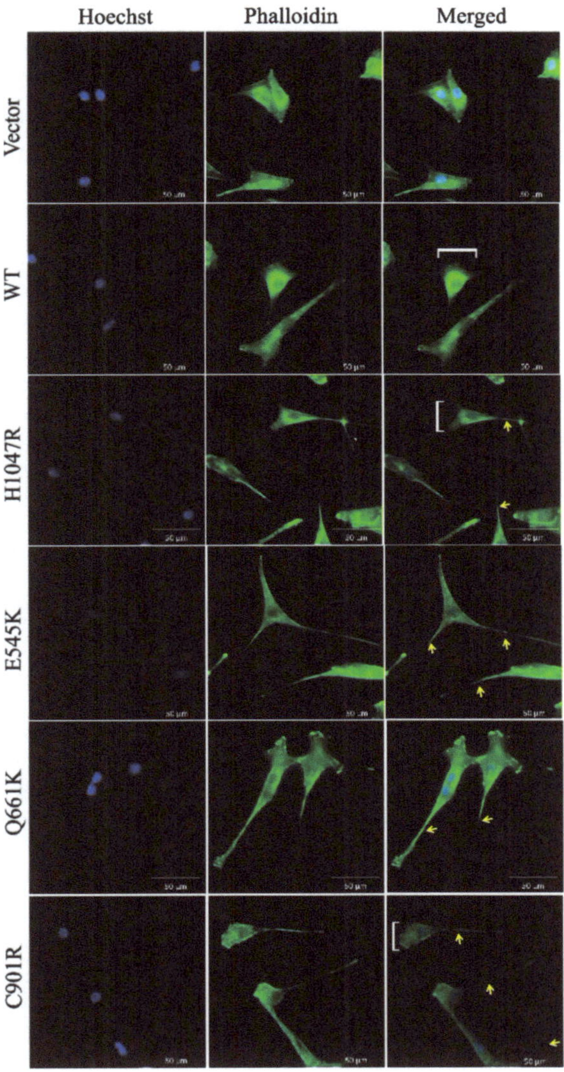

Figure 3. Cytoskeletal reorganization of cells overexpressing wild type, novel, and canonical PIK3CA mutations. White brackets show prominent migrating front. Yellow arrows point to polymerized actin of pseudopod extensions. Scale bar: 50 µm. Data presented are representative of three independent trials in triplicates. WT: wild type.

To evaluate whether the morphological changes induced by the PIK3CA mutants are due, at least in part, to epithelial-to-mesenchymal transition (EMT), the expression levels of E-cadherin and N-cadherin were examined in transfected HCT116 cells, an epithelial cancer cell line. Despite its cancerous cellular background, it is still possible to augment and observe EMT changes in HCT116 after transfection with oncogenic mutants or upregulation of cellular oncogenes [48,49]. Western blot results showed consistent partial EMT only for the E545K mutant, that is, a consistent apparent decrease in E-cadherin, but inconclusive results for N-cadherin. The wild type and other mutant constructs showed inconsistent results across trials, which may suggest no effect on EMT markers (Figure S2). This is not totally surprising given that many aspects that contribute to morphological changes (e.g., cytoskeletal reorganization, elaboration of cellular processes and protrusions, E- to N-cadherin switch) are in fact controlled by distinct signaling pathways [50–53].

3.5. E545K, but not the Other PIK3CA Mutants, Enhanced Migration of NIH3T3 Cells

Given that overexpression of PIK3CA mutations was able to alter cellular morphology and cytoskeletal actin organization, effects of the mutations on cellular migration were also determined. Confluent monolayers were scratched with a sterile pipette tip, and migration of cells into the wound area was monitored for 21 (NIH3T3) or 24 h (HCT116). In NIH3T3 cells, only the E545K canonical mutation was able to significantly promote cellular migration compared to wild type and vector-only controls (Figure 4A,C). In HCT116 cells, no marked differences in migratory capacity were observed across setups at 24 h post-scratch (Figure 4B,D). The different results obtained for E545K may have been due to differences in cellular background. As discussed in Section 3.2 above, NIH3T3 cells are better able to express a transformed phenotype, even in the absence of cooperative complementary mutations [45].

The lack of an effect on cellular migration is not inconsistent with the observed cytoskeletal reorganization induced by all the mutants. There are multiple, complex signaling events integrated to bring about migration. Among the many players involved are the Rho GTPases, integrins, phosphoinositides, RAS proteins, and numerous kinases [54]. Although reorganization of the cytoskeleton is an early event necessary for movement, adhesion-dependent migration is a distinct process altogether.

3.6. E545K Canonical Mutation Enhanced Spheroid Formation in NIH3T3 Cells

Tumors rely on cell–cell and cell–extracellular matrix (ECM) interactions for growth and metastasis [55]. Because there is a significant reduction of these interactions in two-dimensional cultures, it is possible that the phenotype of PIK3CA mutant overexpression could be irreproducible in vivo. To determine whether the results obtained thus far could be replicated in a model that mimics the spatial organization of tumor cells, NIH3T3 cells overexpressing the PIK3CA mutations were grown as three-dimensional spheroids. Only the E545K canonical mutant promoted significant spheroid formation at 96 h post-transfection (Figure 5). The H1047R canonical mutant, as well as the novel mutants Q661K and C901R, did not show any significant differences in size when compared to the wild type and vector-only controls.

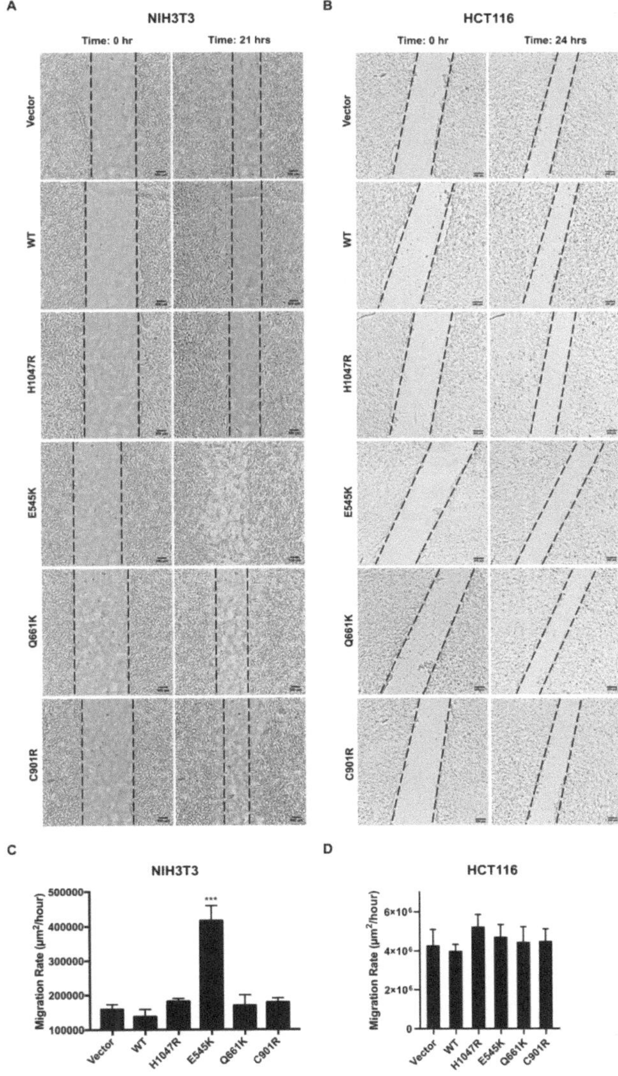

Figure 4. Effect of wild type and PIK3CA mutations on cellular migration in vitro. Confluent monolayers of (**A**) NIH3T3 and (**B**) HCT116 cells were scratched and brightfield images of wound closure were taken for at least a 21 h period post-scratch. Compared with the vector and wild type controls, narrower scratch gaps were noticeable only for NIH3T3 cells overexpressing the PIK3CA E545K hotspot mutant. No significant change was observed for the rest of the PIK3CA mutants in NIH3T3 and HCT116 cells. Scale bars: 100 μm (NIH3T3) and 200 μm (HCT116). Migration rates of (**C**) NIH3T3 and (**D**) HCT116 cells were measured by obtaining the slope of the trend line generated when the wound area was plotted against time for every time point (μm^2/h). Data presented are representative of three independent trials in triplicate, and expressed as mean ± standard deviation. *** $p < 0.001$. WT: wild type; PIK3CA: phosphatidylinositol 3-kinase, catalytic subunit alpha.

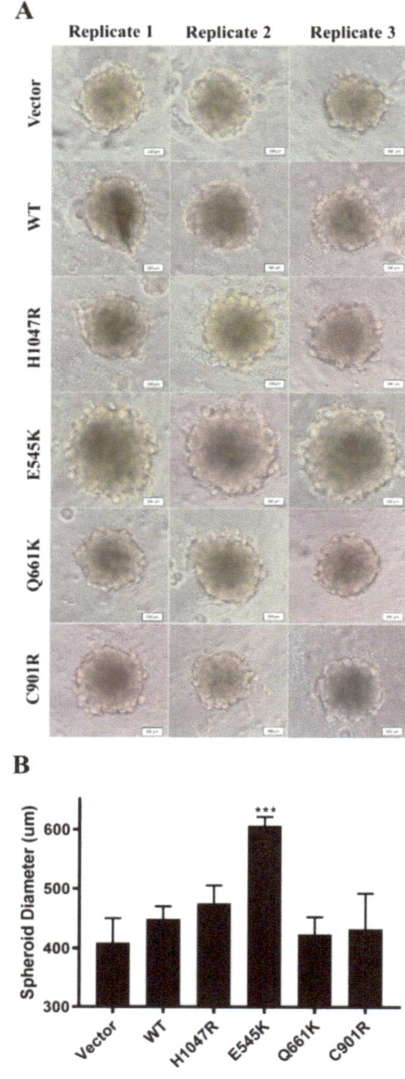

Figure 5. The E545K hotspot mutation, but not the other PIK3CA mutants and wild type, was able to enhance spheroid formation of NIH3T3 cells in vitro. (**A**) Micrographs of the spheroids at 96 h post-transfection show that there was a noticeable increase in the diameter of spheroids overexpressing the E545K hotspot mutation. Scale bar: 200 μm. (**B**) Spheroid diameters were measured at 96 h post-transfection. Data presented are representative of three independent trials in triplicate, and expressed as mean ± standard deviation. *** $p < 0.001$. WT: wild type; PIK3CA: phosphatidylinositol 3-kinase, catalytic subunit alpha.

3.7. In Silico Prediction of the Impact of PIK3CA Mutations on Protein Structure and p85 Alpha-niSH2 Binding

To predict if the mutations can impact protein function, four sequence-based platforms were used (Table 2): polyphen-2, which predicts on the basis of physical and comparative approaches [34];

mutation assessor, which utilizes evolutionary conservation of the affected amino acids in protein homologs [35]; FATHMM, which applies a combination of sequence conservation with hidden Markov models [36]; and SIFT, which relies on sequence homology and the physical properties of amino acids [37]. Apart from the FATHMM prediction, the other three platforms predicted Q661K to be benign and have a minimal effect on protein function. Meanwhile, all four platforms predicted C901R to be oncogenic and have a significant effect on protein function.

Table 2. Functional effects of the Q661K and C901R rare mutations as predicted by multiple bioinformatics platforms.

Mutation	Polyphen-2	Mutation Assessor	FATHMM	SIFT
Q661K [a]	Benign	Neutral	Cancer	Tolerated
C901R [b]	Probably damaging [c]	High	Cancer	Affect protein function

[a] Scores for Q661K: 0.024 (polyphen-2), 0.63 (mutation assessor), −1.63 (FATHMM), 0.46 (SIFT). [b] Scores for C901R: 1.00 (polyphen-2), 3.89 (mutation assessor), −1.81 (FATHMM), 0.00 (SIFT). [c] Introduced substitution is predicted to be damaging with high confidence [56]. Polyphen-2: polymorphism phenotyping 2; FATHMM: Functional Analysis Through Hidden Markov Models; SIFT: Sorting Intolerant From Tolerant; PIK3CA: phosphatidylinositol 3-kinase, catalytic subunit alpha.

Protein models of the mutations were generated with Swiss-model webserver. The models were superimposed to the wild type, and the RMSD and OS were calculated (Figure 6). As expected, the RMSD values for all mutations did not exceed 0.25 Å, as single amino acid changes are not expected to significantly alter protein structure. Among all the mutations, the rare Q661K mutation had the highest RMSD and lowest OS compared to wild type. This suggests that the amino acid change at codon 661 led to the most significant deviation in C-alpha carbon backbone, as well as overall protein conformation, possibly even altering the conformation of residues at the PIK3CA–p85 alpha interface. Nonetheless, the rest of the mutations only slightly differed from Q661K in terms of RMSD and OS, suggesting that they also subtly altered protein structure.

Docking simulations using pyDock webserver were performed to assess whether the mutations affected the ability of the niSH2 domain of p85 alpha to bind to PIK3CA. pyDock is a protein–protein docking algorithm [40] that uses electrostatics, desolvation energy, and a limited van der Waals contribution to score rigid-body docking poses. On the basis of the pyDock scores (Figure 7), E545K had the least favorable interaction with the niSH2 domain of p85 alpha. This was followed by H1047R, Q661K, C901R, and wild type. Altogether, the observed trend was E545K (least favorable interaction) < H1047R < Q661K < C901R < WT (most favorable interaction).

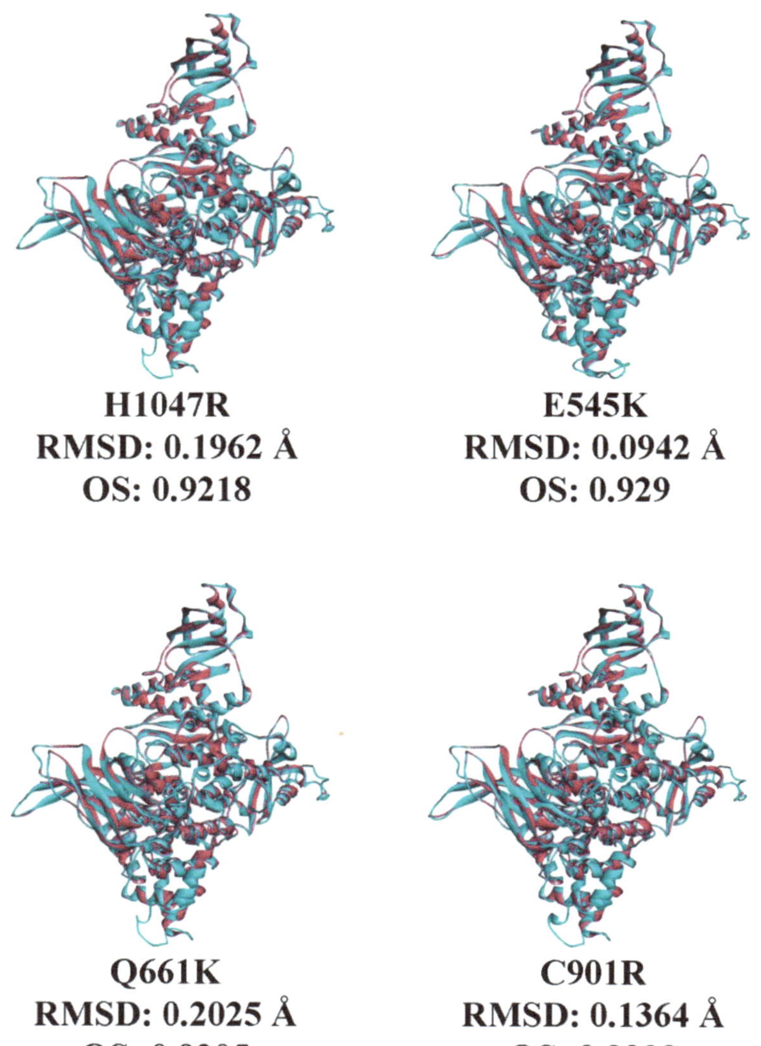

Figure 6. Superposition of modeled PIK3CA mutations (cyan ribbon) with the solved crystal structure of human wild type PIK3CA (PDB ID: 4L1B; red ribbon). Global main chain root mean square distance (RMSD) and overlay similarity (OS) values were also predicted via sequence alignment function for each mutant in comparison with the wild type. RMSD: root-mean-square distance; OS: overlay similarity; PIK3CA: phosphatidylinositol 3-kinase, catalytic subunit alpha.

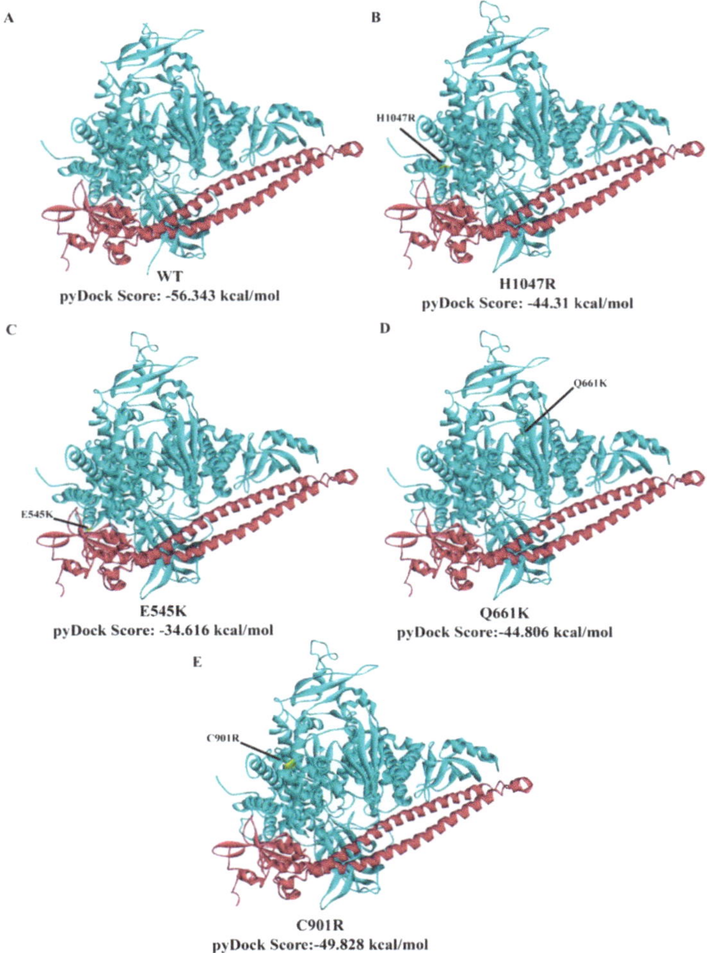

Figure 7. Simulated interactions of niSH2 domain of p85 alpha to wild type and mutant PIK3CA yielded variable binding energies. The solved X-ray crystal structure of human wild type PI3KCA (**A**) PDB 4L1B was used to build mutant homologues of (**B**) H1047R, (**C**) E545K, (**D**) Q661K, and (**E**) C901R. The pose with the most favorable and lowest (i.e., most negative) pyDock score is shown for each interaction. The mutated amino acid is highlighted in yellow and labelled accordingly. WT: wild type; PIK3CA: phosphatidylinositol 3-kinase, catalytic subunit alpha.

To gain some insight into the possible reasons for the differences in docking scores among the PIK3CA mutations, the interactions between interface residues of the PIK3CA and p85 alpha dimers were examined with PDBePISA (Table 3). The E545K-niSH2 p85 alpha dimer, which had the least favorable docking score, had a decrease in both the number of interface residues and buried interface area compared to the PIK3CA wild type-niSH2 p85 alpha dimer. The E545K-niSH2 p85 alpha dimer also gained two additional hydrogen bonds, but also lost four salt bridges that were present in the wild type-niSH2 p85 alpha dimer. This suggests that the salt bridges formed between the E545K–niSH2–p85 alpha interface had a stronger contribution to the binding affinity of the dimer than the hydrogen bonds. This finding is in accordance with the pyDock scoring algorithm, which gives greater value to electrostatics than van der Waals interactions (e.g., hydrogen bonding) in scoring docking poses [41].

Table 3. Analysis of interactions between interface residues of the PIK3CA-niSH2 p85 alpha dimers as predicted by PDBePISA.

	Buried Area, Å2	No. of H-Bonds	No. of Salt Bridges	No. of Interfacing Residues in PIK3CA	%	No. of Interfacing Residues in p85 Alpha	%
WT	3756.4	40	31	106	10.70%	106	38.30%
E545K	3673.5	42	27	102	9.70%	105	37.90%
H1047R	4107.8	48	27	119	11.30%	115	41.50%
Q661K	4097.6	46	26	120	11.40%	116	41.90%
C901R	4078.4	44	28	114	10.80%	115	41.50%

Meanwhile, the H1047R-, Q661K-, and C901R-niSH2 p85 alpha dimers had an increase in both the number of interfacing residues and buried interface area compared to the wild type-niSH2 p85 alpha dimer. Like the E545K-niSH2 p85 alpha dimer, they also had an increase in the number of hydrogen bonds and a decrease in the number of salt bridges between the mutants and the niSH2 domain of p85 alpha.

These results could partially explain the trend observed for the pyDock scores. The E545K dimer had the least favorable pyDock score because in addition to losing four salt bridges in the interface, it also had the least number of interface residues and smallest buried surface area among the PIK3CA variants. On the contrary, the H1047R, Q661K, and C901R dimers also lost a number of salt bridges, but gained in the number of interface residues and buried interface area. Altogether, these results align with the trend E545K (least favorable interaction) < H1047R < Q661K < C901R < WT (most favorable interaction).

4. Discussion

Previous studies suggest that the impact of PIK3CA exon 9 and 20 mutations are cell type- and mutation-specific [10–12]. Many PIK3CA mutant cancer cell lines have also been shown to exhibit transformed phenotypes despite displaying minimal AKT activation, suggesting AKT-independent mechanisms that must be considered [16,18–21]. Indeed, the identification of rare non-hotspot PIK3CA mutations in deep sequencing projects worldwide warrants their functional characterization. Due to multiple signaling pathways affected by PIK3CA, rare mutations may transmit their effects through different effectors, such as PDK1, SGK3, and RAC, among others, to exert their oncogenic outputs [22–25]. Further, the majority of previous sequencing studies have been performed using patient samples from developed nations, and hence the mutational spectrum identified could be biased towards the population under study [14–18]. Ethnic nuances have already been described for KRAS mutants [14,15,57], and it is highly likely that the PIK3CA mutation spectrum may also reveal such differences.

Intratumor heterogeneity, which refers to the coexistence of cell clones harboring variable genetic mutations within a tumor, is another reason for characterizing novel and rare PIK3CA mutations. Intratumor heterogeneity is implicated in failure of some biomarkers to predict tumor response to chemotherapy [58,59]. Because only a small sample from a tumor tissue is used in a biopsy, only the most frequently occurring mutations are detected. Rare mutations present in only a small subset of cells, which could be chemoresistant, remain undetected, and contribute to tumor recurrence after therapy. Further stressing the importance of functionalizing rare mutations is the finding that PIK3CA hotspot mutations differentially impact responses to Mesenchymal Epithelial Transition (MET) receptor tyrosine kinase-targeted therapy [60]. In a similar manner, rare mutations may also differentially impact therapy response.

In this study, two novel non-hotspot mutations, Q661K and C901R, were phenotypically characterized in NIH3T3 and HCT116 cells [28]. PIK3CA mutations in codon 661 were reported to be present in colon cancer, whereas those in codon 901 were reported to be present in colon, breast, and endometrial cancers [61]. Bader et al. [62] reported that in 266 colon cancer samples [26,63], E545 and H1047 mutations occurred at frequencies of 9.8% and 7.1%, respectively, whereas Q661 and C901 mutations both occurred at lower frequencies of 0.4%. Similar trends were observed in breast (580 samples) [26,63–66], brain (382 samples) [26,67,68], liver (73 samples) [69], stomach (291 samples) [26,65,70], lung (253 samples) [26,69], and ovary (489 samples) [63,65,71] tumors, wherein E545K and H1047R mutations occurred at frequencies ranging from 3.3% to 6.2% for E545 and 1.5% to 14.8% for H1047, whereas both Q661 and C901 mutations were almost undetectable. Comparable trends have been reported in the Cancer Genome Atlas (TCGA) database [72]. Interestingly, Q661K was also reported in a genetic screen of sporadic, young-onset CRC patient samples obtained from the Philippine General Hospital, University of the Philippines, Manila [31].

All four sequence-based prediction platforms predicted C901R to be oncogenic and have a significant effect on PIK3CA function, whereas three out of four predicted Q661K to be benign. Although these platforms do not consider the possible effect of the mutations on protein binding partners, these predictions still prompted us to further characterize the rare mutations in detail. Overexpression of the PIK3CA mutants led to distinct oncogenic phenotypes. In all hallmarks tested, the E545K hotspot mutation was consistently the most aggressive. Meanwhile, the H1047R canonical mutation only exhibited certain cancer hallmarks, namely, resistance to apoptosis, cellular proliferation, and cytoskeletal disorganization. These findings are consistent with previous studies, which suggested distinct oncogenic phenotypes of the E545K and H1047R mutations [10–12]. Overexpression of the Q661K and C901R rare mutations also presented distinct cellular phenotypes. The Q661K rare mutation was able to promote cellular proliferation and survival in both NIH3T3 and HCT116 compared to the vector-only and WT controls. The C901R rare mutation, on the other hand, enhanced resistance to apoptosis in NIH3T3 cells compared to vector-only but not WT control, and in HCT116 cells compared to both vector-only and WT controls. The C901R mutant enhanced proliferation only in HCT116 cells relative to vector-only but not WT control. Both Q661K and C901R were able to promote cytoskeletal disorganization.

Although all PIK3CA mutations were able to affect cytoskeletal architecture, only the E545K hotspot mutation was able to enhance cellular migration in wound healing assays, and only in NIH3T3 cells, which are better able to express a transformed phenotype even in the absence of complementary cooperative mutations [45]. Cellular migration is a complex process that involves cytoskeletal reorganization, changes in polarity, formation of cellular extensions, and/or loss of adhesion [54,73]. Hence, the inability of the H1047R hotspot mutation and the Q661K and C901R rare mutations to translate cytoskeletal disorganization to enhanced motility is not unexpected because each mutation may signal through a distinct pathway and may affect different aspects of cellular migration.

Interestingly, these results contradicted the assessment of sequence-based prediction platforms used. Q661K was predicted to be benign and have minimal effect on protein function, and yet was able to show oncogenic properties in a number of cancer hallmarks. These apparently discrepant results are not without precedent. The novel KRAS mutant E31D, which also harbored a conserved amino acid substitution, was predicted to be benign in all in silico prediction platforms used, but showed highly oncogenic properties in all cancer hallmarks tested [57]. In contrast, C901R was predicted to be oncogenic with significant effect on protein function, and yet was only able to alter a subset of the cancer hallmarks assayed. These findings imply the need to consider the cellular context and possible effect on binding partners in predicting the functional impact of protein mutations. Further, the effect of amino acid substitution on folding and protein assembly as well as the position of the altered residue in the folded configuration all merit further analyses.

The present study also aimed to visualize the structural changes induced by the Q661K and C901R rare mutations. Similar to the E545K hotspot mutation, the Q661K rare mutation is found in the helical domain. E545K disrupts the interaction between the helical domain of PIK3CA and the niSH2 domain of p85 alpha (51), and it is hypothesized that Q661K could have the same effect. Meanwhile, the C901R rare mutation is located in the kinase domain. It is hypothesized that C901R enhances the affinity of PIK3CA to the membrane bilayer because kinase domain mutations like H1047R have been found to act in this manner [61]. The basic arginine residue introduced by C901R could possibly increase the affinity of the protein to the negatively charged phospholipid heads in the membrane.

Analysis of RMSD and OS values revealed that the protein models of the Q661K and C901R rare mutations slightly deviated from the wild type structure. Furthermore, docking simulations with pyDock webserver suggested that the binding affinity of both the rare and canonical mutations with niSH2 domain of p85 alpha decreased, due to their more positive pyDock scores compared with the wild type structure. Analysis of the PIK3CA–p85 alpha interface with PDBePISA suggested that the decrease in binding affinity of the mutations with p85 alpha was due to a loss of a number of salt bridges that were originally present in the wild type-p85 alpha dimer. Altogether, these results suggest that the structural basis for the oncogenic activity of the Q661K and C901R mutations is partly due to a relief from the inhibitory control of the p85 alpha regulatory subunit. Future studies could confirm these results via co-immunoprecipitation and Fluorescence Resonance Energy Transfer (FRET) experiments. In addition, it will be interesting to determine the binding affinity of these mutant proteins with lipid membranes of various compositions through biochemical assays [74]. A complete assessment of these mutations can only be achieved through crystallographic techniques checking for the impact of these mutations on the binding of ligands and additional proteins such as KRAS, downstream effectors, scaffolds, and adaptors. The latter can be complemented by cellular assays confirming their phenotypic effects on cancer hallmarks.

The findings in this study, while instructive, should be confirmed in other cellular backgrounds, in vitro using 3D and co-culture experiments, and in vivo using relevant animal models. Correlating these findings with patient outcomes would also be beneficial in establishing these rare mutations as additional prognostic and predictive biomarkers.

Supplementary Materials: The following are available online at http://www.mdpi.com/2073-4409/9/5/1116/s1, Figure S1: Expression of wild type and mutant PIK3CA gene constructs in HCT116 cells, Figure S2: Western blotting detection of epithelial-to-mesenchymal transition markers.

Author Contributions: Conceptualization, A.A.G. and R.L.G.; data curation, A.A.G., J-A.M.T.L., and R.L.G.; formal analysis, A.A.G., J-A.M.T.L., and R.L.G.; funding acquisition, R.L.G.; investigation, A.A.G. and J-A.M.T.L.; methodology, A.A.G., J-A.M.T.L., and R.L.G.; project administration, R.L.G.; resources, R.L.G.; supervision, R.L.G.; writing—original draft, A.A.G. and R.L.G.; writing—review and editing, R.L.G. All authors have read and agreed to the published version of the manuscript.

Funding: This work was supported by grants from the University of the Philippines System (OVPAA-EIDR Code 06-008) and the Philippine Council for Health Research and Development (grant code FP150025).

Conflicts of Interest: The authors declare no conflict of interest. The funders had no role in the design of the study; in the collection, analyses, or interpretation of data; in the writing of the manuscript; or in the decision to publish the results.

References

1. Herreros-Villanueva, M.; Gomez-Manero, N.; Muñiz, P.; García-Girón, C.; Coma del Corral, M.J. PIK3CA mutations in KRAS and BRAF wild type colorectal cancer patients. A study of Spanish population. *Mol. Biol. Rep.* **2011**, *38*, 1347–1351. [CrossRef] [PubMed]

2. Perrone, F.; Lampis, A.; Orsenigo, M.; Di Bartolomeo, M.; Gevorgyan, A.; Losa, M.; Frattini, M.; Riva, C.; Andreola, S.; Bajetta, E.; et al. PI3KCA/PTEN deregulation contributes to impaired responses to cetuximab in metastatic colorectal cancer patients. *Ann. Oncol.* **2009**, *20*, 84–90. [CrossRef] [PubMed]

3. Nosho, K.; Kawasaki, T.; Longtine, J.A.; Fuchs, C.S.; Ohnishi, M.; Suemoto, Y.; Kirkner, G.J.; Zepf, D.; Yan, L.; Ogino, S. PIK3CA Mutation in Colorectal Cancer: Relationship with Genetic and Epigenetic Alterations. *Neoplasia* **2008**, *10*, 534–541. [CrossRef]

4. Prenen, H.; De Schutter, J.; Jacobs, B.; De Roock, W.; Biesmans, B.; Claes, B.; Lambrechts, D.; Van Cutsem, E.; Tejpar, S. PIK3CA Mutations Are Not a Major Determinant of Resistance to the Epidermal Growth Factor Receptor Inhibitor Cetuximab in Metastatic Colorectal Cancer. *Clin. Cancer Res.* **2009**, *15*, 3184–3188. [CrossRef] [PubMed]

5. De Roock, W.; Claes, B.; Bernasconi, D.; De Schutter, J.; Biesmans, B.; Fountzilas, G.; Kalogeras, K.T.; Kotoula, V.; Papamichael, D.; Laurent-Puig, P.; et al. Effects of KRAS, BRAF, NRAS, and PIK3CA mutations on the efficacy of cetuximab plus chemotherapy in chemotherapy-refractory metastatic colorectal cancer: a retrospective consortium analysis. *Lancet Oncol.* **2010**, *11*, 753–762. [CrossRef]

6. Zhao, L.; Vogt, P.K. Helical domain and kinase domain mutations in p110 of phosphatidylinositol 3-kinase induce gain of function by different mechanisms. *Proc. Natl. Acad. Sci. USA* **2008**, *105*, 2652–2657. [CrossRef]

7. Sartore-Bianchi, A.; Martini, M.; Molinari, F.; Veronese, S.; Nichelatti, M.; Artale, S.; Di Nicolantonio, F.; Saletti, P.; De Dosso, S.; Mazzucchelli, L.; et al. PIK3CA Mutations in Colorectal Cancer Are Associated with Clinical Resistance to EGFR-Targeted Monoclonal Antibodies. *Cancer Res.* **2009**, *69*, 1851–1857. [CrossRef] [PubMed]

8. Yang, Z.-Y.; Wu, X.-Y.; Huang, Y.-F.; Di, M.-Y.; Zheng, D.-Y.; Chen, J.-Z.; Ding, H.; Mao, C.; Tang, J.-L. Promising biomarkers for predicting the outcomes of patients with KRAS wild-type metastatic colorectal cancer treated with anti-epidermal growth factor receptor monoclonal antibodies: A systematic review with meta-analysis. *Int. J. Cancer* **2013**, *133*, 1914–1925. [CrossRef]

9. Catasus, L.; Gallardo, A.; Cuatrecasas, M.; Prat, J. PIK3CA mutations in the kinase domain (exon 20) of uterine endometrial adenocarcinomas are associated with adverse prognostic parameters. *Mod. Pathol.* **2008**, *21*, 131–139. [CrossRef]

10. Pang, H.; Flinn, R.; Patsialou, A.; Wyckoff, J.; Roussos, E.T.; Wu, H.; Pozzuto, M.; Goswami, S.; Condeelis, J.S.; Bresnick, A.R.; et al. Differential Enhancement of Breast Cancer Cell Motility and Metastasis by Helical and Kinase Domain Mutations of Class IA Phosphoinositide 3-Kinase. *Cancer Res.* **2009**, *69*, 8868–8876. [CrossRef]

11. Ross, R.L.; Askham, J.M.; Knowles, M.A. PIK3CA mutation spectrum in urothelial carcinoma reflects cell context-dependent signaling and phenotypic outputs. *Oncogene* **2013**, *32*, 768–776. [CrossRef]

12. Meyer, D.S.; Koren, S.; Leroy, C.; Brinkhaus, H.; Müller, U.; Klebba, I.; Müller, M.; Cardiff, R.D.; Bentires-Alj, M. Expression of PIK3CA mutant E545K in the mammary gland induces heterogeneous tumors but is less potent than mutant H1047R. *Oncogenesis* **2013**, *2*, e74. [CrossRef]

13. Hanna, M.C.; Go, C.; Roden, C.; Jones, R.T.; Pochanard, P.; Javed, A.Y.; Javed, A.; Mondal, C.; Palescandolo, E.; Van Hummelen, P.; et al. Colorectal Cancers from Distinct Ancestral Populations Show Variations in BRAF Mutation Frequency. *PLoS ONE* **2013**, *8*, e74950. [CrossRef]

14. Zulhabri, O.; Rahman, J.; Ismail, S.; Isa, M.R.; Wan Zurinah, W.N. Predominance of G to A codon 12 mutation K-ras gene in Dukes' B colorectal cancer. *Singapore Med. J.* **2012**, *53*, 26–31. [PubMed]

15. Tong, J.H.; Lung, R.W.; Sin, F.M.; Law, P.P.; Kang, W.; Chan, A.W.; Ma, B.B.; Mak, T.W.; Ng, S.S.; To, K.F. Characterization of rare transforming KRAS mutations in sporadic colorectal cancer. *Cancer Biol. Ther.* **2014**, *15*, 768–776. [CrossRef] [PubMed]

16. Rudd, M.L.; Price, J.C.; Fogoros, S.; Godwin, A.K.; Sgroi, D.C.; Merino, M.J.; Bell, D.W. A Unique Spectrum of Somatic PIK3CA (p110α) Mutations Within Primary Endometrial Carcinomas. *Clin. Cancer Res.* **2011**, *17*, 1331–1340. [CrossRef] [PubMed]

17. Chen, L.; Yang, L.; Yao, L.; Kuang, X.-Y.; Zuo, W.-J.; Li, S.; Qiao, F.; Liu, Y.-R.; Cao, Z.-G.; Zhou, S.-L.; et al. Characterization of PIK3CA and PIK3R1 somatic mutations in Chinese breast cancer patients. *Nat. Commun.* **2018**, *9*, 1357. [CrossRef] [PubMed]

18. Ikenoue, T.; Kanai, F.; Hikiba, Y.; Obata, T.; Tanaka, Y.; Imamura, J.; Ohta, M.; Jazag, A.; Guleng, B.; Tateishi, K.; et al. Functional Analysis of PIK3CA Gene Mutations in Human Colorectal Cancer. *Cancer Res.* **2005**, *65*, 4562–4567. [CrossRef]

19. Gymnopoulos, M.; Elsliger, M.-A.; Vogt, P.K. Rare cancer-specific mutations in PIK3CA show gain of function. *Proc. Natl. Acad. Sci. USA* **2007**, *104*, 5569–5574. [CrossRef]

20. Zhang, H.; Liu, G.; Dziubinski, M.; Yang, Z.; Ethier, S.P.; Wu, G. Comprehensive analysis of oncogenic effects of PIK3CA mutations in human mammary epithelial cells. *Breast Cancer Res. Treat.* **2008**, *112*, 217–227. [CrossRef]

21. Croessmann, S.; Sheehan, J.H.; Lee, K.; Sliwoski, G.; He, J.; Nagy, R.; Riddle, D.; Mayer, I.A.; Balko, J.M.; Lanman, R.; et al. PIK3CA C2 Domain Deletions Hyperactivate Phosphoinositide 3-kinase (PI3K), Generate Oncogene Dependence, and Are Exquisitely Sensitive to PI3K α Inhibitors. *Clin. Cancer Res.* **2018**, *24*, 1426–1435. [CrossRef]

22. Vasudevan, K.M.; Barbie, D.A.; Davies, M.A.; Rabinovsky, R.; McNear, C.J.; Kim, J.J.; Hennessy, B.T.; Tseng, H.; Pochanard, P.; Kim, S.Y.; et al. AKT-Independent Signaling Downstream of Oncogenic PIK3CA Mutations in Human Cancer. *Cancer Cell* **2009**, *16*, 21–32. [CrossRef]

23. Hu, H.; Juvekar, A.; Lyssiotis, C.A.; Lien, E.C.; Albeck, J.G.; Oh, D.; Varma, G.; Hung, Y.P.; Ullas, S.; Lauring, J.; et al. Phosphoinositide 3-Kinase Regulates Glycolysis through Mobilization of Aldolase from the Actin Cytoskeleton. *Cell* **2016**, *164*, 433–446. [CrossRef] [PubMed]

24. Juvekar, A.; Hu, H.; Yadegarynia, S.; Lyssiotis, C.A.; Ullas, S.; Lien, E.C.; Bellinger, G.; Son, J.; Hok, R.C.; Seth, P.; et al. Phosphoinositide 3-kinase inhibitors induce DNA damage through nucleoside depletion. *Proc. Natl. Acad. Sci. USA* **2016**, *113*, E4338–E4347. [CrossRef] [PubMed]

25. Lien, E.C.; Dibble, C.C.; Toker, A. PI3K signaling in cancer: Beyond AKT. *Curr. Opin. Cell Biol.* **2017**, *45*, 62–71. [CrossRef] [PubMed]

26. Samuels, Y.; Wang, Z.; Bardelli, A.; Silliman, N.; Ptak, J.; Szabo, S.; Yan, H.; Gazdar, A.; Powell, S.M.; Riggins, G.J.; et al. High Frequency of Mutations of the PIK3CA Gene in Human Cancers. *Science* **2004**, *304*, 554. [CrossRef] [PubMed]

27. Cancer Genome Atlas Network. Comprehensive molecular characterization of human colon and rectal cancer. *Nature* **2012**, *487*, 330–337. [CrossRef]

28. Tate, J.G.; Bamford, S.; Jubb, H.C.; Sondka, Z.; Beare, D.M.; Bindal, N.; Boutselakis, H.; Cole, C.G.; Creatore, C.; Dawson, E.; et al. COSMIC: the Catalogue Of Somatic Mutations In Cancer. *Nucleic Acids Res.* **2019**, *47*, D941–D947. [CrossRef]

29. Zhang, M.; Wang, B.; Xu, J.; Wang, X.; Xie, L.; Zhang, B.; Li, Y.; Li, J. CanProVar 2.0: An Updated Database of Human Cancer Proteome Variation. *J. Proteome Res.* **2017**, *16*, 421–432. [CrossRef]

30. Zhao, M.; Kong, L.; Liu, Y.; Qu, H. dbEMT: An epithelial-mesenchymal transition associated gene resource. *Sci. Rep.* **2015**, *5*, 11459. [CrossRef]

31. Sacdalan, D.L.; Uy, C.J.; Cutiongco-de la Paz, E.M.; Garcia, R.L. Next-Generation Sequencing reveals putative novel non-hotspot mutations in EGFR pathway genes in Filipino young-onset sporadic colorectal cancer patients. unpublished.

32. Schneider, C.A.; Rasband, W.S.; Eliceiri, K.W. NIH Image to ImageJ: 25 years of image analysis. *Nat. Methods* **2012**, *9*, 671–675. [CrossRef] [PubMed]

33. Schindelin, J.; Arganda-Carreras, I.; Frise, E.; Kaynig, V.; Longair, M.; Pietzsch, T.; Preibisch, S.; Rueden, C.; Saalfeld, S.; Schmid, B.; et al. Fiji: An open source platform for biological image analysis. *Nat. Methods* **2012**, *9*, 676–682. [CrossRef] [PubMed]

34. Adzhubei, I.A.; Schmidt, S.; Peshkin, L.; Ramensky, V.E.; Gerasimova, A.; Bork, P.; Kondrashov, A.S.; Sunyaev, S.R. A method and server for predicting damaging missense mutations. *Nat. Methods* **2010**, *7*, 248–249. [CrossRef] [PubMed]

35. Reva, B.; Antipin, Y.; Sander, C. Predicting the functional impact of protein mutations: Application to cancer genomics. *Nucleic Acids Res.* **2011**, *39*, e118. [CrossRef]

36. Shihab, H.A.; Gough, J.; Cooper, D.N.; Stenson, P.D.; Barker, G.L.A.; Edwards, K.J.; Day, I.N.M.; Gaunt, T.R. Predicting the Functional, Molecular, and Phenotypic Consequences of Amino Acid Substitutions using Hidden Markov Models. *Hum. Mutat.* **2013**, *34*, 57–65. [CrossRef]

37. Sim, N.-L.; Kumar, P.; Hu, J.; Henikoff, S.; Schneider, G.; Ng, P.C. SIFT web server: Predicting effects of amino acid substitutions on proteins. *Nucleic Acids Res.* **2012**, *40*, W452–W457. [CrossRef]

38. Waterhouse, A.; Bertoni, M.; Bienert, S.; Studer, G.; Tauriello, G.; Gumienny, R.; Heer, F.T.; De Beer, T.A.P.; Rempfer, C.; Bordoli, L.; et al. SWISS-MODEL: Homology modelling of protein structures and complexes. *Nucleic Acids Res.* **2018**, *46*, W296–W303. [CrossRef]

39. Zhao, Y.; Zhang, X.; Chen, Y.; Lu, S.; Peng, Y.; Wang, X.; Guo, C.; Zhou, A.; Zhang, J.; Luo, Y.; et al. Crystal Structures of PI3Kα Complexed with PI103 and Its Derivatives: New Directions for Inhibitors Design. *ACS Med. Chem. Lett.* **2014**, *5*, 138–142. [CrossRef]

40. Jiménez-García, B.; Pons, C.; Fernández-Recio, J. pyDockWEB: A web server for rigid-body protein–protein docking using electrostatics and desolvation scoring. *Bioinformatics* **2013**, *29*, 1698–1699. [CrossRef]

41. Cheng, T.M.-K.; Blundell, T.L.; Fernandez-Recio, J. pyDock: Electrostatics and desolvation for effective scoring of rigid-body protein-protein docking. *Proteins Struct. Funct. Bioinform.* **2007**, *68*, 503–515. [CrossRef] [PubMed]

42. Krissinel, E.; Henrick, K. Inference of Macromolecular Assemblies from Crystalline State. *J. Mol. Biol.* **2007**, *372*, 774–797. [CrossRef] [PubMed]

43. Chalhoub, N.; Baker, S.J. PTEN and the PI3-Kinase Pathway in Cancer. *Annu. Rev. Pathol. Mech. Dis.* **2009**, *4*, 127–150. [CrossRef] [PubMed]

44. Engelman, J.A.; Luo, J.; Cantley, L.C. The evolution of phosphatidylinositol 3-kinases as regulators of growth and metabolism. *Nat. Rev. Genet.* **2006**, *7*, 606–619. [CrossRef]

45. Petty, E.M. Cellular cancer markers. *Am. J. Med. Genet.* **1997**, *68*, 492–493. [CrossRef]

46. Guerrero, S.; Casanova, I.; Farré, L.; Mazo, A.; Capellà, G.; Mangues, R. K-ras Codon 12 Mutation Induces Higher Level of Resistance to Apoptosis and Predisposition to Anchorage-independent Growth Than Codon 13 Mutation or Proto-Oncogene Overexpression. *Cancer Res.* **2000**, *60*, 6750–6756.

47. Pawlak, G.; Helfman, D.M. Cytoskeletal changes in cell transformation and tumorigenesis. *Curr. Opin. Genet. Dev.* **2001**, *11*, 41–47. [CrossRef]

48. Alcantara, K.M.; Garcia, R. MicroRNA-92a promotes cell proliferation, migration and survival by directly targeting the tumor suppressor gene NF2 in colorectal and lung cancer cells. *Oncol. Rep.* **2019**, *41*, 2103–2116. [CrossRef]

49. Alcantara, K.M.M.; Malapit, J.R.P.; Yu, R.T.D.; Garrido, J.A.M.G.; Rigor, J.P.T.; Angeles, A.K.J.; Cutiongco-de la Paz, E.M.; Garcia, R.L. Non-Redundant and Overlapping Oncogenic Readouts of Non-Canonical and Novel Colorectal Cancer KRAS and NRAS Mutants. *Cells* **2019**, *8*, 1557. [CrossRef]

50. Nakhaeizadeh, H.; Amin, E.; Nakhaei-Rad, S.; Dvorsky, R.; Ahmadian, M.R. The RAS-Effector Interface: Isoform-Specific Differences in the Effector Binding Regions. *PLoS ONE* **2016**, *11*, e0167145. [CrossRef]

51. Lauffenburger, D.A.; Horwitz, A.F. Cell migration: A physically integrated molecular process. *Cell* **1996**, *84*, 359–369. [CrossRef]

52. Thiery, J.P. Epithelial–mesenchymal transitions in tumour progression. *Nat. Rev. Cancer* **2002**, *2*, 442–454. [CrossRef] [PubMed]

53. Pećina-Šlaus, N. Tumor suppressor gene E-cadherin and its role in normal and malignant cells. *Cancer Cell Int.* **2003**, *3*, 1–7. [CrossRef]

54. Devreotes, P.; Horwitz, A.R. Signaling Networks that Regulate Cell Migration. *Cold Spring Harb. Perspect. Biol.* **2015**, *7*, a005959. [CrossRef] [PubMed]

55. Quail, D.F.; Joyce, J.A. Microenvironmental regulation of tumor progression and metastasis. *Nat. Med.* **2013**, *19*, 1423–1437. [CrossRef] [PubMed]

56. Adzhubei, I.; Jordan, D.M.; Sunyaev, S.R. Predicting Functional Effect of Human Missense Mutations Using PolyPhen-2. *Curr. Protoc. Hum. Genet.* **2013**, *76*, 7.20.1–7.20.41. [CrossRef]

57. Angeles, A.; Yu, R.; Cutiongco-De la Paz, E.; Garcia, R. Phenotypic characterization of the novel, non-hotspot oncogenic KRAS mutants E31D and E63K. *Oncol. Lett.* **2019**. [CrossRef]

58. Yap, T.A.; Gerlinger, M.; Futreal, P.A.; Pusztai, L.; Swanton, C. Intratumor Heterogeneity: Seeing the Wood for the Trees. *Sci. Transl. Med.* **2012**, *4*, 127ps10. [CrossRef]

59. McGranahan, N.; Swanton, C. Biological and therapeutic impact of intratumor heterogeneity in cancer evolution. *Cancer Cell* **2015**, *27*, 15–26. [CrossRef]

60. Nisa, L.; Häfliger, P.; Poliaková, M.; Giger, R.; Francica, P.; Aebersold, D.M.; Charles, R.-P.; Zimmer, Y.; Medová, M. PIK3CA hotspot mutations differentially impact responses to MET targeting in MET-driven and non-driven preclinical cancer models. *Mol. Cancer* **2017**, *16*, 93. [CrossRef]

61. Miled, N.; Yan, Y.; Hon, W.-C.; Perisic, O.; Zvelebil, M.; Inbar, Y.; Schneidman-Duhovny, D.; Wolfson, H.J.; Backer, J.M.; Williams, R.L. Mechanism of Two Classes of Cancer Mutations in the Phosphoinositide 3-Kinase Catalytic Subunit. *Science (80)* **2007**, *317*, 239–242. [CrossRef] [PubMed]

62. Bader, A.G.; Kang, S.; Zhao, L.; Vogt, P.K. Oncogenic PI3K deregulates transcription and translation. *Nat. Rev. Cancer* **2005**, *5*, 921–929. [CrossRef] [PubMed]

63. Campbell, I.G.; Russell, S.E.; Choong, D.Y.H.; Montgomery, K.G.; Ciavarella, M.L.; Hooi, C.S.F.; Cristiano, B.E.; Pearson, R.B.; Phillips, W.A. Mutation of the PIK3CA Gene in Ovarian and Breast Cancer. *Cancer Res.* **2004**, *64*, 7678–7681. [CrossRef]

64. Bachman, K.E.; Argani, P.; Samuels, Y.; Silliman, N.; Ptak, J.; Szabo, S.; Konishi, H.; Karakas, B.; Blair, B.G.; Lin, C.; et al. The PIK3CA gene is mutated with high frequency in human breast cancers. *Cancer Biol. Ther.* **2004**, *3*, 772–775. [CrossRef]

65. Levine, D.A. Frequent Mutation of the PIK3CA Gene in Ovarian and Breast Cancers. *Clin. Cancer Res.* **2005**, *11*, 2875–2878. [CrossRef] [PubMed]

66. Saal, L.H.; Holm, K.; Maurer, M.; Memeo, L.; Su, T.; Wang, X.; Yu, J.S.; Malmström, P.-O.; Mansukhani, M.; Enoksson, J.; et al. PIK3CA Mutations Correlate with Hormone Receptors, Node Metastasis, and ERBB2, and Are Mutually Exclusive with PTEN Loss in Human Breast Carcinoma. *Cancer Res.* **2005**, *65*, 2554–2559. [CrossRef] [PubMed]

67. Broderick, D.K.; Di, C.; Parrett, T.J.; Samuels, Y.R.; Cummins, J.M.; McLendon, R.E.; Fults, D.W.; Velculescu, V.E.; Bigner, D.D.; Yan, H. Mutations of PIK3CA in Anaplastic Oligodendrogliomas, High-Grade Astrocytomas, and Medulloblastomas. *Cancer Res.* **2004**, *64*, 5048–5050. [CrossRef]

68. Hartmann, C.; Bartels, G.; Gehlhaar, C.; Holtkamp, N.; von Deimling, A. PIK3CA mutations in glioblastoma multiforme. *Acta Neuropathol.* **2005**, *109*, 639–642. [CrossRef]

69. Lee, J.W.; Soung, Y.H.; Kim, S.Y.; Lee, H.W.; Park, W.S.; Nam, S.W.; Kim, S.H.; Lee, J.Y.; Yoo, N.J.; Lee, S.H. PIK3CA gene is frequently mutated in breast carcinomas and hepatocellular carcinomas. *Oncogene* **2005**, *24*, 1477–1480. [CrossRef]

70. Li, V.S.W.; Wong, C.W.; Chan, T.L.; Chan, A.S.W.; Zhao, W.; Chu, K.-M.; So, S.; Chen, X.; Yuen, S.T.; Leung, S.Y. Mutations of PIK3CAin gastric adenocarcinoma. *BMC Cancer* **2005**, *5*, 29. [CrossRef]

71. Wang, Y.; Helland, Å.; Holm, R.; Kristensen, G.B.; Børresen-Dale, A.-L. PIK3CA mutations in advanced ovarian carcinomas. *Hum. Mutat.* **2005**, *25*, 322. [CrossRef] [PubMed]

72. Stephens, B. The Mutational Spectra of Cancer Genes in TCGA Data. Available online: https://www.cancer.gov/research/key-initiatives/ras/ras-central/blog/2017/cancer-mutation-spectra?fbclid=IwAR3MFtXSJ9BlFJtCtEr01p0-CGhaaKKeK8wzC6ZXkTMiyxCNnpkDaQQoBXs (accessed on 21 April 2020).

73. Parsons, J.T.; Horwitz, A.R.; Schwartz, M.A. Cell adhesion: Integrating cytoskeletal dynamics and cellular tension. *Nat. Rev. Mol. Cell Biol.* **2010**, *11*, 633–643. [CrossRef] [PubMed]

74. Mandelker, D.; Gabelli, S.B.; Schmidt-Kittler, O.; Zhu, J.; Cheong, I.; Huang, C.-H.; Kinzler, K.W.; Vogelstein, B.; Amzel, L.M. A frequent kinase domain mutation that changes the interaction between PI3K and the membrane. *Proc. Natl. Acad. Sci. USA* **2009**, *106*, 16996–17001. [CrossRef] [PubMed]

Article

Topical Ascorbic Acid Ameliorates Oxidative Stress-Induced Corneal Endothelial Damage via Suppression of Apoptosis and Autophagic Flux Blockage

Yi-Jen Hsueh [1,2], Yaa-Jyuhn James Meir [2,3], Lung-Kun Yeh [1,4], Tze-Kai Wang [1,2], Chieh-Cheng Huang [2,5], Tsai-Te Lu [2,5], Chao-Min Cheng [2,5], Wei-Chi Wu [1,3] and Hung-Chi Chen [1,2,3,*]

1 Department of Ophthalmology, Chang Gung Memorial Hospital, Linkou branch, Taoyuan 33305, Taiwan
2 Center for Tissue Engineering, Chang Gung Memorial Hospital, Linkou branch, Taoyuan 33305, Taiwan
3 Department of Biomedical Sciences, Chang Gung University College of Medicine, Taoyuan 33305, Taiwan
4 Department of Medicine, Chang Gung University College of Medicine, Taoyuan 33305, Taiwan
5 Institute of Biomedical Engineering, National Tsing Hua University, Hsinchu 30012, Taiwan
* Correspondence: mr3756@cgmh.org.tw; Tel.: +886-3-3281200 (ext. 7855); Fax: +886-3-3287798

Received: 18 February 2020; Accepted: 10 April 2020; Published: 11 April 2020

Abstract: Compromised pumping function of the corneal endothelium, due to loss of endothelial cells, results in corneal edema and subsequent visual problems. Clinically and experimentally, oxidative stress may cause corneal endothelial decompensation after phacoemulsification. Additionally, in vitro and animal studies have demonstrated the protective effects of intraoperative infusion of ascorbic acid (AA). Here, we established a paraquat-induced cell damage model, in which paraquat induced reactive oxygen species (ROS) production and apoptosis in the B4G12 and ARPE-19 cell lines. We demonstrate that oxidative stress triggered autophagic flux blockage in corneal endothelial cells and that addition of AA ameliorated such oxidative damage. We also demonstrate the downregulation of Akt phosphorylation in response to oxidative stress. Pretreatment with ascorbic acid reduced the downregulation of Akt phosphorylation, while inhibition of the PI3K/Akt pathway attenuated the protective effects of AA. Further, we establish an in vivo rabbit model of corneal endothelial damage, in which an intracameral infusion of paraquat caused corneal opacity. Administration of AA via topical application increased its concentration in the corneal stroma and reduced oxidative stress in the corneal endothelium, thereby promoting corneal clarity. Our findings indicate a perioperative strategy of topical AA administration to prevent oxidative stress-induced damage, particularly for those with vulnerable corneal endothelia.

Keywords: ascorbic acid; oxidative stress; apoptosis; autophagic flux blockage; PI3K/Akt; corneal endothelial cells

---

1. Introduction

Phacoemulsification is the most commonly practiced surgery for the treatment of cataracts. Complications of phacoemulsification that compromise visual prognosis remain a concern. These include cystoid macular edema [1], infectious endophthalmitis [2], retinal detachment [3], and corneal edema due to loss of human corneal endothelial cells (HCECs) [4]. The overall rate of HCEC loss after cataract surgery is 2.5% annually [5], while rates of 5.1% [6] to 12.1% [7] have been reported among patients with low HCEC density. HCECs play a cardinal role in the regulation of stromal hydration and corneal transparency [8]. HCECs have been found to exhibit limited proliferative potential, both clinically [9,10] and experimentally [11,12]. We, therefore, wondered whether there were effective and safe strategies to prevent loss of HCECs during and after phacoemulsification.

The current knowledge of the causative mechanisms underlying changes in HCECs following phacoemulsification is insufficient. A clinical study observed that reactive oxygen species (ROS) or free radicals generated by high-intensity ultrasound oscillations in water during phacoemulsification damaged the corneal endothelium [13]. Such phenomena have also been demonstrated in vivo [14,15] and in vitro [16,17]. Furthermore, significant induction of apoptosis has been reported in experimental studies of corneal endothelial cells during phacoemulsification [15,16,18]. Oxidative stress has been reported to induce autophagosome formation [19] in various ocular [20] and systemic [21] diseases. However, until now, there have been no reports demonstrating autophagic flux blockage in ROS-induced corneal endothelial damage.

Conventionally, ascorbic acid is considered an ideal ocular nutritional supplement [22]. Evidence has shown that ascorbic acid inhibits apoptosis of murine [23,24], bovine [16], and human corneal endothelial cells [25], probably through protection against oxidative stress and damage. Likewise, evidence from studies on rabbit eyes has also suggested a protective effect of intraoperative infusion of ascorbic acid during phacoemulsification, due to either free-radical-scavenging properties [14,26] or oxidative stress reduction [15]. Nevertheless, there have been no previous studies focusing on the application of topical ascorbic acid for the corneal endothelium to prevent oxidative stress generated during phacoemulsification.

In this study, we aim to investigate the protective effect of ascorbic acid against oxidative stress in HCEC. Our findings support the hypothesis that ascorbic acid attenuates oxidative stress in HCECs through inhibition of apoptosis and autophagic flux blockage. Furthermore, in a rabbit model, we demonstrate that topical administration of ascorbic acid ameliorates corneal endothelial damage caused by oxidative stress through increasing the concentration of ascorbic acid in the corneal stroma.

2. Materials and Methods

2.1. Materials

Cell culture media and various additives, comprising Dulbecco modified Eagle's medium (DMEM), DMEM/F12, Opti-MEM, human endothelium serum-free medium (HESFM), trypsin-EDTA, fetal bovine serum (FBS), phosphate-buffered saline (PBS), gentamicin, and amphotericin B, were purchased from Invitrogen (Carlsbad, CA, USA), as was Alexa-Fluor-conjugated secondary IgG antibody. Collagenase A was purchased from Roche Applied Science (Indianapolis, IN, USA). Recombinant human FGF-basic (basic-FGF) was purchased from Peprotech (London, UK). FNC coating mix (FNC) was purchased from Athena ES (Baltimore, MD, USA). Ascorbic acid, dimethyl sulfoxide (DMSO), Hoechst 33342 dye, methanol, mitomycin C, penicillin and streptomycin (P/S), Triton X-100, trypan blue, the MTT Cell Growth Assay Kit (MTT assay, CT02), and the ApopTag Plus In Situ Apoptosis Fluorescein Detection Kit (TUNEL assay, S7111) were purchased from Sigma-Aldrich (St. Louis, MO, USA). The Cellular ROS Assay Kit was purchased from Abcam (ab186029, Cambridge, MA, USA). Calcein-AM was purchased from BioVision, Inc. (Mountain View, CA, USA). Antibodies against Bcl2, p-Akt (Ser473), LC3, and Lamin A were purchased from Cell Signaling Technology (Beverly, MA, USA). p62 antibody was purchased from Abcam. Akt, β-actin, and GAPDH antibodies were purchased from Santa Cruz Biotechnology (Santa Cruz, CA, USA). All plastic cell culture dishes and plates were purchased from Corning Incorporated Life Sciences (Acton, MA, USA).

2.2. Tissue Sources

All rabbits were purchased from registered farms and housed at the Animal Care Core Facility of Chang Gung Memorial Hospital in Linkou, Taiwan. All experimental and animal care procedures adhered to the Association for Research in Vision and Ophthalmology (ARVO) Statement for the Use of Animals in Ophthalmic and Vision Research. Corneal tissues were obtained from 4-month-old

New Zealand white rabbits immediately after euthanasia, and the tissues were stored in 50 mL tubes containing DMEM for subsequent examination.

2.3. Cell Culture

The human retinal pigment epithelium cell line ARPE-19 (ATCC, Rockville, MD, USA) and the human corneal endothelial cell line B4G12 (Creative Bioarray, Shirley, NY, USA) were cultured in DMEM/F12 medium supplemented with 10% FBS and 1% P/S, or HESFM supplemented with 2% FBS and 10 ng/mL basic-FGF, respectively. The cells were maintained in a humidified incubator at 37 °C in an atmosphere of 5% CO_2.

The procedures for isolation and culture of rabbit corneal endothelial cells (CECs) were modified from previous methods [12]. Briefly, Descemet's membrane (DM; containing CECs) was stripped from the posterior surface of corneal tissues. The DM fragments were removed and digested at 37 °C for 16 h with 2 mg/mL collagenase A in Opti-MEM containing 50 μg/mL gentamicin and 5 μg/mL amphotericin B. After digestion, the CEC aggregates were collected by centrifugation and then cultured in 24-well plates coated with FNC. The CEC aggregates were maintained in RCEC medium (DMEM supplemented with 10% FBS and 25 μg/mL gentamicin) and subcultured for future investigation.

2.4. Cell Viability

Cell viability was measured by MTT assay. In brief, the culture medium was removed and replaced with 100 μL of fresh medium. Next, 10 μL of the 12 mM MTT stock solution was added to each well, and the plates were incubated at 37 °C for 4 h. Finally, 100 μL of DMSO was added to each well, the well contents were mixed thoroughly by pipette, and the absorbance at 570 nm was read using a Sunrise ELISA reader (Tecan, Salzburg, Austria).

2.5. Reactive Oxygen Species (ROS) Detection

The cells were treated with the ROS inducer (paraquat) for 120 h, followed by incubation with detection reagents included in the Cellular ROS Detection Assay Kit for 30 min. Images were obtained using a Zeiss fluorescence microscope (Oberkochen, Germany).

2.6. Apoptosis Detection

The terminal deoxynucleotidyl transferase (TdT)-mediated dUTP-nick-end-labeling (TUNEL) technique was performed to detect apoptotic cells. After treatment with the ROS inducer (paraquat) for 120 h, cells were fixed with 1% paraformaldehyde and post-fixed with 2:1 ethanol:acetic acid for 5 min at −20 °C and then incubated with TUNEL reagents included in the ApopTag Plus In Situ Apoptosis Fluorescein Detection Kit for 1 h. After counterstaining with Hoechst 33342, cells were examined under a Zeiss fluorescence microscope.

2.7. Autophagosome Formation Detection

LC3 immunofluorescence was performed to detect autophagosome formation. After treatment with the ROS inducer (paraquat) for 120 h, cells were washed three times with PBS (pH 7.2), then fixed in 4% formaldehyde for 15 min at room temperature, rinsed with PBS, permeabilized with 0.2% Triton X-100 for 15 min, and then rinsed again with PBS. After incubation with 2% BSA to block non-specific staining for 30 min, the slides were incubated with the primary antibodies (LC3 at 1:100 dilution) for 24 h at 4 °C. After washing in PBS thrice, the slides were incubated with the corresponding Alexa Fluor-conjugated secondary IgG antibody (1:200 dilution) for 60 min at room temperature. The samples were then counterstained with Hoechst 33342 and examined under a Zeiss fluorescence microscope.

2.8. Western Blotting

Total cell lysates were prepared in RIPA buffer supplemented with 10 mmol/L sodium fluoride, 10 mmol/L sodium orthovanadate, and 1× protease inhibitor cocktail (Sigma-Aldrich). The suspensions were each transferred into a microfuge tube on ice, sonicated to disrupt the cells, and centrifuged for 15 min at 4 °C at maximum speed. The supernatants were then pooled to obtain the total protein extract. The protein extracts were resolved in 10% acrylamide gels and transferred onto polyvinylidene difluoride membranes (MilliporeSigma, Burlington, MA, USA), which were then blocked with 5% (*w/v*) fat-free milk in TBST (50 mmol/L Tris-HCl, pH 7.5, 150 mmol/L NaCl, 0.05% (*v/v*) Tween-20), and probed with the desired primary antibodies at 4 °C overnight, followed by reaction with appropriate horseradish peroxidase–conjugated secondary antibodies. The immunoreactive protein bands were visualized via enhanced chemiluminescence (GE Healthcare, Chalfont St Giles, UK).

2.9. Ex Vivo Rabbit Corneal Damage Model

The rabbit corneal tissues were cultured in tissue culture (TC) medium (DMEM supplemented with 10% FBS and 25 µg/mL gentamicin). The ascorbic acid-pretreatment groups were cultivated in TC medium containing 1 mM of ascorbic acid for two days, followed by the addition of paraquat (25 mM) in the paraquat-treatment groups for 15 min, and then washed with TC medium. Two days later, 1 µmol/L of Calcein-AM was added into TC medium for 15 min, followed by examination under a Zeiss fluorescence microscope to detect the sloughing status of corneal endothelial cells.

2.10. In Vivo Rabbit Corneal Damage Model

Rabbits received administration of 5% ascorbic acid (284 mmol/L in BSS solution) or BSS to the cornea three times a day for two days. Subsequently, 25 mM paraquat (diluted in BSS, total 20 mL) or BSS was slowly injected into the anterior chamber for 15 min using a syringe pump (Harvard Apparatus, Holliston, MA, USA). BSS infusion was performed for another 15 min for those with paraquat infusion. Afterwards, the topical administration of ascorbic acid three times a day was continued for two days. Corneal clarity was recorded using external eye photography. Following the sacrifice of the rabbits, corneal tissues and aqueous humor were obtained for further experiments.

2.11. Ascorbic Acid Measurement

Corneal tissues and aqueous humor were assayed using the OxiSelect™Ascorbic Acid Assay kit (FRASC, Cell Biolabs Inc., San Diego, CA, USA). Briefly, corneas were placed in 1× assay buffer, homogenized, and then centrifuged for 15 min at 4 °C at maximum speed. Subsequently, the samples, including corneal tissue extraction and aqueous humor, were mixed with an ascorbic acid standard and added into 96-well plates. Deionized water (−AO) or 1× ascorbate oxidase (+AO), together with the kit reaction reagents, were added into each well, followed by determination of optical densities at 540 nm. Finally, ascorbic acid concentration (µM) was calculated according to the standard curve.

2.12. Statistics

All data are presented as blots or images from at least three similar experiments or as mean ± S.D. for each group with at least three independent experiments performed. Student's unpaired *t*-tests were performed using SPSS software version 13.0 (SPSS Inc. Chicago, IL, USA). Statistical significance is reported as two-tailed *p*-values, where $p < 0.05$ (*) and $p < 0.01$ (**) are considered statistically significant.

3. Results

3.1. Ascorbic Acid Attenuates Oxidative Stress-Induced Cell Injury

Since phacoemulsification-induced damage with evenly distributed damage areas could not be reproduced for quantification in vitro, paraquat was chosen as an inducer of oxidative stress. In this study, the HCEC cell line (B4G12) was used as a cellular model to examine oxidative stress-induced damage in HCEC. In addition, the retinal pigment epithelium (RPE) cell line (ARPE-19) was used as a parallel control. It has been reported that damage caused by oxidative stress leads to apoptosis [27,28] and autophagic cell death [29] in RPE cells, which support normal functions in the retina.

To further investigate whether ascorbic acid protects cells from oxidative stress, paraquat (an inducer of ROS) was used at various concentrations (B4G12: 0, 0.1, 0.3 mM; ARPE-19: 0, 1, 3 mM). Subsequently, on day two, the viability of the B4G12 and ARPE-19 cells was quantified via MTT assay. Staurosporine, a non-inducer of ROS that incites cell toxicity by protein kinase inhibition, was used as a control. In contrast to the staurosporine group, pretreatment with ascorbic acid (0.25 and 1 mM) in the paraquat group was found to significantly protect cells in a dose-dependent manner ($p < 0.01$, Figure S1).

Cell apoptosis and autophagosome formation were examined via TUNEL assays and immunofluorescence staining for LC3. Whereas the control groups presented normal cellular structures, the paraquat-treated groups revealed cell loss, apoptosis, and autophagosome formation in both B4G12 and ARPE-19 cells on day five (Figure 1). We further examined the time- and dose-dependent protective effects of ascorbic acid by observing cell morphology. B4G12 and ARPE-19 cells were divided into four groups: control, AA (ascorbic acid), paraquat, and AA + paraquat. In the AA and AA + paraquat groups, cells were cultivated in medium containing 1 mM of ascorbic acid for two days. Treatment with paraquat (2 mM for ARPE-19, 0.2 mM for B4G12) was then performed in the paraquat and AA + paraquat groups. As shown in Figure 2A, cell morphology was observed on days 0–5. Compared to the control and AA groups, treatment with paraquat led to reduced cell density at day five, and this cell loss was rescued by pretreatment with 1 mM of ascorbic acid. As shown in Figure 2B, B4G12 and ARPE-19 cells were cultivated in medium containing 0, 0.25, 1, or 2 mM of ascorbic acid for two days, followed by the addition of paraquat (2 mM for ARPE-19, 0.2 mM for B4G12) for five days. According to morphological observations, ascorbic acid (1 and 2 mM) significantly rescued cell loss induced by paraquat.

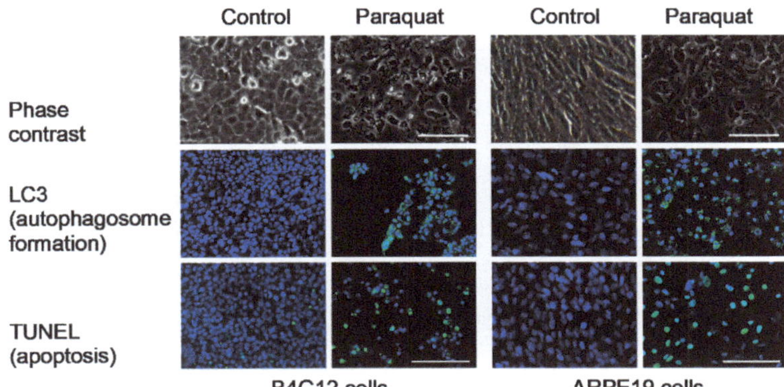

Figure 1. Oxidative stress induces apoptosis and autophagosome formation in B4G12 and ARPE-19 cells. The HCEC cell line (B4G12) and the RPE cell line (ARPE-19) were treated with paraquat (an oxidative stress inducer, 2 mM for ARPE-19, 0.2 mM for B4G12) for five days. Cell morphology was observed using phase-contrast microscopy. TUNEL assay and immunofluorescence of LC3 were performed to examine apoptosis and autophagosome formation (green color). Nuclei were counterstained with Hoechst 33342 (blue color). The scale bars represent 100 μm.

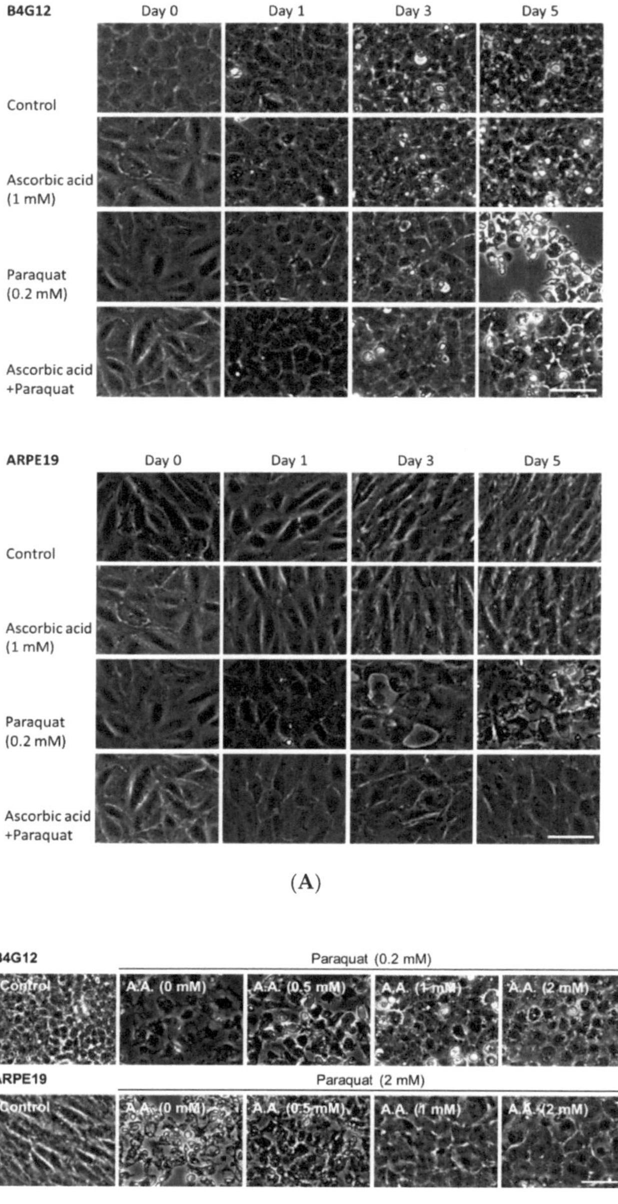

Figure 2. Ascorbic acid protects B4G12 and ARPE-19 cells from oxidative stress. B4G12 and ARPE-19 were pretreated with ascorbic acid (AA) for two days and further treated with paraquat (2 mM for ARPE-19, 0.2 mM for B4G12) for five days. (**A**) The time-dependent protective effect of ascorbic acid (1 mM) on B4G12 and ARPE-19 cells was examined on days 0–5. (**B**) The dose-dependent protective effect of ascorbic acid (0 to 2 mM) on B4G12 and ARPE-19 cells was examined on day five. Cell morphology was observed using phase-contrast microscopy, and images were captured at the same spot on different days. The scale bar represents 100 μm.

3.2. Ascorbic Acid Ameliorates Paraquat-Induced ROS and Oxidative Stress-Induced Apoptosis and Autophagic Flux Blockage in B4G12 and ARPE-19 Cells

Paraquat-induced ROS was detected using ROS fluorescent dye. Paraquat treatment (2 mM for ARPE-19, 0.2 mM for B4G12) significantly increased cellular ROS in both B4G12 and ARPE-19 cells. However, pretreatment with 1.0 mM ascorbic acid significantly rescued paraquat-induced ROS (Figure 3A). Paraquat treatment (2 mM for ARPE-19, 0.2 mM for B4G12) significantly increased apoptosis (B4G12, 14.6% ± 1.4% vs. 67.7% ± 10.2% $p < 0.01$; ARPE-19, 2.4% ± 1.3% vs 87.1% ± 4.4%, $p < 0.01$) and autophagosome formation (B4G12, 2.7% ± 1.0% vs. 70.2% ± 19.9%, $p < 0.01$; ARPE-19, 3.2% ± 1.3% vs. 64.2% ± 13.2%, $p < 0.01$) in the B4G12 and ARPE-19 cells. Likewise, pretreatment with 1.0 mM ascorbic acid significantly attenuated paraquat-induced apoptosis (B4G12, 67.7% ± 10.2% vs. 20.9% ± 4.1% $p < 0.01$; ARPE-19, 87.1% ± 4.4% vs. 6.9% ± 2.6% $p < 0.01$) and autophagosome formation (B4G12, 70.2% ± 19.9% vs. 25.8% ± 9.1% $p < 0.01$; ARPE-19, 64.2% ± 13.2% vs. 6.4% ± 2.3% $p < 0.01$) in both the B4G12 and ARPE-19 cells after five days in culture (Figure 3B,C).

To validate the results of the immunofluorescence assay, an immunoblotting assay was also performed. Bcl2 has previously been regarded as an anti-apoptotic protein, while lamin A cleavage has been considered a marker of apoptosis in the cornea [30]. Recently, between the two forms of LC3, LC3-I, and LC3-II, the autophagosome membrane-bound form (LC3-II) has been viewed as a marker of autophagosome formation. During the process of autophagy, p62 binds to LC3-II to facilitate degradation of ubiquitinated protein aggregates [31] and degrades in autolysosome. Therefore, the protein level of LC3-II and p62 could be regarded as an index of autophagic flux change. As shown in Figure 3D, paraquat treatment significantly suppressed the expression of Bcl2 protein, whereas ascorbic acid in the medium significantly enhanced the expression of Bcl2. In contrast, paraquat treatment significantly elevated the expression of cleaved lamin A, LC3-II and p62, whereas ascorbic acid in the medium significantly suppressed the expression of cleaved lamin A, LC3-II, and p62 in both ARPE-19 and B4G12 cells.

Given that phosphorylation of Akt (Ser473) is a known upstream activator of Bcl-2 and that the PI3K/Akt pathway is involved in ROS-induced autophagic cell death in human CECs [32,33], we investigated the changes in Akt phosphorylation. As shown in Figure 4A, paraquat treatment significantly reduced Akt phosphorylation (p-Akt/Akt), while the addition of ascorbic acid resulted in the reversal of Akt phosphorylation. To further verify the involvement of the PI3K/Akt pathway in the ascorbic acid-mediated protection from ROS-induced cell death, we added LY294002 (PI3K inhibitor, 50 μmol/L) to the medium and observed that this significantly diminished the protective effect (reduced cell loss) of ascorbic acid in B4G12 and ARPE-19 cells (Figure 4B).

3.3. Topical Ascorbic Acid Ameliorates Oxidative Stress-Induced Corneal Endothelial Damage in Rabbits

To investigate whether topical ascorbic acid protects CEC against oxidative stress, we established an in vivo rabbit model of corneal endothelial damage. Accordingly, we first examined the protective effect of pretreatment with ascorbic acid on the primary rabbit corneal endothelial cells (RCECs) in both in vitro and in vivo culture models.

In the in vitro cell culture, confluent primary RCECs (passage 1) were divided into four groups. The A (ascorbic acid) and A + P (ascorbic acid and paraquat) groups were cultivated in medium containing 1 mM of ascorbic acid (A) for two days. This was followed by the addition of 0.2 mM of paraquat in the P (paraquat) and A + P groups for five days. Similar to the results in B4G12 and ARPE-19 cells, paraquat induced cell toxicity in RCECs, while pretreatment with ascorbic acid protected against cell loss (Figure 5A).

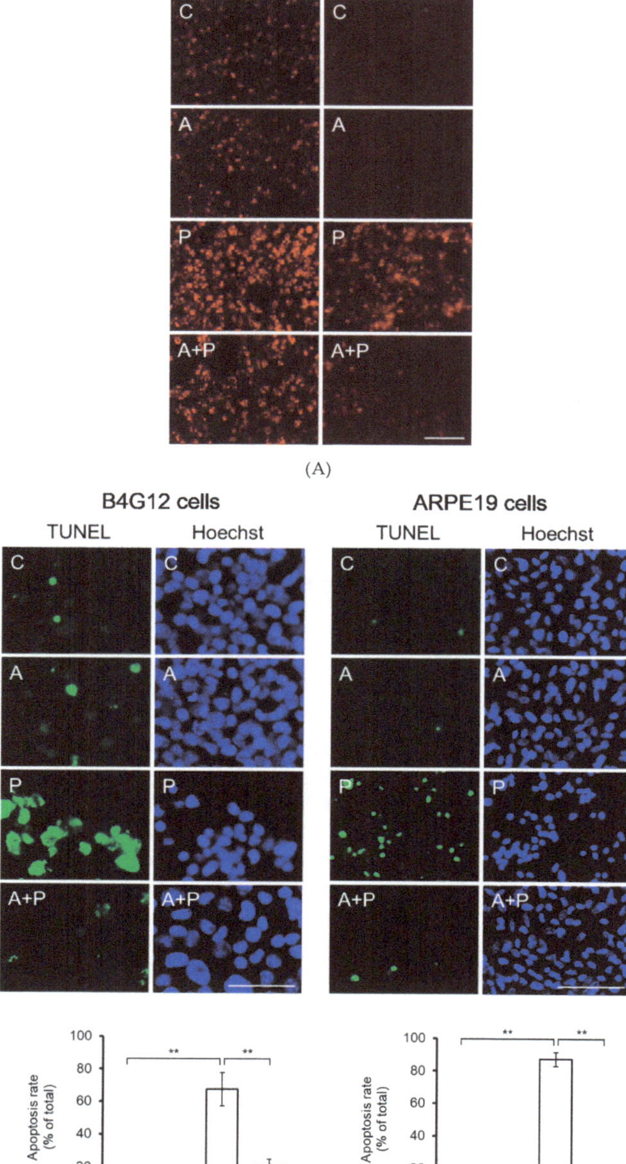

(A)

(B)

Figure 3. *Cont.*

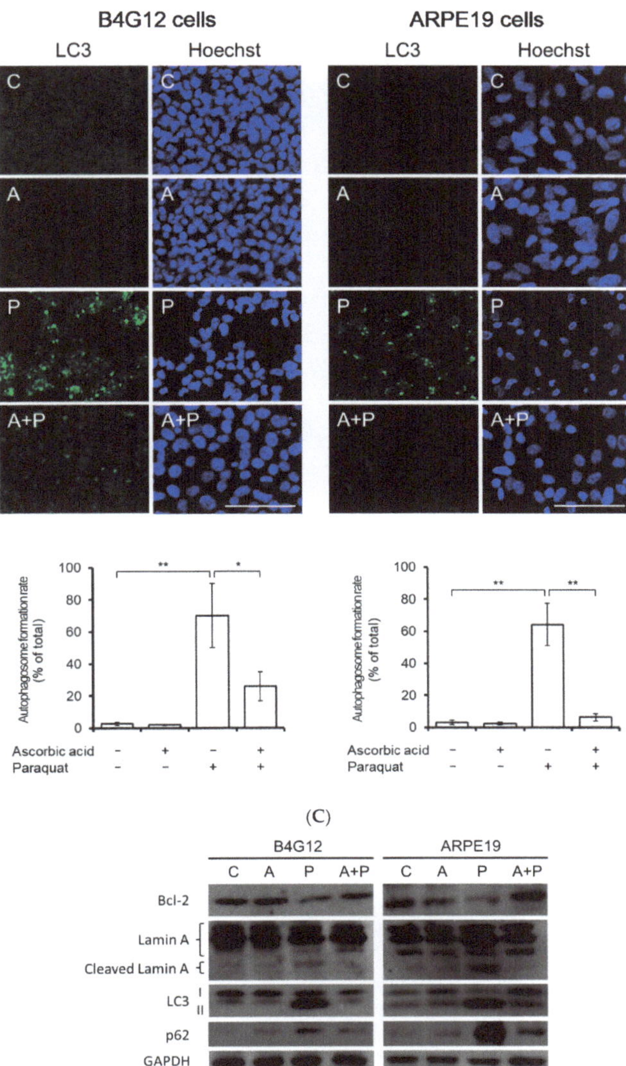

Figure 3. *Cont.*

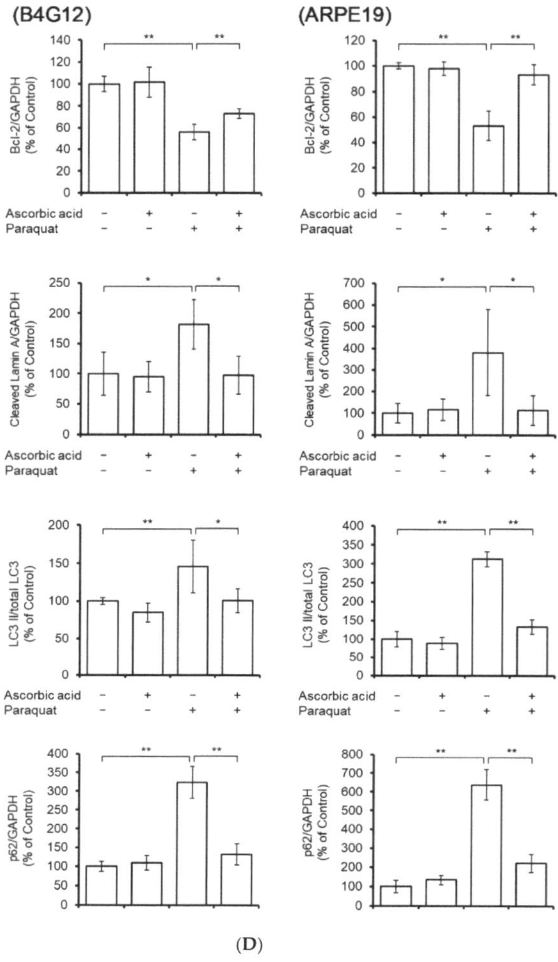

(D)

Figure 3. Pretreatment with ascorbic acid attenuates oxidative stress-induced apoptosis and autophagic flux blockage in B4G12 and ARPE-19 cells. B4G12 and ARPE-19 cells were cultivated in medium with or without 1 mM of ascorbic acid for two days, followed by the addition of paraquat (2 mM for ARPE-19 and 0.2 mM for B4G12) in the paraquat-treated groups (P only and C + P groups) for five days. (**A**) Paraquat-induced cellular accumulation of reactive oxygen species (ROS) and rescue by ascorbic acid was detected using ROS fluorescent dye (red color). ROS was induced by paraquat in both the B4G12 and ARPE-19 cells, and this was ameliorated by treatment with ascorbic acid. (**B**) Paraquat-induced apoptosis and rescue by ascorbic acid was examined using the TUNEL assay (green). (**C**) Paraquat-induced autophagosome formation and rescue by ascorbic acid was examined using immunofluorescence staining for LC3-II (autophagosome formation biomarker; green). (**D**) Effects of the paraquat and ascorbic on the protein expression of anti-apoptosis (Bcl-2), pro-apoptosis (lamin A, including cleaved forms), and autophagic flux (LC3 I/II and p62) biomarkers in B4G12 and ARPE-19 cells were probed using Western blotting. Paraquat induced altered protein expression in both B4G12 and ARPE-19 cells, and this could be reversed by ascorbic acid. The scale bars represent 100 μm (A) and 50 μm (B–C). Nuclei were counterstained with Hoechst 33342 (blue; B–D). (*n* = 3, * *p* < 0.05, ** *p* < 0.01).

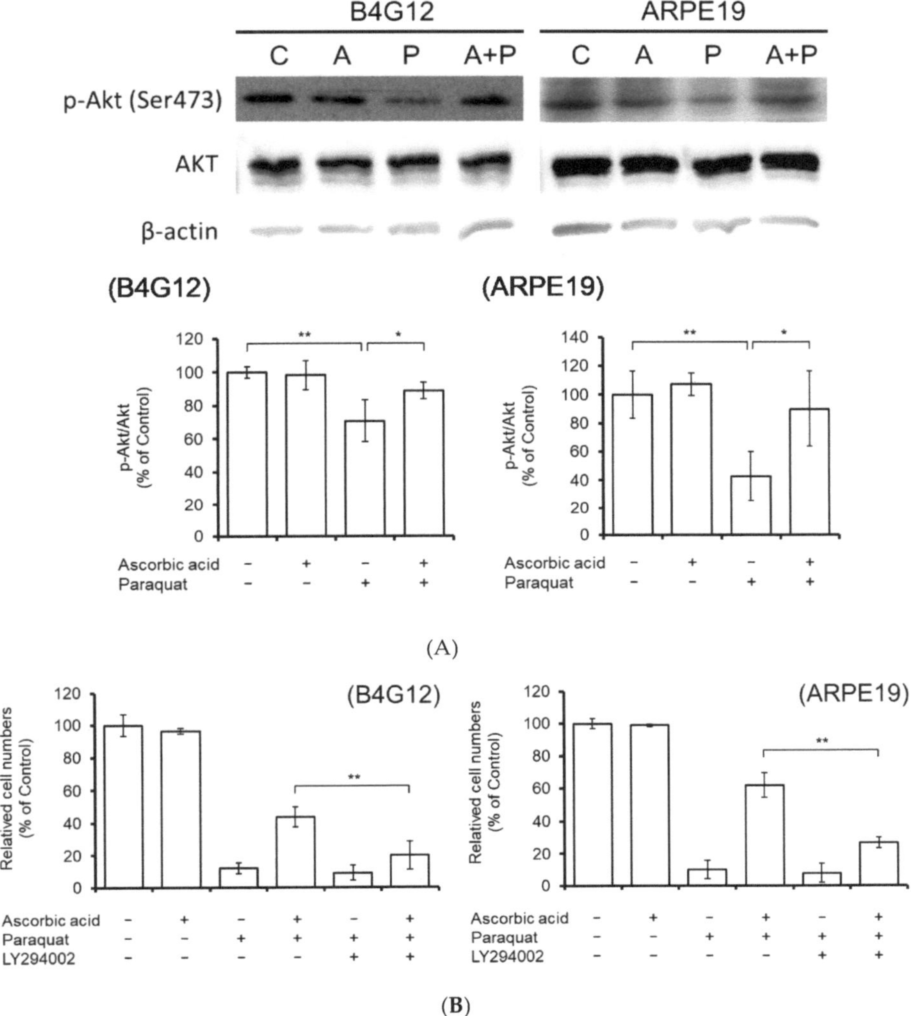

(A)

(B)

Figure 4. The PI3K/AKT pathway is involved in ascorbic acid-mediated cell protection. (**A**) Phosphorylation of Akt (Bcl-2 upstream regulator) was detected by Western blotting. Total Akt and β-actin were used as loading controls. Paraquat-suppressed Akt phosphorylation was rescued by ascorbic acid. (**B**) To examine whether the PI3K/AKT pathway was involved in ascorbic acid-mediated cell protection, LY294002 (a PI3K inhibitor, 50 μmol/L) was added in the culture medium. The cell protection effect was quantified by cell counting. Paraquat-induced cell loss was rescued by pretreatment with ascorbic acid, while the rescue effect of cell loss by ascorbic acid was significantly negated by the addition of LY294002. ($n = 3$, * $p < 0.05$, ** $p < 0.01$).

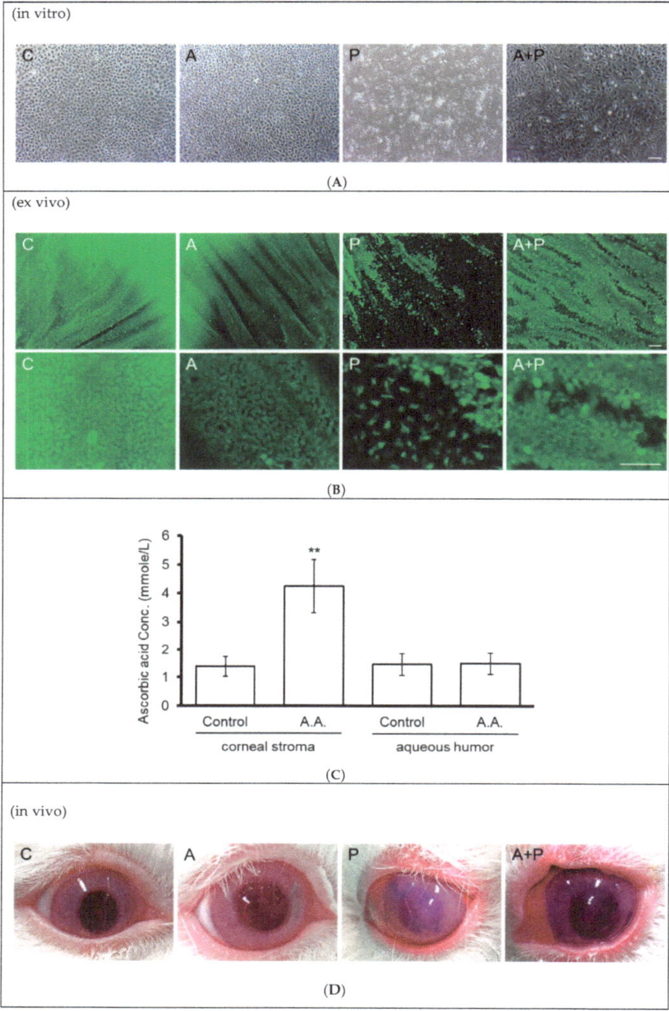

Figure 5. Effect of topical ascorbic acid on oxidative stress-induced corneal endothelial damage in a rabbit model. (**A**) Primary rabbit corneal endothelial cells (in vitro) were cultivated in medium with (A and A + P groups) or without (C and P groups) addition of 1 mM of ascorbic acid for two days, followed by addition of paraquat (0.2 mM) in the paraquat-treated groups (P and A + P groups) for five days. Sloughing of corneal endothelial cells was observed under phase-contrast microscopy. (**B**) Rabbit corneal tissue specimens (ex vivo) were cultivated in medium with (A and A + P groups) or without 1 mM of ascorbic acid for two days, followed by the addition of paraquat (25 mM) to the P and A + P groups for 15 min. Two days later, the sloughing of corneal endothelial cells was observed using Calcein-AM stain. (**C**) Rabbit corneas (in vivo) received an application of ascorbic acid (284 mmol/L in BSS solution) or BSS alone for two days (three times per day). After diffusion, ascorbic acid concentrations in the corneal stroma and anterior chambers were examined using the FRASC assay. ($n = 3$, ** $p < 0.01$). (**D**) Rabbit corneas (in vivo) received an application of ascorbic acid (284 mmol/L in BSS solution) or BSS alone for two days (three times per day), followed by intracameral injection of 25 mM paraquat (diluted in BSS) or BSS alone for 15 min. Corneal transparency was assessed using external eye photography on day two. (The scale bar represents 100 μm).

It is difficult to continuously maintain an elevated paraquat concentration in the anterior chamber in the animal model. This is because systemic toxicity, such as hepatorenal dysfunction, develops in two to three days at a low, poisonous dose of paraquat. Therefore, we chose to establish a model of corneal endothelial damage induced by perfusion of high concentration (25 mmol/L) of paraquat over a short time period (15 min). Hence, we performed ex vivo corneal tissue culture. As in the in vitro experiments, rabbit corneal tissues were divided into four groups. The A and A + P groups were cultivated in medium containing 1 mM of ascorbic acid for two days, followed by the addition of 25 mM of paraquat for the P and A + P groups for 15 min. Two days later, the extent of CEC sloughing was observed using Calcein-AM stain. As shown in Figure 5B, paraquat treatment resulted in the sloughing of RCECs, while pretreatment with ascorbic acid ameliorated loss of RCECs.

Subsequently, after two consecutive days of topical application of ascorbic acid (284 mmol/L) three times a day, diffusion of ascorbic acid into the corneal stroma and anterior chamber was recorded. As shown in Figure 5C, the ascorbic acid concentration was significantly elevated in the corneal stroma (1.39 ± 0.35 vs. 4.24 ± 0.92 mmol/L, $p < 0.01$), but showed no difference in the aqueous humor (1.46 ± 0.39 vs. 1.50 ± 0.38 mmol/L).

Finally, external eye photography was used to monitor the transparency of the cornea in the in vivo rabbit model. CEC damage induced by infusion of paraquat (25 mmol/L) for 15 min resulted in diminished corneal transparency. In contrast, the topical application of ascorbic acid for two consecutive days prior to induction of damage using paraquat improved corneal transparency (Figure 5D).

4. Discussion

Corneal endothelial cell density (ECD) tends to be decreased by phacoemulsification-induced oxidative stress. It has been demonstrated in a canine model that intracameral infusion of ascorbic acid during phacoemulsification minimizes the loss of corneal endothelial cells [34]. Nevertheless, the protective effect of topically administered ascorbic acid on the corneal endothelium has not yet been defined. Clinically, a combined protocol of topical ascorbic acid and steroids has been adopted to treat alkali-induced eye injuries to achieve rapid wound healing and prevent severe ocular sequelae [35]. Recently, we have reported successful phacoemulsification using topical ascorbic acid perioperatively in two patients with low corneal ECD [36]. In the current study, we investigate whether topical ascorbic acid protects HCEC against cell death induced by external oxidative stress.

Phacoemulsification is not the only risk factor for elevated levels of oxidative stress in the anterior chamber. Direct exposure to environmental and solar UV radiation introduces alternative sources of oxidative stress in the cornea, such as photo-oxidative damage [37]. In the corneal endothelium, oxidative stress decreases the levels of cytochrome oxidase and antioxidants; meanwhile, it increases lipid peroxidation, leading to cellular impairment of HCEC [38,39]. In fact, Fuchs endothelial corneal dystrophy (FECD) is characterized by prominent apoptosis resulting from excessive oxidative stress and oxygen-induced DNA damage [39,40]. Hence, it is plausible that topical ascorbic acid could serve a protective role in patient populations with vulnerable corneal endothelium.

As shown in Figure 5C, AA concentration in the corneal stroma significantly increases after a topical application of AA. According to Fernández-Pérez J, AA induces the dendritic morphology, increases the expression of keratocyte markers, and prevents myofibroblast differentiation [41]. Therefore, topical use of AA is not likely to produce side effects on keratocytes. On the other hand, why the AA concentration is not different in the aqueous humor is probably because the aqueous circulation neutralizes the changing pattern of AA concentration. In contrast, there is no dynamic circulation in the interstitial fluid within the corneal stroma, and a long-term reservoir effect can be formed. In addition, Descemet's membrane is AA-penetrable, so that AA from the corneal stroma can exert effects on the corneal endothelium, explaining our observation of improved corneal transparency after topical use of AA (Figure 5D).

In the current study, we have demonstrated that paraquat-provoked oxidative stress induces apoptosis and autophagic flux blockage in corneal endothelial cells, and that inhibition of the PI3K/Akt pathway attenuates the protective effects of ascorbic acid (Figure 4). In addition to the direct reduction of ROS by ascorbic acid, alleviating apoptosis and autophagic flux blockage, the PI3K/Akt pathway has also been known to suppress apoptosis through activation of Bcl-2 [42]. Moreover, previous evidence has shown an autophagosome formation -inhibitory effect of activation of the PI3K/AKT pathway [43]. It remains unclear whether or not the aforementioned pathways are involved in the protective effect of ascorbic acid.

When cells are under oxidative stress, p62/SQSTM1 is induced to express and creates a positive feedback loop by inducing antioxidant response element-driven gene transcription [44]. However, when autophagic defect (blockage of autophagic flux) occurs, the autophagosome fails to fusion with the lysosome, resulting in the accumulation of p62 [45]. Accumulation of p62 subsequently provokes autophagic defect-dependent apoptosis by activating caspase 8 [46,47]. In Figure 3D, paraquat treatment significantly increased the level of LC3-II, indicating an increased formation of the autophagosome. On the other hand, paraquat increases the level of p62, suggestive of autophagic defect induced by paraquat. AA treatment significantly decreased the level of p62, indicating that reducing autophagic defect might be one of the candidate pathways of AA protecting cells from oxidative stress-induced apoptosis. Moreover, crosstalks between autophagy and apoptosis have been shown to regulate cellular survival or death under stress [48]. Therefore, the autophagy-mediated survival pathway could pave the way for therapeutic strategies, for which further investigation is warranted.

As one of the important anti-oxidative constituents in the anterior chamber, levels of ascorbic acid decline as a result of the aging process [49]. In this study, we confirm that the topical administration of ascorbic acid leads to an elevated concentration in the corneal stroma, which subsequently enhances its antioxidant activity against paraquat-induced oxidative stress. Therefore, we propose that topically applied ascorbic acid results in a preventive effect against oxidative stress-associated ocular degeneration.

Supplementary Materials: The following are available online at http://www.mdpi.com/2073-4409/9/4/943/s1, Figure S1: Ascorbic acid improved cell viability from oxidative stress.

Author Contributions: Conceptualization, C.-M.C., H.-C.C., W.-C.W., Y.-J.J.M., L.-K.Y., and Y.-J.H.; Methodology, C.-C.H., C.-M.C., and T.-T.L.; Validation, Y.-J.H.; Investigation, T.-K.W.; Writing—Original Draft Preparation, Y.-J.H.; Writing—Review and Editing, H.-C.C.; Supervision, H.-C.C.; Project Administration, H.-C.C. All authors have read and agreed to the published version of the manuscript.

Funding: This research was funded by the Chang Gung Memorial Hospital (CMRPG3J0511 and CMRPG3G0031~3) and the Ministry of Science and Technology (MOST 107-2314-B-182A-088-MY3).

Acknowledgments: The authors thank the assistance from the Microscope Core Laboratory, Chang Gung Memorial Hospital, Linkou branch.

Conflicts of Interest: The authors declare no conflict of interest.

References

1. Kim, S.J.; Schoenberger, S.D.; Thorne, J.E.; Ehlers, J.P.; Yeh, S.; Bakri, S.J. Topical Nonsteroidal Anti-inflammatory Drugs and Cataract Surgery. *Ophthalmology* **2015**, *122*, 2159–2168. [CrossRef] [PubMed]
2. Pathengay, A.; Flynn, H.W., Jr.; Isom, R.F.; Miller, D. Endophthalmitis outbreaks following cataract surgery: Causative organisms, etiologies, and visual acuity outcomes. *J. Cataract. Refract. Surg.* **2012**, *38*, 1278–1282. [CrossRef] [PubMed]
3. Lois, N.; Wong, D. Pseudophakic retinal detachment. *Surv. Ophthalmol.* **2003**, *48*, 467–487. [CrossRef]
4. Rosado-Adames, N.; Afshari, N.A. The changing fate of the corneal endothelium in cataract surgery. *Curr. Opin. Ophthalmol.* **2012**, *23*, 3–6. [CrossRef] [PubMed]
5. Bourne, W.M.; Nelson, L.R.; Hodge, D.O. Continued Endothelial Cell Loss Ten Years after Lens Implantation. *Ophthalmology* **1994**, *101*, 1014–1023. [CrossRef]
6. Hayashi, K.; Yoshida, M.; Manabe, S.-I.; Hirata, A. Cataract surgery in eyes with low corneal endothelial cell density. *J. Cataract. Refract. Surg.* **2011**, *37*, 1419–1425. [CrossRef] [PubMed]

7.   Yamazoe, K.; Yamaguchi, T.; Hotta, K.; Satake, Y.; Konomi, K.; Den, S.; Shimazaki, J. Outcomes of cataract surgery in eyes with a low corneal endothelial cell density. *J. Cataract. Refract. Surg.* **2011**, *37*, 2130–2136. [CrossRef]

8.   Waring, G.O.; Bourne, W.M.; Edelhauser, H.F.; Kenyon, K.R. The Corneal Endothelium. *Ophthalmology* **1982**, *89*, 531–590. [CrossRef]

9.   Mishima, S. Clinical Investigations on the Corneal Endothelium. *Ophthalmology* **1982**, *89*, 525–530. [CrossRef]

10.   Laing, R.A.; Neubauer, L.; Oak, S.S.; Kayne, H.L.; Leibowitz, H.M. Evidence for Mitosis in the Adult Corneal Endothelium. *Ophthalmology* **1984**, *91*, 1129–1134. [CrossRef]

11.   Zhu, Y.-T.; Chen, H.-C.; Chen, S.-Y.; Tseng, S.C.G. Nuclear p120 catenin unlocks mitotic block of contact-inhibited human corneal endothelial monolayers without disrupting adherent junctions. *J. Cell Sci.* **2012**, *125*, 3636–3648. [CrossRef] [PubMed]

12.   Hsueh, Y.-J.; Chen, H.-C.; Wu, S.-E.; Wang, T.-K.; Chen, J.-K.; Ma, D.H.-K. Lysophosphatidic acid induces YAP-promoted proliferation of human corneal endothelial cells via PI3K and ROCK pathways. *Mol. Ther. Methods Clin. Dev.* **2015**, *2*, 15014. [CrossRef] [PubMed]

13.   Augustin, A.J.; Dick, B.H.; Augustin, A.J. Oxidative tissue damage after phacoemulsification. *J. Cataract. Refract. Surg.* **2004**, *30*, 424–427. [CrossRef]

14.   Nemet, A.Y.; Assia, E.I.; Meyerstein, D.; Meyerstein, N.; Gedanken, A.; Topaz, M. Protective effect of free-radical scavengers on corneal endothelial damage in phacoemulsification. *J. Cataract. Refract. Surg.* **2007**, *33*, 310–315. [CrossRef]

15.   Murano, N.; Ishizaki, M.; Sato, S.; Fukuda, Y.; Takahashi, H. Corneal Endothelial Cell Damage by Free Radicals Associated With Ultrasound Oscillation. *Arch. Ophthalmol.* **2008**, *126*, 816–821. [CrossRef]

16.   Geffen, N.; Topaz, M.; Kredy-Farhan, L.; Barequet, I.S.; Farzam, N.; Assia, E.I.; Savion, N. Phacoemulsification-induced injury in corneal endothelial cells mediated by apoptosis: In vitro model. *J. Cataract. Refract. Surg.* **2008**, *34*, 2146–2152. [CrossRef]

17.   Nishi, Y.; Engler, C.; Na, D.R.; Kashiwabuchi, R.T.; Shin, Y.J.; Cano, M.; Jun, A.S.; Chuck, R.S. Evaluation of phacoemulsification-induced oxidative stress and damage of cultured human corneal endothelial cells in different solutions using redox fluorometry microscopy. *Acta Ophthalmol.* **2010**, *88*, e323–e327. [CrossRef]

18.   Lai, L.J.; Chen, Y.F.; Wu, S.; Tsao, Y.P.; Tsai, R.J. Endothelial cell loss induced by phacoemulsification occurs through apoptosis. *Chang. Gung Med. J.* **2001**, *24*, 621–627.

19.   Filomeni, G.; De Zio, D.; Cecconi, F. Oxidative stress and autophagy: The clash between damage and metabolic needs. *Cell Death Differ.* **2014**, *22*, 377–388. [CrossRef]

20.   Tangvarasittichai, O.; Tangvarasittichai, S. Oxidative Stress, Ocular Disease and Diabetes Retinopathy. *Curr. Pharm. Des.* **2019**, *24*, 4726–4741. [CrossRef]

21.   Navarro-Yepes, J.; Burns, M.; Anandhan, A.; Khalimonchuk, O.; Del Razo, L.M.; Quintanilla-Vega, B.; Pappa, A.; Panayiotidis, M.I.; Franco, R. Oxidative Stress, Redox Signaling, and Autophagy: Cell DeathVersusSurvival. *Antioxid. Redox Signal.* **2014**, *21*, 66–85. [CrossRef] [PubMed]

22.   Bartlett, H.E.; Eperjesi, F. An ideal ocular nutritional supplement? *Ophthalmic Physiol. Opt.* **2004**, *24*, 339–349. [CrossRef] [PubMed]

23.   Serbecic, N.; Beutelspacher, S.C. Vitamins Inhibit Oxidant-Induced Apoptosis of Corneal Endothelial Cells. *Jpn. J. Ophthalmol.* **2005**, *49*, 355–362. [CrossRef] [PubMed]

24.   Serbecic, N.; Beutelspacher, S.C. Anti-oxidative vitamins prevent lipid-peroxidation and apoptosis in corneal endothelial cells. *Cell Tissue Res.* **2005**, *320*, 465–475. [CrossRef]

25.   Shima, N.; Kimoto, M.; Yamaguchi, M.; Yamagami, S. Increased Proliferation and Replicative Lifespan of Isolated Human Corneal Endothelial Cells withl-Ascorbic acid 2-phosphate. *Investig. Opthalmol. Vis. Sci.* **2011**, *52*, 8711–8717. [CrossRef]

26.   Rubowitz, A.; Assia, E.I.; Rosner, M.; Topaz, M. Antioxidant protection against corneal damage by free radicals during phacoemulsification. *Investig. Opthalmol. Vis. Sci.* **2003**, *44*, 1866–1870. [CrossRef]

27.   Park, C.; Lee, H.; Hong, S.-H.; Kim, J.-H.; Park, S.-K.; Jeong, J.-W.; Kim, G.-Y.; Hyun, J.W.; Yun, S.J.; Kim, B.W.; et al. Protective effect of diphlorethohydroxycarmalol against oxidative stress-induced DNA damage and apoptosis in retinal pigment epithelial cells. *Cutan. Ocul. Toxicol.* **2019**, *38*, 298–308. [CrossRef]

28.   Nita, M.; Grzybowski, A. The Role of the Reactive Oxygen Species and Oxidative Stress in the Pathomechanism of the Age-Related Ocular Diseases and Other Pathologies of the Anterior and Posterior Eye Segments in Adults. *Oxid. Med. Cell. Longev.* **2016**, *2016*, 1–23. [CrossRef]

29. Sheu, S.-J.; Chen, J.-L.; Bee, Y.-S.; Lin, S.-H.; Shu, C.-W. ERBB2-modulated ATG4B and autophagic cell death in human ARPE19 during oxidative stress. *PLoS ONE* **2019**, *14*, e0213932. [CrossRef]
30. Mahajan, S.; Thieme, D.; Czugala, M.; Kruse, F.E.; Fuchsluger, T.A. Lamin Cleavage: A Reliable Marker for Studying Staurosporine-Induced Apoptosis in Corneal Tissue. *Investig. Opthalmol. Vis. Sci.* **2017**, *58*, 5802. [CrossRef]
31. Pankiv, S.; Clausen, T.H.; Lamark, T.; Brech, A.; Bruun, J.-A.; Outzen, H.; Overvatn, A.; Bjørkøy, G.; Johansen, T. p62/SQSTM1 Binds Directly to Atg8/LC3 to Facilitate Degradation of Ubiquitinated Protein Aggregates by Autophagy. *J. Boil. Chem.* **2007**, *282*, 24131–24145. [CrossRef] [PubMed]
32. Lin, C.-J.; Chen, T.-L.; Tseng, Y.Y.; Wu, G.-J.; Hsieh, M.-H.; Lin, Y.-W.; Chen, R.-M. Honokiol induces autophagic cell death in malignant glioma through reactive oxygen species-mediated regulation of the p53/PI3K/Akt/mTOR signaling pathway. *Toxicol. Appl. Pharmacol.* **2016**, *304*, 59–69. [CrossRef] [PubMed]
33. Yoon, C.K.; Yoon, S.Y.; Hwang, J.S.; Shin, Y.J. O-GlcNAc Signaling Augmentation Protects Human Corneal Endothelial Cells from Oxidative Stress via AKT Pathway Activation. *Curr. Eye Res.* **2020**, 1–7. [CrossRef] [PubMed]
34. Padua, I.R.M.; Valdetaro, G.P.; Lima, T.B.; Kobashigawa, K.K.; Silva, P.E.S.; Aldrovani, M.; Padua, P.P.M.; Laus, J.L. Effects of intracameral ascorbic acid on the corneal endothelium of dogs undergoing phacoemulsification. *Veter- Ophthalmol.* **2017**, *21*, 151–159. [CrossRef] [PubMed]
35. Brodovsky, S.C.; Mccarty, C.A.; Snibson, G.; Loughnan, M.; Sullivan, L.; Daniell, M.; Taylor, H.R. Management of alkali burns. *Ophthalmol.* **2000**, *107*, 1829–1835. [CrossRef]
36. Lee, C.-Y.; Chen, H.-T.; Hsueh, Y.-J.; Chen, H.-C.; Huang, C.-C.; Meir, Y.-J.J.; Cheng, C.-M.; Wu, W.-C. Perioperative topical ascorbic acid for the prevention of phacoemulsification-related corneal endothelial damage: Two case reports and review of literature. *World J. Clin. Cases* **2019**, *7*, 642–649. [CrossRef]
37. Williams, D.L. Oxidative Stress and the Eye. *Veter. Clin. North Am. Small Anim. Pr.* **2008**, *38*, 179–192. [CrossRef]
38. Shoham, A.; Hadziahmetovic, M.; Dunaief, J.L.; Mydlarski, M.B.; Schipper, H.M. Oxidative stress in diseases of the human cornea. *Free. Radic. Boil. Med.* **2008**, *45*, 1047–1055. [CrossRef]
39. Jurkunas, U.V.; Bitar, M.S.; Funaki, T.; Azizi, B. Evidence of Oxidative Stress in the Pathogenesis of Fuchs Endothelial Corneal Dystrophy. *Am. J. Pathol.* **2010**, *177*, 2278–2289. [CrossRef]
40. Czarny, P.; Kasprzak, E.; Wielgorski, M.; Udziela, M.; Markiewicz, B.; Blasiak, J.; Szaflik, J.; Szaflik, J.P. DNA damage and repair in Fuchs endothelial corneal dystrophy. *Mol. Boil. Rep.* **2012**, *40*, 2977–2983. [CrossRef]
41. Fernández-Pérez, J.; Ahearne, M. Influence of Biochemical Cues in Human Corneal Stromal Cell Phenotype. *Curr. Eye Res.* **2018**, *44*, 135–146. [CrossRef] [PubMed]
42. Pugazhenthi, S.; Nesterova, A.; Sable, C.; Heidenreich, K.A.; Boxer, L.M.; Heasley, L.E.; Reusch, J.E.-B. Akt/Protein Kinase B Up-regulates Bcl-2 Expression through cAMP-response Element-binding Protein. *J. Boil. Chem.* **2000**, *275*, 10761–10766. [CrossRef] [PubMed]
43. Li, X.; Hu, X.; Wang, J.; Xu, W.; Yi, C.; Ma, R.; Jiang, H. Inhibition of autophagy via activation of PI3K/Akt/mTOR pathway contributes to the protection of hesperidin against myocardial ischemia/reperfusion injury. *Int. J. Mol. Med.* **2018**, *42*, 1917–1924. [CrossRef] [PubMed]
44. Jain, A.; Lamark, T.; Sjøttem, E.; Larsen, K.B.; Awuh, J.A.; Øvervatn, A.; McMahon, M.; Hayes, J.; Johansen, T. p62/SQSTM1 Is a Target Gene for Transcription Factor NRF2 and Creates a Positive Feedback Loop by Inducing Antioxidant Response Element-driven Gene Transcription. *J. Boil. Chem.* **2010**, *285*, 22576–22591. [CrossRef]
45. Klionsky, D.J.; Abeliovich, H.; Agostinis, P.; Agrawal, D.K.; Aliev, G.; Askew, D.; Baba, M.; Baehrecke, E.H.; Bahr, B.A.; Ballabio, A.; et al. Guidelines for the use and interpretation of assays for monitoring autophagy in higher eukaryotes. *Autophagy* **2007**, *4*, 151–175. [CrossRef]
46. Yan, X.; Zhong, X.; Yu, S.; Zhang, L.; Liu, Y.; Zhang, Y.; Sun, L.; Su, J. p62 aggregates mediated Caspase 8 activation is responsible for progression of ovarian cancer. *J. Cell. Mol. Med.* **2019**, *23*, 4030–4042. [CrossRef]
47. Young, M.; Takahashi, Y.; Khan, O.; Park, S.; Hori, T.; Yun, J.; Sharma, A.K.; Amin, S.; Hu, C.-D.; Zhang, J.; et al. Autophagosomal Membrane Serves as Platform for Intracellular Death-inducing Signaling Complex (iDISC)-mediated Caspase-8 Activation and Apoptosis. *J. Boil. Chem.* **2012**, *287*, 12455–12468. [CrossRef]

48. Nikoletopoulou, V.; Markaki, M.; Palikaras, K.; Tavernarakis, N. Crosstalk between apoptosis, necrosis and autophagy. *Biochim. Biophys. Acta (BBA) Bioenergy* **2013**, *1833*, 3448–3459. [CrossRef]

49. Canadananovic, V.; Latinovic, S.; Barišić, S.; Babić, N.; Jovanovic, S. Age-related changes of vitamin C levels in aqueous humour. *Vojn. Pregl.* **2015**, *72*, 823–826. [CrossRef]

Article

Galectin-1 Overexpression Activates the FAK/PI3K/AKT/mTOR Pathway and Is Correlated with Upper Urinary Urothelial Carcinoma Progression and Survival

Yu-Li Su [1,2,*] , Hao-Lun Luo [3], Chun-Chieh Huang [4], Ting-Ting Liu [5], Eng-Yen Huang [4], Ming-Tse Sung [5], Jen-Jie Lin [1] , Po-Hui Chiang [3], Yen-Ta Chen [3], Chih-Hsiung Kang [3] and Yuan-Tso Cheng [3]

[1] Division of Hematology Oncology, Department of Internal Medicine, Kaohsiung Chang Gung Memorial Hospital and Chang Gung University, College of Medicine, Kaohsiung 833, Taiwan

[2] Clinical Trial Center, Kaohsiung Chang Gung Memorial Hospital, Kaohsiung 833, Taiwan

[3] Department of Urology, Kaohsiung Chang Gung Memorial Hospital and Chang Gung University, College of Medicine, Kaohsiung 833, Taiwan

[4] Department of Radiation Oncology, Kaohsiung Chang Gung Memorial Hospital and Chang Gung University, College of Medicine, Kaohsiung 833, Taiwan

[5] Department of Pathology, Kaohsiung Chang Gung Memorial Hospital and Chang Gung University, College of Medicine, Kaohsiung 833, Taiwan

* Correspondence: yolisu@mac.com; Tel.: +886-7-731-7123; Fax: +886-7-732-2402

Received: 19 February 2020; Accepted: 25 March 2020; Published: 26 March 2020

Abstract: Galectin-1 (GAL1) is a β-galactoside-binding protein involved in multiple aspects of tumorigenesis. However, the biological role of GAL1 in upper tract urothelial carcinoma (UTUC) has not been entirely understood. Herein, we investigated the oncological effects of GAL1 expression in tumor specimens and identified related gene alterations through molecular analysis of GAL1. Clinical parameter data and tumor specimens were collected from 86 patients with pT3N0M0 UTUC who had undergone radical nephroureterectomy. We analyzed the difference in survival by using Kaplan–Meier analyses and Cox proportional regression models and in GAL1 expression by using immunohistochemical (IHC) methods. Public genomic data from the Cancer Genome Atlas (TCGA) and GSE32894 data sets were analyzed for comparison. Using four urothelial carcinoma (UC) cell lines (BFTC-909, T24, RT4, and J82) as in vitro models, we evaluated the functions of GAL1 in UC cell growth, invasiveness, and migration and its role in downstream signaling pathways. The study population was classified into two groups, GAL1-high (n = 35) and GAL1-low (GAL1 n = 51), according to IHC interpretation. Univariate analysis revealed that high GAL1 expression was significantly associated with poor recurrence-free survival (RFS; $p = 0.028$) and low cancer-specific survival (CSS; $p = 0.025$). Multivariate analysis revealed that GAL1-high was an independent predictive factor for RFS (hazard ratio (HR) 2.43; 95% confidence interval (CI) 1.17–5.05, $p = 0.018$) and CSS (HR 4.04; 95% CI 1.25–13.03, $p = 0.019$). In vitro studies revealed that GAL1 knockdown significantly reduced migration and invasiveness in UTUC (BFTC-909) and bladder cancer cells (T24). GAL1 knockdown significantly reduced protein levels of matrix metalloproteinase-2 (MMP-2) and MMP-9, which increased tissue inhibitor of metalloproteinase-1 (TIMP-1) and promoted epithelial–mesenchymal transition (EMT). Through gene expression microarray analysis of GAL1 vector and GAL1-KD cells, we identified multiple significant signaling pathways including p53, Forkhead box O (FOXO), and phosphoinositide 3-kinase/protein kinase B (PI3K/AKT). We validated microarray results through immunoblotting, thus proving that downregulation of GAL1 reduced focal adhesion kinase (FAK), p-PI3K, p-AKT, and p-mTOR expression. We concluded that GAL1 expression was highly related to oncological survival in patients with locally advanced UTUC. GAL1 promoted UC invasion and metastasis by activating the FAK/PI3K/AKT/mTOR pathway.

Keywords: galectin-1; upper urinary urothelial carcinoma; epithelial–mesenchymal transition; survival; phosphoinositide 3-kinases; focal adhesion kinase; mammalian target of rapamycin

1. Introduction

Upper tract urothelial carcinoma (UTUC) is an aggressive and lethal disease. The incidence of UTUC in Western countries is relatively low (5% of all urothelial carcinoma); however, the prognosis of UTUC is considerably worse than that at the same pathological stage of urothelial carcinoma of the bladder (UCB) [1–3]. The incidence of UTUC in Taiwan is considerably higher than the worldwide incidence (30%–40% of all urothelial carcinoma cells (UCs)), which indicates that some unknown carcinogenic or environmental factors contribute to tumor development and growth [4–7]. Histologically, UCs arising from the upper urinary tract and urinary bladder are grossly identical. However, several comprehensive genomic studies have argued that UTUC and UCB are distinct and that the disease can be characterized by a unique fingerprint mutation signature; this signature takes the form of A:T to T:A transversions induced by aristolochic acid [7]. By contrast, specific insights have been gained by studying an autosomal dominant familial syndrome, namely Lynch syndrome (hereditary nonpolyposis colorectal cancer), which is associated with an increased risk of UTUC [8]. Overall, the carcinogenesis and molecular biology of UTUC and UCB may not involve identical pathways, and both cancers should be treated and discussed separately.

Treatment of patients with UTUC often requires multidisciplinary teams, consisting of urologists, medical oncologists, and radiation oncologists. Radical nephroureterectomy with excision of an ipsilateral bladder cuff is the gold standard treatment for organ-confined UTUC [9]. Early stages of UTUC can be cured by radical surgery; however, disease recurrence and distant metastasis occur commonly in the advanced stages of UTUC (T3 or T4), which is incurable and inevitably results in patient death. Currently, studies with consistent results of adjuvant treatment of locally advanced UTUC and randomized trials to guide postoperative management are not available. Moreover, the comprehensive pathogenesis and molecular features of UTUC are currently under investigation [10]; hence, developing effective treatments is difficult.

Galectin-1 (GAL1) is a β-galactoside-binding protein encoded by *LGALS1* on chromosome 22q12 and participates in multiple aspects of tumorigenesis, including cell proliferation, invasiveness, metastasis, and angiogenesis [11–15]. GAL1 expression has been frequently reported to increase in several types of tumors, including those of the colon, breast, lung, and uterine cervix [16–19] as well as those in Hodgkin lymphoma [20]. Moreover, higher expressions of GAL1 in gastric and cervical cancer have been reported to be positively correlated with advanced tumor stage, tumor invasion, and lymph node metastasis [21,22]. In terms of prognostic effect, several anecdotal studies have demonstrated a consistent relationship between high GAL1 expression and poor survival in patients with cancers of the lung, uterine cervix, and bladder [23–25]. Shen et al. demonstrated that the interplay between GAL1 and bladder cancer invasiveness and that between GAL1 and progression was mediated via the Ras-Rac1-MEKK4-JNK-AP1 signaling pathway [26]. In the lung cancer model, downregulation of GAL1 reduced tumor invasion and migration via the p38 MAPK-ERK and cyclooxygenase-2 (COX2) pathways [27]. Although some previous studies have confirmed the crucial role of GAL1 in tumorigenesis and drug resistance pathways, the role of GAL1 in UTUC remains unknown and has not been investigated thus far.

In the present study, we examined the prognostic role of GAL1 in patients with locally advanced UTUC (pT3). Furthermore, we evaluated the biological roles of GAL1 in UTUC and UCB cell lines and attempted to decipher the GAL1 mediating downstream oncological pathways in UTUC.

2. Materials and Methods

2.1. Antibodies and Reagents

Many reagents, including Dulbecco's modified Eagle's medium (DMEM), McCoy's 5a medium, trypsin-ethylenediaminetetraacetic acid, fetal bovine serum (FBS), and phosphate-buffered saline (PBS), were obtained from Biowest (Nuaillé, France). Polyvinylidene difluoride (PVDF) membranes, and goat anti-rabbit and horseradish peroxidase (HRP)-conjugated immunoglobulin (Ig) G were obtained from Millipore (Billerica, MA, USA). Protease inhibitor cocktail and DMSO were obtained from BioSource International (Camarillo, CA, USA). Cell extraction radioimmunoprecipitation assay (RIPA) buffer was obtained from TOOLS (TOOLS, Taiwan). Enhanced chemiluminescence (ECL) Western blotting reagents were obtained from Pierce Biotechnology (Rockford, IL, USA). Mouse anti-human β-actin antibodies were obtained from Sigma (St Louis, MO, USA). Rabbit anti-human FAK, mTOR, and p-mTOR antibodies were obtained from Epitomics (Burlingame, CA, USA). Rabbit anti-human TIMP-1, AKT, and p-AKT antibodies were obtained from ProteinTech Group (Chicago, IL, USA). Rabbit anti-human MMP-2, MMP-9, PI3K, p-PI3K, and EMT kit (#9782) antibodies were obtained from Cell Signaling Technology (Danvers, MA, USA).

2.2. Patients and Tumor Samples

We enrolled 86 patients with UTUC who had undergone radical nephroureterectomy and bladder cuff excision with final pathologically confirmed as pT3N0 stage between January 2005 and December 2012 in Kaohsiung Chang Gung Memorial Hospital (KSCGMH). Preoperative and pathological features of eligible patients, including age, sex, comorbidity, pathological TNM stage, histopathological subtypes and variants, presence of lymphovascular invasion (LVI) or perineural invasion, pattern of tumor formation (papillary or infiltrative), solitary or multicentric, presence of hydronephrosis, type of disease recurrence (local, regional, distant, or urinary tract recurrence), and date of disease recurrence and death, were recorded in detail. All clinicopathological data were collected retrospectively by accessing an electronic medical record system. All surgical tumor samples were fixed in 10% formalin and embedded in paraffin. All study procedures performed in this research were approved by the Chang Gung Medical Foundation Institutional Review Board (No. 104-5487B). Owing to retrospective nature, human tumor specimens of upper tract urothelial carcinoma were obtained from patients undergoing radical nephroureterectomy at Kaohsiung Chang Gung Memorial Hospital.

2.3. IHC Analysis

Immunohistochemical (IHC) staining for detecting GAL1 was performed in all resected tumor specimens. The paraffin-embedded tumor tissues were cut to obtain 4 mm thick sections. Briefly, after deparaffinization and rehydration, the sections were subjected to heat-induced epitope retrieval in 10 mM citrate buffer (pH 6.0) in a hot water bath (95 °C) for 20 min. After blocking with 1% goat serum for 1 h at room temperature, the sections were incubated with primary antibodies for at least 18 h at 4 °C. GAL1 protein expression was detected using a primary antibody specific to Gal1 (H-45, Santa Cruz Biotechnology, Santa Cruz, CA, USA); subsequently, the sections were incubated with the secondary antibody (Histofine MAX PO, Nichirei, Tokyo, Japan) for 30 min. Immunodetection was performed using the LSAB2 kit (Dako, Carpinteria, CA) followed by 3-3′-diaminobenzidine for color development and hematoxylin for counterstaining. An incubation mixture in which the primary antibody was replaced by PBS was used as a negative control. All sections were scored for GAL1 cytoplasmic expression, and the results were interpreted by two independent pathologists (M.T.S. and T.T.L.) after blinding. We determined 10% expression as the optimal cutoff level. Tumors exhibiting <10% expression of GAL1 were classified as low expression.

2.4. Cell Lines and Culture

Human UC cell lines, namely the BFTC-909 (renal pelvis), J82 (bladder), RT4 (bladder), and T24 (bladder), were purchased from the Food Industry Research and Development Institute (Hsinchu, Taiwan). The BFTC-909 and J82 cells were cultured in DMEM. The T24 and RT4 cells were cultured in McCoy's 5a medium, supplemented with 10% FBS and antibiotics (100 U/mL penicillin and 100 µg/mL streptomycin). All cells were incubated in a humidified atmosphere containing 95% air and 5% CO_2 at 37 °C.

2.5. Immunoblotting

For Western blotting, 5×10^6 BFTC-909 cells were seeded in 10-cm plates and were lysed using a cell extraction RIPA buffer. Proteins (25 µg) extracted from the whole cells were separated through 12.5% SDS gel electrophoresis and then transferred onto a PVDF membrane (Millipore) for 2 h at 400 mA using Transphor TE 22 transfer tank (Hoeffer). The PVDF membranes were then incubated with appropriate rabbit polyclonal antibodies at 4 °C for 2 h or overnight. The membranes were washed five times in PBS buffer containing 0.05% Tween 20 and then probed with goat anti-rabbit HRP-conjugated antibody (1:5000) for 1 h. The blots were then visualized using ECL Western Blotting Reagents (Pierce, Rockford, IL, USA).

2.6. GAL1 Knockdown Cells with shRNA

For shRNA transfection, 1×10^5 cells were seeded on 3-cm plates and incubated for 24 h at 37 °C. Next, the cells were transfected with GAL1 shRNA or respective controls by using lipofectamine 2000 and incubated for 48 h in a serum-free medium. The cells were transferred to a 10 cm dish for growth, addition of antibiotics, and removal of nontransfected cells.

2.7. Stable Knockdown of GAL1 by Using Lentiviruses

We inserted human *LGALS1* (GenBank accession number NM_002305) into the VSV-G pseudotyped lentiviral vectors (Academia Sinica, Taiwan) to silence the expression of GAL1. The two shRNA sequences were as follows:

Gal1sh1 (TRCN0000433733) (5′-CCGGACGGTGACTTCAAGATCAAATCTCGAGATTTGATCTT GAAGTCACCGTTTTTTTG-3′); Gal1sh2 (TRCN0000057425) (5′-CCGGCCTGAATCTCAAACCTGGAG ACTCGAGTCTCCAGGTTTGAGATTCAGGTTTTTG-3′); VSV-G, a pseudotyped lentiviral vector, was constructed to silence the expression of GAL1. Furthermore, a negative control vector containing the cytomegalovirus promoter and expressing high levels of green fluorescent protein was also designed. The negative control was also created using a VSV-G. The lentiviral vectors were transfected into the BFTC-909 cells at a multiplicity of infection ranging from 1 to 10 in the presence of 5 µg/mL polybrene (Sigma-Aldrich, St. Louis, MO, USA).

2.8. Transwell Migration and Invasion Assay

The BFTC-909 cells and knockdown GAL1 cells were seeded into a transwell insert (Neuro Probe, Cabin John, MD, USA) at 1×10^4 cells/well in serum-free media. The cells were incubated at 37 °C for 24 h to allow cell migration. For invasion assay, 20 µL Matrigel (BD Biosciences, MA, USA) was coated onto polycarbonate membrane filters of 8 µm pore-size, and BFTC-909 and knockdown GAL1 cells were plated in the upper chamber of the Matrigel-coated transwell insert. The migrated and invaded cells on the lower chamber were fixed with 100% methanol and stained with 0.1% crystal violet. Cell numbers were counted using a 100× light microscope.

2.9. RNA Isolation and Quantitive PCR

Total RNA was extracted from the cell lines using QIAGEN RNA purification kit. The total RNA (5 µg) was then reverse transcribed using RevertAidTM H Minus Reverse Transcriptase (Fermentas,

Waltham, MA, USA). Real-time PCR was performed using SYBR Green PCR master mix (Life Technologies, Carlsbad, CA, USA) and ABI 7500 sequence detection system (Life Technologies).

Real-time PCR primers used in this study were as follows:

GAL1 forward: 5′- AGCAGCGGGAGGCTGTCTTTC-3′;

GAL1 reverse primer: 5′- ATCCATCTGGCAGCTTGACGGT-3′.

GAPDH forward: 5′-GTCTCCTCTGACTTCAACAGCG-3′;

GAPDH reverse primer: 5′-ACCACCCTGTTGCTGTAGCCAA-3′.

All primers were purchased from OriGene (Rockville, MD, USA) and checked for specificity using BLAST (NCBI). Exon and intron junctions were spanned.

2.10. Gene Expression Microarray Assay

Total RNA extraction from peripheral blood mononuclear cells (PBMCs) was performed using a miRNeasy mini kit following the manufacturer's protocol (Qiagen GmbH, Hilden, Germany). Furthermore, cRNA preparation, sample hybridization, and scanning were performed following the protocols provided by Affymetrix (Affymetrix, Santa Clara, CA, USA) and Cogentech Affymetrix microarray unit (Campus IFOM IEO, Milan, Italy). All samples were hybridized on a Human Clariom D (Thermo Fisher Scientific) gene chip and were analyzed using the Transcriptome Analysis Console 4.0 software (Applied Biosystem, Foster City, CA, USA by Thermo Fisher Scientific, Waltham, MA, USA). Human Clariom D arrays enable investigation of more than 540,000 transcripts sourced from the largest public databases starting from as little as 100 pg of total RNA. Relative gene expression levels of each transcript were validated by applying a one-way analysis of variance ($p \leq 0.01$) and multiple testing corrections. Coding genes and lncRNAs that displayed an expression level at least 1.5-fold different in the test sample versus control sample ($p \leq 0.01$) were carried forward in subsequent analyses.

2.11. Bioinformatics Analysis

We analyzed comprehensive TCGA cancer genome expression data by using UALCAN platform (http://ualcan.path.uab.edu/index.html). UALCAN is a publicly online website which can deeply analyses gene expression, promoter methylation, and correlation across defined clinicopathological features [28]. The mRNA expression level between normal tissue and cancer were analyzed from the Gene Expression Omnibus (GEO) data sets GSE32894 by using ShinyGEO online tool (https://gdancik.github.io/shinyGEO/). ShinyGEO is a web-based platform implemented using R package for building interactive web applications to analyze the difference of gene expression and survival outcome [29].

2.12. Statistical Analysis

IHC staining of tumor specimens was performed using anti-GAL1 antibody. The UTUC cell line (BFTC-909) was used for in vitro study of tumor invasiveness and migration. Kaplan–Meier analyses and Cox proportional regression models were used for univariate and multivariate survival analyses.

3. Results

3.1. LGALS1 mRNA Expression Increased Significantly in Advanced UC

To determine the extent of GAL1 expression in UC and normal tissues, we first examined the *LGALS1* mRNA levels in human bladder cancer tissues from the TCGA cohort and GSE32894 data set and our in-house q-PCR analysis of 55 paired normal tissue and cancer specimens (KSCGMH cohort). We found that the levels of *LGALS1* transcripts were significantly lower in tumor tissues than in the normal urothelium in KSCGMH cohort ($p < 0.001$; Figure 1A); however, this trend was not observed in the TCGA cohort ($p = 0.74$; Figure 1B); the difference in trend was possibly attributable to the small number of normal tissue samples (n = 19) in the TCGA cohort. Furthermore, high levels of *LGALS1* mRNA were expressed at the advanced stage of bladder cancer (stages 3 and 4) and not at the early stage (stages 1 and 2; Figure 1C) in the TCGA cohort. We further examined *LGALS1* expression in bladder cancers of different levels of invasiveness from the GSE32894 data set; *LGALS1* mRNA levels increased significantly in muscle invasion bladder tumors ($p < 0.001$; Figure 1D). Similarly, q-PCR analysis of all stages of bladder cancer revealed that GAL1 expression was higher in muscle invasion tumors than in non-muscle invasion tumors ($p = 0.03$; Figure 1E). Overall, GAL1 expression levels are strongly associated with bladder tumor stage and invasiveness.

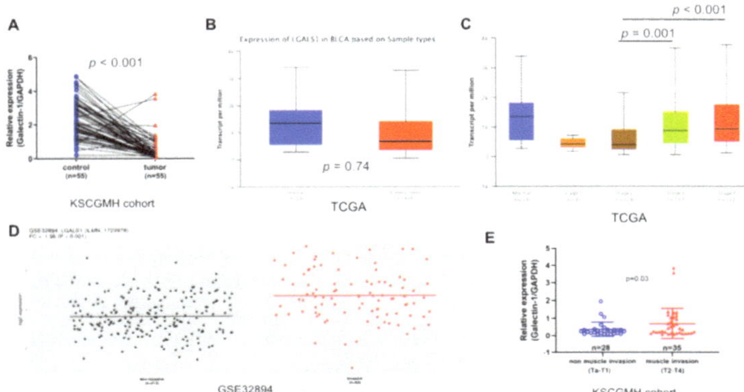

Figure 1. Galectin-1 (GAL1) expression was highly correlated with invasiveness of bladder cancer: (**A**) q-PCR analysis of GAL1 in paired normal bladder and cancer tissues in the Kaohsiung Chang Gung Memorial Hospital (KSCGMH) cohort. (**B**) Analysis of *LGALS1* expression in normal bladder tissues and cancer tissues by using the TCGA bladder cancer cohort (BLCA). (**C**) *LGALS1* expression in all stage of the BLCA cohort. (**D**) Analysis of LGALS1 expression in non-muscle invasion (Ta and T1) and muscle invasion (T2, T3, and T4) tumors by using GSE32894 data sets. (**E**) q-PCR analysis of GAL1 expression stratified by tumor invasiveness in the KSCGMH cohort.

3.2. High Expression of GAL1 is Associated with Poor Disease Recurrence and CSS

To understand the clinical effect of GAL1 expression in UC, we further investigated the differences in survival based on GAL1 expression levels in the TCGA BLCA and GSE32894 data sets. The overall survival of patients with tumors with high GAL1 expression was significantly poorer than that of patients with tumors with low GAL1 expression in the TCGA ($p = 0.01$) and GSE32894 cohorts ($p = 0.0011$; Figure 2A,B). To validate the findings and to explore the clinical importance of GAL1 expression in UTUC, we enrolled 86 patients with pT3 UTUC from the KSCGMH cohort for demographic and immunohistochemical (IHC) analysis. The median age was 71 years (interquartile range (IQR), 64–77). Among the 86 patients, 49 (57%) patients were female and 60% of patients had primary tumor located in the renal pelvis. Representative micrographs of GAL1 immunostaining are shown in Figure 2C–F. The study population was classified into two groups, namely GAL1-high (n = 35) and GAL1-low (n = 51) groups. The basic clinicopathological characteristics were comparable between the two groups (Table 1), and significant intergroup differences were not observed.

Within the median follow-up time of 40.4 months, 33 among 86 patients (38.4%) experienced disease recurrence and 26 (30.2%) patients died of disease. High GAL1 expression was significantly associated with a poor recurrence-free survival (RFS; $p = 0.028$; Figure 2G) and cancer-specific survival (CSS; $p = 0.025$; Figure 2H); hence, RFS and CSS were lower in the GAL-high group than in the GAL1-low group. After adjusting for all possible covariates by using Cox regression hazard models, high expression levels of GAL1 in UTUC tumors was an independent predictive factor for RFS (hazard ratio (HR) 2.43; 95% CI 1.17–5.05, $p = 0.018$) and CSS (HR 4.04; 95% CI 1.25–13.03, $p = 0.019$; Table 2). The other independent predictive factor for RFS and CSS was presence of lymphovascular invasion (LVI) (HR 2.41 for RFS; HR 3.56 for CSS).

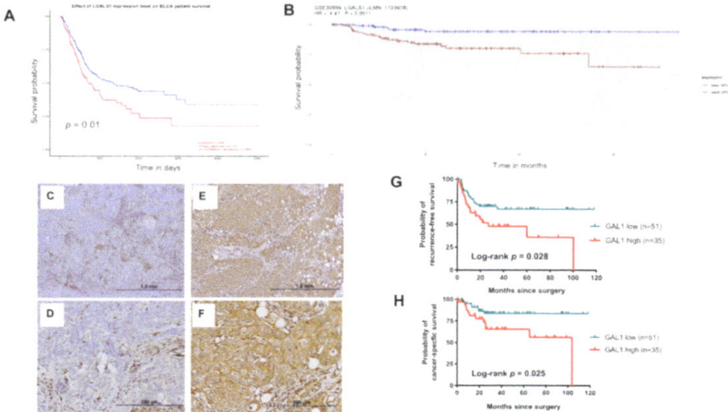

Figure 2. Evaluation of the association between GAL1 expression and oncological outcomes in urothelial carcinoma (UC): High expression of GAL1 reduced overall survival in patients with bladder cancer from the TCGA (**A**) and GSE32894 (**B**) cohorts. Representative immunohistochemical (IHC) images of low GAL1 (**C,D**) and high GAL1 (**E,F**) expression levels. Kaplan–Meier analysis of 86 patients with pT3 upper tract urothelial carcinoma (UTUC) for determining recurrence-free survival (RFS) (**G**) and cancer-specific survival (CSS) (**H**).

Table 1. Clinicopathological features of 86 patients with pT3 upper tract UC stratified by GAL1 expression.

	GAL1-Low	GAL1-High	*p*-Value
	(N = 51, %)	(N = 35, %)	
Age (years)	70.5 ± 10.8	68.0 ± 9.8	0.27
Gender			
Female	31 (60.8)	18 (51.4)	0.39
Male	20 (39.2)	17 (48.6)	
Smoking			
Yes	8 (15.7)	4 (11.4)	0.75
No	43 (84.3)	31 (88.6)	
Primary site			
Renal pelvis	26 (51.0)	26 (74.3)	0.09
Ureter	19 (37.3)	7 (20.1)	
Both	6 (11.8)	2 (5.7)	
Grade			
Low	0 (0)	1 (2.9)	0.41
High	51 (100)	34 (97.1)	
Histopathologic variant			
Presence	20 (39.2)	14 (40)	0.94
Absence	31 (60.8)	21 (60)	
CIS			
Presence	17 (33.3)	13 (37.1)	0.72
Absence	34 (66.7)	22 (62.9)	
LVI			
Presence	20 (39.2)	14 (40)	0.94
Absence	31 (60.8)	21 (60)	

Table 1. *Cont.*

	GAL1-Low	GAL1-High	*p*-Value
	(N = 51, %)	(N = 35, %)	
Tumor necrosis			
Presence	18 (35.3)	16 (45.7)	0.33
Absence	33 (64.7)	19 (54.3)	
Multicentricity			
Yes	12 (23.5)	5 (14.3)	0.29
No	39 (76.5)	30 (85.7)	
Papillary feature			
Presence	31 (60.8)	15 (42.9)	0.10
Absence	20 (39.2)	20 (57.1)	

Abbreviations: CIS, carcinoma in situ; LVI, lymphovascular invasion.

Table 2. Univariate and multivariate analyses of RFS and CSS.

	RFS			CSS		
Characteristics	Univariate	Multivariate		Univariate	Multivariate	
	p-Value	HR (95% CI)	*p*-Value	*p*-Value	HR (95% CI)	*p*-Value
Age						
≥65 vs. <65	0.99			0.67		
Gender						
Male vs. female	0.17			0.41		
Smoking history						
Yes vs. no	0.16			0.08		
Primary site						
Kidney vs. ureter	0.24	2.03 (0.96–4.28)	0.06	0.23		
Histological variant						
Presence vs. absence	0.74			0.26	3.65 (1.20–11.12)	0.023
LVI						
Presence vs. absence	0.043	2.41 (1.15–5.05)	0.02	0.18	3.56 (1.06–11.88)	0.039
Tumor necrosis						
Presence vs. absence	0.25			0.35		
CIS						
Presence vs. absence	0.97			0.96		
Papillary feature						
Presence vs. absence	0.039			0.62		
Multicentricity						
Yes vs. no	0.74			0.82		
Galectin-1						
High vs. low	0.028	2.43 (1.17–5.05)	0.018	0.025	4.04 (1.25–13.03)	0.019

Abbreviations: CI, confidence interval; CIS, carcinoma in situ; HR, hazard ratio; CSS, cancer-specific survival; LVI, lymphovascular invasion; RFS, recurrence-free survival.

3.3. Downregulation of GAL1 in UC Cell Lines

To determine the biological role of GAL1 in UC, we evaluated the expression levels of GAL1 in four UC cell lines (BFTC-909, T24, J82, and RT4) through Western blot and q-PCR analyses. As indicated in Figure 3A,B, the mRNA expression levels of GAL1 were highly upregulated in the BFTC-909, T24, and J82 cells and were positively correlated with high levels of GAL1 protein expression. The RT4 cell line exhibited low expression levels of GAL1 protein and mRNA. We selected the BFTC-909 and T24 cell lines, which exhibited the highest expression levels of GAL1, to construct GAL1 knockdown cell lines (sh-Gal1). As shown in Figure 3C,D, the levels of GAL1 protein and mRNA were significantly reduced in the sh-GAL1-T24 and sh-GAL1-BFTC-909 cells, indicating the efficiency of silencing

LGALS1 gene function. Furthermore, we established stable knockdown cell lines through the lentivirus transfection assay. As shown in Figure 3E–H, the amount of GAL1 protein and mRNA were significantly downregulated in the sh-GAL1-RNAi-A and sh-GAL1-RNAi-B BFTC-909 and T24 cells compared with shLuc and Mock. The knockdown efficiency of RNAi-A was higher than that of RNAi-B.

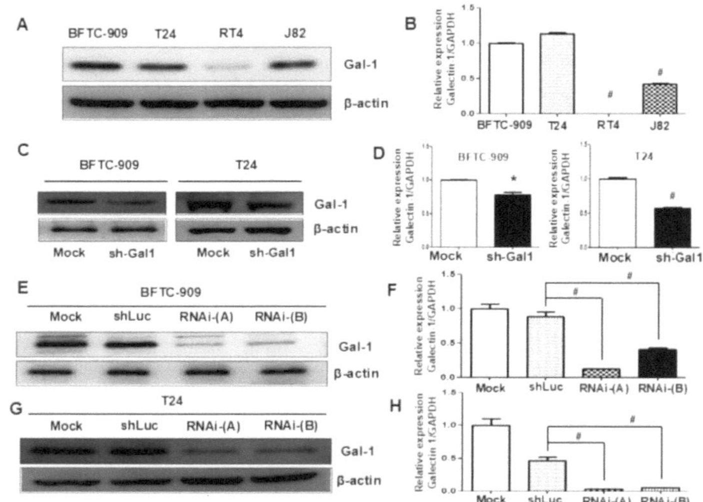

Figure 3. Evaluation of GAL1 expression in UC cell lines and verification of downregulation efficiency through Western blotting and q-PCR analyses: Data are presented as mean ± standard error of the mean from three independent experiments. (**A,B**) High expression of GAL1 in the BFTC-909 and T24 cell lines and low GAL1 expression in the J82 and T24 cell lines were observed. (**C,D**) The efficiency of GAL1 knockdown by using shRNA in the BFTC-909 and T24 cell lines was assessed through Western blotting and q-PCR analyses. (**E,F**) The efficiency of GAL1 knockdown by using lentiviruses was assessed through Western blotting and q-PCR analyses in the BFTC-909 cell line. (**G,H**) The efficiency of GAL1 knockdown by using lentiviruses was assessed through Western blotting and q-PCR analyses in the T24 cell line. * $p < 0.05$, # $p < 0.001$.

3.4. GAL1 Expression Increased Tumor Invasiveness and Migration

Next, we investigated tumor invasiveness of the GAL1 knockdown cells produced using lentivirus through transwell migration and invasion analysis. The results showed that lentivirus knockdown of GAL1 by using either RNAi-A or RNAi-B significantly reduced migration and invasion in the BFTC-909 and T24 cells (Figure 4A,B). Subsequently, we added GAL1 recombinant protein to the J82 cells for 24 h to verify the difference in migration and invasion assessed using the transwell method. The results showed the GAL1 recombinant protein (2.5–3 µg/mL) exhibited cytotoxicity (Figure 4C). Therefore, we selected 1 or 2 µg/mL as the concentration of GAL1 recombinant protein for subsequent analysis. The results showed that, after treatment with GAL1, recombinant protein, migration, and invasion increased significantly in the J82 cells (Figure 4D).

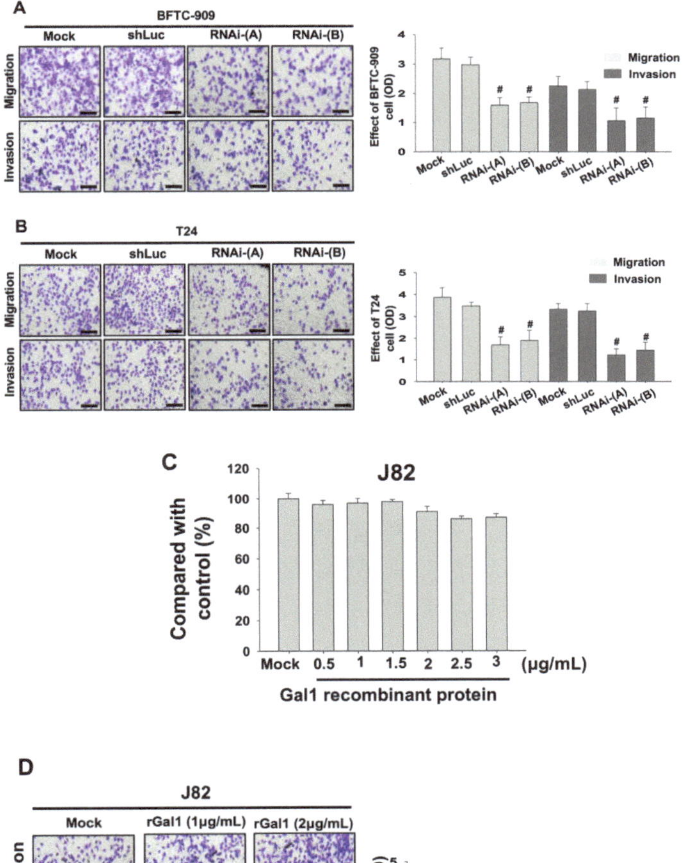

Figure 4. Downregulation of GAL1 affected cancerous behavior in the BFTC-909 and T24 cells and effects of different concentrations of GAL1 recombinant protein (0–3 µg/mL) on the J82 cells. GAL1 knockdown cells produced using lentiviruses exhibited a reduction in migration and invasiveness in the BFTC-909 (**A**) and T24 cells (**B**). (**C**) Viability of the J82 cells at different concentrations of GAL1 recombinant protein (0–3 µg/mL) after incubation for 24 h. (**D**) After 24 h of treatment with GAL1 recombinant protein (1–2 µg/mL), the percentages of migration of and invasion by the J82 cells significantly increased compared with the controls (Mock: cells treated with vehicle DMSO). Scale bar = 20 µm. * $p < 0.05$, # $p < 0.001$.

3.5. GAL1-Mediated Epithelial–Mesenchymal Transition in UC

MMP-2 and MMP-9 are well-known extracellular matrix (ECM)-degrading enzymes that have been reported to play crucial roles in cancer cell metastasis and invasion. We used Western blot analysis to investigate the effect of GAL1 knockdown or overexpression on associated protein levels of migration and invasion. The results showed that knockdown of GAL1 expression in the BFTC-909 and T24 cells significantly reduced the protein levels of MMP-2 and MMP-9 and increased TIMP-1 protein expression (Figure 5A), whereas increasing GAL1 expression in J82 cells by adding recombinant GAL1 protein reversed the phenomenon. EMT is one of the crucial mechanisms by which tumor cells detach from their primary site and invade surrounding tissues as well as the vascular system. We selected six representative molecules and assessed their expression levels in the BFTC-909 and T24 cells after GAL1 knockdown or overexpression. The results showed downregulation of N-cadherin, vimentin, β-catenin, and snail and upregulation of E-cadherin and ZO-1 after knockdown GAL1 in the BFTC-909 and T24 cells. The results of the J82 cells were contrary to the aforementioned results after the addition of GAL1 recombinant protein (Figure 5B).

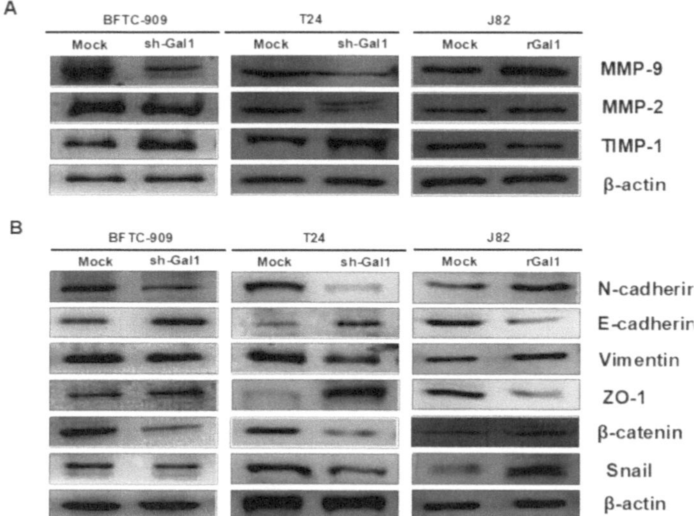

Figure 5. Effect of GAL1 on MMPs and EMT pathway proteins assessed through Western blotting analysis. Mock: cells treated with DMSO vehicle only. β-Actin was used as the protein loading control. (**A**) Total cell lysates of the BFTC-909, T24, and J82 cells treated with shRNA or GAL1 recombinant protein (1 μg/mL) were analyzed in terms of expression levels of MMP-2, MMP-9, and TIMP-1 by Western blotting. (**B**) Total cell lysates of the BFTC-909, T24, and J82 cells treated with shRNA or GAL1 recombinant proteins (1μg/mL) were analyzed in terms of expression levels of EMT pathway through Western blotting analysis.

3.6. Gene Expression Variations Among Samples

To determine GAL1-associated gene alteration and downstream signal transduction pathways, we performed a high-throughput analysis by using Clariom D microarray to analyze two replicate samples (GAL1 vector and GAL1-KD). GAL1-mediated upregulation and downregulation of genes was depicted in a heatmap (Figure 6A), and all differentially expressed genes between the samples GAL1 vector and GAL-KD with at least fold change ≥2 and a nominal significance level of 0.01 are shown in Supplementary File S1 (Table S1). Subsequent KEGG pathway analysis was used to identify the top five gene enrichment pathways including p53, FOXO, cell cycle, PI3K/AKT, and ECM receptor signaling pathways (Figure 6B). We focused on gene alteration in the PI3K/AKT pathway; the results showed that *THBS1, CCND1, VEGFA, IL7R, MYC, PP2R2A, COL6A3,* and *CCNE2* were upregulated in GAL1-KD whereas *SPP1, RRAGD, PRKAA2,* and *IL7* were downregulated (Figure 6C).

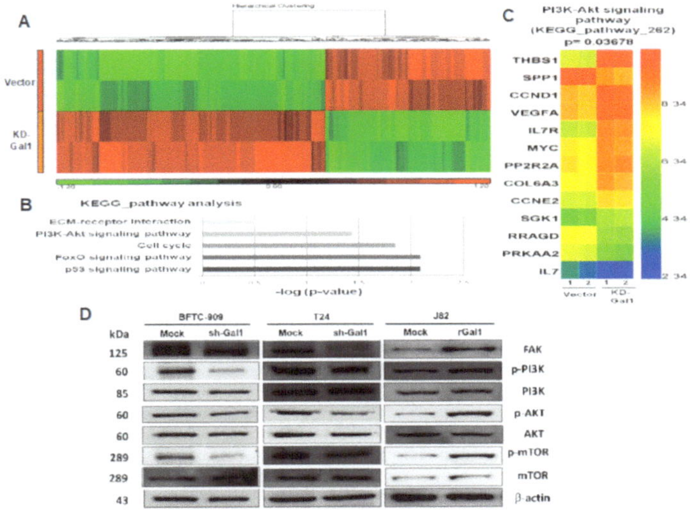

Figure 6. GAL1-mediated PI3K/AKT/mTOR pathway alteration assessed through microarray and Western blotting analyses: (**A**) A heatmap revealed upregulation and downregulation of genes with at least 2-fold difference between the GAL1-vector and GAL1-KD cells. (**B**) Top five gene enrichment pathways in KEGG pathway analysis. (**C**) Heatmap of the 13 altered genes in the PI3K/AKT signaling pathway in the BFTC-909 cells (GAL-vector and GAL1-KD). (**D**) Western blotting of GAL1 knockdown or overexpression in the FAK/PI3K/AKT/mTOR signaling pathways in the BFTC-909, T24, and J82 cells. Mock: Cells were treated with DMSO vehicle only. β-Actin was used as the protein-loading control.

3.7. Downregulation of GAL1 Suppressed FAK and Phosphorylated PI3K-AKT-mTOR

We further investigated the effects of GAL1 on the FAK/PI3K/AKT/mTOR pathways in vivo. The results showed that, in the GAL1 knockdown BFTC-909 and T24 cells, the phosphorylation of FAK, PI3K, AKT, and mTOR decreased. Moreover, in the J82 cells with GAL1 overexpression, the phosphorylation of FAK, PI3K, AKT, and mTOR increased. The protein expression of PI3K, AKT, and mTOR did not change after GAL1 knockdown or overexpression (Figure 6D).

4. Discussion

For treating patients diagnosed with pathological stage T3 UTUC, a consensus is not currently available for guiding clinicians to provide adjuvant therapy to prevent disease recurrence. Although several clinicopathological factors, such as LVI and tumor growth pattern, significantly affect prediction of disease recurrence, more precise and reliable biological markers are required to decipher mechanisms underlying UTUC progression and to guide medical practitioners. In the present study, GAL1 protein expression was highly associated with RFS and CSS in patients with UTUC. To our best knowledge, this is the first report to evaluate the prognostic value of GAL1 expression in patients with UTUC. We also demonstrated that downregulated GAL1 reduced tumor migration and invasion whereas addition of recombinant GAL1 protein resulted in increased malignant behavior in the J82 cells. Furthermore, we found that GAL1 mediated an increase in EMT, increased in MMP2/MMP9 activity, and altered the FAK/PI3K/AKT/mTOR pathway.

GAL1 is a homodimeric, β-galactoside-binding protein composed of 14.5 kDa subunits. GAL1 is abundantly expressed in various types of malignant tumors, including those in colorectal cancer [16], breast cancer [17], lung cancer [18], cancer of the uterine cervix [19], Hodgkin lymphoma [20], melanoma [30], ovarian cancer [31], and glioblastoma multiforme [32]. Regarding the clinical role of GAL1 expression in bladder cancer, Wu et al. demonstrated that overexpression of GAL1 in bladder tumors was significantly associated with higher pathological T grade and nodal stage as well as an increased risk of disease recurrence [25]. In the present study, we observed that 40% of pT3 UTUC tumors exhibited high expression of GAL1 (cutoff level was 10%). The clinicopathological features of UTUC with high or low GAL1 expression were not significantly different from each other. We further evaluated the prognostic role of GAL1 expression in UTUC patients through Kaplan–Meier analysis and by using Cox regression model. We found that increased expression of GAL1 protein by the IHC method was significantly associated with relapse or recurrence survival and CSS in both univariate and multivariate analyses, thus suggesting that GAL1 was an independent prognostic factor in determining oncological outcomes. Given that we selected patients with UTUC at the same pathological stage (pT3), the effect of the confounding factor of T stage was diminished, which reinforced the strength of the prognostic role of GAL1 in patients with pT3 stage UTUC.

In humans, GAL1 is involved in many biological processes including cell adhesion, cell proliferation, invasion, migration, tumor angiogenesis, and immune escape [12–15]. Shen et al. elucidated the underlying molecular pathways of GAL1-mediated tumorigenesis in bladder cancer, mainly through the RAS-Rac1-MEKK4-JNK-AP1 pathway [26]. By using a comprehensive proteomic approach to investigate GAL1-regulated proteins, Li et al. showed that deregulated proteins are involved in several biological pathways, including lipid, amino acid, and energy metabolism; cell proliferation and apoptosis; cytoskeleton functions; cell–cell interaction; metastasis; and protein degradation [33]. Not only were several key proteins, such as fatty acid-binding protein 4 (FABP4), glutamine synthetase, toll interacting protein, and alcohol dehydrogenase NADP+ (AKR1A1), validated functionally by using immunoblotting methods but also their prognostic values were confirmed in a cohort study [33]. In the present study, we confirmed the results of the study by Shen et al., and our results showed that GAL1 expression was associated with tumor invasiveness and migration ability in UTUC (BFTC-909) and UCB (T24). We could successfully knockdown GAL1 expression in the BFTC-909 and T24 cell lines by using the shRNA and lentivirus methods, resulting in significant inhibition of tumor invasion and migration. By contrast, by adding recombinant GAL1 protein to cells with relatively low GAL1 expression (J82), tumor invasiveness and migration ability increased significantly, which indicated that GAL1 is a crucial factor for tumor aggressiveness and malignant behavior.

The main intracellular pathway for GAL1 is through protein–protein interaction with H-RAS and activation of downstream MEK/ERK signaling [34]. GAL1 binds to Gemin4, which mediates the biogenesis of microRNA ribonucleoprotein and regulates pre-RNA splicing modulation [35]. However, the underlying signaling transduction pathway of GAL1-mediated UTUC carcinogenesis remains unknown. Through gene microarray analysis of parental and knockdown GAL1 cells, multiple crucial signaling pathways are involved in alteration of GAL1 expression, including the p53, FOXO, cell cycle, and PI3K/AKT pathways. We further validated microarray results through immunoblotting analysis, which showed that GAL1 regulated downstream FAK/PI3K/AKT/mTOR protein expression. In the previous study, Zhang et al. also showed that GAL1 induced hepatocellular carcinoma metastasis and resistance to sorafenib by upregulation of αvβ3-integrin and activated the PI3K/AKT pathway [36]. Our findings demonstrated comparable and consistent results, which indicated that GAL1-mediated tumor invasion and metastasis occurs through the FAK/PI3K/AKT/mTOR pathway.

5. Conclusions

In brief, our study implied that the GAL1 protein is highly associated with oncological outcomes of UTUC by promoting tumor invasion, metastasis, and epithelial–mesenchymal transition, possibly through the FAK/PI3K/AKT/mTOR pathway (Figure 7).

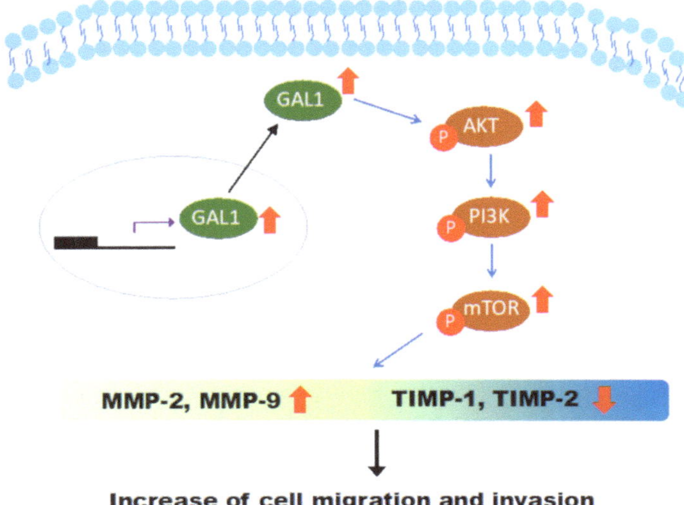

Figure 7. Representative Figure of GAL1-induced cancerous behavior in UTUC cells: Based on the results of our study, the cancerous behavior caused by GAL1 is mediated by the PI3K/AKT/mTOR pathway.

Supplementary Materials: The following are available online at http://www.mdpi.com/2073-4409/9/4/806/s1. Supplementary File 1: Table S1. All differentially expressed genes (DEGs) between GAL1-vector and GAL1-KD UC cells.

Author Contributions: Conceptualization and design of the study, Y.-L.S., J.-J.L., and H.-L.L. Clinical data curation, C.-H.K., P.-H.C., and Y.-T.C. IHC scoring and interpretation, T.-T.L. and M.-T.S. In vitro cell culture, transfection, and transwell procedure, J.-J.L., E.-Y.H., and C.-C.H. Statistical analysis and validation, J.-J.L. and Y.-L.S. Manuscript writing, Y.-L.S., J.-J.L., and H.-L.L. All authors have read and agreed to the published version of the manuscript.

Funding: The study was supported partly by a grant from Chang Gung Memorial Hospital, Kaohsiung, Taiwan (CMRPG8F0581, CMRPG8F1431, and CMRPG8G1431).

Acknowledgments: We thank the multidisciplinary team of the genitourinary cancer service at our hospital for their generous assistance and cooperation. This manuscript was edited by Wallace Academic Editing.

Conflicts of Interest: The authors declare no conflict of interest.

Abbreviations

CIS:	carcinoma in situ;
CSS:	cancer-specific survival;
ECM:	extracellular matrix;
EMT:	epithelial-mesenchymal transition;
FAK:	focal adhesion kinase;
GAL1:	galectin-1;
HR:	hazard ratio;
IHC:	immunohistochemical;
IQR:	interquartile range;
LVI:	lymphovascular invasion;
MMP:	matrix metalloproteinase;
mTOR:	mammalian target of rapamycin;

PI3K: phosphoinositide 3-kinase;
q-PCR: quantitative polymerase chain reaction;
RFS: recurrence-free survival;
TCGA: The Cancer Genome Atlas;
TIMP: tissue inhibitors of metalloproteinases;
UC: urothelial carcinoma;
UCB: urothelial carcinoma of the bladder;
UTUC: upper tract urothelial carcinoma.

References

1. Siegel, R.; Naishadham, D.; Jemal, A. Cancer statistics. *Cancer J. Clin.* **2013**, *63*, 6311–6330.
2. Munoz, J.J.; Ellison, L.M. Upper tract urothelial neoplasms: Incidence and survival during the last 2 decades. *J. Urol.* **2000**, *164*, 1523–1525. [CrossRef]
3. Raman, J.D.; Messer, J.; Sielatycki, J.A.; Hollenbeak, C.S. Incidence and survival of patients with carcinoma of the ureter and renal pelvis in the USA 1973–2005. *BJU Int.* **2011**, *107*, 1059–1064. [CrossRef] [PubMed]
4. Li, C.C.; Chang, T.H.; Wu, W.J.; Ke, H.L.; Huang, S.P.; Tsai, P.C.; Chang, S.J.; Shen, J.T.; Chou, Y.H.; Huang, C.H. Significant predictive factors for prognosis of primary upper urinary tract cancer after radical nephroureterectomy in Taiwanese patients. *Eur. Urol.* **2008**, *54*, 1127–1134. [CrossRef] [PubMed]
5. Chou, Y.H.; Huang, C.H. Unusual clinical presentation of upper urothelial carcinoma in Taiwan. *Cancer* **1999**, *85*, 1342–1344. [CrossRef]
6. Cukuranovic, R.; Ignjatovic, I.; Visnjic, M.; Velickovic, L.J.; Petrovic, B.; Potic, M.; Stefanovic, V. Characteristics of upper urothelial carcinoma in an area of Balkan endemic nephropathy in south Serbia. A fifty-year retrospective study. *Tumori* **2010**, *96*, 674–679. [CrossRef]
7. Poon, S.L.; Pang, S.T.; McPherson, J.R.; Yu, W.; Huang, K.K.; Guan, P.; Weng, W.H.; Siew, E.Y.; Liu, Y.; Heng, H.L.; et al. Genome-wide mutational signatures of aristolochic acid and its application as a screening tool. *Sci. Transl. Med.* **2013**, *5*, 197ra101. [CrossRef]
8. Roupret, M.; Yates, D.R.; Comperat, E.; Cussenot, O. Upper urinary tract urothelial cell carcinomas and other urological malignancies involved in the hereditary nonpolyposis colorectal cancer (lynch syndrome) tumor spectrum. *Eur. Urol.* **2008**, *54*, 1226–1236. [CrossRef]
9. Roupret, M.; Babjuk, M.; Comperat, E.; Zigeuner, R.; Sylvester, R.J.; Burger, M.; Cowan, N.C.; Bohle, A.; Van Rhijn, B.W.; Kaasinen, E.; et al. European Association of Urology Guidelines on Upper Urinary Tract Urothelial Cell Carcinoma: 2015 Update. *Eur. Urol.* **2015**, *68*, 868–879. [CrossRef]
10. Patel, N.; Arya, M.; Muneer, A.; Powles, T.; Sullivan, M.; Hines, J.; Kelly, J. Molecular aspects of upper tract urothelial carcinoma. *Urol. Oncol.* **2014**, *32*, e11–e20. [CrossRef]
11. Camby, I.; Le Mercier, M.; Lefranc, F.; Kiss, R. Galectin-1: A small protein with major functions. *Glycobiology* **2006**, *16*, 137R–157R. [CrossRef]
12. Adams, L.; Scott, G.K.; Weinberg, C.S. Biphasic modulation of cell growth by recombinant human galectin-1. *Biochim. Biophys. Acta.* **1996**, *1312*, 137–144. [CrossRef]
13. Wells, V.; Mallucci, L. Identification of an autocrine negative growth factor: Mouse beta-galactoside-binding protein is a cytostatic factor and cell growth regulator. *Cell* **1991**, *64*, 91–97. [CrossRef]
14. Thijssen, V.L.; Postel, R.; Brandwijk, R.J.; Dings, R.P.; Nesmelova, I.; Satijn, S.; Verhofstad, N.; Nakabeppu, Y.; Baum, L.G.; Bakkers, J.; et al. Galectin-1 is essential in tumor angiogenesis and is a target for antiangiogenesis therapy. *Proc. Natl. Acad. Sci. USA* **2006**, *103*, 15975–15980. [CrossRef] [PubMed]
15. Liu, F.T.; Rabinovich, G.A. Galectins as modulators of tumour progression. *Nat. Rev. Cancer* **2005**, *5*, 29–41. [CrossRef] [PubMed]
16. Barrow, H.; Rhodes, J.M.; Yu, L.G. The role of galectins in colorectal cancer progression. *Int. J. Cancer* **2011**, *129*, 1–8. [CrossRef]
17. Dalotto-Moreno, T.; Croci, D.O.; Cerliani, J.P.; Martinez-Allo, V.C.; Dergan-Dylon, S.; Mendez-Huergo, S.P.; Stupirski, J.C.; Mazal, D.; Osinaga, E.; Toscano, M.A.; et al. Targeting galectin-1 overcomes breast cancer-associated immunosuppression and prevents metastatic disease. *Cancer Res.* **2013**, *73*, 1107–1117. [CrossRef]

18. Szoke, T.; Kayser, K.; Baumhakel, J.D.; Trojan, I.; Furak, J.; Tiszlavicz, L.; Horvath, A.; Szluha, K.; Gabius, H.J.; Andre, S. Prognostic significance of endogenous adhesion/growth-regulatory lectins in lung cancer. *Oncol.* **2005**, *69*, 167–174. [CrossRef]

19. Huang, E.Y.; Chanchien, C.C.; Lin, H.; Wang, C.C.; Wang, C.J.; Huang, C.C. Galectin-1 is an independent prognostic factor for local recurrence and survival after definitive radiation therapy for patients with squamous cell carcinoma of the uterine cervix. *Int. J. Radiat. Oncol. Biol. Phys.* **2013**, *87*, 975–982. [CrossRef]

20. Ouyang, J.; Plutschow, A.; Pogge von Strandmann, E.; Reiners, K.S.; Ponader, S.; Rabinovich, G.A.; Neuberg, D.; Engert, A.; Shipp, M.A. Galectin-1 serum levels reflect tumor burden and adverse clinical features in classical Hodgkin lymphoma. *Blood* **2013**, *121*, 3431–3433. [CrossRef]

21. Chen, J.; Zhou, S.J.; Zhang, Y.; Zhang, G.Q.; Zha, T.Z.; Feng, Y.Z.; Zhang, K. Clinicopathological and prognostic significance of galectin-1 and vascular endothelial growth factor expression in gastric cancer. *World J. Gastroenterol.* **2013**, *19*, 2073–2079. [CrossRef] [PubMed]

22. Kim, H.J.; Do, I.G.; Jeon, H.K.; Cho, Y.J.; Park, Y.A.; Choi, J.J.; Sung, C.O.; Lee, Y.Y.; Choi, C.H.; Kim, T.J.; et al. Galectin 1 expression is associated with tumor invasion and metastasis in stage IB to IIA cervical cancer. *Hum. Pathol.* **2013**, *44*, 62–68. [CrossRef]

23. Schulkens, I.A.; Heusschen, R.; Van den Boogaart, V.; van Suylen, R.J.; Dingemans, A.M.; Griffioen, A.W.; Thijssen, V.L. Galectin expression profiling identifies galectin-1 and Galectin-9Delta5 as prognostic factors in stage I/II non-small cell lung cancer. *PLoS ONE* **2014**, *9*, e107988. [CrossRef]

24. Punt, S.; Thijssen, V.L.; Vrolijk, J.; de Kroon, C.D.; Gorter, A.; Jordanova, E.S. Galectin-1, -3 and -9 Expression and Clinical Significance in Squamous Cervical Cancer. *PLoS ONE* **2015**, *10*, e0129119. [CrossRef] [PubMed]

25. Wu, T.F.; Li, C.F.; Chien, L.H.; Shen, K.H.; Huang, H.Y.; Su, C.C.; Liao, A.C. Galectin-1 dysregulation independently predicts disease specific survival in bladder urothelial carcinoma. *J. Urol.* **2015**, *193*, 1002–1008. [CrossRef] [PubMed]

26. Shen, K.H.; Li, C.F.; Chien, L.H.; Huang, C.H.; Su, C.C.; Liao, A.C.; Wu, T.F. Role of galectin-1 in urinary bladder urothelial carcinoma cell invasion through the JNK pathway. *Cancer Sci.* **2016**, *107*, 1390–1398. [CrossRef]

27. Chung, L.Y.; Tang, S.J.; Sun, G.H.; Chou, T.Y.; Yeh, T.S.; Yu, S.L.; Sun, K.H. Galectin-1 promotes lung cancer progression and chemoresistance by upregulating p38 MAPK, ERK, and cyclooxygenase-2. *Clin. Cancer Res.* **2012**, *18*, 4037–4047. [CrossRef]

28. Chandrashekar, D.S.; Bashel, B.; Balasubramanya, S.A.H.; Creighton, C.J.; Rodriguez, I.P.; Chakravarthi, B.V.S.K.; Varambally, S. UALCAN: A portal for facilitating tumor subgroup gene expression and survival analyses. *Neoplasia* **2017**, *19*, 649–658. [CrossRef]

29. Dumas, J.; Gargano, M.A.; Dancik, G.M.; Shiny, G.E.O. A web-based application for analyzing gene expression omnibus datasets. *Bioinformatics* **2016**, *32*, 3679–3681. [CrossRef]

30. Tinari, N.; Kuwabara, I.; Huflejt, M.E.; Shen, P.F.; Iacobelli, S.; Liu, F.T. Glycoprotein 90K/MAC-2BP interacts with galectin-1 and mediates galectin-1-induced cell aggregation. *Int. J. Cancer* **2001**, *91*, 167–172. [CrossRef]

31. Zhang, P.; Zhang, P.; Shi, B.; Zhou, M.; Jiang, H.; Zhang, H.; Pan, X.; Gao, H.; Sun, H.; Li, Z. Galectin-1 overexpression promotes progression and chemoresistance to cisplatin in epithelial ovarian cancer. *Cell Death Dis.* **2014**, *5*, e991. [CrossRef] [PubMed]

32. Chou, S.Y.; Yen, S.L.; Huang, C.C.; Huang, E.Y. Galectin-1 is a poor prognostic factor in patients with glioblastoma multiforme after radiotherapy. *BMC Cancer* **2018**, *18*, 105. [CrossRef] [PubMed]

33. Li, C.F.; Shen, K.H.; Chien, L.H.; Huang, C.H.; Wu, T.F.; He, H.L. Proteomic Identification of the Galectin-1-Involved Molecular Pathways in Urinary Bladder Urothelial Carcinoma. *Int. J. Mol. Sci.* **2018**, *19*, E1242. [CrossRef] [PubMed]

34. Paz, A.; Haklai, R.; Elad-Sfadia, G.; Ballan, E.; Kloogm, Y. Galectin-1 binds oncogenic H-Ras to mediate Ras membrane anchorage and cell transformation. *Oncogene* **2001**, *20*, 7486–7493. [CrossRef]

35. Vyakarnam, A.; Dagher, S.F.; Wang, J.L.; Patterson, R.J. Evidence for a role for galectin-1 in pre-mRNA splicing. *Mol. Cell Biol.* **1997**, *17*, 4730–4737. [CrossRef]

36. Zhang, P.F.; Li, K.S.; Shen, Y.H.; Gao, P.T.; Dong, Z.R.; Cai, J.B.; Zhang, C.; Huang, X.Y.; Tian, M.X.; Hu, Z.Q.; et al. Galectin-1 induces hepatocellular carcinoma EMT and sorafenib resistance by activating FAK/PI3K/AKT signaling. *Cell Death Dis.* **2016**, *7*, e2201. [CrossRef]

*cells*

Article

Elevated PDK1 Expression Drives PI3K/AKT/MTOR Signaling Promotes Radiation-Resistant and Dedifferentiated Phenotype of Hepatocellular Carcinoma

Oluwaseun Adebayo Bamodu [1,2,†] , Hang-Lung Chang [3,4,†], Jiann-Ruey Ong [5,6], Wei-Hwa Lee [7] , Chi-Tai Yeh [1,2,8] and Jo-Ting Tsai [9,10,11,*]

[1] Department of Hematology and Oncology, Cancer Center, Taipei Medical University-Shuang Ho Hospital, New Taipei City 235, Taiwan; dr_bamodu@yahoo.com (O.A.B.); ctyeh@s.tmu.edu.tw (C.-T.Y.)
[2] Department of Medical Research and Education, Taipei Medical University-Shuang Ho Hospital, New Taipei City 235, Taiwan
[3] Department of General Surgery, En Chu Kong Hospital, New Taipei City 237, Taiwan; changhl0321@gmail.com
[4] Department of Health Care Management, Yuanpei University of Medical Technology, Hsinchu 300, Taiwan
[5] Department of Emergency Medicine, School of Medicine, Taipei Medical University, Taipei City 110, Taiwan; 12642@s.tmu.edu.tw
[6] Department of Emergency Medicine, Taipei Medical University-Shuang Ho Hospital, New Taipei City 235, Taiwan
[7] Department of Pathology, Taipei Medical University-Shuang Ho Hospital, New Taipei City 235, Taiwan; whlpath97616@s.tmu.edu.tw
[8] Department of Medical Laboratory Science and Biotechnology, Yuanpei University of Medical Technology, Hsinchu City 300, Taiwan
[9] Department of Radiology, School of Medicine, College of Medicine, Taipei Medical University, Taipei City 110, Taiwan
[10] Department of Radiology, Taipei Medical University-Shuang Ho Hospital, New Taipei City 235, Taiwan
[11] Graduate Institute of Clinical Medicine, College of Medicine, Taipei Medical University, Taipei City 110, Taiwan
* Correspondence: 10576@s.tmu.edu.tw
† These authors have contributed equally to this work.

Received: 21 January 2020; Accepted: 13 March 2020; Published: 18 March 2020

Abstract: Resistance to radiotherapy (IR), with consequent disease recurrence, continues to limit the efficacy of contemporary anticancer treatment for patients with hepatocellular carcinoma (HCC), especially in late stage. Despite accruing evidence implicating the PI3K/AKT signaling pathway in cancer-promoting hypoxia, cancerous cell proliferation and radiotherapy-resistance, it remains unclear which molecular constituent of the pathway facilitates adaptation of aggressive HCC cells to tumoral stress signals and drives their evasion of repeated IR-toxicity. This present study investigated the role of PDK1 signaling in IR-resistance, enhanced DNA damage repair and post-IR relapse, characteristic of aggressive HCC cells, while exploring potential PDK1-targetability to improve radiosensitivity. The study employed bioinformatics analyses of gene expression profile and functional protein–protein interaction, generation of IR-resistant clones, flow cytometry-based ALDH activity and side-population (SP) characterization, siRNA-mediated loss-of-PDK1function, western-blotting, immunohistochemistry and functional assays including cell viability, migration, invasion, clonogenicity and tumorsphere formation assays. We showed that the aberrantly expressed PDK1 characterizes poorly differentiated HCC CVCL_7955, Mahlavu, SK-HEP1 and Hep3B cells, compared to the well-differentiated Huh7 or normal adult liver epithelial THLE-2 cells, and independently activates the PI3K/AKT/mTOR signaling. Molecular ablation of PDK1 function enhanced susceptibility of HCC cells to IR and was associated with deactivated PI3K/AKT/mTOR signaling. Additionally, PDK1-driven IR-resistance positively correlated with activated PI3K signaling, enhanced HCC cell motility and invasiveness, augmented EMT, upregulated stemness markers ALDH1A1, PROM1, SOX2, KLF4 and POU5F1, increased tumorsphere-formation efficiency and suppressed biomarkers of DNA damage—RAD50, MSH3, MLH3 and ERCC2. Furthermore, the acquired IR-resistant phenotype of Huh7 cells was strongly associated with significantly increased ALDH activity, SP-enrichment, and direct ALDH1-PDK1 interaction. Moreover, BX795-mediated pharmacological inhibition of PDK1 synergistically enhances the radiosensitivity of erstwhile resistant cells, increased Bax/Bcl-2 apoptotic ratio, while suppressing oncogenicity and clonogenicity. We provide preclinical evidence implicating PDK1 as an active driver of IR-resistance by activation of the PI3K/AKT/mTOR signaling, up-modulation of cancer stemness signaling and suppression of DNA damage, thus, projecting PDK1-targeting as a putative enhancer of radiosensitivity and a potential new therapeutic approach for patients with IR-resistant HCC.

Keywords: hepatocellular cancer; HCC; LIHC; PDK1; PI3K/AKT/mTOR pathway; BX795; selective inhibitor; radiotherapy; radioresistance; combination therapy; stemness; DNA damage

---

1. Introduction

Liver cancer with 841,080 new cases and 781,631 disease-specific deaths in 2018 alone, ranks as the 6th most diagnosed malignancy, and 4th commonest cause of cancer-related mortality globally [1]. Histologically, liver cancer is subclassified as focal nodular hyperplasia (FNH), cholangiocarcinoma (CC), hepatocellular adenoma (HCA), hepatocellular carcinoma (HCC) and combined HCC-CC [2]. Hepatocellular carcinoma, with an increasing annual incidence and arising mostly (90%) in the context of chronic liver disease, such as underlying liver cirrhosis and chronic hepatitis B or C, accounts for a significant 75% of all liver cancer incidence and is associated with very poor survival rates, especially as patients present in late stage, with comorbidities, micro- and/or macrovascular invasion, multicentric or multifocal large tumors, organ shortage and are thus inoperable [2–4]. The post-diagnosis median survival time of patients with inoperable disease is 6–20 months, while the 5-year survival remains <5% [1].

Given the complex nature of HCC, therapeutic decisions in HCC clinics are dependent on disease staging, location and size of tumor, presence and extent of extra-hepatic spread and underlying hepatic function. Currently, the preferred curative modalities for patients with HCC are surgical resection and orthotopic liver transplantation (OLT), however, for patient not meeting the criteria for curative therapy, such as those with unresectable tumors, treatment options include systemic chemotherapy, molecularly targeted therapies, transarterial chemoembolization (TACE), percutaneous ethanol injection (PEI), cryoablation and various forms of IR, namely, microwave ablation (MWA), radiofrequency ablation (RFA), radioembolization, stereotactic body radiotherapy (SBRT) and external beam radiation therapy (EBRT) [4–6]; these are fraught with enhanced risk of severe drug-related adverse events (AEs), acquired resistance to anticancer therapeutics and IR, and relatively dismal survival benefits [5,6], thus, necessitating concerted screening for or development of novel highly efficacious therapeutics and/or the discovery of new actionable molecular oncotargets, which inhibit disease progression, alleviate resistance to treatment and improve prognosis in patients with pancreatic ductal adenocarcinoma (PDAC).

The role of IR in the treatment of HCC continues to evolve, especially with regards to technological advancement and in the context of combinatorial therapy, aimed at enhancing IR safety and efficacy, nevertheless, the anti-HCC efficacy of IR is non-apparent in the intrinsically IR-resistant cells or blunted over time by the acquisition of IR-resistance, subsequently resulting in disease relapse and poor prognosis [7,8]. Thus, the need for continued unraveling of the mechanistic underlining of IR-resistance in HCC, identification of reliable molecular targets and development of more effective therapies.

In the last decade, there has been increased implication of the phosphatidylinositol-3-kinase (PI3K)/protein kinase B (PKB, AKT)/mammalian target of Rapamycin (mTOR) signaling in the acquisition of an IR-resistant phenotype by cancerous cells of different histological origin [9–12]. Accruing evidence suggest a role for activated AKT in the prediction of sensitivity to anticancer chemo- or radiotherapy. Mechanistically, AKT plays a vital role in the PI3K/AKT/mTOR signaling cascade, however, while PI3K binds stably to the pleckstrin homology domain (PHD) finger of AKT without fully activating AKT, evidence abound that the activation of AKT through phosphorylation of the threonine residue (Thr)-308 in AKT activation loop by 3-phosphoinositide-dependent protein kinase-1 (PDK1/PDPK1) enhances AKT activity by over 100-fold, and when followed by phosphorylation of Serine (Ser)-473 at the C terminus, which is the AKT hydrophobic motif, it induces AKT by an additional 7–10-fold, and stabilizes AKT active conformation [13]. Consistent with these, PDK1 has been shown to be dysregulated in some malignancies, and accruing evidence suggesting that the underexplored PDK1 may serve as a therapeutic target and is a probable modulator of sensitivity to cancer therapy [14]. Thus, this present study investigated the role of PDK1 signaling in IR-resistance, enhanced DNA damage repair and post-IR relapse, characteristic of aggressive HCC cells, while exploring potential PDK1-targetability to improve radiosensitivity.

2. Materials and Methods

2.1. Ethics Approval and Consent to Participate

Clinical samples were collected from Taipei Medical University-Shuang Ho hospital (Taipei, Taiwan). All enrolled patients gave written informed consent for their tissues to be used for scientific research. The study was approved by the Institutional Review Board (IRB) of the Taipei Medical University-Shuang Ho hospital (Taipei, Taiwan), consistent with the recommendations of the declaration of Helsinki for biomedical research (Taipei Medical University-Shuang Ho hospital, Taiwan) and followed standard institutional protocol for human research. This study was approved by the Institutional Human Research Ethics Review Board (TMU-JIRB No. 201302016) of Taipei Medical University.

2.2. Access and Analysis of Public Cancer Datasets

The public online cancer data repositories used in this study include Oncomine, The Cancer Genome Atlas (TCGA), Gene Expression Omnibus (GEO) and Broad Institute Cancer Cell Line Encyclopedia (CCLE). We probed the Wurmbach liver (HG-U133_Plus_2) Affymetrix Human Genome U133 Plus 2.0 Array dataset ($n = 75$) using the Oncomine platform (https://www.oncomine.org/resource/main.html#v:18). We also used the Affymetrix Human Genome U133 Plus 2.0 Array dataset GSE6465/GPL570 analyzing the high-throughput gene expression profile of hepatocellular carcinoma xenografts ($n = 53$ samples, 54,675 genes), from the Gene Expression Omnibus (GEO) using the National Center for Biotechnology Information (NCBI) GEO Data Browser (https://www.ncbi.nlm.nih.gov/geo/geo2r/?acc=GSE6465 &platform= GPL570).

2.3. Drug and Reagents

BX-795 hydroxide (#SML0694, HPLC ≥ 98%) was purchased from Sigma Aldrich Co. (St. Louis, MO, USA). Stock solutions of 1 mM were dissolved in dimethyl sulfoxide (DMSO) at 15 mg/mL, and stored in dark room at −20 °C. Phosphate buffered saline (PBS, #P7059), dimethyl sulfoxide (DMSO, #D2650), sulforhodamine B (SRB) reagent (#230162), trypsin/ethylenediaminetetraacetic acid (Trypsin-EDTA, #T4049) solution, trisaminomethane (Tris) base (#93352) and acetic acid (#695092) were purchased from Sigma Aldrich Co. (St. Louis, MO, USA), while Gibco™ Dulbecco's modified Eagle's medium (DMEM) was purchased from Invitrogen (#11966025, Invitrogen Life Technologies, Carlsbad, CA, USA).

2.4. Cell lines and Culture

The human HCC SK-HEP1 (ATCC® HTB-52™) and normal adult liver epithelial THLE-2 (ATCC® CRL-2706™) cells were obtained from American Type Culture Collection (ATCC. Manassas, VA, USA), Huh7 (JCRB0403) from the NIBIOHN ((National Institute of Biomedical Innovation, Health and Nutrition, Japanese Collection of Research Bioresources (JCRB) Cell Bank, Japan)), while FOCUS, Mahlavu, Hep3B cells were also purchased from ATCC. All cells were cultured in DMEM (Invitrogen Life Technologies, Carlsbad, CA, USA), supplemented with 10% fetal bovine serum (FBS, #16140071) and 1% penicillin-streptomycin (Invitrogen, Life Technologies, Carlsbad, CA, USA) in 5% humidified CO_2 incubator at 37 °C. Cells were subcultured at full confluence or media changed every 48–72 h. The cell lines were identified and authenticated based on karyotype and short tandem repeat analyses by the vendors and were regularly checked and confirmed free from any mycoplasma contamination. The cells were subjected to treatment with indicated IR dosage and/or concentrations of BX795.

2.5. Immunohistochemistry (IHC) Analysis

For immunohistochemistry (IHC), tissue microarray (TMA) slides of the TMU-SHH HCC cohort were established, then heat-based antigen retrieval was performed in EDTA-containing buffer, sections blocked with 5% bovine serum albumin (BSA)/1% HISS/0.1% Tween20 solution and incubated with primary recombinant antibody against PDK1 (1:400 dilution; Anti-PDK1 antibody, ab90444) overnight, at 4 °C. PDK1 immunoreactivity/positivity was detected using the mouse IgGk light chain binding protein conjugated to horseradish peroxidase m-IgG BP-HRP (#sc-516102; Santa Cruz Biotechnology, Inc., Santa Cruz, CA, USA) and the EXPOSE mouse and rabbit specific HRP/DAB detection IHC kit (#ab80436, Abcam plc., Cambridge, MA, USA). This study was approved by the Institutional Human Research Ethics Review Board (TMU-JIRB No. 201302016) of Taipei Medical University.

2.6. Establishment of IR-Resistant HCC Cell Lines

In preliminary studies to determine optimal IR dose, Mahlavu, Hep3B and Huh7 cell lines were exposed to IR of 2-10 Gy for 5 consecutive days to determine the maximum tolerated dose (MTD). Based on the cell dysmorphia, cytoplasmic vacuolization, nuclei pleomorphism and cell hyperplasia in irradiated HCC cells compared to the control group, MTD was determined to be 2 Gy/per day for the five consecutive days in all 3 HCC cell lines. Thus, to establish IR-resistant cell lines, the cells were subsequently exposed to 2 Gy at 130 KV, 5.0 mA, every 48 h for 30 cycles (i.e., 60 Gy cumulative dose in 2 months), using the Faxitron® CellRad X-ray cell irradiator (Precision X-ray Irradiation, North Branford, CT, USA). The viable HCC cells after the 30 IR cycles were designated IR-resistant—Mahlavu-R, Hep3B-R and Huh7-R. Culture media was changed every 48–72 h or cells subcultured if confluent. To confirm IR-resistance, the Mahlavu-R, Hep3B-R and Huh7-R alongside their control counterparts were exposed to 0.5–2 Gy single-doses of IR, then evaluated using functional assays, including cell viability and clonogenic-survival assay. The HCC-R cell survival fractions, clonogenicity and tumorsphere-formation efficacy were markedly higher compared to the HCC control cells.

2.7. Western Blot Analysis

After separation of 20 μg protein samples using 10% sodium dodecyl sulfate-polyacrylamide gel electrophoresis (SDS-PAGE) gel, the protein blots were transferred onto polyvinylidene fluoride (PVDF) membranes in the Bio-Rad Mini-Protein electro-transfer system (Bio-Rad Laboratories, Inc., CA, USA), followed by membrane-blocking in 5% skimmed milk in Tris-buffered saline with Tween 20 (TBST) for 1 h. Thereafter, membranes were incubated overnight at 4 °C with primary antibodies against p-PI3 Kinase p85 (Tyr458)/p55 (Tyr199; #4228S; 1:1000, Cell Signaling Technology Inc., Danvers, MA, USA), PI3 Kinase p110α (#4292S; 1:1000, Cell Signaling Technology), p-PDK1 (Ser241; #3061L; 1:1000, Cell Signaling Technology), PDK1 (#3062S; 1:1000, Cell Signaling Technology), p-AKT (Ser473; #9271L; 1:1000, Cell Signaling Technology), AKT (#4691L; 1:1000, Cell Signaling Technology), p-mTOR (Ser2448; #2971L; 1:1000, Cell Signaling Technology), mTOR (#2972S; 1: 1000, Cell Signaling Technology), E-cadherin (#3195S; 1:1000, Cell Signaling Technology), N-cadherin (#13116S; 1:1000, Cell Signaling Technology), Vimentin (#5741S; 1:1000, Cell Signaling Technology), Snail (#3879S; 1:1000, Cell Signaling Technology), Bcl-2 (#15071S; 1:1000, Cell Signaling Technology), Bax (#5023S 1:1000, Cell Signaling Technology) and β-actin (sc-69879; 1:500, Santa Cruz Biotechnology, Santa Cruz, CA, USA) in Supplementary Table S1.

This was followed by incubation of membranes in appropriate secondary antibodies conjugated with horseradish peroxidise (HRP) at room temperature for 1 h, washed carefully with PBS thrice, and then protein band detection performed using the enhanced chemiluminescence (ECL) detection system (Thermo Fisher Scientific Inc., Waltham, MA, USA), and band densitometry-based quantification done using the ImageJ software (https://imagej.nih.gov/ij/).

2.8. Sulforhodamine B Cytotoxicity Assay

Using 96-well plates, 3×10^3 HCC and/or HCC-R cells were seeded per well in quadruplicates and cultivated for 24 h. The cells were thereafter exposed to 0.5–2 Gy IR and/or indicated concentrations of BX795 for 48 h, subjected to 10% trichloroacetic acid (TCA) fixation, carefully washed with double-distilled water (ddH$_2$O), then stained with 0.4% 0.4:1 (*w/v*) SRB/acetic acid solution. The unbound SRB dye was removed by carefully washing the cells with 1% acetic acid thrice before air-drying the plates. Thereafter, bound SRB dye was solubilized in 10mM Tris base, and absorbance, which is strongly correlated to the number of viable stained cells over a wide range, was read at a wavelength of 570 nm in the Molecular Devices Spectramax M3 multimode microplate reader (Molecular Devices LLC., San Jose, CA, USA).

2.9. Tumorsphere Formation Assay

In non-adherent 6-well plates, 5×10^4 Mahlavu, Mahlavu-R, Huh7 or Huh7-R cells were seeded per well (Corning Inc., Corning, NY, USA) containing DMEM supplemented with Gibco™ B-27™ supplement (#17504044, Invitrogen, Carlsbad, CA, USA), 20 ng/mL basic fibroblast growth factor (bFGF; #13256029, Invitrogen, Carlsbad, CA, USA) and 20 ng/mL epidermal growth factor (EGF; #PHG0311, Invitrogen, Carlsbad, CA, USA). Cells were cultured for 12 days and formed tumorspheres ≥150 μm were counted under inverted phase contrast microscopy.

2.10. Colony Formation Assay

Clonogenicity was assessed as previously described [15]. Briefly, 1×10^3 Mahlavu, Hep3B or Huh7 cells, with or without their IR-resistant counterparts were pre-exposed to indicated treatment regimen for 24 h were seeded per well in 6-well plates and incubated in 5% humidified CO$_2$ incubator at 37 °C for 15 days. The colonies formed (>50 cells/colony) were then stained with crystal violet dye, photographed and counted.

2.11. Immunofluorescence (IFC) Staining

In Nunc™ Lab-Tek™ II 8-well chamber slides, 2×10^4 Mahlavu, Mahlavu-R, Huh7 or Huh7-R cells pre-exposed to indicated IR dosage or transfected with shPDK1 were seeded (#154534, Thermo Fisher Scientific Inc., Waltham, MA, USA) for 24 h. For SOX2, or OCT4A staining, the seeded cells were fixed with Image-iT™ fixative solution (4% paraformaldehyde; #FB002, Thermo Fisher Scientific Inc., Waltham, MA, USA) at room temperature for 20 min, washed with 1X PBS, permeabilized with 0.1% Triton X-100 (#28314, Thermo Fisher Scientific Inc., Waltham, MA, USA) in 0.01 M PBS (pH 7.4) for 5 min, blocked with 0.2% bovine serum albumin for 1 h, air-dried and rehydrated in 1X PBS. The cells were then incubated with rabbit monoclonal antibody against Sox2 (#3579S, Cell Signaling Technology Inc., Danvers, MA, USA) or Oct-4A (#2840S, Cell Signaling Technology Inc., Danvers, MA, USA) diluted 1:500 in 1X PBS containing 3% normal goat serum at room temperature for 2 h, washed thrice in 1X PBS for 10 min each, and then incubated with goat anti-rabbit fluorescein isothiocyanate (FITC) IgG—conjugated secondary antibody (Jackson ImmunoResearch Inc., West Grove, PA, USA) diluted 1:500 in 1X PBS at room temperature for 1 h. Thereafter, the cells were washed in 1X PBS, mounted using Vectashield® antifade mounting medium with 4′,6-diamidino-2-phenylindole (DAPI; #H-1200, Vector Laboratories, Burlingame, CA, USA) for nuclear staining. Cell images were taken under a Zeiss Axiophot fluorescence microscope (Carl Zeiss Microscopy LLC., Thornwood, NY, USA).

2.12. Transwell Matrigel Invasion Assay

For the evaluation of cell invasion, the modified Boyden chambers consisting of Corning®
Transwell® membrane filter inserts (8-μm pore size; Corning Costar Corp., Cambridge, MA, USA) in
24-well tissue culture plates were used. For an invasion assay, 2×10^5 Mahlavu, or Mahlavu-R cells
were seeded into 200 μL serum-free DMEM medium in the upper surface of membranes coated with
100 μL Matrigel (BD Biosciences, San Jose, CA, USA) and allowed to invade toward the underside of
the membrane in 24-well tissue culture plates containing 500 μL complete growth media with 10%
FBS for 24 h. Non-invaded cells were removed by carefully wiping the upper side of the membrane
with sterile cotton buds while the invaded cells on the underside of the membrane were fixed with
ice-cold methanol, and then stained with 0.5% crystal violet dye in 20% ethanol for 30 min. Thereafter,
the invaded cells were counted under a light microscope in 5 randomly selected visual fields at a
magnification of ×400.

2.13. Scratch-Wound Healing Migration Assay

To evaluate cell migration, we used well-established protocol. Briefly, Mahlavu, or Mahlavu-R
cells were seeded into 6-well plates (Corning Inc., Corning, NY, USA) containing complete growth
media with 10% FBS, cultured to 98–100% confluence, then the median axes of the cell monolayers
were denuded with sterile yellow pipette tips. The scratch-wound healing cum cell migration was
monitored over time and images captured under a light microscope with 10× objective lens at the 0
and 48 h time-points after denudation, and then images were analyzed with the NIH ImageJ software
(https://imagej.nih.gov/ij/download.html).

2.14. Small Interfering RNA (siRNA) Transfection

For transient silencing of PDK1, the PDK1-specific siRNA ((PDPK1 siRNA (h), sc-29448)) was
purchased from Santa Cruz (Santa Cruz Biotechnology, Santa Cruz, CA, USA). Lipofectamine
2000 transfection reagent (#11668019, Thermo Fisher Scientific Inc., Waltham, MA, USA) was used for
the transfection of the siRNA following the manufacturer's protocol. Total protein extracted 48 h after
transfection was used for Western blot analyses.

2.15. Flow Cytometry-Based Side Population Analyses

For identification of the side population (SP), after washing the HCC cells in warm DMEM
supplemented with 3% FBS and 10 mmol/L Gibco™ 4-(2-hydroxyethyl)-1-piperazineethanesulfonic
acid (HEPES) buffer (#15630106, Invitrogen-Life Technologies, Carlsbad, CA, USA), 1×10^6 cells were
resuspended per mL of DMEM supplemented with 3% FBS, 10 mmol/L HEPES buffer and 5 μg/mL
Hoechst 33342 dye, then incubated for 1.5 h at 37 °C with gentle agitation. Hoechst dye excitation was
at 350–356 nm, while fluorescence was quantified at 424/44 nm or 620 nm for Hoechst blue or Hoechst
blue, respectively. Dead cells or doublets were gated out. SP phenotype sorting gates were defined by
100 μM/L reserpine, an ABC-transporter inhibitor (#R0875, Sigma-Aldrich Corp., St. Louis, MO, USA).
Single cell suspension was obtained by filtering cells through a 70 μm filter, sorted into SP and non-SP
cell fractions and then analyzed in the BD FACSAria II System (BD Biosciences, San Jose, CA, USA).
SP cell purity was ≥98%.

2.16. ALDH Aldefluor Activity

The Aldefluor™ kit (#01700, Stem Cell Technologies) was used for the profiling and
isolation of Huh7 or Huh7-R cells with high or low ALDH activity following the manufacturer's
instruction. Briefly, after incubating the cells in Aldefluor™ assay buffer containing ALDH substrate,
BODIPY-aminoacetaldehyde (BAAA) for 45 min at 37 °C. ALDH$^+$ cells were delineated by the ability to
catalyze BAAA to its fluorescent product, BODIPY-aminoacetate (BAA), and ALDH enzymatic activity
was blocked by Aldefluor™ DEAB reagent (#01705), a specific ALDH inhibitor. Fluorescence-activated

cell sorting (FACS) gates were defined relative to the DEAB-treated samples-based baseline fluorescence. The cells were resuspended in fresh assay buffer after incubation. ALDH$^+$ and ALDH$^-$ cells were sorted in the BD FACSAria II System (BD Biosciences, San Jose, CA, USA).

2.17. Determination of Synergism of Combinatorial Therapy

The synergistic effect of combining IR and BX795 was evaluated by adapting the Chou-Talalay algorithm of multiple drug combination. CompuSyn software (ComboSyn Inc., Paramus, NJ, USA) was used following the guideline of two therapies combination analysis. All combination dose-points falling within the right-angled 'isobologram' triangle, was defined as synergism, if the dose-points laid on the hypotenuse, additivity was considered, and when the dose-points fell outside the isobologram, the combination was designated as antagonistic.

2.18. Statistical Analysis

All results represent the mean ± SD of assays performed at least three times in triplicates. The 2-sided Student's *t*-test was used for intergroup comparison, while 1-way ANOVA with a Tukey's post-hoc test was used for comparisons between multiple groups. All statistical analyses were performed using the GraphPad Prism version 7.0 for Windows (GraphPad Software, Inc., La Jolla, CA, USA). *p*-value < 0.05 was considered statistically significant.

3. Results

3.1. PDK1 Is an Independent Driver of the PI3K/AKT/mTOR Signaling Pathway and Its Aberrant Expression Characterizes Poorly Differentiated Aggressive HCC Cells

Following our in-house mining of high through-put gene expression data from public repositories, to unravel the role of the PI3K/AKT/mTOR signaling axis, and more specifically PDK1 signaling in HCC, we performed in silico proteotranscriptomic analyses of selected molecular components of the pathway in question. Results of our differential expression profiling of the Wurmbach liver cohort ($n = 75$), with 45 context-relevant sample, revealed concomitant significant up-regulation of PI3K (1.22-fold, $p = 0.030$), PDK1 (2.17-fold, $p = 8.95 \times 10^{-6}$), mTOR (1.37-fold, $p = 6.26 \times 10^{-4}$) and statistically insignificant down-regulation of AKT mRNA (-1.14-fold, $p = 0.847$) in HCC samples ($n = 35$) compared to the normal liver tissues ($n = 10$; Figure 1A). Protein–protein interaction (PPI) enrichment analyses using the STRING-db (https://string-db.org) platform for PPI network prediction revealed very strong association between PI3K, PDK1, AKT, mTOR and other indicated effectors and/or mediators of the PI3K/AKT/mTOR signaling axis, as expressed by the relatively high mean local clustering coefficient of 88.3% ($p < 1.0 \times 10^{-16}$ (Figure 1B)). Furthermore, results of our Western blot analyses showed that compared to its non-expression in the normal adult liver epithelial THLE-2 cells, the expression of p-PI3K p85 (Tyr458)/p55 (Tyr199), PI3K, p-PDK1 (Ser 241), PDK1, p-AKT (Ser 473), AKT, p-mTOR (Ser 2448) and mTOR proteins were highly or moderately highly enhanced in the poorly differentiated SK-HEP1 and Mahlavu, and Hep3B cell lines, moderately enhanced in poorly differentiated FOCUS (CVCL_7955) cells, but mildly or not expressed in the well-differentiated Huh7 cells (Figure 1C). Moreover, coupled with the consistent high mRNA intensity and protein expression levels of PDK1 in earlier data, from IHC-based proteome analyses using the human protein atlas platform version 18.1 (https://www.proteinatlas.org/ENSG00000152256-PDK1/pathology/tissue/liver+cancer#img) we observed a significant positive correlation between PDK1 immunoreactivity/expression and disease stage/progression (Figure 1D). These data did indicate, at least in part, that PDK1 was an independent driver of the PI3K/AKT/mTOR signaling pathway and its aberrant expression characterized poorly differentiated aggressive HCC cells (Figure 1D).

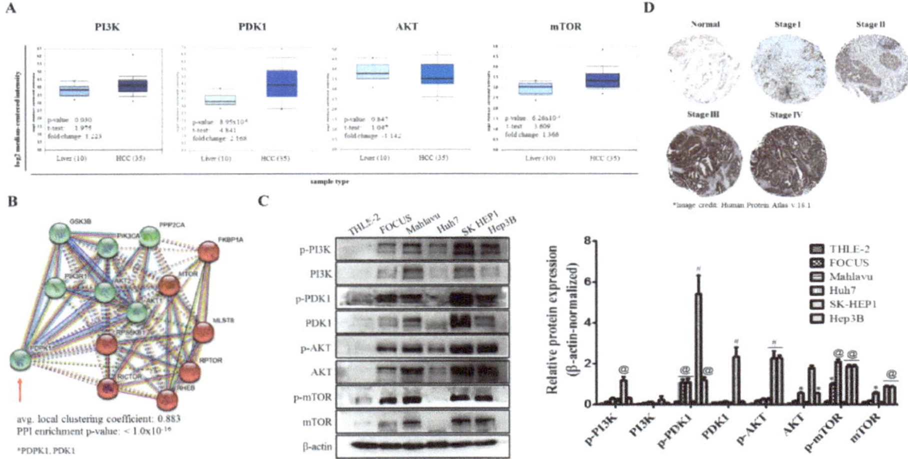

Figure 1. PDK1 is an independent driver of the PI3K/AKT/mTOR signaling pathway and its aberrant expression characterizes poorly differentiated aggressive hepatocellular carcinoma (HCC) cells. (**A**) Box and whisker plots of the differential expression of PI3K, PDK1, AKT and mTOR in HCC or normal liver tissues from the Wurmbach liver cohort. (**B**) STRINGdb-generated visualization of the protein–protein interaction between PDK1/PDPK1, and molecular components of the PI3K/AKT/mTOR signaling pathway. (**C**) Representative Western blot images and histograms showing the differential expression of p-PI3K p85 (Tyr458)/p55 (Tyr199), PI3K, p-PDK1 (Ser 241), PDK1, p-AKT (Ser 473), AKT, p-mTOR (Ser 2448) and mTOR proteins in THLE-2, CVCL_7955, Mahlavu, Huh7, SK-HEP1 or Hep3B cell lines. (**D**) Representative IHC images of PDK1 immunoreactivity in Stages I–IV HCC, compared to normal liver tissue. β-actin served as loading control. * $p < 0.05$, @ $p < 0.01$, # $p < 0.001$ vs. THLE-2; Green nodes, PI3K/AKT signaling; red nodes, mTOR signaling; PPI, protein–protein interaction.

3.2. Altered PDK1 Expression Is Sufficient for the Deactivation of the PI3K/PDK/AKT/mTOR Oncogenic Signaling and Sensitizes Aggressive HCC Cells to Radiotherapy

Having shown that enhanced PDK1 expression plays an essential role in the PI3K/AKT/mTOR signaling axis and characterizes poorly differentiated aggressive HCC cells, to gain some insight into the mechanistic underlining of PDK1 on HCC oncogenicity and therapy response, we evaluated the effect of increasing doses of IR on different HCC cells. We observed that while 24 h exposure to 0.5–2 Gy IR significantly reduced the viability of PDK1^{low} Huh7 (46.1%, $p < 0.05$ at 2 Gy) and PDK1$^{low/moderate}$ FOCUS cells (38.4%, $p < 0.05$ at 2 Gy), its effect on the PDK1^{moderate} Hep3B cell was mild (~24% at 2 Gy), and it had no apparent or insignificant cytocidal effect on the PDK1^{high} SK-HEP1 and Mahlavu cells, respectively (Figure 2A). Furthermore, we demonstrated that aside from eliciting a 59% decline in the viability of PDK1^{high} Mahlavu cells, siRNA-mediated loss-of-PDK1 function (siPDK1) enhanced their sensitivity to 2 Gy IR by a significant 80%; similarly in synergism with siPDK1, the cytocidal effect of 2 Gy IR was increased to 97% from 42% alone in PDK1^{low} Huh7 cells, and to 93% from 26% alone in PDK1^{moderate} Hep3B cells (Figure 2B), thus, indicating a vital role for PDK1 expression and/or activity in the radiosensitivity of HCC cells. In addition, compared to the 27%, 14% or 51% loss of clonogenicity elicited by exposure to 2 Gy IR in Mahlavu, Hep3B or Huh7 cells, respectively, siPDK1 inhibited the ability of Mahlavu, Hep3B or Huh7 cells to form colonies by 91%, 96% or 99.8%, respectively (Figure 2C). Combining siPDK1 with 2 Gy IR was incompatible with clonal survival in all three cell lines (data not provided). Interestingly, the observed inhibition of cell viability and attenuation of clonogenicity by siPDK1 alone or in synergism with 2 Gy IR, were associated with down-regulated p-PDK1, PDK1, p-PI3K p85 (Tyr458)/p55 (Tyr199), PI3K, p-AKT, AKT, p-mTOR and mTOR in the PDK1^{high} Mahlavu cells (Figure 2D). These results do indicate that altered PDK1 expression via specific targeting of PDK1 was sufficient to deactivate the PI3K/AKT/mTOR oncogenic signaling in HCC cells and sensitized the aggressive cells to IR.

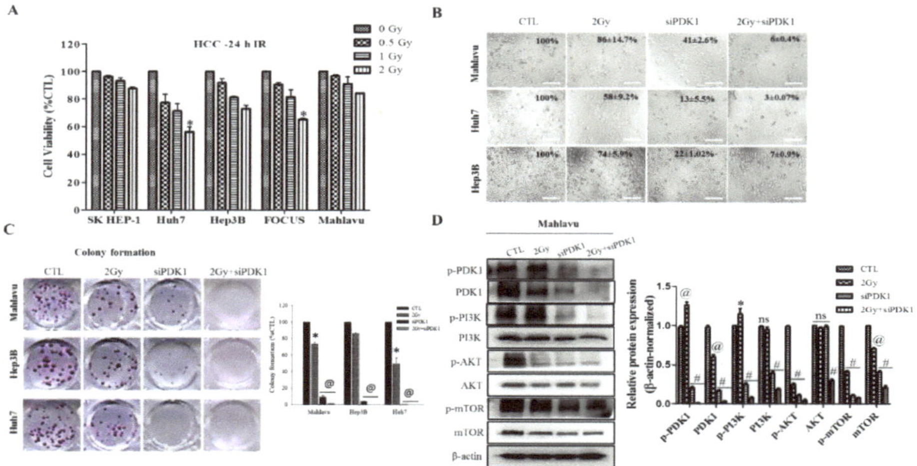

Figure 2. Altered PDK1 expression is enough for the deactivation of the PI3K/PDK/AKT/mTOR oncogenic signaling and sensitizes aggressive HCC cells to radiotherapy. (**A**) Histograms of the effect of 24 h exposure to 0.5–2 Gy IR on the viability of SK-HEP1, Huh7, Hep3B, CVCL_7955 or Mahlavu cell lines. (**B**) Representative photo-images of the effect of 2 Gy, siPDK1 or 2 Gy+siPDK1 on the viability/proliferation, and morphology of Mahlavu, Huh7 or Hep3B cells. Scale bar: 100 μm, 10× objective. (**C**) Representative photo-images showing the effect of 2 Gy or siPDK1 on clonogenicity of Mahlavu, Huh7 or Hep3B cells. Scale bar: 100 μm, 10× objective. (**D**) Representative western blot images and histograms showing the effect of 2 Gy, siPDK1 or 2 Gy+siPDK1 on the expression levels of p-PI3K p85 (Tyr458)/p55 (Tyr199), PI3K, p-PDK1 (Ser 241), PDK1, p-AKT (Ser 473), AKT, p-mTOR (Ser 2448) and mTOR proteins in Mahlavu cells. * $p < 0.05$, @ $p < 0.01$, # $p < 0.001$ vs. CTL; ns, not significant.

3.3. Aberrant PDK1 Expression Is Implicated in the Acquisition of Radioresistance and Evasion of DNA Damage by HCC Cells

Due to the implication of DNA damage in the death of cancerous cells exposed to IR and the therapeutic benefit of exploiting the reduced resolution of IR-induced clustered DNA damage [16] and altered DNA damage repair (DDR) genes in liver cancer [17], firstly, we probed and reanalyzed Huynh H et al.'s E-GEOD-6465, A-AFFY-44, AFFY_HG_U133_PLUS_2 data set on the array expression profiling of xenografts of HCC (n = 53 samples, 54,675 genes) (https://www.ebi.ac.uk/arrayexpress/experiments/E-GEOD-6465/). Generated expression-based heat-map revealed a dichotomization of our selected gene-sets, such that the PDK1, ALDH1A1, CD133/PROM1, OCT4A/POU5F1, SOX2 and KLF4 genes clustering with TP53 were up-regulated, while the DDR genes RAD50, MSH3, MLH3, ERCC2,and BLM gene cluster were down-regulated (Figure 3A). Additional analyses using the STRING-db platform (https://string-db.org) for PPI network prediction further confirmed earlier results, as we observed a very strong association between components of the PI3K/AKT signaling, namely PDK1/PDPK1, AKT1, MTOR and stemness marker complicit in IR-resistance ALDH1Al, CD133/PROM1, OCT4A/POU5F1, SOX2 and KLF4 pooled together, while DNA damage markers RAD50, MSH3, MLH3, ERCC2 and BLM pooled together (Figure 3B). The average local clustering coefficient for the clustered proteins was 0.827 and PPI enrichment p-value was $p < 2.45 \times 10^{-10}$ (Figure 3B). Due to the suggested implication of cancer stem cell (CSCs) markers in PDK1-induced IR-resistance, we further examined if and to what extent PDK1-induced IR-resistance affects the side population (SP), which is representative of the CSCs pool in vitro. Comparative analyses of the HCC wild type and PDK1-rich IR-resistant clones

(HCC-R) revealed a 1.40-fold increase in the SP in Mahlavu-R compared to its wild type counterpart; similarly compared to Hep3B or Huh7 cells, the SP was increased by 2.14-fold or 7.03-fold in Hep3B-R or Huh7-R cells, respectively (Figure 3C). More so, because of the documented implication of ALDH1 in IR-resistance [18], we probed the GDC TCGA liver cancer (LIHC, n = 469) for probable relationship between PDK1 and ALDH, and showed a positive correlation between ALDH1A1 and PDK1 (R = 0.27, $p = 1.6 \times 10^{-8}$; Figure 3D). Consistent with the 'coexpression - function similarity' paradigm, and in conformity with conventional knowledge that when an inactive enzyme, otherwise known as an 'apoenzyme', (in this case, apoALDH1A1) binds with an organic or inorganic helper-molecule/cofactor, a complete and catalytically active form of the enzyme called an 'holoenzyme' is formed, we generated a spatiotemporal visualization of the probable interaction between PDK1 and ALDH1 using the Schrödinger's PyMOL molecular graphics system (https://pymol.org/2/), and demonstrated that the catalytic domain of PDK1 (protein data bank, PDB: 1H1W) binds directly with human apoALDH1A1 (PDB: 4WJ9) with an interaction score of 17.5, an atomic contact energy (ACE) of -50.05 kcal/mol, and root-mean-square deviation (RMSD) of 23.52Å (Figure 3E, also see Supplementary Figure S1). In parallel assays, we observed significantly enhanced ALDH1 activity in the PDK1-rich Huh7-R cells, as demonstrated by a 24.16-fold increase in ALDH activity in the former PDK1^{low} Huh7 cells when they acquire an IR-resistant phenotype (Huh7-R; Figure 3F), which is consistent with our predicted PDK1-ALDH1 interaction probability of 0.80 or 0.99 using the random forest (RF) or support-vector machine (SVM) classifier algorithm, respectively, as shown in Figure 3D. PDK1 interacts with ALDH and directly modulate the expression and/or activity of ALDH in HCC cells. Representative Western blot image and histograms of the differential expression of PDK1 and ALDH1 in adherent wild-type Mahlavu cells or their tumorsphere counterparts (Supplementary Figure S2). These data indicate, at least in part, that PDK1 directly interacts with and activates ALDH1A1, and implicates the aberrant PDK1 expression in the acquisition of IR-resistance and evasion of DNA damage by HCC cells.

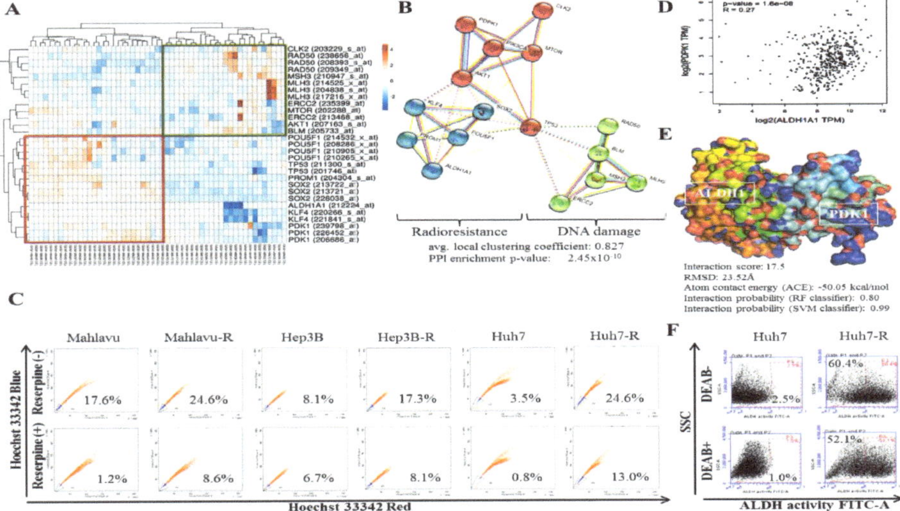

Figure 3. Aberrant PDK1 expression is implicated in the acquisition of radioresistance and evasion of DNA damage by HCC cells. (**A**) Heatmap of gene-set expression in the E-GEOD-6464, AFFY_HG_U133_PLUS_2 data set. Rows are centered; unit variance scaling is applied to rows. All 29 rows and 53 columns were clustered using correlation distance and average linkage. (**B**) STRINGdb-generated visualization of the

protein-protein interaction between PDK1/PDPK1, PI3K/AKT/mTOR signaling pathway (red nodes), CSCs markers (blue nodes) and DNA damage markers (green nodes). (**C**) Graphical representation of the differential SP in Mahlavu, Mahlavu-R, Hep3B, Hep3B-R, Huh7 or Huh7-R cells, in the presence or absence of reserpine. (**D**) Graphical representation of the correlation between ALDH and PDK1 expression in the GDC TCGA liver cancer cohort, *n* = 469. (**E**) Schrödinger's PyMOL molecular graphics system-generated molecular docking of PDK1 and ALDH1. (**F**) Representative images showing the differential ALDH activity in Huh7 or Huh7-R cells in the presence or absence of DEAB. DEAB, Aldefluor inhibitor; DEAB+, negative control; Reserpine served as inhibitor of Hoechst 33342 dye efflux; PPI, protein–protein interaction; RSMD, root-mean-square deviation; TPM, transcript per million.

3.4. PDK1-Dependent Radioresistance Is Associated with the Enhanced Metastatic and Cancer Stem Cell-Like Phenotypes of HCC Cells

Having implicated the aberrant expression of PDK1 in the acquisition of IR-resistance by HCC cells, we sought to rule out its complicity in the enhanced oncogenicity and CSCs-like phenotypes of the HCC cells. We demonstrated that the Mahlavu-R cells were significantly more mobile than the Mahlavu cells, as indicated by a 2.15-fold ($p < 0.001$) enhanced cell migration in the Mahlavu-R compared with the Mahlavu cells (Figure 4A). Additionally, we observed that compared to the Mahlavu cells, a significant time-dependent increase was induced in the number of invaded Mahlavu-R cells (wild-type (WT) vs. R: 24 h, 1.48-fold, $p < 0.01$; 48 h, 1.47-fold, $p < 0.001$; Figure 4B). In addition, compared to their wild type counterpart, the clonogenicity in the Huh7-R (6.21-fold, $p < 0.001$), Mahlavu-R (13.18-fold, $p < 0.01$) and Hep3B-R (5.30-fold, $p < 0.01$) was significantly enhanced (Figure 4C). We also demonstrated that the observed enhanced migration, invasion and colony formation potentials in the PDK1-rich Hep3B-R, Mahlavu-R and Huh7-R cells were concomitantly associated with marked up-regulation of p-PDK1, PDK1, N-cadherin, Vimentin and Snail protein expression levels, with converse down-regulation of E-cadherin, compared to their expression in the HCC-WT cells (Figure 4D). Moreover, the Huh7-R or Mahlavu-R cells exhibited a marked increase in the tumorsphere sizes (Huh7 vs. Huh7-R: 2.01-fold, $p < 0.001$; Mahlavu vs. Mahlavu-R: 2.48-fold, $p < 0.001$), number (Huh7 vs. Huh7-R: 3.29-fold, $p < 0.001$; Mahlavu vs. Mahlavu-R: 4.72-fold, $p < 0.001$) and formation efficiency (Huh7 vs. Huh7-R: 4.14-fold, $p < 0.001$; Mahlavu vs. Mahlavu-R: 1.83-fold, $p < 0.01$; Figure 4E). For tumorsphere formation efficiency (TFE), the formula used was $TFE = \frac{Nx}{N1} - 1$, where N_1 is the number of formed tumorspheres ≥ 150 mm in primary or first generation and N_x is the number of formed tumorspheres ≥ 150 mm in the subsequent generation. Expectedly, nuclear OCT4A and SOX2 immunoreactivity were enhanced in the Mahlavu-R and Huh7-R cells compared with their wild type counterparts (Figure 4F). These results are indicative of existent association between the PDK1-dependent IR-resistance, increased epithelial-to-mesenchymal transition (EMT) and the enhanced metastatic and CSCs-like phenotypes of HCC cells.

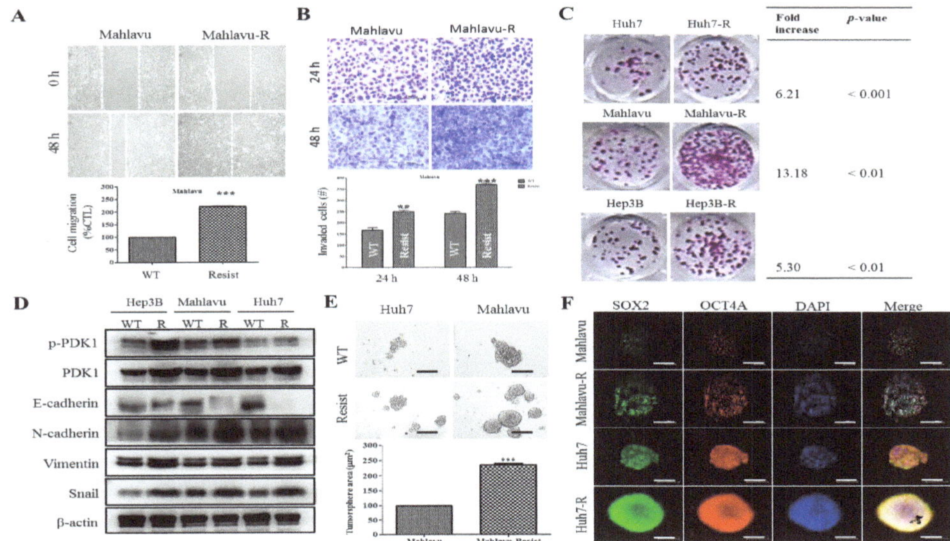

Figure 4. PDK1-dependent radioresistance is associated with the enhanced metastatic and cancer stem cell-like phenotypes of HCC cells. Representative photo-images (upper panel) and histograms (lower panel) comparing (**A**) migration and (**B**) invasion potential between Mahlavu and Mahlavu-R cells. Scale bar: 100 μm, 10× objective. (**C**) Representative photo-images (left panel) comparing clonogenicity between Huh7-R, Mahlavu-R, Hep3B-R and their wild-type counterparts. Scale bar: 100 μm, 10× objective. (**D**) Representative Western blot images of the differential expression of p-PDK1, PDK1, E-cadherin, *N*-cadherin, Vimentin or Snail protein in wild-type and IR-resistant Hep3B, Mahlavu or Huh7 cells. β-actin was used as a loading control. (**E**) Representative photo-images (upper panel) and histograms (lower panel) comparing the tumorsphere formation potential between wild-type (WT) and IR-resistant Huh7 or Mahlavu cells. Scale bar: 100 μm, 10× objective. (**F**) Representative immunofluorescence staining images of SOX2 and OCT4A expression in WT and IR-resistant Huh7 or Mahlavu cells. DAPI served as nuclear marker. Scale bar: 100 μm, 10× objective. * $p < 0.05$, ** $p < 0.01$, *** $p < 0.001$.

3.5. Pharmacologic Targeting of PDK1 Resensitizes HCC Cells to Radiotherapy-Induced Apoptosis Signals Dose-Dependently, and Significantly Suppress Their Oncogenicity

In the light of our accruing evidence implicating PDK1 in the enhanced IR-resistance and associated CSCs-linked oncogenicity of HCC cells, we evaluated the translatability and probable clinical feasibility of our findings by investigating the likely efficacy of pharmacologically targeting PDK1 on the IR-resistant phenotype of HCC cells, using BX795, a potent small molecule ATP-competitive inhibitor of PDK1. Our results indicate that in the presence of 1.25–10 μM enhanced the cytocidal effect of 0.5–2 Gy IR on the Mahlavu-R cells, dose-dependently (Figure 5A, upper panel). Interestingly, this synergistic effect was akin to that observed in the wild type Mahlavu subjected to same combinatorial therapeutic regimen, howbeit greater in the wild type cells (Figure 5A, lower panel). Furthermore, we performed a synergy evaluation of the BX795/IR combinatorial therapy using the Chou-Talalay isobologram algorithm, and the results were corroboratory of our earlier data, as demonstrated by all twelve BX795/IR combination dose-points lying within the isobologram (Figure 5B). In addition, we demonstrated that the observed therapeutic effect was associated with enhanced Bax/Bcl-2 apoptotic index as shown by the moderate or strong expression intensity of Bax protein in Mahlavu-R cells treated with 1.25 μM BX795 alone or 1.25 μM BX795/2 Gy IR combined, respectively, compared to the null/mild Bax protein expression in the 2 Gy IR alone or untreated control group; conversely, compared to the strong expression of Bcl-2 protein in the control or 2 Gy IR alone group, Bcl-2 protein expression was mild or null in the 1.25 μM BX795 alone or 1.25 μM BX795/2 Gy IR combined group, respectively (Figure 5C), suggesting an apoptosis-dependent mechanism for the synergistic therapeutic effect. Consistent with the above, we also observed that in comparison to the control group, treatment with 2 Gy IR, 1.25 μM BX795 or 1.25 μM BX795/2 Gy IR combined elicited a 39% ($p < 0.05$), 48.3% ($p < 0.01$) or 97% ($p < 0.001$) reduction in the number of invaded wild type Mahlavu cells, respectively, and a 20% ($p < 0.05$), 27.6% ($p < 0.05$) or 71.9% ($p < 0.001$) reduction in the number of invaded Mahlavu-R cells (Figure 5D). A similar effect trend was also observed for the effect of 2 Gy IR, 1.25 μM BX795 or 1.25 μM BX795/2 Gy IR combined on the ability of Mahlavu and Mahlavu-R cells to form colonies (Figure 5E). These data do indicate that the BX795-mediated pharmacological targeting of PDK1 resensitizes HCC cells to radiotherapy-induced apoptosis signals dose-dependently, and significantly suppress their oncogenicity.

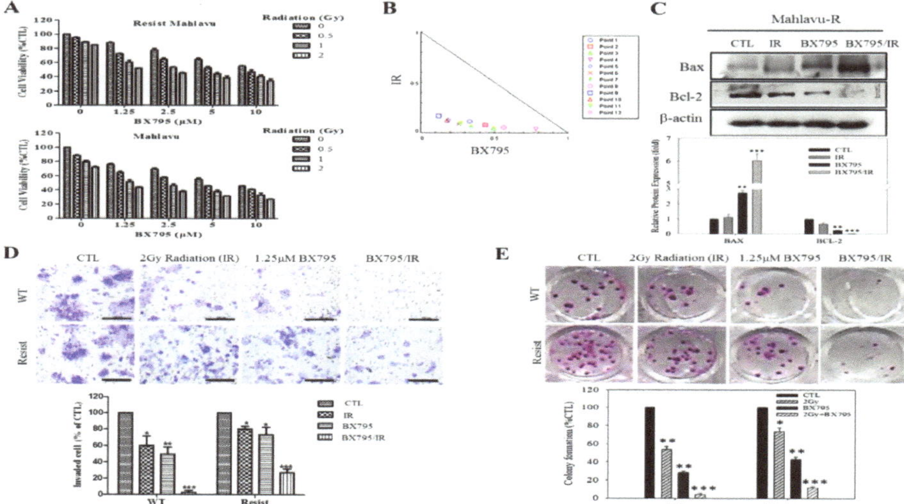

Figure 5. BX795-mediated pharmacologic targeting of PDK1 resensitizes HCC cells to radiotherapy-induced apoptosis signals dose-dependently and significantly suppresses their oncogenicity. (**A**) Histograms of the effect of 24 h exposure to 0.5–2 Gy IR on the viability of IR-resistant (upper panel) or wild-type (lower panel) Mahlavu cells. (**B**) Isobologram indicating synergism between different doses of IR and BX795. (**C**) The effect of exposure to 2 Gy IR, 1.25 μM BX795 or 2 Gy IR+1.25 μM BX795 on the expression level of Bax and Bcl-2 proteins in Mahlavu-R cells as shown by Western blot analyses. Photo-images (upper panel) and histograms (lower panel) comparing the effect of exposure to 2 Gy IR, 1.25 μM BX795 or 2 Gy IR+1.25 μM BX795 on the (**D**) invasion, and (**E**) colony formation potential of WT or IR-resistant Mahlavu cells. β-actin was used as a loading control. Scale bar: 100 μm, 10× objective; * $p < 0.05$, ** $p < 0.01$, *** $p < 0.001$; Resist, IR-resistant; WT, wild-type; CTL, control.

4. Discussion

While the last 3 decades has been characterized by the broadening of our understanding of HCC pathobiology, and advances in diagnostic and therapeutic strategies for managing patients with HCC, therapeutic success and clinical outcome remain poor, especially as the incidence of resistance to conventional anticancer therapies and cancer recurrence after an initial response to treatment, including IR, continue to rise [1–6]. Given the critical role of oncogene addiction and/or aberrant oncogene expression in the initiation, metastasis, resistance to therapy and recurrence of HCC, oncogene addiction constitutes an 'Achilles heel' in any successful molecular anticancer therapy, and its targetability represents a putative therapeutic strategy with high curative efficacy and strong antirelapse potential. This necessitates the discovery of novel onco-addictive molecular targets and development of new therapeutic strategies that are efficacious against HCC oncogenicity, biomolecular drivers of HCC development, dissemination and the acquisition of a therapy-resistant phenotype, and therapy failure in patients with PDAC [19–22]. Against this background, this study unravels a role for PDK1 as one such onco-addictive driver of oncogenicity and IR-resistance.

In the present study, we demonstrated that (i) PDK1 is an independent driver of the PI3K/AKT/mTOR signaling pathway and its aberrant expression characterizes poorly differentiated aggressive HCC cells, with (ii) altered PDK1 expression being sufficient for the deactivation of the PI3K/PDK/AKT/mTOR oncogenic signaling, and sensitization of aggressive HCC cells to radiotherapy. Mechanistically, we also showed that while (iii) aberrant PDK1 expression is implicated in the acquisition of IR-resistance and evasion of DNA damage by HCC cells, (iv) PDK1-dependent IR-resistance is associated with the enhanced metastatic and cancer stem cell-like phenotypes of HCC cells and that the (v) pharmacologic targeting of PDK1 resensitizes HCC cells to radiotherapy-induced apoptosis signals dose-dependently and significantly suppresses their oncogenicity. As with all our works, these findings add to the current repertoire of knowledge on HCC pathobiology, and are posited to help shape potential therapeutic decision-making for managing patients with HCC, in the context of the contemporary clinical challenge of IR-resistance, therapy failure, disease recurrence and poor prognosis amongst patients with HCC.

In the present study, for the first time, to the best of our knowledge, we demonstrated that PDK1 is an independent driver of the PI3K/AKT/mTOR signaling pathway and its aberrant expression characterizes poorly differentiated aggressive HCC cells (Figure 1). This is consistent with reports indicating that while PDK1 oncogenic activities are not dependent on PI3K/AKT signaling, the later more often than not is modulated by PDK1 expression [23], and as rightly put by Tan J et al., while PDK1 is almost always linked with the PI3K/AKT signaling pathway, evidence abound that it does also induce other efferent oncogenic signaling, such as demonstrated by its unmediated induction of Polo-like kinase 1 (PLK1) phosphorylation, with subsequent activation and nuclear accumulation of MYC, resulting in the growth and survival of cancerous cells, induction of an embryonic stem cell (ESC)-like gene-signature, which is associated with aggressive cancer traits and robust CSC-driving signaling, as well as resistance to mTOR-targeted therapy [24]. More so, the characterization of poorly differentiated aggressive HCC cells by aberration in PDK1 expression is of clinical relevance, since it is broadly understood that poorly differentiated cancerous cells exhibit greater degree of resistance to therapy, and are strongly associated with increased metastasis and poor prognosis [25], thus, highlighting a critical role for PDK1 as a putative molecular target in HCC.

Our data indicating that the aberrant expression of PDK1 characterizes poorly differentiated aggressive HCC cells is of translational relevance, considering that PDK1 facilitates the phosphorylation of its protein substrates, and phosphorylation has been shown to modulate the maintenance or repression of pluripotency, which is the ability of individual cells to differentiate into any somatic cell lineage [26,27]. Understanding that malignant cells differ from normal cells based, among other traits, on their propensity to proliferate without terminal differentiation; we posit that the observed aberration in PDK1 expression and/or activity permits and facilitates the occurrence of perpetual/unlimited proliferation of HCC lineage-committed progenitors while deterring terminal differentiation of the cancerous cells [28]. In fact, differentiation-failure and the degree of such differentiation-failure, as reflected by the undifferentiated or poorly differentiated cell status, distinguish benign from malignant tumors and dictate the degree of malignant transformations [28]. Our findings thus indicate the complicity of PDK1 in differentiation-failure oncogenic transformation, and the therapy-resistant phenotype of HCC cells, especially as cellular differentiation depends on the activity of master pluripotency transcription factors (TFs) such as MYC, OCT4, or SOX2 and cofactors like PDK1 invariably serving as important transcriptional coactivator or corepressor that use adenosine triphosphate (ATP) for chromatin remodeling, which 'switches on' or 'switches off' substrates/target genes, enhancing the proliferative index and consequently eliciting treatment failure and poor prognosis [27,28].

In addition, we demonstrated that altered PDK1 expression is enough for the deactivation of the PI3K/PDK/AKT/mTOR oncogenic signaling and sensitizes aggressive HCC cells to radiotherapy (Figure 2). This finding aligns with our evolving understanding of the critical role of PDK1 as a cancer addictive oncogene. In the light of our findings, we posited a probable dependence of disease progression in patients with HCC on expression and/or activity aberrations in oncogene PDK1, which is reminiscent of oncogene addiction, such that the therapeutic targeting of PDK1 or selective blocking of its activities was enough and sufficient to suppress cell viability, impede proliferation, inhibit colony formation and deactivate the PI3K/AKT/mTOR signaling cascade. Having said this, our finding is consistent with the recently demonstrated role of PDK1 as a crucial regulator of cancerous cell migration, invasion and dissemination, while concomitantly modulating the activation status of several oncogenic proteins, including PI3K and AKT/PKB [29], as well as evidence indicating that altered PDK1 expression is a crucial component of the oncogenic PI3K/AKT signaling in breast cancer, and that the targeting of PDK1 sensitizes cancerous cells to therapy [30].

In addition, we demonstrated that aberrant PDK1 expression is implicated in the acquisition of IR-resistance and evasion of DNA damage by HCC cells (Figure 3). This is particularly important considering the role of dysregulated DNA damage/DNA damage repair in the death of cancerous cells exposed to IR and the exploitable therapeutic benefit of reduced resolution of IR-induced clustered DNA damage [16] and altered DNA damage repair genes in liver cancer [17]. The last 3 decades has seen a surge in advocacy for and research into the exploitation of DNA damage response proteins as potential molecular targets to enhance the anticancer effect of radiotherapy, and numerous agents or molecular events that impair key response proteins continue to be combined with radiation in clinical trials [31]. Our data demonstrating an inverse correlation between PDK1 clustered with CSCs/IR-resistance markers and DNA damage repair gene cluster is of translational relevance. This is particularly so, considering the therapeutic challenge of innate or acquired resistance to genotoxic agents or events such as topoisomerase (TOP) inhibitors and ionizing radiation, which are known to induce some of the most toxic genomic lesions—double-strand DNA breaks (DSBs) [31]. Against the background that 1 Gy IR elicits 1000 single strand breaks (SSBs) and 35 DSBs per cell [31], and extrapolating from our presented data, it is probable that the aberrant expression of PDK1 facilitates replication forks, up-regulates DSB religation while suppressing the conversion of IR-induced primary DNA damage/lesions into fatal DSBs through the deregulation of non-homologous end joining (NHEJ), which occurs during late S and G2 cell cycle phases [31]. Intriguingly, we demonstrated for the first time to the best of our knowledge that PDK1 not only interacts with ALDH1A1, but that it also activates it.

Intriguingly, our demonstration of direct interaction between the catalytic domain of PDK1 and human apoALDH1A1 is the first evidence-based documentation of the role of PDK1 in the activation of ALDH1, to the best of our knowledge (Figure 3). Herein, PDK1 serving as a cofactor, covalently binds to the inactive ALDH1A1 apoenzyme (apoALDH1A1) to form a complete and catalytically active ALDH1A1 holoenzyme (holoALDH1A1). This in part, explains the observed clustering of PDK1 with CSC markers, and their association with enhanced oncogenicity, suppressed DNA damage genes RAD50, MSH3, MLH3, ERCC2, CLK2 and BLM, with resultant resistance to radiotherapy.

We also showed that PDK1-dependent IR-resistance is associated with the enhanced metastatic and CSCs-like phenotypes of HCC cells (Figure 4 and Supplementary Figure S3). This is consistent with contemporary knowledge that CSCs-like cells exhibit increased resistance to chemo- and radiotherapy compared to their non-CSCs counterparts in same tumor niche [32], even as cells undergoing EMT have been shown to exhibit resistance to genotoxic stress mediated by conventional radio- and chemotherapy [33,34], thus, linking resistance with CSCs-like and EMT-phenotypes. In fact, it was recently suggested that the acquisition of EMT and CSCs phenotypes is linked with the activation of PI3K/AKT/mTOR signaling in IR-resistant prostate cancer cells [11]. Furthermore, we also demonstrated that the BX795-mediated pharmacologic targeting of PDK1 resensitizes HCC cells to IR-induced apoptosis signals dose-dependently, and significantly suppress their oncogenicity (Figure 5), which is consistent with findings showing that PDK1 signaling is critical for the growth and survival of cancerous cells, and that small-molecule inhibition of PDK1/PLK1 activity effectively targeted and impaired MYC dependency with its associated therapy resistance [24].

5. Conclusions

This present study as we depicted in our schematic abstract (Figure 6), demonstrated that aberrantly expressed PDK1 independently drives PI3K/AKT/mTOR oncogenic signaling, characterizes poorly differentiated cells, activates ALDH1 and is associated with the desensitization of aggressive hepatocellular carcinoma cells to radiotherapy. Thus, we provide preclinical evidence for the pleiotropic contribution of PDK1 to the IR-resistant phenotype of aggressive HCC cells by modulating oncogenic, stemness, metastatic and DNA damage repair signaling. These findings provide important mechanistic insights into HCC biology and have significant therapeutic implications.

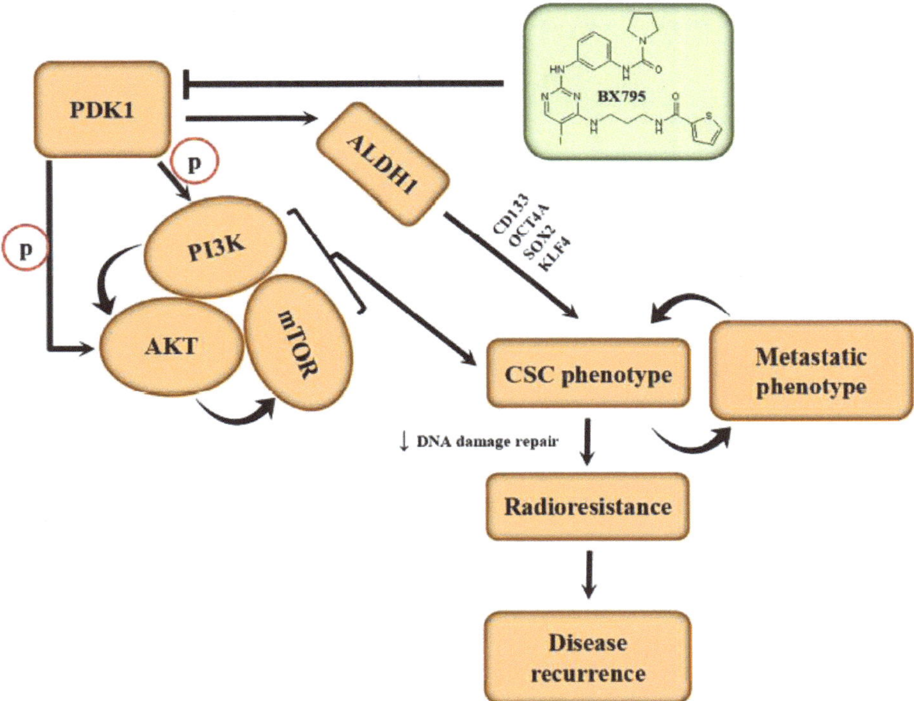

Figure 6. Schematic abstract of the aberrantly expressed PDK1 independently drives PI3K/AKT/mTOR oncogenic signaling in HCC. This present study demonstrated that aberrantly expressed PDK1 independently drives PI3K/AKT/mTOR oncogenic signaling, characterizes poorly differentiated cells, activates ALDH1 and is associated with the desensitization of aggressive hepatocellular carcinoma cells to radiotherapy.

Supplementary Materials: The following are available online at http://www.mdpi.com/2073-4409/9/3/746/s1, Table S1. Antibody list. Supplementary Figure S1. PDK1 directly binds to and activates ALDH1. (A) ALDH1-PDK1 protein interaction matrix (upper panel) with local and global RMSD data indicated (lower panel). (B) ALDH1-PDK1 protein sequence alignment confirming complementarity. Supplementary Figure S2. PDK1 interacts with ALDH and directly modulate the expression and/or activity of ALDH in HCC cells. (A) Representative Western blot image and histograms of the differential expression of PDK1 and ALDH1 in adherent wild-type Mahlavu cells or their tumorsphere counterparts. (B) Graph showing the effect of siPDK1 on the expression level of ALDH1 mRNA in Mahlavu-R cells. (C) Representative Western blot image and histograms showing the effect of siPDK1-1 and siPDK1-2 on the expression levels of PDK1 or ALDH1 protein in Mahlavu-R cells. $* p < 0.05, ** p < 0.01, *** p < 0.001$. Mahlavu-R, radioresistant Mahlavu cells; WT, wild-type; NC, negative control. Supplementary Figure S3. PDK1 is associated with the modulation of cellular pluripotency and proliferation. Graphical representation of the correlation between PDK1 and (A) c-MYC/MYC, or (B) Ki-67/MKI67 in the TCGA-liver hepatocellular carcinoma (LIHC) cohort, $n = 374$. (C) 2D visualization of the protein–protein interaction network between PDK1, MYC, and Ki-67. Figure S4. Full-size blots of Figure 1C. Figure S5. Full-size blots of Figure 2D. Figure S6. Full-size blots of Figure 4D. Figure S7. Full-size blots of Figure 5C. Figure S8. Full-size blots of Figure S3A,B.

Author Contributions: O.A.B. and H.-L.C.: Study conception and design, collection and assembly of data, data analysis and interpretation, and manuscript writing. O.A.B. and C.-T.Y.: Study conception and design, collection and assembly of data, data analysis and interpretation, manuscript writing and Revision. J.-R.O., W.-H.L.: Data analysis and interpretation. J.-T.T.: Study conception and design, data analysis and interpretation. All authors read and approved the final manuscript.

Funding: This study was also supported by grants from Taipei Medical University (102TMU-SHH-02) to Wei-Hwa Lee.

Acknowledgments: The authors thank all research assistants of the Cancer Translational Research Laboratory and Core Facility Center, Taipei Medical University - Shuang Ho Hospital, especially Wan-Chen Tsai, for her assistance with the flow cytometry, molecular and cell-based assays.

Conflicts of Interest: The authors declare that they have no potential financial competing interests that may in any way gain or lose financially from the publication of this manuscript at present or in the future. Additionally, no non-financial competing interests are involved in the manuscript.

Ethical Approval: This study was approved by the Institutional Human Research Ethics Review Board (TMU-JIRB No. 201302016) of Taipei Medical University and performed in accordance with the Institutional Guide for the Care and Use of Laboratory Animals of the Taipei Medical University.

Availability of Data and Materials: The datasets used and analyzed in the current study are publicly accessible as indicated in the manuscript.

Abbreviations

epithelial-to-mesenchymal transition (EMT), focal nodular hyperplasia (FNH), transarterial chemoembolization (TACE), percutaneous ethanol injection (PEI), microwave ablation (MWA), radiofrequency ablation (RFA), stereotactic body radiotherapy (SBRT) and external beam radiation therapy (EBRT).

References

1. Ferlay, J.; Ervik, M.; Lam, F.; Colombet, M.; Mery, L.; Piñeros, M.; Znaor, A.; Soerjomataram, I.; Bray, F. Global Cancer Observatory: Cancer Today. Lyon, France: International Agency for Research on Cancer. 2018. Available online: https://gco.iarc.fr/today (accessed on 3 May 2019).

2. Walther, Z.; Jain, D. Molecular pathology of hepatic neoplasms: Classification and clinical significance. *Patholog Res. Int.* **2011**, *2011*, 403929. [CrossRef]

3. Akinyemiju, T.; Abera, S.; Ahmed, M.; Alam, N.; Alemayohu, M.A.; Allen, C.; Al-Raddadi, R.; Alvis-Guzman, N.; Amoako, Y.; Artaman, A.; et al. The Burden of Primary Liver Cancer and Underlying Etiologies From 1990 to 2015 at the Global, Regional, and National Level: Results from the Global Burden of Disease Study 2015. *JAMA Oncol.* **2017**, *3*, 1683–1691.

4. Villanueva, A. Hepatocellular Carcinoma. *N. Engl. J. Med.* **2019**, *380*, 1450–1462. [CrossRef]

5. Lurje, I.; Czigany, Z.; Bednarsch, J.; Roderburg, C.; Isfort, P.; Neumann, U.P.; Lurje, G. Treatment Strategies for Hepatocellular Carcinoma- A Multidisciplinary Approach. *Int. J. Mol. Sci.* **2019**, *2*, 1465. [CrossRef]

6. Crissien, A.M.; Frenette, C. Current management of hepatocellular carcinoma. *Gastroenterol. Hepatol. (N. Y.)* **2014**, *10*, 153–161.

7. Kalogeridi, M.A.; Zygogianni, A.; Kyrgias, G.; Kouvaris, J.; Chatziioannou, S.; Kelekis, N.; Kouloulias, V. Role of radiotherapy in the management of hepatocellular carcinoma: A systematic review. *World J. Hepatol.* **2015**, *7*, 101–112. [CrossRef]

8. Klein, J.; Dawson, L.A. Hepatocellular Carcinoma Radiation Therapy: Review of Evidence and Future Opportunities. *Int. J. Radiat Oncol. Biol. Phys.* **2013**, *87*, 22–32. [CrossRef]

9. Xu, S.; Li, Y.; Lu, Y.; Huang, J.; Ren, J.; Zhang, S.; Yin, Z.; Huang, K.; Wu, G.; Yang, K. LZTS2 inhibits PI3K/AKT activation and radioresistance in nasopharyngeal carcinoma by interacting with p85. *Cancer Lett.* **2018**, *420*, 38–48. [CrossRef]

10. Horn, D.; Hess, J.; Freier, K.; Hoffmann, J.; Freudlsperger, C. Targeting EGFR-PI3K-AKT-mTOR signaling enhances radiosensitivity in head and neck squamous cell carcinoma. *Expert Opin. Ther. Targets* **2015**, *19*, 795–805. [CrossRef]

11. Chang, L.; Graham, P.H.; Hao, J.; Ni, J.; Bucci, J.; Cozzi, P.J.; Kearsley, J.H.; Li, Y. Acquisition of epithelial-mesenchymal transition and cancer stem cell phenotypes is associated with activation of the PI3K/Akt/mTOR pathway in prostate cancer radioresistance. *Cell Death Dis.* **2013**, *4*, e875. [CrossRef]

12. Szymonowicz, K.; Oeck, S.; Malewicz, N.M.; Jendrossek, V. New Insights into Protein Kinase B/Akt Signaling: Role of Localized Akt Activation and Compartment-Specific Target Proteins for the Cellular Radiation Response. *Cancers (Basel)* **2018**, *10*, 78. [CrossRef]

13. Park, J.; Feng, J.; Li, Y.; Hammarsten, O.; Brazil, D.P.; Hemmings, B.A. DNA-dependent Protein Kinase-mediated Phosphorylation of Protein Kinase B Requires a Specific Recognition Sequence in the C-terminal Hydrophobic Motif. *J. Biol. Chem.* **2009**, *284*, 6169–6174. [CrossRef]

14. Gagliardi, P.A.; Puliafito, A.; Primo, L. PDK1: At the crossroad of cancer signaling pathways. *Semin. Cancer Biol.* **2018**, *48*, 27–35. [CrossRef]

15. Franken, N.A.; Rodermond, H.M.; Stap, J.; Haveman, J.; van Bree, C. Clonogenic assay of cells in vitro. *Nat. Protoc.* **2006**, *1*, 2315–2319. [CrossRef]

16. Lomax, M.E.; Folkes, L.K.; O'Neill, P. Biological Consequences of Radiation-induced DNA Damage: Relevance to Radiotherapy. *Clin. Oncol. (R. Coll. Radiol.)* **2013**, *25*, 578–585. [CrossRef]

17. Lin, J.; Shi, J.; Guo, H.; Yang, X.; Jiang, Y.; Long, J.; Bai, Y.; Wang, D.; Yang, X.; Wan, X.; et al. Alterations in DNA Damage Repair Genes in Primary Liver Cancer. *Clin. Cancer Res.* **2019**, *25*, 4701–4711. [CrossRef]

18. Cojoc, M.; Peitzsch, C.; Kurth, I.; Trautmann, F.; Kunz-Schughart, L.A.; Telegeev, G.D.; Stakhovsky, E.A.; Walker, J.R.; Simin, K.; Lyle, S.; et al. Aldehyde Dehydrogenase Is Regulated by β-Catenin/TCF and Promotes Radioresistance in Prostate Cancer Progenitor Cells. *Cancer Res.* **2015**, *75*, 1482–1494. [CrossRef]

19. Li, Z.; Huang, X.; Zhan, H.; Zeng, Z.; Li, C.; Spitsbergen, J.M.; Meierjohann, S.; Schartl, M.; Gong, Z. Inducible and repressable oncogene-addicted hepatocellular carcinoma in Tet-on xmrk transgenic zebrafish. *J. Hepatol.* **2012**, *56*, 419–425. [CrossRef]

20. Luo, J.; Solimini, N.L.; Elledge, S.J. Principles of cancer therapy: Oncogene and non-oncogene addiction. *Cell* **2009**, *136*, 823–837. [CrossRef]

21. Weinstein, I.B.; Joe, A. Oncogene Addiction. *Cancer Res.* **2008**, *68*, 3077–3080. [CrossRef]

22. Villanueva, A.; Lilvet, J.M. Targeted Therapies for Hepatocellular Carcinoma. *Gastroenterol* **2011**, *140*, 1410–1426. [CrossRef]

23. Emmanuouilidi, A.; Falasca, M. Targeting PDK1 for Chemosensitization of Cancer Cells. *Cancers (Basel)* **2017**, *9*, 140. [CrossRef]

24. Tan, J.; Li, Z.; Lee, P.L.; Guan, P.; Aau, M.Y.; Lee, S.T.; Feng, M.; Lim, C.Z.; Lee, E.Y.J.; Wee, Z.N.; et al. PDK1 signaling toward PLK1-MYC activation confers oncogenic transformation, tumor-initiating cell activation, and resistance to mTOR-targeted therapy. *Cancer Discov.* **2013**, *3*, 1156–1171. [CrossRef]

25. Gaianigo, N.; Melisi, D.; Carbone, C. EMT and Treatment Resistance in Pancreatic Cancer. *Cancers* **2017**, *9*, 122. [CrossRef]

26. Fernandez-Alonso, R.; Bustos, F.; Williams, C.A.C.; Findlay, G.M. Protein Kinases in Pluripotency–Beyond the Usual Suspects. *J. Mol. Biol.* **2017**, *429*, 1504–1520. [CrossRef]

27. Fedorov, O.; Müller, S.; Knapp, S. The (un)targeted cancer kinome. *Nat. Chem. Biol.* **2010**, *6*, 166–169. [CrossRef]

28. Enane, F.O.; Saunthararajah, Y.; Korc, M. Differentiation therapy and the mechanisms that terminate cancer cell proliferation without harming normal cells. *Cell Death Dis.* **2018**, *9*, 912. [CrossRef]

29. Di Blasio, L.; Gagliardi, P.A.; Puliafito, A.; Primo, L. Serine/Threonine Kinase 3-Phosphoinositide-Dependent Protein Kinase-1 (PDK1) as a Key Regulator of Cell Migration and Cancer Dissemination. *Cancers* **2017**, *9*, 25. [CrossRef]

30. Raimondi, C.; Falasca, M. Targeting PDK1 in cancer. *Curr. Med. Chem.* **2011**, *18*, 2763–2769. [CrossRef]

31. Goldstein, M.; Kastan, M.B. The DNA damage response: Implications for tumor responses to radiation and chemotherapy. *Annu. Rev. Med.* **2015**, *66*, 129–143. [CrossRef]

32. Shibue, T.; Weinberg, R.A. EMT, CSCs, and drug resistance: The mechanistic link and clinical implications. *Nat. Rev. Clin. Oncol.* **2017**, *14*, 611–629. [CrossRef] [PubMed]

33. Santamaria, P.G.; Moreno-Bueno, G.; Cano, A. Contribution of Epithelial Plasticity to Therapy Resistance. *J. Clin. Med.* **2019**, *8*, 676. [CrossRef] [PubMed]

34. Wang, N.; Wang, S.; Li, M.Y.; Hu, B.G.; Liu, L.P.; Yang, S.L.; Yang, S.; Gong, Z.; Lai, P.B.; Chen, G.G. Cancer stem cells in hepatocellular carcinoma: An overview and promising therapeutic strategies. *Ther. Adv. Med. Oncol.* **2018**, *10*, 1758835918816287. [CrossRef] [PubMed]

MDPI AG
Grosspeteranlage 5
4052 Basel
Switzerland
Tel.: +41 61 683 77 34

Cells Editorial Office
E-mail: cells@mdpi.com
www.mdpi.com/journal/cells